Scholastic WORLD GEOGRAPHY

Scholastic WORLD GEOGRAPHY

Robert A. Harper

Professor of Geography, University of Maryland

Joseph P. Stoltman

Chair, Department of Geography, Western Michigan University

Scholastic Inc.

Robert A. Harper is Professor of Geography and former chairperson of the Geography Department at the University of Maryland, College Park. A former president of the National Council for Geographic Education, Dr. Harper received his Ph.D. in geography from the University of Chicago. He is the author of several geography texts at elementary, secondary, and college levels.

Joseph P. Stoltman is Professor of Geography and former chairperson of the Geography Department at Western Michigan University, Kalamazoo. He received his Ed.D. in geographic education from the University of Georgia and taught secondary schools for several years. Dr. Stoltman has written numerous articles for geographic journals.

CURRICULUM CONSULTANTS

Miller L. Barron
Supervisor of Social Studies Education
Cobb County Public Schools, Marietta, Georgia

Todd I. Berens
Social Studies Department Chairman
Lexington Junior High School, Cypress, California

Sister Marijon Binder, C.S.J.
Educational Consultant
Global Concerns Center, Chicago, Illinois

Loyal L. Darr
Supervisor of Social Studies
Denver, Colorado, Public Schools

Marsha Delfino
Gahanna Middle School
Gahanna, Ohio

Larry D. Hatke
Slinger Middle School
Slinger, Wisconsin

Dr. Richard M. Haynes
Assistant Superintendent
Tarboro, North Carolina, County Schools

Margy Nurik
Lakeside Middle School
Pompton Lakes, New Jersey

Jo Beth Oestreich
Oliver Wendell Holmes High School
San Antonio, Texas

Bruce E. Tipple
Director, Minnesota Curriculum Laboratory
Minneapolis, Minnesota

Roy Tison
Board Member, National Council for Geographic Education

Joyce Williamson
Social Studies Coordinator
Marshall High School, Marshall, Texas

REGIONAL CONSULTANTS

Latin America and the Caribbean:
Dr. Ernst Griffin, Chair, Department of Geography
San Diego State University, California

South and Southeast Asia, East Asia and the Pacific:
Dr. Laurence J.G. Ma, Professor of Geography
University of Akron, Ohio

Tropical and Southern Africa:
Dr. Jane J. Martin, U.S. Educational and Cultural
Foundation; Monrovia, Liberia

FIELD TESTERS

Nevil Barr
Elkins Junior High School
Elkins, West Virginia

Dennis Gorecki
Central Junior High School
Evergreen Park, Illinois

Michael A. Mazella, Jr.
St. Ann's School
Staten Island, New York

Staff

Publisher: Eleanor Angeles
Editorial Director: Carolyn Jackson
Project Editor: William Johnson
Skills Editor: Mollie L. Cohen
Contributing Editors: Norman L. Lunger, Ira Peck
Art Design and Director: Irmgard Lochner
Design of Book, Thematic Maps, and Charts: Hal Aber
  Graphics, Inc.
Regional, National, and Atlas Maps: Donnelley
  Cartographic Services
Photo Research: Roberta Guerette
Revision and Teacher's Annotated Edition: E.L. Wheetley
  and Associates, Inc.
Cover Image: C.P. Jones
Cover Design: Hal Kearney
Cover Picture Research: Connie McCullom

ISBN 0-590-35407-8
Copyright © 1989 by Scholastic Inc.
All rights reserved.
12 11 10 9 8
Printed in the U.S.A.

4/9

23

CONTENTS

Western Europe, 215

The Soviet Union and Eastern Europe, 277

The Middle East and North Africa, 325

Charts, Graphs, and Diagrams

You and Your World

*L*adies and gentlemen, welcome to the fair! We have some really exciting events for you today. There'll be contests in swimming, ice skating, and surfing on our lake. On the same track, there'll be a downhill ski race and an attempt to break the world's record for the mile run. You'll also see displays of our finest local crops — bananas, coffee, and tomatoes. You'll see our local wildlife — polar bears, camels, alligators, and walruses."

Take a careful look at this announcement. Is there anything wrong with it?

The answer is *yes*, nearly everything in it is wrong. There's no place on Earth where all those events and displays could be brought together at one time. Swimming needs water above freezing, while skating needs water frozen into ice. Surfing needs big, rolling ocean waves. Downhill skiing needs a snowy slope, while a mile run needs a smooth flat track. Bananas grow in hot, steamy lowlands. Coffee plants do best in tropical highlands. Tomatoes prefer cooler places. Polar bears and walruses live in cold regions, while camels and alligators need warmth.

The **environment** (surroundings) in which people and animals live sets limits on what they can do. Within those limits, however, people have a wide range of choices. In many places, for example, they can choose to ski *and* skate. In many places, they can grow tomatoes along with apples, cabbages, and countless other food crops.

Often they can find ways around the limits. They can build an indoor skating rink for times when there's no natural ice. They can breed a plant that will grow outside its normal environment. They can change a dry environment by bringing in water.

All over the world, there is give and take between the environment and the people who live in it. The environment shapes people's lives — but people also reshape the environment. This happens where you live too. Check for yourself by answering the following questions.

■ Is there a river near where I live? If so, is there a bridge that lets me cross it? How else could I cross?

■ Is it easy to travel to and from my neighborhood? Are there highways, railroads, or air routes going in many directions?

■ Do I ever have the chore of sweeping up dead leaves or clearing away snow? Or aren't there any dead leaves or snow where I live?

■ Do I often have to think about wearing boots or a raincoat? Or can I nearly always be sure it won't rain?

■ What kinds of animals live in my area? Are some of them wild? Are others looked after by people?

■ What outdoor sports are possible? Can people go swimming, surfing, ice skating, or skiing?

■ Suppose I want to stay in my neighborhood when I finish my schooling. Is there a wide choice of jobs? Or is there just one kind of job for most people?

These questions deal with only part of your environment. But they are enough to give you an idea of how it influences your life.

What makes up the environment? How and why does it differ from place to place? Just how important a part does it play in people's lives? How do they change it? Answering these questions is the task of **geography.**

The word *geography* means "writing about Earth." It is first of all the study of the **land** and the various forms or shapes that the land can take around the world. Geography is also the study of **climate** (the kind of weather a place has over a period of time). It is the study of the land's **resources** (the water we drink, the plants and animals we grow for food, the minerals we dig out of the ground, and other useful things). Geography is also the study of the **people** who live on the land. It shows how they use, depend on, and change the land. In short, geography shows how land and people affect each other.

As you can see, geography covers a lot of ground. Sometimes special names are given to its different areas. The study of land and water is called **physical geography.** The study of people and their ways of life in different places is called **human** or **cultural geography.** The study of the way people use Earth's resources is called **economic geography.** There are other kinds of geography too, but these three cover most of the field.

Why Geography?

The purpose of this book is to give you a basic understanding of the geography of today's world. This understanding can be helpful to you in many ways.

Look back at the questions on page 12. They show how geography can affect the place where you live, the way you live, and your future choice of jobs. Geography also affects the things you need and use in your daily life. Food, water, clothing, housing, lighting, heating, fuel for transportation, and many other items depend in some way on Earth's resources — they depend on Earth's geography.

Geography also helps you to understand news events and how they may affect you. In today's world, things that happen thousands of miles away may affect you as much as events in your own community. Very often the land and its resources are involved. For example, the U.S. needs to buy oil from other nations. Suppose a nation doesn't want to sell oil to the U.S., or sets a very high price. You will soon see the result in long lines or higher prices at your neighborhood gas station.

There are many other examples. Earthquakes, floods, and other natural disasters may lead to worldwide calls for help. A natural disaster may also reduce the supply of an important food crop or mineral resource. On the other hand, the discovery of a new source of minerals in a far-off part of the world may boost trade in our nation. Conflicts between distant nations may threaten world peace. Cooperation (nations working together) may strengthen it. Both conflicts and cooperation often arise over the land and its resources. One way or another, geography plays a large part in your life.

This Book

The main part of this text is divided into 10 units, with two to five chapters in each unit. The first unit looks at Earth as a whole. It takes you into space to see the surface of Earth. Then it brings you closer to Earth to discover Earth's resources. You will find out how different environments on Earth can influence the people who live there. You will also learn the basic ways in which people make use of their environments.

Units 2 through 9 take you to different regions of the world. (A **region** is an area of land where most people have enough in common to be considered as a group, different from other peoples.) In these regional units, you will see how the basic ideas of Unit 1 work in practice for different people in different places. You will also learn how these people and places are connected with the U.S. and your own way of life.

The final unit, Unit 10, once again looks at Earth as a whole. It focuses on some important changes, problems, and hopes that you have come across in your journey from region to region. It helps you to shape your own ideas on where Earth — and all of us who live on it — may be headed.

Making the Most of It

If you flip through the pages of this book, you'll see that it contains a large number of photos, maps, charts, and diagrams. These illustrations are some of the tools that will help you to understand geography. They give certain kinds of information that can't be put into words easily. But like any other tool, they are helpful only when you know how to use them. That's why this text also includes a series of "Geography Skills" lessons. There are two or more lessons with each chapter in Unit 1, and one lesson with each chapter in the rest of the book. These skills lessons explain the tools of geography, step by step.

At the end of every chapter, you'll find a section called "Your Local Geography." Here you'll find suggestions for comparing and contrasting other places around the world with your own community.

In each unit, you'll find a section called "Sidelights," which is a kind of surprise package. It may be part of a diary, a song, a novel, a letter, or other kind of writing. These sidelights have one thing in common: They all involve geography. Reading them, you'll see how geography has a way of throwing light on unexpected areas of life.

Two features of this book will help you find your way through it:
■ The Contents at the front of the book lists first the titles of each unit and chapter, and tells you the page on which each begins. It also lists the Geography Skills lessons. Then there are separate listings for the maps and charts.
■ The Index at the back of the book helps you to find any place or topic dealt with in the text. All the places and topics are listed in alphabetical order, and the numbers of the pages on which they are discussed are shown to the right of the listing. The Index also tells you how to pronounce each place-name.

Near the back of this book, you'll find many pages of reference material:
■ The Atlas consists of maps covering the whole world.
■ The Glossary explains the meaning of key words used in the book. The words are listed in alphabetical order.
■ The Checklist of Nations gives many basic facts and figures about the countries in each region of the world.

Until you're familiar with all of the book's features, you can check back to this page. You're now ready to discover more about the planet you live on.

1

Planet Earth

Chapter 1

A Special Planet

A spaceship speeds through the universe. Its crew members are alien beings whose home star is a long way behind. They are on a voyage of exploration. Their mission: to search for planets where there is life.

Now the spaceship is approaching a star — a huge and powerful source of energy. The crew members peer at the new bright spot on their view screen. Will this star have any **planets** (masses of cooler substances turning around it)? If so, will there be any life on these planets?

Suddenly there is excitement aboard the craft. Instruments have spotted a small planet quite close to the spaceship. The instruments sweep the darkness ahead, searching for more planets. One by one, the total builds. There are seven, eight, nine planets in all. They are many million miles apart, but that seems close in the vast distances of space. Each planet spins like a top while moving in a huge oval around the star.

The planets vary greatly in size. The biggest is about 50 times larger than the smallest, but it is still much smaller than the star. Each planet moves around the star at a different distance. The one farthest from the star is more than three billion miles (*five billion kilometers*) away. The planet closest to the star is only about 30 million miles (*50 million kilometers*) away.

The spaceship has come upon our solar system. (A **solar system** is a group of planets and other bodies revolving around a star.) The star is our Sun, and the planets that turn around it are Earth and its neighbors.

Around the Sun

The planet farthest from the Sun is Pluto, the smallest of all. It is dark and cold — its surface seems to be covered with a frozen poisonous gas. The space travelers find that Pluto is too cold for life. They glide around to check the other planets.

To do this, the travelers have to follow a winding course. The planets circle the Sun on roughly the same plane, or level. But they do not go around side by side, like the front line of a marching band. Instead they move at different speeds, like racing cars around a track. Thus the space travelers find the planets scattered at different points around the Sun.

After Pluto the spaceship swings past two larger planets, Neptune and Uranus. Then comes Saturn, even bigger, and Jupiter, the biggest of all the planets. All four are still much too cold for life. Saturn and Jupiter are completely covered with poisonous gases.

The spaceship keeps moving closer to the Sun and soon passes tiny Mars. This planet looks as lifeless as the rest.

Seen from an approaching spacecraft, Earth appears quite small. Our planet looks like a bright, blue-and-white ball rolling through space.

Our Solar System

Pluto · Neptune · Uranus · Saturn · Jupiter · Mars · Earth · Venus · Mercury · Sun

The next planet is about twice as big as Mars. It is 93 million miles (*150 million kilometers*) from the Sun. At once the travelers notice something different about this planet.

Its size isn't impressive. Jupiter, Saturn, Uranus, and Neptune are much larger. Its shape isn't different. Like the other planets, it is shaped like a **sphere** (ball).

It does have a **moon** — a planetlike mass that turns around it. But this isn't so special either. Saturn has 17 moons. Even Mars has two small ones.

The unusual thing about this planet is its appearance. The other planets were white, gray, reddish, or brown. This planet looks like a blue marble with swirls of white in it.

Moving closer, the travelers can see patches of green and brown among the blue and white. These patches are land surfaces. The green comes from growing things — different kinds of trees and grasses. The visitors from space have had their first view of life in our solar system. They have come upon our planet, Earth.

This picture shows the relative sizes of the planets in our solar system. The Sun itself is so big that only a tiny part can be included. The picture also shows the order in which the planets circle the Sun. (Pluto's path is off-center, so occasionally it moves closer to the Sun than Neptune.) Relative distances are *not* shown. If the distance of Earth from the Sun were used as the basis for relative distances, Pluto would have to be placed an arm's length off the page!

SECTION REVIEW

1. What is a solar system?

2. Which of the following makes Earth different from other planets in our solar system? (a) It has a moon. (b) It can support life. (c) It is sphere-shaped.

3. Pluto and the other outer planets are unable to support life. What is the main reason for this?

4. What can the visitors from space tell about Earth from viewing it at a distance? In what way does Earth appear able to support some kinds of life?

Understanding Globes

Imagine that you are a traveler approaching Earth from outer space. The photograph below shows what you would see from your spacecraft. This photograph of Earth was taken from more than 100,000 miles (*160,000 kilometers*) out in space. It shows that Earth is shaped like a *sphere* (ball). But it shows only one side of the planet — a **hemisphere** (half a sphere).

For another view of Earth, look at a classroom globe such as the one shown in Drawing A. An actual globe shows a whole sphere. It is a mini-Earth, a model of the entire planet. A globe, however, does not look exactly like Earth as seen by a space traveler. In some ways, the globe is clearer. For example, the cloud swirls that surround Earth don't surround the globe. Also, you can turn a globe to view different hemispheres.

Photograph of Earth

A Classroom Globe

Drawing A

As you turn a globe, you can see how major land and water areas are arranged on Earth's surface. Study the two views of a globe on page 20 in Drawings B and C. Each of these drawings shows a hemisphere of Earth. In each drawing, the largest land areas, called **continents**, and the largest water areas, called **oceans**, are labeled by name.

(Turn page.)

Use both drawings to answer the following questions. Write the answers on a separate sheet of paper.

1. Three continents are shown in the half of Earth called the **Western Hemisphere**. Which continents are they?
2. How many continents are in the half of Earth called the **Eastern Hemisphere?** Which continent is shown partly in both the Western and Eastern hemispheres?
3. The globe drawings show Earth's four oceans. What are their names? Which ocean is located only in the Eastern Hemisphere?

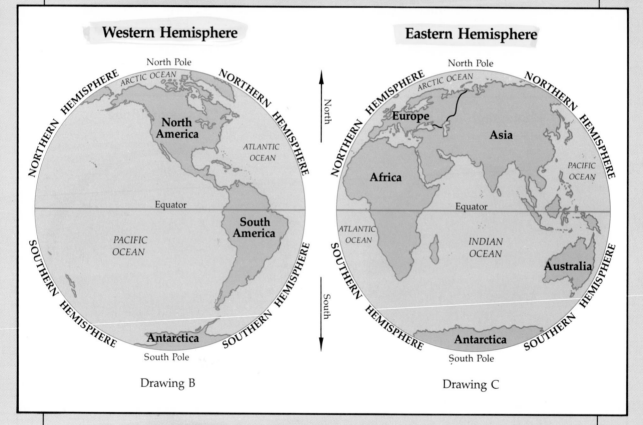

Drawing B

Drawing C

Find the **North Pole** and the **South Pole** on each globe drawing. The poles are fixed points at opposite ends of the planet. They are used for finding direction on Earth's surface. North is always toward the North Pole. South is always toward the South Pole. Whenever you move toward the North Pole from another place on Earth, you are going north. Whenever you move toward the South Pole from any place on Earth, you are going south.

The **Equator** is also shown on both globe drawings. The Equator is a line drawn around a globe, halfway between the North and South poles. The Equator divides Earth into a **Northern Hemisphere** and a **Southern Hemisphere.**

Use both globe drawings to answer the following questions. Write the answers on your separate sheet of paper.

4. Which of the two poles is located over an ocean? Which ocean?
5. Which of the two poles is located over a continent? Which continent?
6. In what general direction, north or south, would you be traveling if you went: (a) from North America to South America? (b) from Australia to Asia? (c) from Europe to Africa?
7. Which continents are entirely in the Northern Hemisphere?
8. Which continents are entirely in the Southern Hemisphere?
9. Which continents are in both hemispheres?
10. Which ocean is in only one of these hemispheres?

GEOGRAPHY SKILLS 2

Locating Places on a Globe

You are circling Earth in a high-speed spacecraft. Except for the clouds, the planet below looks like a giant classroom globe. Mostly you see oceans, which cover more than two thirds of the surface. Of course, the largest land areas are the continents. You can also spot islands here and there. **Islands** are smaller land areas surrounded by water.

Every point on the surface beneath you can be located on a globe. It may be the town or city where you live, or it may be a tiny island in the vast Pacific Ocean. Any place can be pinpointed exactly — if you know its latitude and longitude.

Understanding Latitude

Latitude is the position of a place north or south of the Equator. Lines of latitude are drawn around a globe, as shown on page 22. In Drawing A, you see a side view of a globe with the Equator in the middle. In Drawing B, your view shifts north of the Equator. You can see more clearly what happens to lines of latitude near the poles.

(Turn page.)

Lines of latitude never meet on a globe. They are **parallel** — that is, they are always the same distance from one another. For this reason, they are often called parallels of latitude. As you study Drawings A and B, answer the following questions on a separate sheet of paper.

1. Are parallels of latitude on a globe straight lines or circles?
2. Do lines of latitude become shorter or longer near the North and South poles?
3. Which is the longest line of latitude?

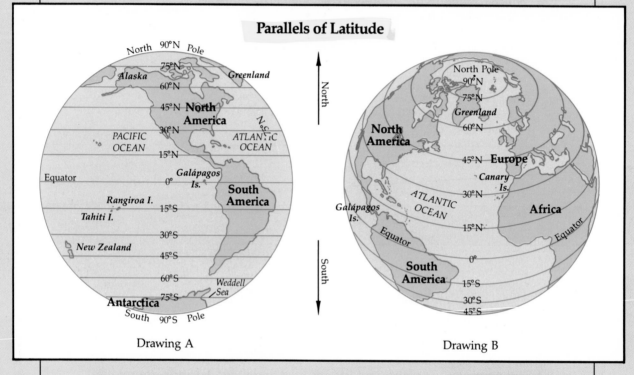

Parallels of Latitude

Drawing A

Drawing B

Notice how each line of latitude is numbered. Latitude is measured in **degrees**. The symbol for degrees is a small circle, like this °, to the right of each number. The Equator is numbered zero degrees (0°). There are 90 degrees between the Equator and the North Pole. So the North Pole has a latitude of *90 degrees north*, which is written like this: 90°N. There are also 90 degrees between the Equator and the South Pole. The latitude of the South Pole is 90°S.

You can use the globe drawings to find the latitudes of places. For example, Rangiroa Island in Drawing A has a latitude of about 15°S. On your separate sheet of paper, write the latitudes of the following places.

4. Southern tip of Greenland.
5. Galápagos Islands.
6. Weddell Sea.
7. Southern New Zealand.

Understanding Longitude

Longitude is measured east and west around the globe. For someone on Earth facing north, *east* is on the right and *west* is on the left.

Lines of longitude, also called **meridians,** are drawn on a globe, as is shown in Drawings C and D. Meridians of longitude are *not* parallel. They meet at the North Pole and the South Pole. Drawing D shows how they meet at the North Pole.

Like parallels of latitude, meridians of longitude are numbered in degrees. Unlike parallels of latitude, meridians of longitude have no obvious starting point for numbering. By international agreement, the meridian that runs through Greenwich, England, was chosen as the **Prime Meridian** (0°).

You can find the Prime Meridian in the center of Drawing C. Other meridians, as shown in Drawing D, are numbered by degrees either east (E) or west (W) of the Prime Meridian to 180°. You can use these lines to find the longitudes of places. For example, in Drawing C, Madagascar is at 45°E.

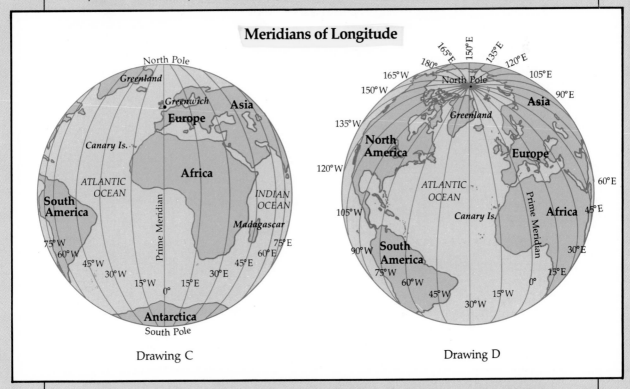

Drawing C

Drawing D

Use the globe drawings to decide whether each of the following statements is true or false. Write the answers on your separate sheet of paper.

8. The longitude of the Canary Islands is about 15°W.
9. The eastern tip of Africa is about 45° west of the Prime Meridian.
10. In Drawing D, the meridian numbered 0° circles the entire globe over the North Pole.

(Turn page.)

Using a Global Grid

Parallels of latitude and meridians of longitude cross each other around the globe to form a **grid.** With this global grid, you can give the exact location of any place on Earth.

For example, imagine that you are on a ship in the Pacific Ocean off New Zealand. The ship is sinking. To get help, you must radio your location to another ship. You are at parallel of latitude 30 degrees south of the Equator. You are at a meridian of longitude 165 degrees east of the Prime Meridian. You would radio your location as 30°S, 165°E. Latitude is always given first, longitude second. This marks your exact location.

You can find the location for 30°S, 165°E in Drawing E, which shows the Western Hemisphere. Drawing F shows part of the Eastern Hemisphere too.

Use both drawings to locate the grid points and answer these questions.

11. What places do you find at these locations: (a) 15°S, 150°W? (b) 0°, 90°W?

12. At what grid points would you locate the Canary Islands?

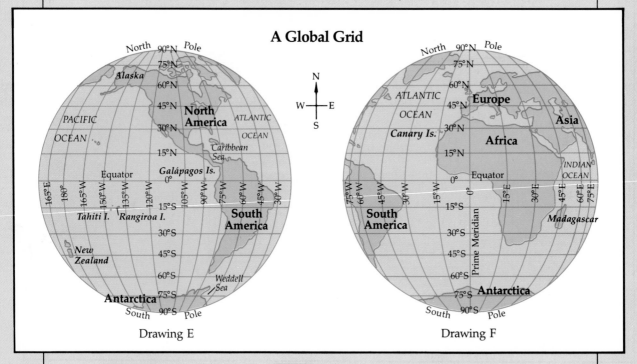

A Global Grid

Drawing E   Drawing F

Near the globe drawings is a **compass rose**, which points out the directions north (N), south (S), east (E), and west (W). Use these directions and the global grid in Drawings E and F to answer the following questions.

13. To get from 30°N, 75°W to 60°N, 0°, you must cross which ocean?

14. If you sailed east from 45°N, 165°W, you would land on which continent?

15. To reach the coast of South America from 15°S, 15°W, you would sail in what direction?

Earth, a Good Place To Live

Earth is the only planet in our solar system that supports life as we know it. Its distance from the Sun allows the right temperatures for life. The other planets that the travelers passed are too far from the Sun and do not receive enough heat. Venus and Mercury, the planets closer to the Sun, are *too* close. On their surfaces, living things would burn up.

The unusual appearance of Earth gave the visitors two more hints that they would find life. Can you tell what those hints are?

First there's the blue that covers much of this planet. The visitors' instruments show that the blue areas are water, which is necessary for life as we know it. Of our Sun's planets, only Earth has a large supply of water.

Then there are the swirls of white, which are clouds. These are part of the **atmosphere** — the layer of gases that covers Earth like a transparent (see-through) wrapping. The instruments show that one of the gases is **oxygen**, which is also necessary for life as we know it. Only Earth has a large supply.

A Profile of Earth

The visitors from space want to record what they see. One of them takes notes while the others call out observations.

■ Planet Earth: A sphere about 25,000 miles *(40,000 kilometers)* around, the fifth largest in this solar system.

■ Its surface: About 70 percent of Earth is covered by large bodies of salty water called *oceans.* These oceans are never still, but flow and change all the time. About 30 percent of Earth is dry land. The great landmasses of Earth are called *continents.* They are surrounded by the oceans.

Now the observers are ready to learn more about planet Earth. They descend for a closer look at the signs of life they saw before. These are the green patches of **vegetation** (plant life) on the land.

In some places, there are huge forests of tall trees. Large areas of grasses cover other parts of Earth. The visitors marvel at the rich green of forests near the Equator. As the spaceship moves around Earth, the observ-

In this satellite view of the Florida area, blue water and hazy air show that Earth is able to support life.

ers can see treeless areas of rock and sand. They note too that there are huge ice-covered areas around the poles.

Another Kind of Life

The visitors in the spacecraft glide even closer to Earth's surface. Now they can see plants growing in straight rows. The rows form squares that look like a checkerboard on Earth's surface. It seems as if the plants are arranged this way on purpose. What could have done this?

The visitors talk excitedly. Then a short distance away, they notice something else. Gray ribbons stretch across the land. Beside these ribbons there are gray and brown and red structures of different sizes. The visitors are seeing highways and streets and buildings.

The visitors realize that there is another kind of life on Earth. Intelligent beings live here. These beings can change the land to suit their needs. The rows of plants are their farmlands; the buildings make up the villages and cities where they live. Earth is indeed very different from the other planets in this solar system.

For the visitors from space, this trip is only a beginning. What they have seen so far raises many new questions. For example:

- Why is there plant life on some parts of Earth and not on others?
- Why do those intelligent beings — people — live in some places and not in others?
- How does Earth affect the way people live?
- How do people make use of Earth?

As you read in the Introduction, geography is the study of planet Earth. In this chapter, you have looked at Earth through the eyes of visitors from space. You have learned what makes Earth different from the other planets in our solar system.

In the following chapters of this unit, you will take a closer look at Earth. You will begin to answer the questions raised by the visitors from space. You will also begin to understand how your own life shapes and is shaped by the planet you live on.

SECTION REVIEW

1. Although Earth's temperatures are right for life, these alone could not support life as we know it. What gas and liquid are also necessary?

2. What two kinds of masses make up the surface of Earth? Which of them covers about 30 percent of the surface?

3. Describe two kinds of vegetation that the visitors from space observed on Earth.

4. From their survey of Earth so far, would the space visitors be able to tell that all areas of Earth do not have the same temperatures? Why, or why not?

YOUR LOCAL GEOGRAPHY

1. Using a classroom globe or wall map (or the map on page 118), find the state where you live. Then as closely as you can, estimate where your community is located. Find its latitude and longitude. Now find where your community stands in relation to the rest of Earth. Is your community closer to the Equator or to a pole? (In other words, is your community's latitude more or less than 45 degrees?) Find three other places around the world that are on the same latitude as your community. Then find two other places on the same longitude. Which continent is farthest from your community? Which ocean is nearest? Which ocean is the farthest?

2. Suppose the visitors' spaceship travels right over your community. They are looking down from a height at which they can see objects or patterns on the ground that are bigger than 100 yards (90 meters) square — about the size of a football stadium. Could they see any signs of human life in your community? If so, give three examples. What could they tell about your community from those features?

Understanding Scale

Imagine that you are approaching North America from outer space. High above Earth's surface, you see an entire continent framed in your spacecraft window. The image looks as small as in the **map** below, which shows North America reduced, or "scaled down," to fit on a page.

Suppose you compare North America's size on the map with its actual size on Earth. The proportion between them is called **scale.** The map is a **scale model** of a section of Earth, just as a globe is a scale model of the entire planet.

Every inch or centimeter on a map or globe represents a much bigger distance on Earth. For example, one inch on the map of North America below represents about 45 million inches on Earth. Therefore we say that the map's scale is about one to 45 million. We can also show this as 1:45,000,000.

Coming closer to Earth's surface from outer space, you would no longer see the entire continent. You would see, instead, just a part of North America, enlarged and in greater detail. The view from your spaceship window now might be of the Florida area, shown in the map on the next page.

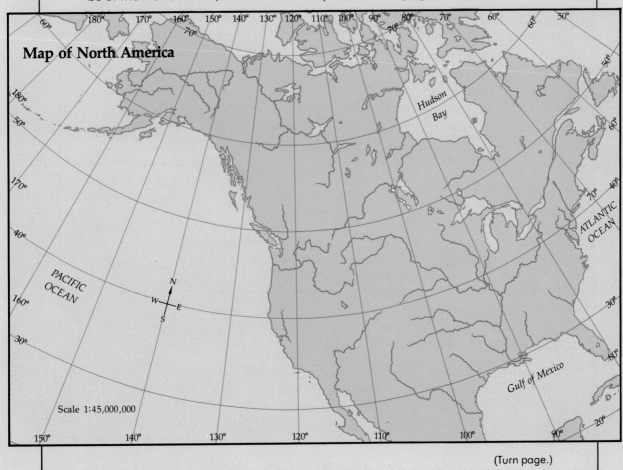

Map of North America

Scale 1:45,000,000

(Turn page.)

27

Compared to the map on page 27, this view of the Florida area is in much larger scale. Notice the scale printed to the left of the large-scale map. One inch on this map represents how many inches on Earth? There are also **scale lines** with a series of marks and numbers. These lines show what distances on the map represent in *miles* and *kilometers* on Earth. Use the scale lines to answer the following questions on a separate sheet of paper.

1. The markings and numbers on the top scale line refer to miles. What do markings and numbers on the bottom scale line refer to?

2. On the top scale line, the distance between the longer marks stands for 50 miles on Earth. On the bottom scale line, the distance between longer marks stands for how many kilometers?

A scale line is used for figuring real distances on Earth between places shown on a map. Here's how to figure real distances. Put the edge of a sheet of paper along one scale line. With your pencil, mark off points on the edge of your paper to match the longest marks above or below the line. Number the pencil marks in miles or kilometers to match the numbers on the scale line. You now have a movable ruler that you can use to measure distances on the map.

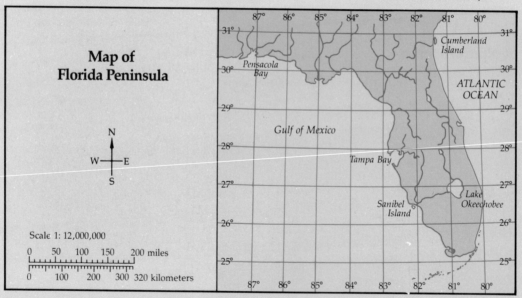

Use your movable scale line to measure straight-line distances between the following points on the map above. First find each distance in miles. Then find each distance in kilometers. Write your answers for the distances below.

3. From the western rim of Lake Okeechobee to Sanibel Island.

4. Across Florida from the Atlantic coast to the Gulf coast, along 28°N.

5. From Pensacola Bay to Cumberland Island.

6. From 25°N, 81°W to Cumberland Island.

7. From 30°N, 87°W to 25°N, 80°W.

Understanding Map Projections

A globe is the most accurate way to represent Earth. Like the planet itself, a globe is a sphere with a curved surface. A flat map of the world always distorts the surface — stretching or splitting it in some way. To see why, cut an orange or grapefruit in half. Scoop out the fruit from one half. Then try to flatten the peel. You simply cannot make a curved surface flat without stretching or splitting some part of it.

Nevertheless, a flat map is usually more convenient to use than a globe. So mapmakers have worked out various ways to represent a round Earth on a flat surface. These flat maps are called **projections** — as if different views of the globe were projected or thrown onto a flat screen.

Different projections are useful for different purposes. One may distort shape but show distance accurately. Another may distort distance but show direction accurately. The **Robinson projection** (page 573), which has minor distortions around the poles, is often used for world maps.

The map below is called an **interrupted projection.** Do you think this is a good name for it? On an interrupted projection, most continents are the right shape. You are also shown quite accurately how the continents compare with each other in size. Refer to this map to answer the questions below.

Interrupted Projection

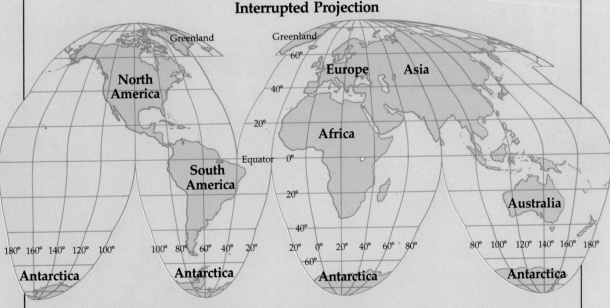

1. Are the interruptions on this map made mostly over continents or over ocean areas?
2. Would this interrupted projection be a useful map for ship navigators? Explain why, or why not.

(Turn page.)

The map on this page is called a **Mercator projection**. It is named for its inventor, Gerardus Mercator, a European mapmaker who lived in the 1500's. How does a Mercator projection compare to a globe?

On a Mercator projection, parallels and meridians *do* meet at right angles as they do on a globe. But meridians of longitude do *not* come together at the poles as they do on a globe. On a Mercator projection, they are pulled apart and drawn as straight lines. Moreover, parallels are *not* the same distance apart on a Mercator projection as they are on a globe. They are spaced farther and farther apart north and south of the Equator. This is why land areas, especially toward the poles, look much bigger than they really are.

Nevertheless, the Mercator projection is useful. Ocean navigators use it because the direction of a ship's course can be plotted as a straight line.

Use the Mercator projection to decide whether each of the following statements is true or false. Write your answers on a separate sheet of paper.

3. Meridians of longitude on this map are *not* drawn parallel.
4. Parallels of latitude are drawn 20 degrees apart on this map.
5. On a Mercator projection, parallels of latitude are spaced farther apart near the poles.
6. Greenland has only about one eighth the area of South America. On a Mercator projection, Greenland looks much too large.
7. If you wanted to compare information about world land areas, the Mercator projection would be more useful than the interrupted projection.

Mercator Projection

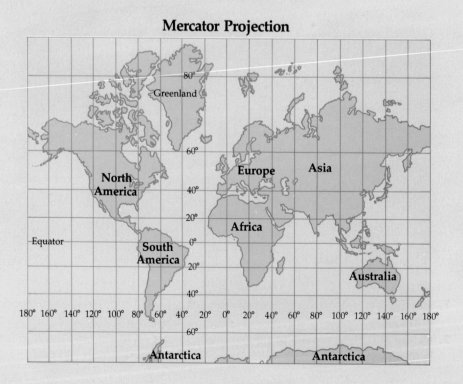

CHAPTER REVIEW

A. Words To Remember

In your own words, define each of the following terms.

atmosphere   Equator   latitude
continent   hemisphere

B. Check Your Reading

1. What is the solar system? How many planets do we have in our solar system?

2. How does Earth differ from the other planets when seen from space?

3. What three things does Earth have that enable it to support life?

4. If you approached Earth from space, what are the first signs of intelligent life that you would see?

5. Which contains the most land area, the Northern Hemisphere or the Southern Hemisphere? How can you tell?

C. Think It Over

1. The main color of Earth as seen from space comes from the water on its surface. Why does water set the color, rather than land or ice?

2. Arrange these three planets in order from coldest to hottest: Earth, Venus, Pluto. On which of these planets would the Sun appear biggest in the sky? Why?

3. Why would the space visitors notice plant life before animal or human life?

D. Things To Do

Prepare a bulletin-board display of our solar system. Use an encyclopedia to make lists of basic facts about each planet: size, appearance, etc. If possible, find actual pictures of the planets in current magazines. You could use different colors of construction paper to represent the Sun and the nine planets revolving around it.

Chapter 2

Earth as a Life-Support System

Have you ever said thank you to a plant? This thought is not as silly as it sounds. Plant life is of the first importance to those who live on Earth. Without it none of us could survive.

Think back to Chapter 1. The color of leaves and grass was one of the first things the visitors from space noticed about Earth. That's because no other planet around the Sun has plant life. The many shades of green told the visitors that life exists here.

This vegetation is a source of life for Earth's people and animals. It provides a gas called oxygen, which we need to breathe. It also provides food for growth and energy. People and animals eat plants such as vegetables and fruits. People and animals also eat other animals that feed on plants, such as cattle and hogs. In addition to oxygen and food, vegetation supplies people with many other important things — fuel, clothing, paper, medicine, and material for buildings. Plants and trees offer shade, homes for animals, and protection from wind. They also provide pleasure and beauty.

Through the centuries, people have learned more and more uses for Earth's plant life. To suit their needs and wants, people have changed much of the **natural vegetation** (the land's original plant life). They have done this by putting the land un-der **agriculture** (farming). Where wild grasses once grew, people have planted corn or wheat. Forests have been cleared for fields of rice. Deserts have been turned into orchards.

Still, there's a limit to these changes. Although a few plants will grow almost anywhere on Earth, most plants grow well only in certain areas, and grow poorly or not at all elsewhere.

Why won't all plants grow everywhere on Earth? The answer is that plant growth depends on two vital conditions that vary from place to place: energy from the Sun's rays (warmth) and moisture (water).

These two conditions create an area's climate. (*Climate* refers to the kind of weather conditions that are found in an area over a period of time.) Different places around Earth can have very different climates. In other words, the amount of the Sun's energy and moisture can vary greatly from place to place. Let's see why, beginning with the Sun's energy.

Earth and Sun

"Whew! The Sun's hot today!"

People often talk as if the Sun itself changes from hot to cold. But the Sun produces the same amount of energy all the time. What changes is the amount of the

---

The rainbow is a sign of sunlight and moisture — the two basic needs of Earth's plant life. Plants, in turn, provide food and oxygen to support human and animal life.

Sun's energy that different places on Earth receive at different times. When the place we're in receives more of that energy, we feel hot. When it receives less, we feel cold. We measure these differences in degrees of temperature. Degrees are shown as °F on the Fahrenheit (FAR-uhn-hiet) scale and °C on the Celsius (SEHL-see-uhs) scale.

Since the Sun gives out the same energy all the time, why do different places on Earth receive different amounts? There are two reasons. First, the amount of energy falling on a place depends on the period of sunlight — the length of time the Sun is above the horizon. Second, the amount of energy depends on the angle of the Sun's rays as they fall on the place.

Let's look first at the length of the day. What causes the change from daylight to darkness? Earth makes one complete **rotation** (turn) every 24 hours. When the part of Earth you live on faces the Sun, it is daytime. You watch the Sun "come up" in the morning, "rise" higher and higher in the sky as the day goes on, and then "go down" in the evening. (Of course, the Sun only *appears* to move. Actually Earth is turning.) When the part of Earth you live on faces away from the Sun, it is night. Then you do not get any direct heat or light from the Sun.

At some times of the year, the Sun may come up early in the morning and go down late in the evening. At other times, the daylight hours are shorter. In fact, the period of daylight changes constantly, day by day. In the next section, you will see why.

SECTION REVIEW

1. Plants supply Earth with a gas that is necessary for life. What is this gas called?

2. What is natural vegetation?

3. What do we call the kind of weather conditions that are found in an area over a period of time?

4. The amount of the Sun's energy falling on a place depends on two things. What are they?

Understanding a Functional Diagram

Earth is a planet in motion. Spinning like a top, it turns around completely every 24 hours. This motion, which produces day and night, is called **rotation**. At the same time, Earth is traveling in a giant yearly **revolution** (circular movement) around the Sun. This motion leads to the different seasons.

Earth's motions can be illustrated by **diagrams** (drawings), like the ones in this lesson. We call these drawings **functional diagrams**. *Functional* comes from a Latin word which means "to perform." Functional diagrams are action diagrams. They show how something works. They show movement.

Earth's Rotation

Diagram A shows Earth rotating on its axis. Earth's **axis** is like an imaginary rod through the center of the planet, from the North Pole to the South Pole. In the diagram, curving arrows at the poles show the direction of Earth's rotation.

You can see from the diagram that only one side of the planet can face the Sun at a time. This side is in daylight. The other side is in darkness. On a rotating Earth, places are always moving from daylight to darkness — from day to night — and back again.

Study Diagram A. Then decide whether each of the following statements is true or false. Write your answers on a separate sheet of paper.

1. Earth rotates on its axis from west to east.
2. It is noontime along the part of Earth that is closest to the Sun.
3. When it is noontime anywhere on Earth, it is midnight on the side of the planet directly opposite.
4. When it is midnight anywhere on Earth, the Sun is rising on the side of the planet directly opposite.

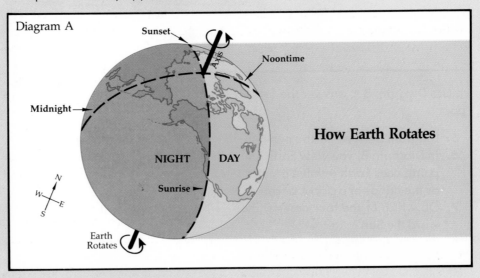

Diagram A — How Earth Rotates

(Turn page.)

Earth's Revolution

Diagram B shows Earth's yearly revolution around the Sun. In this diagram, you can see Earth's position on four different dates of the year. The large curved arrows show Earth's path around the Sun. Notice how Earth's axis is **inclined** (tilted).

Diagram B

Earth's Revolution Around the Sun

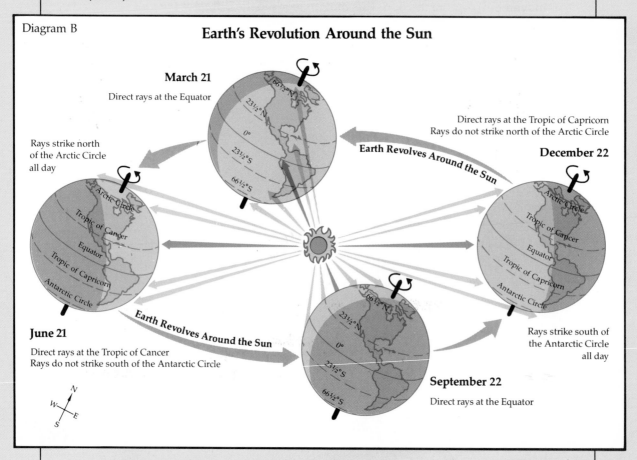

Use Diagram B to answer the following questions on your separate sheet of paper.

5. Diagram B shows Earth's position on which four dates of the year?

6. In Diagram B, you view Earth from above the North Pole. From this viewpoint, does Earth revolve around the Sun *clockwise* (in the same direction as the hands of a clock), or *counterclockwise* (in the opposite direction)?

7. On which of the four dates shown is the Northern Hemisphere tilted toward the Sun?

8. On which of the four dates is the Northern Hemisphere tilted away from the Sun?

Special Parallels

Because Earth is tilted as it revolves around the Sun, the direct rays of the Sun strike different parts of the planet at different times of the year. This causes the change in seasons (*see page 38*).

Look again at Diagram B on the opposite page. Thin arrows in the diagram show where the Sun's rays strike Earth at different times. These places on Earth are marked by the Equator and four other special parallels of latitude. The **Tropic of Cancer** (at $23\frac{1}{2}°N$) is the most northerly parallel where the Sun's rays strike from directly overhead. The **Tropic of Capricorn** (at $23\frac{1}{2}°S$) is the most southerly parallel where the Sun's rays strike from directly overhead. The **Arctic Circle** (at $66\frac{1}{2}°N$) is the farthest parallel that the Sun's rays reach over the North Pole in the northern summer. The **Antarctic Circle** (at $66\frac{1}{2}°S$) is the farthest parallel that the Sun's rays reach over the South Pole in the southern summer.

Refer to Diagram B to write your answers to the following questions.

9. The Sun's *direct* rays strike Earth at which special parallel on: (a) June 21?   (b) December 22?   (c) March 21?   (d) September 22?
10. On June 21, the Sun's rays reach: (a) all of Earth north of which special parallel?   (b) none of Earth south of which special parallel?

The angle of the Sun's rays affects the temperatures and hours of daylight on Earth at different latitudes. For example, the Sun's rays strike the planet *most* directly in the **low latitudes** or **tropics.** These areas lie between the Tropic of Cancer and the Tropic of Capricorn, with the Equator in the middle. The Sun's rays strike Earth *least* directly in the **high latitudes** near the North and South poles. These include the *Arctic zone* between the Arctic Circle and the North Pole, and the *Antarctic zone* between the Antarctic Circle and the South Pole. The areas between the low and high latitudes in the Northern and Southern hemispheres are known as the **midlatitudes** or **temperate zones.** (*Temperate* means moderate — without extremes of heat or cold.)

Refer to Diagram B again to answer the following questions.

11. In which latitudes would you expect to find: (a) the warmest temperatures?   (b) the coldest temperatures?
12. Which zone has 24 hours of daylight on: (a) June 21?   (b) December 22?
13. On which dates of the year do the Arctic and Antarctic zones have equal hours of daylight?

Around the Year

Why aren't there 12 hours of daylight everywhere on Earth, all year round? The reason is that Earth is inclined, or tilted, on its axis. The axis is an imaginary axle around which Earth revolves.

As you can see in Diagram B on page 36, either the Northern or the Southern Hemisphere can be tilted toward the Sun. The other hemisphere then is tilted away. The hemisphere that is inclined toward the Sun receives more sunlight, so daylight there lasts longer than 12 hours. The other hemisphere, of course, is inclined away from the Sun and receives less than its share. There daylight lasts less than 12 hours.

Does this mean that one hemisphere always has more daylight than the other? No, because even as Earth is rotating (spinning) on its axis, it is also revolving (moving in a circle) around the Sun. As Earth revolves, it stays tilted at the same angle. This means that the Northern and Southern hemispheres each have their turn receiving a larger share of sunlight.

Earth makes one complete revolution around the Sun in a year. Look at Diagram B again, and you will see that on June 21, the Northern Hemisphere is tilted closer to the Sun. On or around this date, places in the Northern Hemisphere have their longest period of daylight. It is the beginning of the season called summer.

What is happening to the Southern Hemisphere at this time? You can see that it is tilted away from the Sun. So places in the Southern Hemisphere are having their shortest period of daylight. There June 21 marks the season called winter.

Six months later, on December 22, Earth has traveled halfway around the Sun and is on the opposite side. But as you can see from the diagram, Earth is still tilted in the same direction. This means that the Northern Hemisphere is now tilted *away* from the

Sun. So the Northern Hemisphere is having its shortest daylight period of the year, and the season beginning now is winter.

In the Northern Hemisphere, the daylight period gradually becomes shorter from June 21 to December 22, and then starts lengthening again. In the Southern Hemisphere, the change is the other way around. Each of these dates is known as a **solstice** (SOL-stis), from Latin words meaning "Sun at standstill." The date when daylight is at its longest is called the **summer solstice**, while the shortest day is the **winter solstice**.

Twice a year — around September 22 and March 21 — both hemispheres have days and nights that are the same length. Each of these dates is known as an **equinox** (EE-kwuh-noks), from Latin words meaning "equal night." These dates mark the beginning of two other seasons. In the Northern Hemisphere, fall begins around September 22 and spring around March 21. In the Southern Hemisphere, it's the other way around.

On March 21 and September 22, every place on Earth has about the same period of daylight — 12 hours. Whether you live in Alaska, Puerto Rico, or anywhere in between, you'll have 12 hours of daylight.

Temperature and Latitude

The amount of the Sun's heat reaching a place depends on more than the length of the day. It also depends on the angle of the Sun's rays falling on the area.

You can see how this angle varies by looking at the diagram on the opposite page. Notice how the Sun's rays fall directly on areas near the Equator but strike areas near the poles at a slant. Thus the same amount of energy is spread over a wider area near the poles than near the Equator and cannot warm Earth so much.

For the same reason, the Sun warms Earth less in winter than in summer. In most places on Earth, the Sun is higher in the sky

The Sun's Rays on Earth

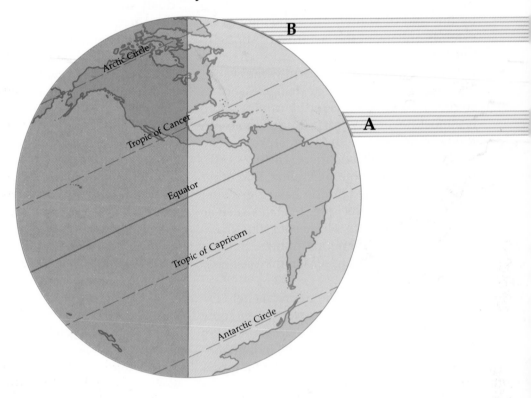

The Sun's rays warm the low latitudes more than the high latitudes. The bands of sunlight shown falling on A and B are the same width and therefore have the same amount of energy. But that energy is spread much more thinly over Earth's surface at B than at A.

in summer than in winter. So the Sun's rays also bring more heat in summer because they strike Earth more directly.

The area of Earth that changes least throughout the year is the tropics. There the Sun is high in the sky all the time. There too, as you learned earlier, the daylight period stays about the same throughout the year. In much of the tropics, it's hard to see any difference between the four seasons. This zone has been called "the land where winter never comes."

Growing Seasons

Energy from the Sun is important for plant growth. All plants need some warmth. Most grow quickly when it's hot, slowly when it's cool, and not at all when it's freez-

ing cold. In the U.S., for instance, plant growth starts slowly in the spring. By summer when the temperature reaches 80°F (28°C), farmers sometimes say, "You can sit and watch the corn grow." That's because it grows so fast then. In the fall, when temperatures drop, the growth slows.

The period when it is warm enough for a plant to begin and complete its growth is called the **growing season**. This season varies sharply around the world. The tropics are nearly always warm and have no winter

39

to interrupt the growing season. Plants in the tropics can grow all year round.

Now look at the higher latitudes, outside the tropics. Here plant growth is interrupted by the cold of winter. The growing season extends only from the last frost of spring to the first frost of fall.

The length and harshness of winter varies widely, and so does the growing season. In southern Florida, close to the tropics, there is almost never any frost. In fact, there is a town in Florida named Frostproof. The growing season is long there. Farther north, most of Georgia, South Carolina, and other states on the same latitude have a growing season eight months long. In states such as Maine and northern Minnesota, the growing season is less than four months. In parts of Alaska, the growing season may last only a few weeks.

These different growing seasons help explain why many plants grow well in some places but not in others. Some plants may grow quickly enough to do well in high latitudes. For example, potatoes need a growing season of only three months or less. On the other hand, the cotton plant needs well over six months. You can see why a farmer in Maine would not try to grow cotton. Would that farmer be able to grow potatoes?

SECTION REVIEW

1. When the Sun's rays are striking an area *most* directly, what season is it there?

2. What is the growing season?

3. The length of the growing season varies from place to place. How long does it last in the tropics?

4. Celery and lettuce are cool-weather crops. They grow well when the temperature is above freezing, but not too high. (a) Name some places where celery and lettuce would probably not do well at any time of year. (b) Name some places where celery and lettuce could be planted only part of the year.

Water and Climate

A tropical Sun burns down on dry and dusty soil. Brown stalks of grass lie flat and dead across the village fields. Looking at this scene in central India, you might think that no plants could ever grow there again. But then the sky fills with dark storm clouds, and torrents of rain flood the land. Before long, new shoots of wheat are peeping out of the mud.

Though energy from the Sun is vital for plant growth, it is not enough by itself. Let's look at the other important need of plant life — water.

In Chapter 1, you read that much of Earth's surface is covered by water. Water from Earth's oceans, lakes, and rivers provides the moisture needed for plant growth. But how does water get to the plants?

As you know, water can exist not only as a liquid but also as a solid, when frozen, and as a **vapor** (gas) when warmed. You can see it as vapor when it comes from a boiling kettle, or when it rises from the ground after a shower on a hot day. Warmth causes the water to **evaporate** (change from liquid to gas). There is always some water vapor in the air, even when you can't see it. Air that contains a lot of water vapor is said to be **humid**, or to have high **humidity**.

Just as liquid water changes to vapor when warmed, water vapor changes back to liquid when cooled. The vapor is then said to **condense**. If you take a container out of the refrigerator, you will see tiny drops of water collect on the outside. The container has cooled some of the water vapor in the air and made it condense.

Now look at the diagram opposite. It shows how water goes from Earth's surface to its atmosphere and back again to the surface. This is called Earth's **water cycle**.

The warmth of the Sun causes water on Earth's surface to evaporate into the air. At the same time, the Sun warms the air near the surface, causing it to expand. This makes the mixture of air and water vapor lighter than the surrounding air, and so it rises.

Temperatures in the air become lower as **altitude** (height) increases. So as the vapor rises, it begins to cool. This makes it condense and form water droplets that you see as clouds. If the clouds are cooled more, the water droplets become bigger and fall back to Earth. Usually they fall as rain or snow, but there are other forms in which water droplets may reach the ground. (Can you think of some?) All these forms of falling water are known as **precipitation** (pri-sip-uh-TAY-shuhn).

Wet Lands or Dry Lands?

"Ballgame stopped by rain." . . . "Snow blocks highway." . . . "Hail damages crops." . . . Moisture from the clouds can cause trouble. But most of the time it quietly provides the water needed by plant life on Earth. However, if there is a **drought** (an extra-long period without precipitation), plants may wither and die.

Some places on Earth are dry nearly all the time. Some are wet nearly all the time. Most

Moisture evaporates mainly from oceans, but some rises from lakes, rivers, and moist land. The vapor condenses and falls as precipitation — which evaporates again. "Drift" is moist air moving from oceans to land; "runoff," water flowing from land to oceans.

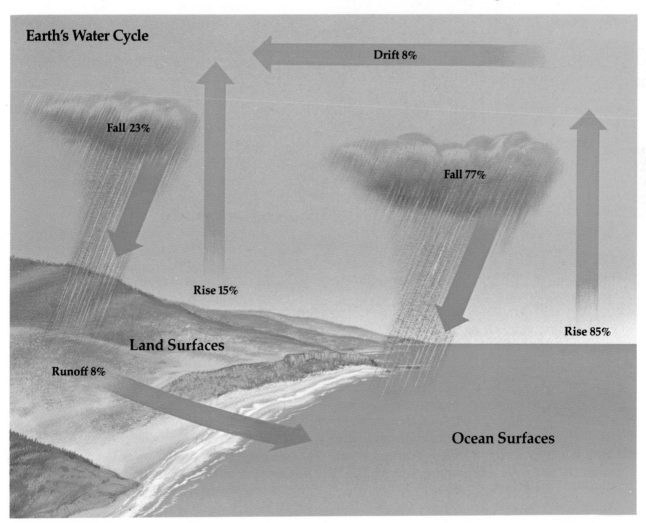

Earth's Water Cycle

Drift 8%

Fall 23%

Fall 77%

Rise 15%

Rise 85%

Land Surfaces

Runoff 8%

Ocean Surfaces

Earth's Moving Air

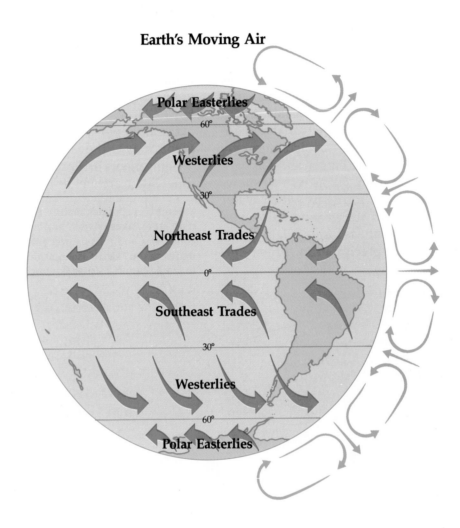

Polar Easterlies

60°

Westerlies

30°

Northeast Trades

0°

Southeast Trades

30°

Westerlies

60°

Polar Easterlies

places on Earth fall somewhere between these two extremes. Why do different places receive different amounts of rain?

In most places, there is more rain in the summer than in the winter. This is because warm air can hold more moisture than cold air. Hot summer air of 100°F (*38°C*) can hold twice as much moisture as cold winter air of 32°F (*0°C*). So any rainfall from that summer air can be twice as heavy. Since places in the tropics have warm weather all year long, many of them get a lot of rain.

Some warm areas on Earth are very dry, however. Many of the world's biggest des-

This is the basic pattern of Earth's winds. Arrows at right show how air rises at the Equator, moves out, and falls (causing similar movements at higher latitudes). It then moves back at ground level. Earth's rotation sets these winds at an angle (heavy arrows).

erts lie just on the edge of the tropics. (A **desert** is a dry area where ordinary plants cannot grow.) What happens here? For one reason or another, the air has not been able to collect enough water vapor from Earth's surface. For example, if the air has come a long distance across land, or traveled high above Earth's surface, it will remain dry.

Thus precipitation depends on the way the air moves across Earth's surface. This movement of air is known as wind. What makes the wind blow? The basic reason is that warm air rises. In an area of Earth heated by the Sun, the air becomes warm and rises from the surface. As it rises, cooler air moves in from surrounding areas to take its place. This cooler air becomes warm and rises in turn, while the warm air cools and falls back toward the surface. As long as this movement continues, a wind blows toward the heated area.

This happens on a large scale all around Earth. Air rises at the Equator, and cooler air moves in from north and south. Because Earth rotates, the air moves at an angle across its surface. As you can see in the diagram opposite, the heated air sinks to Earth around 30° latitude and then blows back toward the Equator. At the same time, some of the air at 30° is pushed the opposite way. Thus there are other movements of the air in the higher latitudes.

The diagram shows what are called **prevailing winds**. These are the winds that blow most often in the different parts of Earth. Most of the U.S. is in the belt of the **westerlies** — that is, winds which blow *from* the west. (Winds are always named for the direction from which they come.) However, winds in most parts of the U.S. may blow from any direction at one time or another. The diagram shows how the winds would blow all the time if Earth's surface were perfectly smooth. Since the surface varies, so do the winds.

Why Moisture Becomes Rain

Rain falls when warm, moist air rises. But what makes the moist air rise and drop rain on one place rather than another? There are three main causes.

In or near the tropics, Earth's surface is often hot enough to warm the air above it and make the air rise. If the air is moist enough,

rain will fall. If there is no strong wind, the water may evaporate, rise, and fall in another shower on the same place.

Another cause of rain involves larger masses of air. A warm, moist mass of air may run up against a mass of colder air. The two masses may interact in various ways that bring rain. For example, since the cold air is heavier, it may stay down low while the warm air is forced to move up and over it. Thus the rising warm air is cooled and forms rain clouds. Because large masses of air are involved, the rain may last for a long time.

The shape of the land can force moist air upward in a similar way. Suppose winds blow across the ocean, becoming moist, and then reach land. If there are mountains rising high above sea level, the moist air will run into them and be forced upward. As the air cools, it will form rain clouds. Mountain slopes facing winds from the oceans are some of the wettest places on Earth.

You have seen that rainfall depends on the temperature of the air, how much moisture the air is carrying, the direction in which the air is moving, and the shape of the land. In many places, the first three conditions vary at different times of the year. In these places, the amount of rainfall can also vary around the year.

The shape of the land may be quite different in two places that are close together. As a result, these places may receive completely different amounts of rain. In the state of Washington, for example, there is a dramatic contrast between the western and eastern sides of the Cascade Range. The western slopes, which face the Pacific Ocean, are on the **windward** side of the mountains — the side toward which the wind usually blows. The eastern slopes are on the **ieeward** side — the side that is sheltered from the wind.

As a result, the western slopes may be drenched with more than 100 inches (*250 centimeters*) of rain each year. By the time the

air from the Pacific reaches the eastern slopes, however, it has lost most of its moisture. On the eastern side of the mountains, there are dry areas with an annual rainfall of less than five inches (*13 centimeters*). These areas are said to be in the **rain shadow** of the mountains.

Plants and Moisture

On the wet side of the Cascade Range, you can find Douglas fir trees and red cedars that stretch hundreds of feet into the sky. On parts of the dry side, you can find only small scrublike plants. Here, as in most of the world, the amount of moisture determines the type of vegetation that will grow. There are four basic types of vegetation.

Forests need the most moisture, because trees are much bigger than most other plants. Trees cannot grow in places where the ground dries completely for part of the year.

Grasslands need less moisture than forests. Grasses are smaller than trees and grow much more quickly. Even in places that have a long, dry season, grasses can spring up again as soon as the rainy season begins.

Desert plants can get by with very small amounts of moisture. Perhaps you live near a desert or have had a desert cactus as a houseplant. If so, you know it has a tough "skin" that lets it store moisture with little evaporation. It can survive without water for a long time.

Tundra, the fourth type of vegetation, is determined by the Sun's heat, not by moisture. It is found in the cold, high latitudes. Tundra consists of mosses and small flowering plants. Parts of the tundra zones may be moist enough for grasslands or even forests, but there is just too little sunlight and warmth for these other types of vegetation to grow.

From Plants to Other Life

The different kinds of plant life on Earth cover enormous areas of its land surface. That's why the visitors from space could not miss seeing it, even from a height of many miles. But what about the other forms of life on Earth? To see those directly, the visitors would have to descend much closer.

As the visitors skim low over Earth's surface, they begin to see for themselves how vegetation supports other forms of life. They see antelopes eating grass, and giraffes stripping twigs and leaves from trees. These animals live directly off the plants around them. Then the visitors see a lion kill and eat an antelope. Animals like the lion live *indirectly* off the plants that other animals eat. The visitors can guess that most land animals are found where there is the most natural vegetation.

What about human life? Where do most of Earth's people live? Do they also settle where there is the most natural vegetation?

The answers to these questions aren't quite so simple. There are reasons besides food that determine why people settle where they do. You will discover some of these reasons in the next chapter.

These pictures show the environments of four basic types of vegetation. Above left: grasslands in Africa. Above right: tundra in North America. Below left: a desert area in Asia. Below right: a forest in Europe.

SECTION REVIEW

1. Describe the process through which water from oceans, lakes, and rivers provides the moisture needed for plant life. What is this called?

2. In what direction do winds mostly tend to blow in the low latitudes north of the Equator? Why?

3. Why do most places on Earth usually get more rain in summer than in winter?

4. Name two factors that determine how much rain will fall in any one place.

Reading Map Symbols

Climate refers to a place's weather conditions over long periods of time. You already know how temperatures on Earth vary at different latitudes. To find out about differences in rainfall — or rainfall patterns — you could use a map like the one below. Like many maps, this one uses **symbols** (images) to stand for real objects. Some maps use small pictures as symbols. Others, like this one, use lines, dots, or other simple shapes.

A map's **key**, also called a **legend**, is a list of its symbols. Refer to the key for this map. Answer the following questions on a separate sheet of paper.

1. Which symbol in the key stands for: (a) heavy rainfall all year?   (b) moderate rainfall with dry summers?   (c) light rainfall?

2. Which rainfall pattern is represented by light slanted lines?

Notice how symbols are arranged on the map. They show only general patterns of rainfall. They do not show any gradual changes from one pattern to another that you would find on Earth. Answer the following questions.

3. Near which latitude would you find most places with heavy rainfall all year round? Which rainfall pattern would you find in the bordering regions?

4. Are areas of moderate rainfall found mostly in Earth's low, middle, or high latitudes? In which latitudes would you find areas with low rainfall?

Earth's Yearly Rainfall Patterns

KEY

HEAVY RAINFALL	MODERATE RAINFALL	LOW RAINFALL
All Year	All Year	Light Rainfall
Dry Winters	Dry Summers	Little or No Rainfall

Comparing Maps

The map on this page shows the general patterns of vegetation around Earth. By comparing it to the rainfall map, you can draw conclusions about the effect of rainfall on vegetation. This map uses different colors as symbols instead of lines and shapes. The key tells what each color symbol stands for. Study the map and key. Answer these questions on a separate sheet of paper.

1. How many different types of forest vegetation are represented in the key? What types are represented?
2. Which color represents: (a) tall grasses? (b) desert shrubs and plants?

Now use the key to find patterns on the map. Answer these questions.

3. What is the main type of vegetation in northern Africa?
4. What type of forest vegetation is found in land areas near the Equator?
5. What is the main type of vegetation north of the Arctic Circle?

Does rainfall affect Earth's natural vegetation? Compare this map with the map on the opposite page. Answer the following questions.

6. What kind of vegetation is found in areas with: (a) heavy rainfall all year? (b) dry winters? (c) dry summers? (d) low rainfall?
7. What effect does rainfall seem to have on Earth's vegetation patterns?

Earth's Natural Vegetation Patterns

KEY

FORESTS

- Broadleaf Evergreen (Rain Forest)
- Mixed Forest
- Needleleaf (Coniferous) Forest
- Scrub and Thorn Forest

GRASSLANDS

- Tall Grasses
- Short Grasses

OTHER

- Desert Shrubs and Plants
- Tundra (Low Plants and Mosses)
- Little or No Vegetation (Mountains, Ice, etc.)

YOUR LOCAL GEOGRAPHY

1. Are there big differences between the seasons where you live? Or are the differences only slight? Make a list of items in your environment that change throughout the year. Beside the list, draw four columns headed Spring, Summer, Fall, and Winter. In each column, note the differences in each item.

Here are some of the items you might want to consider: Sun's position at midday; temperature at midday; length of daylight; amount of precipitation; type of precipitation (rain, snow, etc.); condition of trees and other plants. For temperatures, length of daylight, and precipitation, you can simply make comparisons (more, longer, hotter, etc.). To check these, you can ask at your school or local library where to find weather records.

2. Suppose the visitors from space landed in your community. They want to find out more about your environment by studying plant life. What kinds of plant life would they find? Which of these would not be natural but planted by humans? How could the visitors tell? (In Chapter 1, for example, they knew that crop fields were planted by humans because the fields formed neat squares and rectangles. See if you can think of other clues.)

There are four basic types of natural vegetation: forests, grasslands, desert plants, and tundra. Suppose the visitors wanted to know which type best described the area where you live. Which would you choose? Why?

Is this natural vegetation? There are signs of human life in the squared-off fields separated by hedges. Furthermore, the buildings do not look like natural objects, and the grouping of animals seems planned.

CHAPTER REVIEW

A. Words To Remember

From the following list, select the term that best completes each of the sentences below. On a separate sheet of paper, write your answer next to the number of each sentence.

climate   revolution   tropics
high latitudes   rotation   water cycle
midlatitudes   scale   westerlies
precipitation   symbols

1. _____ is the kind of weather conditions that are found in an area over a period of time.

2. Earth makes one complete _____ (turn) every 24 hours.

3. The zone around the Equator between the Tropic of Cancer and the Tropic of Capricorn is known as the _____ .

4. The process by which moisture goes from Earth's surface to its atmosphere and back again to the surface is known as the _____ .

5. Images used on maps to stand for real objects are known as _____ .

B. Check Your Reading

1. Plant life supplies us with oxygen and what other basic need?

2. The climate of a place depends on two things. One is energy from the Sun. What is the other?

3. On part of Earth, the length of day stays much the same all year round. Is this true of the high latitudes, the midlatitudes, or the low latitudes?

4. What is the name given to the period of time when it is warm enough for a plant to begin and complete its growth?

5. What is a prevailing wind?

C. Think It Over

1. Where on Earth would the sky be dark at noon in June? Why? What would the sky be like in that same place at midnight in December?

2. Compared with Florida, is the growing season in Minnesota longer, shorter, or the same? Why? The Equator runs through the nation of Zaire in Africa. What kind of growing season does Zaire have?

3. Suppose you live in an area near the coast where there are mountains. Winds often blow from the ocean over the land. What would the rainfall be like on the ocean side of the mountains? What would it be like on the inland side?

D. Things To Do

Make a list of your three favorite fruits, vegetables, or other food crops. Are they available throughout the year in your community? Can you guess why, or why not? Look up each one in an encyclopedia at the library. Note where each grows best and at what time of year. Report your findings to the class.

Chapter 3

Oceans and Land

"All right!" says the spaceship captain. "Let's see what we know about this planet so far."

"The blue areas of its surface are water," says one crew member. "The areas of other colors are land."

"Right," says another crew member. "We also know that the green on the land is vegetation. And the puffs of white up in the air are clouds of water vapor. They carry the moisture that the vegetation needs."

The captain scratches one of his foreheads. "But we still don't know much about the land itself. What shape is it? What's it made of?"

The crew looks at the viewing screen, which shows the land they are flying over. When the spaceship comes down low, the visitors can see that Earth's surface is molded into many different shapes.

Geographers measure the height of Earth's land surfaces from the point where land and sea meet. They call the height of land above sea level **elevation**. When geographers talk about *changes* in elevation, they use the term **relief**. From the low-flying spaceship, the visitors can see areas of **high relief** (big changes in elevation), such as

mountains, and areas of **low relief** (flat areas), such as sandy coastlines. (You will read more about elevation and relief in Geography Skills 8.)

The depth of the ocean floor changes too. The edges of the continents slope gently under the oceans for about 100 miles (*160 kilometers*), and in some places, for as far as 500 miles (*800 kilometers*). These sloping edges are called **continental shelves**. Beyond them the ocean floor drops steeply. In some places, the floor is more than six miles (*10 kilometers*) below the surface.

The land and water of Earth's surface affect our lives in two basic ways.

■ They can hinder us by setting up *barriers*. A mountain is a barrier if it forces people to travel around it.

■ They can help us by providing *resources*. A resource is anything in the environment that people can use for their own benefit. Coal dug out of the earth and fish caught in a lake are two kinds of resources.

This chapter shows the most important ways in which Earth's surface helps and hinders us. First you will look at the part of the surface covered by water, and then in the following section, at the land.

Water meets land on an island in the Mediterranean (meh-di-tuh-RAY-nee-uhn) Sea. The many shapes taken by land and water affect all living things on Earth.

The Oceans

Oceans have been important throughout history. At first they seemed to affect human life mostly as barriers that kept people on different continents apart.

About 3,000 years ago, peoples such as the Chinese, Arabs, Greeks, and Vikings began sailing on long ocean journeys. But they usually kept as close as they could to land. It was the Polynesians (pol-uh-NEE-zhuhns) of the Pacific who first traveled regularly across the open ocean. They sailed hundreds of miles in small boats from one island to another.

Over the centuries, the oceans became highways rather than barriers. Traveling by water could be easier than by land. Sailors did not have to clear a path or build a road. There were no deserts to cross or mountains to climb. What's more, ships could use an energy source that was as free as air. This was the wind, which moved ships by blowing against their sails.

Of course, sailors had problems. Some-

Few people live on Earth's water, but many make their living on it. Fishing provides a major food supply. Some fishing people use big ocean boats. Others cast their nets from rowboats, as here in Haiti (HAY-tee).

times the wind died, and ships couldn't move. At other times, the wind was too strong, blowing ships off course or wrecking them. All the same, sailors could do better than land travelers. Christopher Columbus took 10 weeks on his first voyage from Spain to the Caribbean, some 4,000 miles (*7,000 kilometers*) away. Compare this with the land journey made by Meriwether Lewis and William Clark from 1804 to 1806. They were sent by Thomas Jefferson to explore the lands west of the Mississippi River known as the Louisiana Territory. The distance from the Mississippi to the Pacific Ocean is only half as far as the distance Columbus sailed across the Atlantic. But Lewis and Clark took seven times as long to cover it!

Even today water transportation remains very important. It's still a cheap way to move cargo. If the water is deep enough, a ship can go anywhere on the open seas. Ships can also be very large. An oil supertanker has a deck big enough to cover three football fields with room to spare. Under that deck, it can carry as much cargo as a freight train 100 miles (*160 kilometers*) long!

So although oceans are barriers, people have learned to get over them. People have learned to use the ocean's resources too.

Ocean Resources

You already know about one important ocean resource — moisture. As you read in Chapter 2, it is rain that provides the moisture needed by all life on Earth. And most of the moisture in Earth's atmosphere comes from the oceans. Although ocean water is salty, the salt remains behind when the water evaporates.

Another vital resource is oxygen. The oceans contain billions of tiny plants, similar to the green algae that you may have seen in ponds or aquariums. Like the trees and grasses of the land, these ocean plants give off oxygen. In fact, because the oceans cover so much of Earth's surface, their plants produce far more oxygen than do land plants.

Another important resource from the oceans is food. The oceans are full of life. Thousands of kinds of fish, shellfish, and plants live there. In a few nations, such as Japan, seaweed is a common ingredient of soups and other dishes. In nations such as Japan, Peru, Norway, and parts of China and the Soviet Union, people eat more fish than meat.

Some of the world's best fishing grounds lie in the shallow waters over the continental shelves. There some sunlight reaches below the ocean's surface. Large numbers of sea plants and tiny animals live on these shelves. Small fish such as herring feed on these plants and animals. Larger fish such as cod feed on the herring. And humans in fishing fleets come to catch all kinds of fish, large and small.

Although the oceans are a source of food, they do not provide nearly as much food as land does. That's because the oceans are not farmed the way land is. For example, dairy farmers feed their cattle and know where to find them at milking time. Most people who fish simply go to where the fish are likely to be and try to catch them. In this way, fishing is more like hunting.

In some places, lobsters, oysters, and other shellfish *are* farmed. They are bred in underwater "pens" off the coast. But the farming of ocean fish is still in the future.

SECTION REVIEW

1. Name one landform that could give an area high relief.

2. What is the name for things in our natural environment that we can use for our own benefit?

3. Ocean plants provide Earth with a vital resource. What is it?

4. In what way is the ocean both a help and a hindrance to humans?

Reading an Elevation Map

Suppose you are on a ship in a calm, flat ocean when you see an island. Because you see the island in profile (from the side), you can tell at a glance whether or not the island rises high above the ocean. But how can you get a more exact picture of the island's surface?

Below is a **profile drawing** (a side view) of the island of Hawaii, in the Pacific Ocean. Bands of color on the profile drawing represent levels of *elevation* or *depth* — that is, how far places are above or below sea level. Next to the drawing, a key explains what each color means. The left side of the key gives distances measured in *feet*. The right side gives the same distances in *meters*.

Find the dark green band on the key. Notice how the bottom edge of this color is marked "Sea Level" and the top edge is marked for 1,000 feet. This tells you that dark green on the profile drawing represents all places on the island of Hawaii that are between sea level and 1,000 feet above sea level.

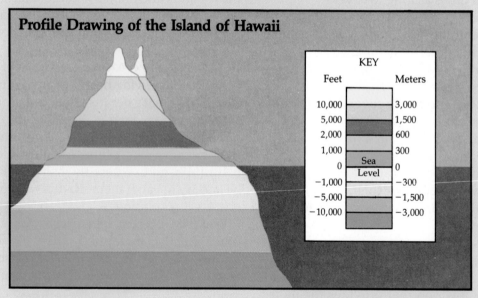

Profile Drawing of the Island of Hawaii

KEY		
Feet		Meters
10,000		3,000
5,000		1,500
2,000		600
1,000		300
0	Sea Level	0
−1,000		−300
−5,000		−1,500
−10,000		−3,000

Study the key and the profile drawing together. Then answer the following questions on a separate sheet of paper.

1. How many color bands does the key show for: (a) elevation *above* sea level? (b) depth *below* sea level?
2. Light green represents elevations between 1,000 and 2,000 *feet* above sea level. What does it represent in *meters*?
3. What range of elevation or depth (in feet) is represented by: (a) yellow? (b) brown? (c) lightest blue? (d) darkest blue?
4. The highest peaks on the island of Hawaii are above what elevation: (a) in feet? (b) in meters?

If you were flying over Hawaii in a plane, you could not see the island in profile. Instead you would have to use a flat map to figure out the island's elevation and the depth of the surrounding sea.

The map below shows the island as if viewed from above. Notice that there is no compass rose on this map. One arrow points "north." Can you figure out which way south, east, and west are on this map?

Bands of color represent elevation and depth on this map, just as they did on the profile drawing. Use the map and its key to complete these statements.

5. The northwest corner of the island rises to an elevation of about (2,000; 5,000; 10,000) feet.

6. Ten miles off the northwest corner of the island, the ocean floor slopes to a depth below (−1,000; −5,000; −10,000) feet.

On page 51, you read about *relief* — the difference in elevation between the low points and high points of an area. The pattern of color bands on a map can tell you which areas have high relief or low relief.

Compare the map and the profile drawing. Notice where the bands of color on the map are narrow. There the land rises steeply over a short distance, from one elevation to a higher one. You can see clearly from the profile drawing that these are areas of high relief. Where the map shows wider bands of color, the rise is more gradual. These areas of the island have lower relief.

Refer to the map to answer the following questions.

7. Which is the best example of a high-relief area: the area east of Kilauea Crater or north of Mauna Kea Volcano? Give reasons for your choice.

8. Is the sharpest drop in the ocean floor north of the island or south of it?

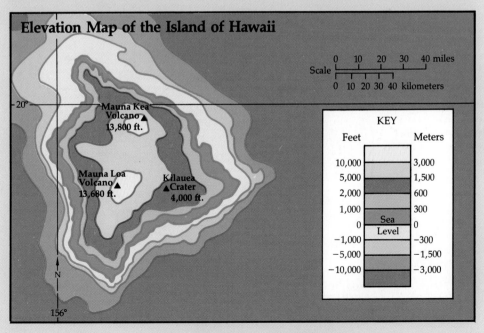

The Changing Land

"It's good to be back on solid land again!"

People often say this at the end of a bumpy ride by boat or airplane. They don't like being bounced up and down by waves or by currents in the air. Landing is like coming home. Land is where we live, and most of us think of it as solid and unmoving.

Yet Earth's surface is not always like that. Forces from deep within Earth are always causing changes on or near its **crust** (surface). Earth's crust varies in thickness, but under the continents, it is about 20 miles (*30 kilometers*) deep.

Nobody has ever traveled to the bottom of the crust, but there are some clues to what's inside Earth. (Look at the diagram below.) For example, we know that some rocks below the crust are hot enough to flow like molten (melted) metal. How do we know? Sometimes the fiery rocks burst through

gaps in the crust onto Earth's surface. This is what happens when a **volcano** erupts.

One such eruption took place in Washington State in 1980, when Mount St. Helens literally blew its top. Other eruptions soon followed. Sixty people were killed, along with thousands of animals, birds, and fish. Trees were destroyed over an area of 150 square miles (*400 square kilometers*). A vast crater was left where the top of the mountain had been.

Earthquakes also bring fast changes to the land surfaces. Like volcanic eruptions, earthquakes are caused by pressures building up in Earth's crust. In fact, most quakes

---

This is a *structural diagram*—a drawing that shows how the inside of something is structured or made up. A section of Earth is cut away to reveal its interior. How is this known? Scientists can tell partly from volcanic eruptions and partly from the different ways in which earthquake shocks travel through Earth.

Inside Planet Earth

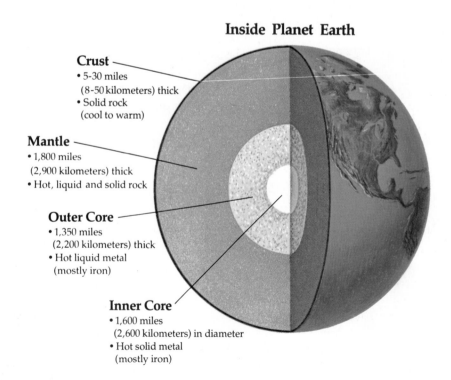

Crust
- 5-30 miles
  (8-50 kilometers) thick
- Solid rock
  (cool to warm)

Mantle
- 1,800 miles
  (2,900 kilometers) thick
- Hot, liquid and solid rock

Outer Core
- 1,350 miles
  (2,200 kilometers) thick
- Hot liquid metal
  (mostly iron)

Inner Core
- 1,600 miles
  (2,600 kilometers) in diameter
- Hot solid metal
  (mostly iron)

and active volcanoes are found in the same areas (see the map on page 483).

With earthquakes the pressures cause cracks and movements along weak sections in Earth's surface. These weak sections are known as **faults**. The best-known fault in the U.S. is the San Andreas Fault, which runs through California. Pressures on this weak spot caused quakes that destroyed most of San Francisco in 1906 and did widespread damage in Los Angeles in 1971.

Long-Term Changes

The forces at work in volcanoes and earthquakes today were even more active in the past. Molten rock from beneath the crust pushed the land surface upward, or poured out on top and then hardened like cement. Many scientists believe that the continents as we know them today were once joined together. The continents lie on hard sections of crust known as **plates**. Ages ago, these scientists believe, the plates were gradually pushed apart by pressures from the molten rocks, and the continents drifted to their present positions.

Pressures in the crust twisted or split the surface of the continents in many different ways. To see one of these ways, place a piece of paper flat on your desk. With your fingers resting on each end, push gently inward. If the piece of paper represents a large section of Earth's surface, what have you produced?

Land surfaces also change in gentler ways. Have you ever seen mud washed to the bottom of a hill after a heavy rain? Have you seen sand dunes shifted by the wind? Have you seen overhanging cliffs where waves have worn away the rocks below?

These are some examples of **erosion** (the wearing away of the land). Water and wind can erode soil or rock and carry it from one place to another. So can **glaciers** (slowly moving masses of ice). Several times in the distant past, Earth has gone through cold periods known as ice ages. During those ages, vast glaciers have changed the shape of much of Earth's land surface.

Over long periods of time, erosion by water, wind, or ice can scrape away mountain tops, gouge out valleys, build up hills, and shift coastlines. For example, the Grand Canyon was carved out by erosion — and it is more than one mile (*1½ kilometers*) deep.

Because of all these forces, the surface of Earth has a tremendous variety of shapes. In the next section, you'll look at some of the most important shapes on Earth's surface.

Landforms

If you pushed on a piece of paper as suggested in the previous section, you created a bulge in the middle. Similar bulges in Earth's crust are **mountains**. Often they come in large groups known as **mountain ranges**.

Mountains have been, and in many places still are, barriers to human travel. They are also barriers to human settlement. For one thing, temperatures drop steadily as you go up a mountain. Since you are moving very slightly closer to the Sun, you might expect to get warmer. But most of the Sun's rays pass through the air without warming it. Their heat collects in and under Earth's surface, which then warms the air above it. Thus the farther you rise above the main mass of Earth, the cooler the air becomes. Temperatures fall about 3°F for each 1,000 feet of elevation (*1°C for each 180 meters*).

This drop in temperature affects the kinds of plants that can grow on a mountain. First the environment becomes too cold for oaks, maples, and other broad-leafed trees. Then it becomes too cold for hemlocks, spruces, and other needle-leafed trees. In the end, it becomes too cold for any plant life.

The air becomes thinner too. Above 5,000 feet (*1,500 meters*), you may find yourself panting a little. Above 10,000 feet (*3,000 meters*), you will probably be panting hard, your head will ache, you'll feel dizzy and exhausted, and you may throw up. This is why mountaineers wear oxygen masks when they tackle the world's highest peaks.

Plateaus and Plains

Not all of Earth's highlands rise to peaks and sharp ridges, like mountains. Some are flat on top and may spread over vast distances. These high flatlands are known as **plateaus**. The edges of a plateau may be steep and act as a barrier to travel and farming. But the top may be pleasant for living and suitable for farming.

In and near the tropics, plateaus are cooler than the lowlands but still warm enough for most plants. The land is flat enough to allow both travel and farming. So people do live on tropical plateaus, as in East and South Africa and in central India. In higher latitudes, however, plateaus are often too dry and cold for much life.

Now let's look at Earth's **lowlands**. The most important lowland areas are **plains**. These are broad lands that are flat or gently rolling. They are the easiest kind of land to travel across. In most places, they are also the best kind of land for farming. This means that most of the world's food is grown on plains. Most of the world's cities and towns are built on them too.

Mountain ranges, plateaus, and plains are the major landforms, since they may cover vast areas. There are countless other types of landforms. Can you think of any different shapes that the land takes where you live? Perhaps there are **hills**. These may be mountains that have been worn down by erosion, or rocks piled up by moving water or ice. Between hills and mountains there may be **valleys** (low folds in the land). The drawing on the opposite page shows all these and other landforms.

In general, as you have read, people live in the broad lowlands. But their choice depends on more than the shape of the land. The next section shows how people are also interested in what is on and in the land.

1. What is a plateau?

2. Name and describe two other landforms.

3. Give one example of a violent natural force that can change the shape of Earth's surface.

4. Give one example of a gentler natural force that can change the shape of Earth's surface.

Natural Resources

A gun shot rings out. There is shouting, a clatter of hooves and wheels, and a great cloud of dust. A line of wagons lurches madly across the rolling plain.

This is the beginning of one of the 19th-century land rushes that is opening some of the U.S. plains to settlers. It is a race among families to stake a claim on farming land. The government is selling large areas at low prices to people who claim the land and promise to stay there and work it. In choosing a spot, what do you think a family would be most likely to look for first?

The best places to settle would be close to a source of water, which the settlers would need for themselves and their animals to drink. They would also be more likely to find trees for lumber along a stream or river. In dry grasslands, water can spread through the ground from such a source and enable trees to grow nearby.

You read earlier that water is a vital resource of the oceans. It is also a vital resource of the land. As you know, the water in rivers, lakes, and streams comes from the oceans in the first place. You might think of these inland bodies of water as stopover

Above right, the drawing brings together shapes that can be found on Earth's surface. All the land and water forms in the drawing are defined in the list below it. Of course, you would not find a real place with all these forms together. Can you guess why not?

Major Land and Water Forms

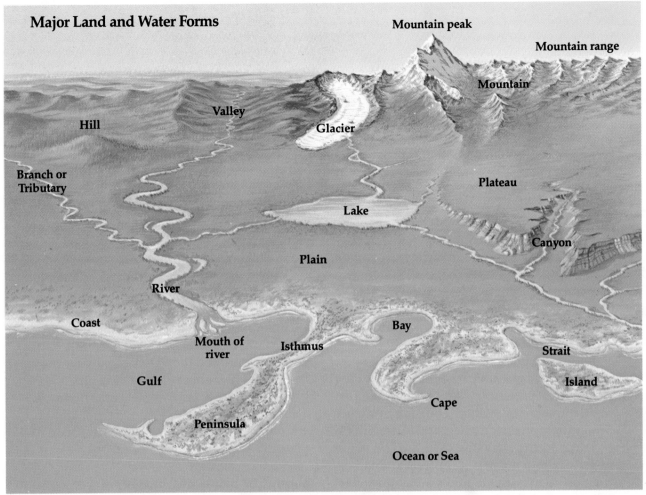

Mountain peak

Mountain range

Mountain

Valley

Glacier

Hill

Plateau

Branch or Tributary

Lake

Canyon

Plain

River

Coast

Bay

Mouth of river

Isthmus

Strait

Gulf

Island

Cape

Peninsula

Ocean or Sea

Shapes on Earth's Surface

Here is a list of terms used for some of Earth's most common land and water forms.

Bay — Curved part of a coastline where the water juts into the land.

Body of water — Any area of water, large or small, such as an ocean, sea, or lake.

Branch or Tributary — Stream or river that flows into a larger stream or river.

Canyon — Steep-sided valley cut in rock by a river.

Cape — Curved or pointed part of a coastline where the land juts into the water.

Coast — Land beside a sea or an ocean.

Coastline — Border between land and sea or ocean.

Continent — Large mass of land surrounded by oceans.

Glacier — Slowly moving mass of ice. It is usually formed in high mountains where snow cannot melt and run off as a river.

Gulf — Broadly curved coastline where the water juts into the land. A gulf is usually larger than a bay.

Hill — Raised area of land, usually with a rounded top. A hill is smaller than a mountain.

Island — Body of land entirely surrounded by water.

Isthmus — Narrow strip of land joining two larger areas of land.

Lake — Area of water surrounded entirely by land. Its water is usually fresh, not salt.

Mountain — Raised area of land, usually with steep, rocky sides. A mountain is higher than a hill.

Mountain peak — Highest point of a mountain.

Mountain range — Chain of mountains.

Mouth of a river — Area where a river flows into a larger body of water.

Ocean — Large body of water separating continents.

Peninsula — Narrow piece of land jutting out into water and almost surrounded by it.

Plain — Broad area of low, mostly flat land.

Plateau — Broad area of high, mostly flat land.

River — Large stream of water that flows through the land to a sea, lake, or ocean.

Sea — Large body of water partly or entirely surrounded by land. A sea is usually larger than a lake, and its water is salt.

Strait — Narrow strip of water joining two larger bodies of water.

Tributary — see Branch.

Valley — Area of low land, usually longer than it is wide, between hills or mountains.

places in Earth's water cycle. Rainwater that falls on mountainsides or hilltops runs downward and forms streams. These streams join together to form rivers. The rivers empty into other bodies of water — lakes, ponds, swamps, larger rivers. Nearly all river water sooner or later flows back into the oceans.

A lake may be small enough for you to walk around in a few minutes. Or it may be huge like Lake Superior, which is larger than 11 of the states of our nation. Rivers also vary in size, ranging from a few miles in length to more than 4,000 miles (*7,000 kilometers*) — the distance from southern Florida to northern Alaska.

Big or small, these lakes and rivers are very useful to people. Like the oceans, they are sources of fish. Unlike the salty oceans, they contain fresh water. Thus they provide water for homes, industry, and farmlands.

Water from rivers and lakes can be made to flow where it will turn wheels and generate electricity. This is known as **hydroelectric power.** Inland waterways also serve as important transportation routes for boats and barges. Many towns and cities are located along the banks of rivers and lakes.

Plants and Animals

You have already read about another of Earth's important resources — vegetation (see page 33). Can you name some kinds of vegetation? Can you tell some of the ways people use Earth's vegetation?

Much of the vegetation that is farmed grows on Earth's plains. As you read earlier, plains can provide good **arable land** (land suitable for farming). They are not too steep and uneven like mountains. They are not too low and wet like swamps. They are not too dry like deserts. On most of the world's plains, farmers can use machines to plow, plant, and harvest huge fields of corn, wheat, and other crops.

Vegetation depends not only on water but also on another resource of the land — **soil**. This is found where the rocks on Earth's surface have been broken into tiny pieces by wind and weather. Soils vary from place to place and make a difference to plant growth.

What makes a good soil? One important ingredient is material from dead plants and animals. This provides many **nutrients** (foods) that living plants need. Farmers can improve poor soils by adding **fertilizer** (manure or chemical nutrients).

Animals are used as resources by people. They provide food in the form of meat and dairy products. They also provide other useful items, such as wool and leather for clothing. Many of these animals are kept on the plains, where the food they eat grows best. But some, like sheep, do well on the short grasses of higher land — hills and plateaus. This is true in New Zealand, for example, where there are 20 times as many sheep as people.

Resources Below the Surface

Flying low across Earth's surface, the visitors from space have been able to see the oceans, the different shapes of the land, the inland bodies of water, and the changes in vegetation. But to find other resources, the visitors will need more than good vision, because these other resources are hidden under the land.

Beneath the soil there are resources known as **minerals**. These are rocks or rocklike materials that people can dig up and put to use. Some minerals can be used directly, the way they are dug up. Coal is one example. It can be taken straight from the mine and burned as a fuel. But most mineral resources are found mixed with other kinds of rocks, and have to be **extracted** (taken out) by special processing. Some mixtures are richer in the mineral resource than others. Those that are rich enough to be worth digging up are called **ores**. Some ores are mostly waste, but the mineral is valuable

Resources are any things on Earth that humans can put to use. They include animal resources, like the wool or meat of New Zealand sheep (top). There are also mineral resources, like salt in Mexico (bottom).

enough to mine anyway. For instance, miners may dig up a ton of ore to get a few ounces of gold.

A mineral becomes a resource only when humans discover ways to use it. People learned to use copper and iron thousands of years ago, turning them into knives, necklaces, and other tools and ornaments. It was only about 200 years ago that people discovered how to use coal to power steam boilers. The heavy demand for oil did not begin until the gasoline engine was invented about 100 years ago. Before that, people who dug for water and found oil instead might have cursed their bad luck.

The most important minerals in today's world are those used to produce energy. These are called **fuels**. Coal, oil, and natural gas are three of these fuels. (Since there are many different oils, the one used most widely as a fuel is also referred to as **petroleum**.) These fuels actually come from dead plants or tiny animals that have been changed by lying underground for millions of years. They are sometimes called **fossil fuels**.

These fuels also have many other uses. They go into the making of such things as matches, perfumes, soap, printing ink, synthetic (human-made) fabrics, dyes, aspirin, and plastics. Today oil is the most widely used of these mineral fuels. Not only can it be burned to provide heating, but it powers all kinds of machinery — from mopeds to jetliners.

Another important group of minerals consists of **metals**. These are usually smooth, tough, and easily shaped. Of all the metals, iron is the most important. Iron ore is the basis of steel, the metal used to build skyscrapers, bridges, railroads, machinery, trucks, cars, and airplanes. Can you think of other things you use that are made of steel?

So far the space visitors have been looking at the people of Earth from a distance. They have seen how Earth's surface offers different places where people can live and different resources that people can use. But what are the people themselves like? Do they all eat the same food? Do they all speak the same language? Do they all use the same resources? If not, how and why do they differ?

You will find answers to these questions in Chapter 4.

SECTION REVIEW

1. Give two reasons why plains offer better farming conditions than other kinds of land.

2. Vegetation is one of Earth's most important resources. Name two resources on which vegetation depends.

3. What is a mineral?

4. When is a mineral considered a resource? Which kind of minerals are considered the most important resources today? Why?

YOUR LOCAL GEOGRAPHY

1. What kinds of land and water forms are there in and around your community? Would they have been barriers to travel before there were highways, bridges, airfields, and other modern aids to transportation? Do they still act as barriers in any way? If so, how?

Suppose you had to travel to a point five miles (*eight kilometers*) from your school, on foot and/or in a rowboat. Which direction would be the easiest? Which would be the most difficult? Explain your answers.

2. Is there an active volcano in your state, or does your state ever have earthquakes? What other signs are there that the shape of the land may be changing by natural forces? Here's a checklist of some of the things to look for: rock falls; mud slides; sand or dirt shifted by the wind; riverbanks shifted by the water flow; shores washed away by waves. Do you live in a place with many natural changes or very few? How do these changes — or lack of them — affect your life?

Reading a Bar Graph

The modern world's most important fuel is petroleum (oil). Because people use oil in many ways, they need to know how much is produced, or pumped out of the ground, at any time.

The graph on this page shows how much petroleum was produced in seven major world regions during one recent year. **Graphs** show **quantities** (amounts) of things by means of lines, shapes, colors, and other symbols. A **bar graph** uses bars of various lengths. These bars let you see at a glance how amounts compare with one another. You can find out the subject of a bar graph from its title.

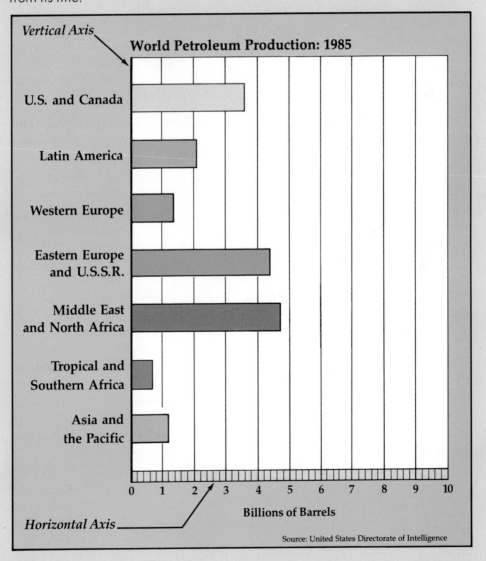

Vertical Axis

World Petroleum Production: 1985

U.S. and Canada

Latin America

Western Europe

Eastern Europe and U.S.S.R.

Middle East and North Africa

Tropical and Southern Africa

Asia and the Pacific

0 1 2 3 4 5 6 7 8 9 10

Billions of Barrels

Horizontal Axis

Source: United States Directorate of Intelligence

(Turn page.)

The line up and down the left side of the graph is called the **vertical axis**. In this bar graph, information along the vertical axis tells what each bar represents. You read down the axis to see what is being compared.

The line along the bottom of the graph is called the **horizontal axis**. Along this axis is a number scale. You read along the horizontal axis to see what *amount* each bar represents.

Study all parts of the bar graph. Make sure you understand what is being compared, and in what amounts. Then answer the following questions on a separate sheet of paper.

1. The graph shows petroleum production during what year?
2. What is the source of the information on the graph?
3. The graph gives information for which seven world regions?
4. Read the label below the horizontal axis. What amounts do numbers on this axis represent — thousands, millions, or billions of barrels of oil?
5. Read along the horizontal axis and count the tiny marks between numbers. How many barrels does each mark represent?

Find the bar for Asia and the Pacific. Compare the length of this bar with the number scale on the horizontal axis. The bar shows that Asia and the Pacific produced less than two billion barrels of oil in 1985. Now use the bar graph to find more facts and make some comparisons. Tell whether each of the following statements is true or false.

6. In 1985 the U.S. and Canada produced slightly more than four billion barrels of oil.
7. Tropical and Southern Africa produced about one barrel of oil less than Asia and the Pacific.
8. The largest amount of oil was produced in the U.S. and Canada.
9. The smallest amount of oil came from Tropical and Southern Africa.
10. The Middle East produced more oil than the U.S. and Canada and Latin America combined.

Note that bar graphs can be drawn so that the bars run vertically, instead of horizontally as in the petroleum graph on page 63. Suppose this graph had been drawn with vertical bars. Where would the names of the oil-producing regions be placed? Where would the number scale for oil produced be placed? Can you suggest any reason for choosing one arrangement of the graph rather than the other?

CHAPTER REVIEW

A. Words To Remember

Three of the following terms are defined by the numbered phrases below. On a separate sheet of paper, write each term next to the number of its definition. Then write the two terms that are *not* defined and give a definition for each.

continental shelf   elevation   mountain
crust   erosion

1. Raised area of land, usually with steep, rocky sides.

2. The action of water, wind, and ice wearing away soil or rock in one place and carrying it elsewhere.

3. The sloping edge of a landmass near the coastline.

B. Check Your Reading

1. What word is used to describe changes in elevation?

2. Name three landforms mentioned in the chapter and tell what they are.

3. Which landform is best suited for farming? Why?

4. What word is used to describe minerals used to produce energy? Give three examples of such minerals.

5. Are Earth's landforms and water forms more of a help or a hindrance? Explain.

C. Think It Over

1. Why did explorers in the past find it quicker to travel over oceans than over land? Is ocean travel still quicker today?

2. Volcanoes and earthquakes both involve violent movements in Earth's crust. How do they differ?

3. Suppose you were leading an expedition to an Earthlike planet in another solar system. Name three things you should look for when deciding where to settle. Explain.

D. Things To Do

Do some research on the eruption of Mount St. Helens in Washington State in 1980. Perhaps your school or local librarian can help you locate a few magazine and newspaper articles about it. See if you can find pictures showing how Mount St. Helens appeared before, during, and after the eruption. How much did the eruption change the land surface? Is the land surface still changing?

Chapter 4

People on Earth

"We have a problem," says the spaceship captain. "How are we going to find out what the people on Earth are like?"

"Well," says a crew member, "why don't we choose a few places at random and send down a rocket with a TV camera and microphone? That will give us an idea of any differences there are between one place and another. Then we can pick two very different places for a closer look."

"Good!" says the captain. "Get the first rocket probe ready!"

For the first sample, the visitors return to the place where they first saw signs of human life. As you read on, can you guess what nation they have picked?

The rocket settles gently on top of a farmhouse at the edge of a square field. The field runs for about half a mile (*almost one kilometer*) alongside a paved road. Corn grows in long parallel lines that stretch as far as the camera can see. In the distance, huge machines move slowly along the lines, harvesting the ears of corn. Sleek new automobiles are traveling speedily along the road.

One car turns into a driveway below, and a woman in jeans steps out. She carries bags of groceries into the farmhouse. Music and singing are coming from a machine of some kind. A voice says, "Hi, Mom, did you remember to get the hot dogs?"

You may have guessed by now that this scene is in the U.S. The farm is somewhere in the northern part of our nation's central plain.

Now the space visitors move around Earth to Egypt — a nation in North Africa. Here the rocket lands beside the Nile River. The fields are square but much smaller than in the U.S. Midwest. Instead of machines, there is a camel pulling a plow. A man in a loose white robe walks beside the camel, shouting at it from time to time. A ditch runs nearby, carrying water from the Nile River. Beyond the field, the ground is bare and sandy. The desert stretches to the horizon.

The farmer's hut stands at the edge of the desert. It is built of blocks of dried mud. Here too there are none of the machines found around the U.S. farmhouse. There's no automobile and no stereo.

Inside the hut, a woman can be heard singing, and there is a clatter of pots. The woman comes out carrying a dish. On it are some cooked beans and a piece of wheat bread. It is the midday meal for her husband in the field.

"Well, that's certainly different," says the captain. "Now let's move to a third location."

The next stop for the space visitors is India. Here the rocket lands beside a village in

Earth's humans behave in many different — and similar — ways. The same enthusiasm shown here by Argentine soccer fans may be shown by different peoples for different reasons.

the northern plains. The fields are small, like those in Egypt, but many have irregular shapes. They are covered with water, because the rice planted here grows best in flooded soil. A man holds a wooden plow that is being pulled by a water buffalo. The man wears a cotton cloth wrapped around his waist and between his legs.

The villagers' homes are made of thick clay walls with straw roofs. There is no glass in the windows, just wooden shutters to keep out the heat of the sun. Outside the nearest hut sits the village blacksmith. He is hammering the blade of a sickle (a curved knife used for cutting the rice plants).

A broad, muddy path runs past the rice field. This is the main road to the village. Three women walk along it, carrying clay jars on their heads. They are going to fetch water from the village well.

Here are two different ways of performing the same human task — growing food. In Egypt (left), a farmer uses a hand tool and muscles to work the land. In the U.S. (below), crops are tended by machinery.

Different Ways of Life

Some of the differences among these three places come as no surprise to the space travelers. They already know that places on Earth can vary widely in climate, vegetation, and landforms. They are discovering other differences, however, for the first time.

Most obvious, perhaps, are the different ways these farmers work their fields. In the U.S. scene, there is big, complicated machinery. In the Egyptian and Indian fields, the farmers rely on different kinds of animals and on human muscles.

This is only the beginning. The closer the travelers look at the three scenes, the more differences they will find.

For example, there are differences among the people themselves. All humans belong to the same family, called *homo sapiens* (Latin for "thinking human"). However, there is tremendous variety within that family. Humans can be divided into many **ethnic groups** — that is, large groups of people who have more in common with each other than they do with other peoples.

Some ethnic groups differ physically from others. However, all humans are born with certain physical differences. They differ in height and weight; skin, eye, and hair color; features of the face; and other ways. Often there are no overall physical differences between one ethnic group and another.

Other differences between human groups are acquired after birth. These are differences in **culture** (customs and ways of doing things). The space travelers quickly notice one example: the different methods of farming. They also see people wearing jeans in the U.S., a loose white robe in Egypt, and a cotton cloth in India. They hear English spoken in the U.S., Arabic in Egypt, and Hindi in India. There is rock music from a stereo in the U.S. and a folk song sung by a woman in Egypt.

How did such cultural differences come about? How important a part do they play in today's world? Are there any cultural similarities among Earth's peoples that balance or outweigh the differences? The following sections will look for answers to these questions.

SECTION REVIEW

1. Are clothing, hair styles, and language examples of physical or cultural characteristics?

2. In what ways are the Egyptian and Indian farms more similar to each other than to the U.S. farm?

3. In what ways are all three farm settings similar?

4. Name two kinds of cultural differences existing among the groups of people discussed in this section.

Why Cultures Differ

Today it is easy to get in touch with far-off places. In the past, however, it was difficult or impossible. Fifty years ago, there were telephones, but there was no long-distance dialing. There were airplanes, but no fast jets. Television programs did not exist. One hundred years ago, there were no airplanes, and the telephone was a new invention. Two hundred fifty years ago, there were no powered machines for traveling or sending messages. The fastest ride was on a galloping horse. (See the chart on page 70 for more on developments in travel.)

In Chapter 3, you saw how the oceans and the land could act as barriers to people. Through most of human history, there were no easy ways of crossing those barriers. Most people did not travel much, or very far. They had little or no contact with people outside their own small area. They did not always know what their neighbors 10 miles (*15 kilometers*) away were like. They could have wild ideas about people farther away. Even in more recent times, there were ru-

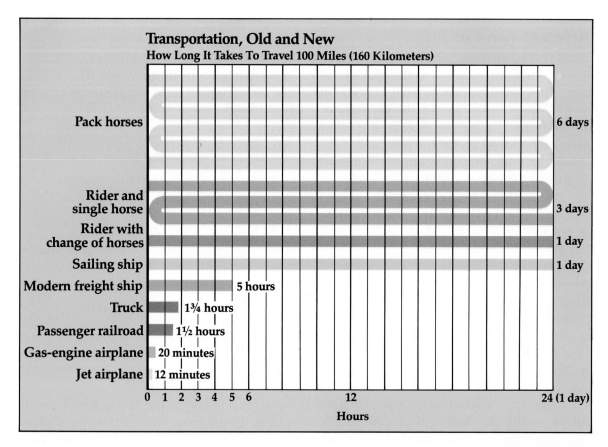

Transportation, Old and New
How Long It Takes To Travel 100 Miles (160 Kilometers)

Pack horses	6 days
Rider and single horse	3 days
Rider with change of horses	1 day
Sailing ship	1 day
Modern freight ship	5 hours
Truck	1¾ hours
Passenger railroad	1½ hours
Gas-engine airplane	20 minutes
Jet airplane	12 minutes

0 1 2 3 4 5 6   12   24 (1 day)

Hours

mors of giants 50 feet (*15 meters*) tall, or people whose heads grew under their arms.

While Earth's surface sets up barriers, it also offers plants, animals, and minerals that people can use as resources. These resources vary from place to place. In the distant past, each group of people had to rely on the resources in their own area. They had to develop their own ways of living.

How would different resources lead to different ways of living? One area might have antelope, wild pigs, and other animals. The people there would take up hunting as their everyday "job." For them food would mean meat. Another area might be beside a river teeming with trout or salmon. The people there would work at fishing, and food for them would mean fish. Still another area might be full of coconut palms, and the people there would live mainly off the meat and milk of the coconut.

Until the 19th century, horses and sailing ships were the fastest means of travel. Transporting goods by land was extremely slow, and sailing ships were often delayed by bad weather. Today goods and people can be carried easily and rapidly over long distances. A jet plane could fly three times around the Equator in the time it takes pack horses to cover 100 miles.

The hunters would wear clothing made of animal skins. The fishing people would wear the skins of water animals such as otters and beavers. The coconut eaters would make their clothes out of leaves and grasses.

Their homes would vary in much the same way. The hunters might make tents of animal skins. The fishing people might make huts of mud from the riverbanks. The coconut eaters might make huts of palm leaves. Even the purpose of the homes would differ from one group to another. In some places, people would want mainly to keep out the cold. In other places, they would want shelter from hot sunlight or heavy rain.

People's ways of life varied beyond the basic needs of food, clothing, and shelter. Each group would find more ways of using the resources in its area. There were different rocks that could be shaped into tools for cutting or scraping. There were different animals that could be **domesticated** (tamed to do work), like the camel in Egypt and the water buffalo in India.

Other Cultural Differences

So far you have looked at ways of life that were linked directly to the land and its resources. But there were also other ways that did not have this direct link. Language is one example.

People around the world needed language for various reasons. For example, language helped them work together and pass on knowledge to their children. People needed words to name their plants, animals, tools, and other resources. But the words themselves had no natural connection with these objects. They were simply sounds or marks that each group of people happened to choose. When a Chinese says *ma*, a German says *Pferd*, and an Arab says

A *pictograph* uses symbols to show amounts, with a key to the symbols. Here, to find how many people speak a language, multiply the number of whole symbols by the key amount. Estimate the fraction of any incomplete symbols, and add that fraction of the key amount. For example, Russian has two symbols and two thirds of a symbol—a total of 266 million.

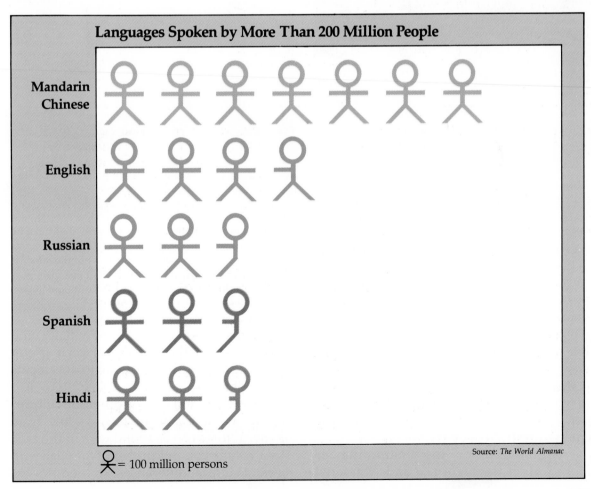

Languages Spoken by More Than 200 Million People

Mandarin Chinese

English

Russian

Spanish

Hindi

Source: *The World Almanac*

= 100 million persons

71

Music is one of many cultural features that have no direct link with the environment. The instruments *might* be made from local resources — but the music is shaped by humans. At left, two Chinese girls play lutes; at right, Mexican men play a violin and guitar.

hosan, they are each talking about the same kind of animal. It's what you call a *horse*.

Like language, art could vary from place to place. Eskimos would carve a face on a piece of driftwood to be used as a fishing float. For a ceremonial dance, West Africans would carve a mask out of wood and decorate it with copper. American Indians would wear ornaments they made out of shells, feathers, and animal bones. The objects out of which different peoples made their art depended on local resources. But the designs could come from people's imagination.

From the tropics to the tundra, humans looked up at the night sky and wondered where the stars came from, and what would happen when they died. As each group faced these questions, it began to develop its own religion.

There are hundreds of different cultures in the world today. In fact, people from many cultures can be found here in the U.S. As a result, you are probably familiar with several cultural differences. In food, for example, you may have tasted chili (Mexican), pizza (Italian), chow mein (Chinese), and fish and chips (English). In language you may use such words as *parade* (from Spanish), *chocolate* (from Mexican Indian), *raccoon* (from North American Indian), *boss* (from Dutch), and many others.

What Cultures Have in Common

Smoke rises from a special area in the dry land. Across the ground, some men are spreading hot ashes from a fire. Others are whirling bull roarers around their heads. A *bull roarer* is a piece of wood or bone attached to a cord; as it whirls, it makes a noise.

A group of 13-year-old boys is led into the area. For the past few days, the men have taught these boys the myths and traditions of their people. Now the boys must run through the hot ashes in their bare feet, to show that they can stand pain. Afterward they will be treated as men.

Nearly all human groups have held special ceremonies for young people becoming adults. This happens today in the U.S. with confirmation among Christians and bar mitzvah among Jews. Sometimes the details of a ceremony in one group are very similar to those in another group. The scene described above could be found in places 8,000 miles (*13,000 kilometers*) apart. Both the original inhabitants who lived in southeast Australia and the Indians who lived in central California used bull roarers and hot ashes in their manhood rites.

What could have caused these similarities between groups living far apart? One reason was given earlier in this chapter — the fact that all humans belong to the same family. This basic human similarity can show up in very different cultures.

For one thing, all human cultures have found the same general ways of living on the land. They have developed some kind of clothing, shelter, and methods of preparing food. They all have a spoken language —

even though it differs from place to place. Nearly all human groups have developed some kind of art, music, and religion.

Often many groups also found the same way of solving a problem. How could they store and carry water? If there was clay in their area, they probably made large pots. How could they hunt animals? If there were trees in their area, they probably used a straight branch with a sharpened tip.

Although most people did not travel outside their own small area, some had to. A group might lack a vital resource in its area, such as salt. Someone would have to go out and find salt. They would probably meet and trade with other groups on the way. Or a group might grow too big for the resources in its area. Then some or all of the group would have to **migrate** (move from their homeland) to another place.

Thus different groups of people came into contact with one another. Not only goods but also ideas were exchanged. If one group found a better way of making knives or curing a fever, other groups nearby would learn about it and use it themselves. As time went by, the new idea could spread farther and farther.

SECTION REVIEW

1. Would it be true to say that the cultures of the world are: basically different; basically similar; or a mixture of basic differences and similarities?

2. Name one way in which coconut palms could serve as a local resource.

3. Various kinds of cultural differences arose between different peoples. Which of the following were *not* directly linked to the land and its resources: homes, food, language, tools, or pets? Explain.

4. Which of these statements is true of all human groups: They eat meat; they ride animals; they speak a language; or they hunt for their food?

Reading a Table

People around the world follow a variety of religions. What are the major religions? Does each one have more followers in some areas of the world than in others?

You can answer these questions from the **table** shown below. A table lists groups of facts so they can be compared easily. Facts are grouped together up and down the table in *columns*, and across the table in *rows*. Along the top of the table are *headings* that tell you what kinds of facts are listed in each column. Other headings on the left side of the table tell what kinds of facts are grouped in each row.

Study the title and headings of this table. To complete each statement, write the best choice on a separate sheet of paper.

1. The table has facts about members of (all; most; major) religions.
2. The column headings include (six; seven; eight) world areas.
3. The far-left column lists (religions; continents; numbers).

Membership of Major World Religions

RELIGION	NORTH AMERICA	SOUTH AMERICA	EUROPE	ASIA	AFRICA	PACIFIC	TOTALS
Christianity	235,109,500	177,266,000	342,630,400	95,987,240	129,717,000	18,063,500	998,773,640
Judaism	6,155,340	635,800	4,061,620	3,212,860	176,400	76,000	14,318,020
Islam	371,200	251,500	14,145,000	427,266,000	145,214,700	87,000	587,335,400
Shintoism	60,000	92,000	—	57,003,000	200	—	57,155,200
Taoism	16,000	10,000	—	31,261,000	—	—	31,287,000
Confucianism	97,100	70,150	—	157,887,500	1,500	80,300	158,136,550
Buddhism	171,250	192,300	192,000	254,241,000	14,000	30,000	254,840,550
Hinduism	88,500	849,300	350,000	473,073,000	1,079,800	499,000	475,939,600
TOTALS	242,068,890	179,367,050	361,379,020	1,499,931,600	276,203,600	18,835,800	2,577,785,960

Source: *Encyclopaedia Britannica Book of the Year, 1980*

Facts listed down the columns and across the rows form an information *grid*. To learn how many members of a *religion* live on a certain *continent*, find where numbers for that row and column meet. To find out how many Taoists live in Asia, for example, read down the left column to the name of that group. Then read across the row of numbers to the column for Asia. Now find these facts.

4. How many followers of Christianity are in: (a) North America? (b) Asia?
5. In Asia how many people believe in: (a) Buddhism? (b) Hinduism?
6. In Africa which religion has the largest number of members?

Cultural Regions

A song recorded in Los Angeles or Nashville today can be heard next week on radios and stereos all over the U.S. Not only songs but also other cultural features can spread rapidly across whole continents.

In the past, ideas moved much more slowly, but in time they could still spread across large areas. Even today groups who live in the same general area often have more in common with each other than with groups who live farther away.

You can see this most easily in an area we call a **nation** (a group of people who share the same political system, or type of government). Most people in a nation eat the same kinds of food, follow the same kinds of sport, and celebrate the same holidays. Can you think of any other cultural features

these people are likely to have in common?

In Canada you'd find that many Canadians speak English, are familiar with hamburgers and hot dogs, and follow football and baseball. Of course, you'd notice cultural differences too. But similarities can be found in areas larger than a single nation. For example, look at the languages spoken in Western Europe. English and French both belong to the same language family. Many words in the two languages are similar, and so is the way they are put together in sentences. In France you would also play a *disque*, if you wanted to *danser* to *musique*.

Another cultural similarity found in large areas of the world is religion. In Western Eu-

Canada and the U.S. belong to one cultural region. One sign of this is the number of similar English and Indian place-names found on both sides of the border.

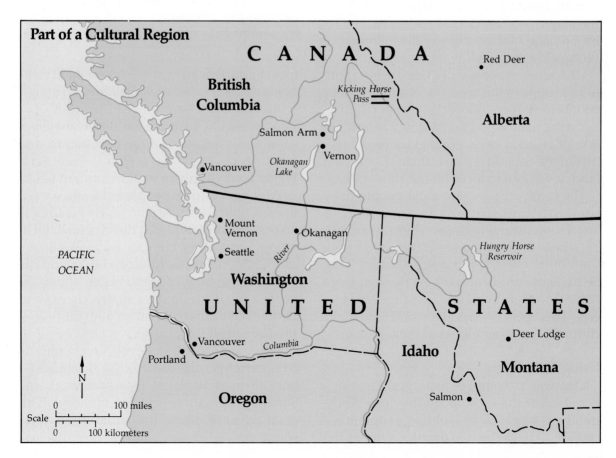

Part of a Cultural Region

Technology is any skill or tool that humans develop to help them live better. Above, in the Philippines, two examples of technology meet: a complex flying machine and a simple device for watering crops.

rope, the Christian religion is most widespread. In North Africa and the Middle East, the majority of people are Moslems. The table on page 74 shows where the followers of the world's major religions are found.

Of course, these similarities do not apply to all groups in the area. For example, many people in Europe follow the Jewish religion, as do the majority in the Middle Eastern nation of Israel. However, to get a broad picture of the ways in which humans live on the land, it's useful to focus on the similarities.

The large areas where most people have some cultural features in common are known as **cultural regions**. These areas may be whole continents, separated by oceans from most other landmasses. Or they may be parts of continents, cut off by such large natural barriers as mountain ranges or deserts. This book divides the world into eight cultural regions and looks at them in turn.

Using Technology

One way to compare nations and regions is to look at the technology they use. **Technology** is any skill or tool that people have created to help them live better. One of the most important advances ever made in technology was the rise of *agriculture* (farming). Instead of searching for wild plants to eat, people started planting seeds in one area. They chose seeds from the best plants, and kept weeds away. As a result, they got a bigger food supply in one convenient place.

With agriculture, people led more settled lives. Civilizations began. (A **civilization** is an organized society that develops art, technology, and other cultural features.) Many more advances in technology followed. The most far-reaching of these began about 250 years ago. It brought so many changes to people's lives that it is called a revolution — the **Industrial Revolution**.

Before the Industrial Revolution, winds or river currents were used to run machines — but only in a few places. Nearly all work was done by human or animal muscles. The great advance of the Industrial Revolution was a powerful new way of running machinery — the steam engine fueled by coal. The steam engine was followed by even more efficient engines run by oil, electricity, and other sources of power. This new machinery made it possible for humans to find and use much more of Earth's resources.

Different Stages of Development

The U.S. is one part of the world where people's lives show the full effects of the Industrial Revolution. Think of the field that the space visitors looked at in the Midwest. Although it is large, machines quickly move through it to plow, plant, and harvest the corn. Technology has also improved the quality of the seeds and provided fertilizer for the soil. Fields like this produce much more food than the same kind of land farmed by older methods.

In some parts of the world, technology is not so fully developed. There are areas that the Industrial Revolution has barely touched. This is true of the fields that the visitors looked at in Egypt and India. While some farmers in these countries do use modern technology, much farming is done the same today as it was 250 years ago.

Thus advances in technology made in the last two centuries have created a gap between the peoples of Earth today. On one side are the people who live in what might be called the modern way, using the technology of the Industrial Revolution. On the other side are the people who live in the traditional way, relying mainly on the technology of 250 years ago. To understand how people live on Earth today, the visitors must look at places on each side of the gap.

As you'll see in the next chapter, the gap is not always clear-cut. In some places, the two different ways are found side by side, or both exist together as people change from one to the other. For the most part, however, the contrast between these two ways is the most obvious and dramatic difference between the peoples of Earth.

SECTION REVIEW

1. Why are there cultural similarities between peoples who live in the same region?

2. What is technology? Give an example.

3. What advance in technology led to the start of the Industrial Revolution?

4. Why have advances in technology since the Industrial Revolution created a gap between the peoples of the world today?

YOUR LOCAL GEOGRAPHY

1. The names of places in the U.S. often give clues to the culture of the people who settled there in the past. For example, Chicago comes from the original Indian name of that locality. Settlers gave the name of an English town to what is now the city of Boston. Spanish missionaries named San Miguel (Spanish for "St. Michael").

Do the names in your locality give similar cultural clues? If you are not sure of the origins, ask at the library for a book such as *American Place-Names*, by George R. Stewart. If the names do not refer to cultural groups, what are they based on? For example, they may describe the area, like Hot Springs or Anchorage. Or they may be taken from a person, like Houston (named for Sam Houston, the Texas soldier and statesman) or Jefferson City (named for Thomas Jefferson, the third President of the U.S.). Do such names still give any hint about the kind of people who chose them?

2. Have you ever traveled outside the U.S.? If so, what cultural differences made the biggest impression on you? Did they have any connection with landforms, climate, or local resources?

Have you ever met visitors from a foreign country? If so, did they mention any cultural differences in the U.S. that made a big impression on them? Did these differences have any connection with landforms, climate, or local resources?

Reading a Line Graph

With the spread of technology in the past 200 years, Earth's population has increased dramatically. To show such a change that takes place over a period of time, one of the most useful types of graphs is a **line graph**. Like a bar graph (see Geography Skills 9), a line graph has a horizontal axis and a vertical axis. However, *time periods* are marked along one axis, usually the horizontal axis. Number amounts are marked along the vertical axis.

Study the information along both axes of the line graph at right. (*Axes* is the plural of *axis*.) Write the best choice for each statement below on a separate sheet of paper.

1. The graph shows population growth for (the United States; North America; the world).

2. The horizontal axis represents a period of time between the year 1600 and the year 2000.
   (a) Along this axis, there are numbers for every (10; 50; 100) years.
   (b) Between the numbered years, there are marks along the axis for every (year; 10 years; 50 years).

3. Read the label next to the vertical axis.
   (a) Each number on the axis stands for (thousands; millions; billions) of people.
   (b) Between the numbers on this axis, each mark stands for (four; one; one quarter) billion people.

Lines from the horizontal and vertical axes cross each other on the graph to form a grid. Dots are drawn on the grid to show the total population at the time of each numbered year. An **indicator line** connects the points. As you read the graph from left to right, a rising indicator line shows increase. A downward indicator line would show decrease. The increase or decrease is called a **trend**.

Study the trend on this line graph and answer the questions below on your answer sheet.

4. On the vertical axis, find the number representing one billion people. Read across the graph to where the line for this number is crossed by the indicator line. In about what year, shown on the horizontal axis, was the world population one billion people?

5. In 1900 there were about how many billions of people in the world? Fifty years later, by how many billions had world population increased?

6. Just 35 years later, in 1985, world population was close to how many billion people?

7. If this rate of growth continues, world population is expected to be about how large in the year 2000?

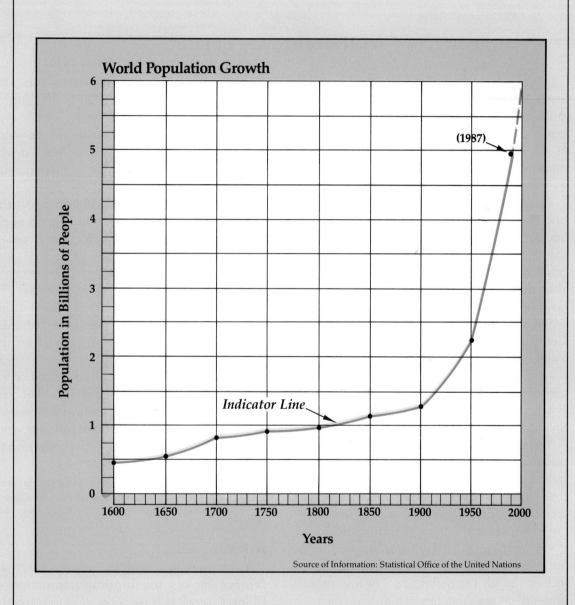

World Population Growth

Population in Billions of People

(1987)

Indicator Line

Years

Source of Information: Statistical Office of the United Nations

A Different Way of Life

Lillian Carter, mother of former President Jimmy Carter, did volunteer work in India when she was nearly 70. She had never been out of the U.S. before. Here are some sections of her letters home. As you read them, note what cultural differences made the biggest impression on her.

This is my first Christmas away from home, and the most exciting I have ever had. A little boy came to our flat [apartment] and washed the floors with a rag. Aloo and her husband gave us a jelly dish and a pair of napkin rings. They are Hindu, but wanted to share our holiday with us. . . .

We finally went to the big bazaar [market]. Got the necessities (iron, sheets, towels) and some groceries. But there wasn't much selection, so we will just have to eat whatever we can afford, and forget about balanced diets. There is so much poverty, I am losing my appetite anyway, and I know we live in luxury.

Yesterday I watched the workmen paving a place for classes to sit. They used cow dung, mixed with water. These classes are for illiterates to learn Hindi and Marathi [languages spoken in India]. I guess I'm one of "them," as I'm trying so hard to learn to communicate.

I'm learning to . . . drink from a bottle by holding it up and pouring it in my mouth at six inches. (So far I'm getting half in my mouth, and half on my dress front!)

We are able to get beans, cauliflower, and sweet potatoes, but meat is just out. The women at the Settlement grind wheat and make chapatis [chuh-PAH-tees] for the workmen. They squat all day, cooking. . . .

What is a chapati? It's the life of India. Flat bread the size of a dessert plate, made of flour, water, salt, and grease. It's rolled paper-thin and cooked on big grills until it's slightly brown. . . .

If I had one wish, I would wish for some clean rags, and if I had two wishes, I'd want some cheddar cheese! I'd rather have a chunk of cheese than diamonds.

Mabel's bag came, and she had a few cans of food in it. I had forgotten how good applesauce was! I'd give anything for a jar of peanut butter. . . .

I've been barefoot all day — anyone can go barefoot any time here, outside or inside. . . .

It now rains constantly, and where I once looked at cracked, parched earth, there is beautiful green grass. The trees are green after looking dead for months.

I have tried and tried to think, but I cannot remember. Will you please describe cold weather for me? When Mabel's vacation ends, we'll have one more monsoon [annual winds bringing rain], and then I'll be coming home. That's how I keep track of time here.

— From *Away from Home: Letters to My Family*, by Lillian Carter and Gloria Carter Spann. New York: Simon & Schuster, 1977.

Ask Yourself . . .

1. In the letters selected, what cultural difference makes the biggest impression on Lillian Carter? Is it housing, clothing, food, or religion?

2. Name three types of food that are eaten in India. Name three types of food that Lillian Carter could *not* find in India.

3. Do you think that in some ways Lillian Carter found living in India easier than in the U.S.? Give examples to support your answer.

A. Words To Remember

From the following list, select the term that best completes each sentence below. Write your answers on a separate sheet of paper.

bar graph   ethnic group   migrate
cultural region Industrial Revolution nation
culture   line graph   technology

1. _____ is a people's customs and ways of doing things.

2. A(n) _____ is a group of people who share the same political system, or type of government.

3. Any skills or tools that people have created to help them live better are known as _____.

4. A(n) _____ is a group of people who have more in common with each other than they do with other peoples. They may have physical similarities.

5. A chart that lets you see a change that takes place over a period of time is known as a _____.

B. Check Your Reading

1. The space visitors looked at farms in three different countries. What did the Egyptian and Indian farmers rely on for work: water power, animal and human muscles, or machinery?

2. The chapter mentions three modern inventions that overcame barriers to travel and communication. Name two of them.

3. Throughout most of human history, did different groups of people have a lot, a little, or no contact with one another?

4. In the past, the way people met certain basic needs depended on their local resources. Give two examples of those needs.

5. About 250 years ago, machines driven by new energy sources began to replace animal and human muscles. What name is given to that change?

C. Think It Over

1. Even in the U.S. today, there are cultural differences among people. Of the cultural ways mentioned in the chapter, name three in which Americans may differ from one another.

2. The U.S. and Canada are said to belong to the same cultural region. What does this mean?

3. What was the most important result of the Industrial Revolution?

D. Things To Do

In the past, people living in different places developed different languages. Today people living in the same place but belonging to different generations have their own jargon or slang. Write as many words you can think of that are used by students in the U.S. but not by their parents' generation.

Chapter 5

Two Ways of Life

"Prepare to land and watch out for the cow!"

The visitors are about to set their spacecraft down in a field outside a small village in Latin America. It is the first of two stops they have decided to make. In their explorations so far, the visitors have noticed that some settlements on Earth make great use of modern technology. Others do not.

The space visitors want to look at these two ways of life more closely. They begin their observation in the village, where life is very simple. Later they will go to a very different place — a big, modern city. By comparing what they see in these two places, they hope to have a rough idea of what life may be like anywhere else on Earth.

The Traditional World

As the space visitors descend, they see a group of houses made of rough stone. They also notice that the village has no railroad, no seaport, and no airport. The road that leads from the village is not a highway, but a dirt path. The visitors also note that there are no telephone lines, no electric wires, and no television antennas. The village is still largely cut off from the outside world.

One link it does have with other places is the radio. A few villagers own battery-operated transistor radios, and people gather around them to listen to the one radio station they can receive. This station is run by the government. It broadcasts school lessons to the children. It teaches people about cooking and farming. It tells them about health practices, and gives them some news of outside events.

The dirt road offers the villagers some contact with nearby places. Every few days, a truck brings a load of supplies for the village store. Once a week, a doctor comes in a jeep to treat any sick people. And once a week, there is a bus that goes to the next town.

For the most part, life centers around the local environment. Villagers walk to the nearby fields where their crops are planted and their animals graze. They also walk to the forest where they cut their firewood. The villagers eat what they grow and build with what they find in the forest. They carry their crops and their wood on their backs or on their animals. There are few wagons or carts.

Children may go to a government-run school for a few years and learn to read and write a bit. But they may have little interest in becoming **literate** — able to read and write. (A person who cannot read or write is **illiterate**.) Parents teach their children the "subjects" that are important for life in the village — how to plant and harvest crops;

In today's world, some ways of life have changed little in centuries, while others change almost daily. This African computer operator wears clothes of an age-old tradition.

how to prepare meals and to weave cloth. Families often look to a village leader to solve their problems and give advice.

The people of this village have learned the best ways to live in their separate world. Sometimes the teacher in the school, the public health nurse, or the radio bring new ideas to the village. Sometimes they suggest new ways of doing things. But the people in the village are not always sure that the new ways are best. What about the old ways that the villagers have followed for centuries? These people do not change their ways of life easily. Their customs are important to them.

A village like this lives in its own local world. It uses the resources of that particular place. Sometimes these are not enough. For instance, the people in this village use a well for their water. When the dry season comes, the well goes dry. The crops are stunted and the pastures turn brown. Then the whole village suffers. The technology of the village isn't advanced enough to overcome this resource problem.

Life in this village is **traditional** (following old ways). Life is much the same now as it has been for hundreds of years. Changes come slowly. Few inventions of modern technology have reached here.

In Chapter 4, you read that the Industrial Revolution brought major changes to some people. These changes have not happened in the traditional village. Life here is much like life anywhere before the Industrial Revolution.

Many Local Worlds

There are millions of these traditional villages on Earth today. In India alone, there are more than 500,000 farm villages. China probably has more.

Because these villages rely on their local resources, they can look very different in different parts of the world. In south China, for example, the climate is hot and wet. Here villagers grow rice in small fields that are flooded with water brought from rivers and streams. The villagers raise fish and ducks in the flooded rice fields as another source of food. They have small garden plots for vegetables. They raise a few chickens and hogs, but not cattle. The land cannot be spared for livestock feed. It is needed for human food.

Life is quite different in a hot, dry village on the edge of the Arabian Desert. Few crops will grow, so people here depend on flocks of sheep and goats. They drink goat milk, eat lamb, and use the animals' wool for their clothing and tents. These villagers move from place to place during the year to find new pastures for the animals they depend on.

There are traditional villages in northern Norway, where the forest ends and the tundra begins. Crops will not grow in the cold here, so villagers herd reindeer, which survive well in the cold environment. Like the sheep in the desert, the reindeer are a source of clothing and shelter as well as food.

In northern Canada, there are Eskimos who live neither by farming nor by herding. They live by hunting such wild animals as polar bears and seals. They also gather berries from the tiny Arctic plants of the tundra.

Each of these traditional villages has its own customs and life-style. Each looks very different from the Latin American village where the space visitors chose to land.

At the same time, all of these villages have a lot in common. The people in each place rely on things they can find on the spot. They rely on their muscles and skills to make most things they need. Although the space visitors have seen only one village, they have some idea of life in the millions of other traditional villages on Earth.

Since traditional villages rely on local resources, they vary widely around the world. In dry Afghanistan (bottom), vegetation is sold as firewood. It is too rare to use on homes, as in rainy West Africa (top).

Modern ways of life are seen most clearly in a big city like Chicago. Cities depend on energy for lighting (above) and other purposes. They also need transportation routes (left) in order to carry resources in and out.

The Modern World

To see the modern way of life, the space visitors want to find a place that looks as different as possible from an isolated village. They want to see a big, sprawling, bustling modern city. The city they choose is Chicago, the business and industrial center of the U.S. Midwest.

One of the first things the visitors notice as they approach Chicago is its vast **transportation system**. Roads, highways, and superhighways lead into, through, and around the city. These highways are filled with trucks, cars, buses, and taxicabs.

Railroads with thousands of freight cars link Chicago with other parts of the continent. An elevated railroad loops through the middle of the city. At least three airports can be seen. You may have heard of O'Hare Airport. It covers 10 square miles (*25 square kilometers*) and is one of the world's busiest airports.

The observers can also see that Chicago lies beside a huge lake. This is Lake Michigan, one of the Great Lakes. The city is a bustling port with ships of every size.

Compare all these transportation routes with the single dirt road leading out of the Latin American village. Of course, about 10,000 times as many people live in Chicago as in the village. But even 10,000 dirt roads would not be enough for Chicago, which could not keep going without its paved highways, railroads, docks, and airports. These are Chicago's lifelines.

Some of the most vital resources for a city such as Chicago are those that provide **energy** (the power to do work). Highways, railroads, docks, and airports are useless without gasoline and other oil fuels to drive the cars, trucks, trains, ships, and planes. Steel factories cannot work without special coal for their furnaces. Offices, stores, and homes need oil or natural gas for heating, and electricity for lighting.

People use some kind of energy resource whenever they pick up a phone, switch on the TV, cook a meal, or take a hot shower. Everything they buy has taken energy to make and bring to the store. And these energy resources themselves are brought from other parts of the U.S. or from other nations.

Like any other city, Chicago cannot grow enough food for all of its people. Therefore, it has to import food along its transportation lifelines. It also imports many other resources from different parts of the U.S. and from nations around the world. In return, it produces goods that go out along its transportation lifelines to other parts of the U.S. and the world. For example, two of Chicago's biggest industries are steel-making and printing. Iron ore and paper are two resources that come into Chicago. Steel girders and magazines are two products that are sent to the outside world.

Here you see the main differences between the traditional and modern worlds. Latin American villagers use local resources to produce things mainly for themselves. Chicago workers use resources from many parts of the world to produce things for people in many parts of the world.

SECTION REVIEW

1. What do we call the way of life that is much the same now as it was hundreds of years ago?

2. Which of these places would probably *not* have modern technology: the capital of France; a Japanese industrial center; a forest village in Southeast Asia; the largest city in South Africa? Explain your answer.

3. Which of the following is most characteristic of the modern way of life: eating home-grown food; talking by telephone; making tools by hand; traveling on foot?

4. A big city like Chicago has to bring in several kinds of resources to meet its people's daily needs. Name one of these kinds of resources.

Interpreting Photographs

On a hillside in northern Chile, an Indian farmer harvests oregano leaves by beating them loose from their stems. (*Oregano* is an herb used in pizza and other foods.) This scene is captured in the photograph on the opposite page. Notice the tool that the farmer is using. Is it a modern tool? Do you think the farmer bought the tool or made it?

You can find out many things about a place and its people from photographs, if you know what kinds of clues to look for. Some details in a photograph may tell you about the land and climate of a place — its *physical geography*. Other details may tell you about the people and their way of life or *culture*.

Study the photo on the next page. What clues can you find about the area's land and climate? Answer the following questions on a separate sheet of paper.

1. Which of the following do you see in this photo: (a) a lush, green forest? (b) snow-covered mountains? (c) grassy plains? (d) short, stubby plants? (e) rolling hills? (f) sandy desert?
2. What else do you see in the photo that tells you about natural features of the land?
3. From the details you have noted, which type of climate do you think you would find in this place: (a) tropical climate — hot and rainy all year round? (b) highland climate — with little rain? (c) midlatitude climate — warm and cool rainy seasons?

Can you find any details in the photo that show something about the people and their life-style? Answer the following questions on the same sheet of paper.

4. The photo shows which of the following examples of culture? (a) clothing; (b) art; (c) houses; (d) methods of farming; (e) language; (f) religion.
5. How do people in this place use local resources? List as many clues from the photograph as you can.
6. Do people in this place use any resources from outside their own village? List any clues you find.
7. Which of the following statements best describes the way of life shown in the photo? People in this village: (a) have no contact with the modern world at all. (b) have a traditional way of life for the most part. (c) depend on many outside resources and modern technology.

Reading a Pie Graph

The car or bus you may ride to school runs on gasoline. The TV set you watch uses electricity. Electricity, in turn, may be produced by coal, oil, gas, nuclear power, or water power. The many resources that we use to make things work or move are called **energy resources**.

Nations in the modern world use much more of the world's energy resources than do nations with more traditional life-styles. The **pie graph** on this page shows the use of the world's energy resources by different nations for a recent year. In this type of graph, the "pie" always represents a whole amount of something. We use the term *100 percent* (100%) to mean a whole amount. The pie is then divided into "slices" or **percentages** (shares of the whole).

Use of World Energy Resources, 1983

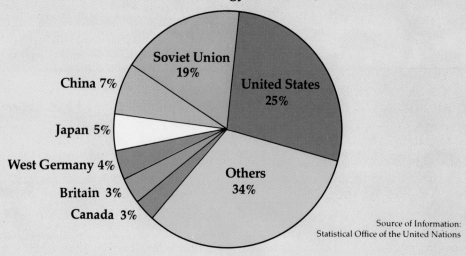

Source of Information:
Statistical Office of the United Nations

Answer these questions on a separate sheet of paper.

1. The whole pie graph represents the world's use of energy resources during which year?

2. Where did the information in the graph come from?

3. Which nation used seven percent of the world's energy resources?

4. What percentage of energy resources was used by the United States?

Compare the different shares of the pie graph with each other. Then decide whether each of the following statements is true or false.

5. Japan used more energy than any other nation.

6. All together the nations represented by "others" in the graph used more energy than the United States alone.

7. The United States used about as much energy as China and the Soviet Union combined.

Modern Ways

Suppose you don't live in a big city. Suppose you live in a small city, a suburb, a small town, or in a farmhouse in the country. Life is much quieter there, without the noise, bustle, and crowds of a city of millions of people. Does this mean that you *don't* live in the modern world?

Think of the place where your family buys its groceries. Do all the products for sale there come from your community? Or are there items like packaged cereals and soft drinks, which are produced in other parts of the country? Where does your neighborhood service station get its gasoline?

Although these areas of the U.S. differ in many ways from the big cities, they still belong to the modern world. So too do mining settlements and lumber camps — even though they are often found in out-of-the-way places. All of these areas bring in resources from other parts of the U.S. and the world. All of these areas send products to other parts of the U.S. and the world.

Think of the farm areas of the U.S. Each farmer does not simply produce food for his or her family, like the villagers you read about earlier. Instead, most modern farmers grow only one major crop that is well suited to the latitude and climate. As a result, there are wheat farmers, corn farmers, cotton farmers, and so on. Each is a specialist in one major crop. These farmers use scientific methods and knowledge. They have modern equipment, specially prepared seeds, and fertilizers. Their crops are carried away in trucks, on railroad cars, and in ships.

Like other people in the modern world, farmers rely not only on transportation but also on **communications** (ways of exchanging information). They get weather forecasts and other news from radio, television, and newspapers. They use the phone for ordering supplies or making sales. What means of communication do you rely on in your life?

In our modern world, people count on many services that people in a traditional village rarely have. Even in rural areas, there is usually water on tap and electricity available at the flick of a switch. There are food stores, gas stations, and all kinds of repair services. At a nearby town or city, people in the modern world can get meals in restaurants, goods of all kinds in department stores, medical care in hospitals, information in libraries, entertainment in theaters and sports stadiums. Can you think of other services you use?

Earlier you saw that traditional villages can look very different in different parts of the world. Suppose you traveled from Chicago to Toronto, Canada. Here too you would find tall office buildings in the downtown section. You would see roads and superhighways leading into and around the city. You would find railroads, an international airport, and a modern subway. Like Chicago, Toronto is on one of the Great Lakes — Lake Ontario — and is a busy port for shipping.

Now suppose you went to Tokyo, Japan. As in Chicago and Toronto, you could travel by bus or cab from the airport to a big downtown hotel. You would find big stores, theaters, sports stadiums, and museums. You would also find restaurants where you could eat anything from hamburger to raw fish.

Your next stop might be São Paulo, Brazil. Here too you would find the same kind of busy downtown area with the same kinds of big hotels, stores, and restaurants.

Of course, you would also find many differences among these big cities. The money used in Toronto or Tokyo is different from the U.S. dollar bills used in Chicago. People speak Japanese in Tokyo and Portuguese in São Paulo. Palm trees grow in São Paulo's parks, while there are oaks and maples in Toronto. But the basic pattern of life is much the same from one city to another.

Because cities use resources from all over the world, they can look much the same anywhere. Compare the traffic, buildings, street lamps, and clothes in Istanbul, Turkey (top), and Bangkok, Thailand (bottom).

People in the Modern World

It's not surprising that both the culture and problems of modern life can be seen most clearly in or near the cities, because that's where most people live. In the U.S., more than 70 percent of the population lives in **metropolitan areas** (cities and their suburbs). About the same percentage is found in many other parts of the modern world.

Where is the modern world? As you've seen, nearly everyone in the U.S. belongs to it, no matter where they live. The same is true of people in Canada, Europe, Japan, Australia and New Zealand, and the Soviet Union. In other regions, modern and traditional ways are found side by side. Some parts of Latin America, Africa, the Middle East, Asia, and the Pacific follow modern ways. Other parts follow traditional ways. Still other areas of these regions are divided between the two.

In these mixed nations, modern ways are found in the big cities. Once you leave the cities, you enter the traditional world of small villages living mainly on their own resources. In most of these nations, however, modern ways are spreading. The governments are trying to bring modern farming, industry, transportation, and communications into the traditional areas. These nations are known as **developing areas**.

The Rest of This Book

In this first unit, you have looked at Earth and its people as visitors from space might have seen them. You first saw Earth as one of nine spheres revolving around the Sun. Then you saw Earth as a planet that could support life. Looking closer, you saw how Earth could support human life. You discovered the wide variety of human cultures around the world. Finally in this chapter, you saw how these cultures can be grouped into two basic ways of living on Earth's resources.

You should now have a general picture of human life on Earth. Most of the time, however, people don't think of the whole Earth as their home. They are aware of only part of it — the part where they spend their lives. In the units that follow, you will explore those smaller parts of Earth. You will discover how these parts fit the general picture of human life on Earth, and how they add new and often surprising details to it.

SECTION REVIEW

1. Are farming communities in the United States considered part of the modern world? Why, or why not?

2. One of the following items does not belong with the rest: telephone, TV, department store, newspaper, mail service. Which item is it? What do the rest have in common?

3. What is a developing area?

4. If you wanted to visit the traditional world, would you be most likely to find it: (a) in a rural area in Japan? (b) in a rural area in the Middle East? (c) in an urban area in Africa? Give reasons for your choice.

YOUR LOCAL GEOGRAPHY

1. Suppose you want to travel from your community to a place in another state. Make a list of the different transportation methods you could use: car, bus, railroad, plane, boat. You might draw a simple sketch map showing the different routes, using a different color pencil for each type of transportation.

Make a similar list of the different communication links between your community and other parts of the U.S. Which of these links are two-way (allowing you to send *and* receive messages)? Which are only one-way?

2. Think of the things you have eaten, worn, and used today. Were any of them produced entirely in your community? Make a list of three things that came from outside your community. Which of them came from outside the U.S.?

Finding Information in an Atlas

Imagine that you are a runner in an international track meet. You have made friends with a runner from the city of Adelaide (AD-uh-layd) in Australia. There is much that you don't know about your new friend's city. You don't even know exactly where it is located. How could you find out? The place to start is a world atlas.

An **atlas** is a series of maps, bound together, which show different parts of the world. The map section at the back of this textbook is a mini-atlas. Most world atlases are larger. Some are huge books in themselves. The librarian in your school or local library can show you many different atlases that are shelved in the reference section.

The maps in an atlas are usually arranged by continents. In a large atlas, each continent map is followed by larger scale maps showing different areas or countries of that continent. The major maps generally combine two types of information: They show **physical features** like river systems and land elevations. They also show **political features** like cities, states, and national boundaries. Such maps are sometimes called **geopolitical** or **general-purpose maps**.

Most large atlases also contain **special-purpose maps** with information on topics such as climate, vegetation, rainfall, population, or industrial products of a region. Atlases often contain tables, charts, and graphs with geographic or economic information as well.

All maps in an atlas are listed with page numbers in the *contents* at the front of the book. However, the quickest way to find a particular place like Adelaide is to turn to the *index* at the back of the atlas. There you will find, listed in alphabetical order, the names of most places shown on general-purpose maps in the atlas.

On the opposite page is a section from an atlas index. The information is arranged somewhat like a table with column headings. Use the sample index to answer the following questions on a separate sheet of paper.

1. What information is given in parentheses ( ) after each place-name?
2. What kind of information is listed for each place in the other two columns of the index?
3. Different places sometimes have the same name. The index lists three places named "Albany." How can you tell which one is in Australia?
4. On what page of the atlas would you find a map showing: (a) Acapulco, Mexico?   (b) the country of Algeria?   (c) Adelaide, Australia?

Index

Places	Page No.	Map Key

A

You know from the index that Adelaide, Australia, can be found in a map on page 90 of the atlas. Besides giving a map page number for each place, the index also has a key with **letter/number coordinates** to help you find each place on the map. For example, the index gives the letter/number coordinates of Adelaide, Australia, as D3.

(Turn page.)

Now imagine you have turned to page 90 of the atlas. There you find a general-purpose map of Australia, like the one shown below. Look at this map. Notice how letters go across the top and bottom; numbers, along each side. The letter D refers to the column of boxes down the grid between 130°E and 140°E. The number 3 refers to the row of boxes across the map between 30°S and 40°S. You can find Adelaide in the square where Column D meets Row 3.

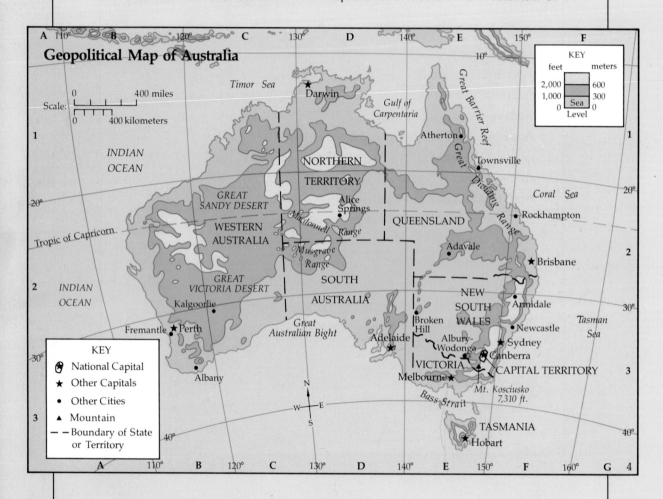

Use the index and the map of Australia to do the following exercises.

5. List Albany, Alice Springs, Adavale, and Albury-Wodonga. Next to each place-name: (a) Write the letter/number coordinates for that place from the index. Use the coordinates to find each place on the map.   (b) Write the numbers of the nearest lines of latitude and longitude.

6. Australia has six states and two territories. Find them on the map and list their names. Beside the name of each, write the name of the capital city and its letter/number coordinates. Include the name of the national capital and its coordinates.

A. Words To Remember

Three of the following terms are defined by the numbered phrases below. On a separate sheet of paper, write each term next to the number of its definition. Then write the two terms that are *not* defined and give a definition for each.

atlas   metropolitan areas   traditional
general-purpose maps   special-purpose maps

1. Much the same now as it has been for hundreds of years.

2. Maps with information on topics such as the climate, vegetation, rainfall, population, or industrial products of an area.

3. Cities and their suburbs.

B. Check Your Reading

1. Do people in a traditional village rely mainly on local resources, resources from outside, or an equal mixture of both?

2. Do traditional villages look much the same in different parts of the world? Why, or why not?

3. In any nation, are you most likely to find modern ways in villages, in cities, or in heavily populated farming areas?

4. Name one of the vital resources that a modern city like Chicago has to import.

5. Name two ways in which Chicago and Tokyo are similar. Name two ways in which they are different.

C. Think It Over

1. What is the biggest disadvantage of relying on local resources?

2. Why do big cities tend to look alike?

3. Why is transportation vital to people in the modern world?

D. Things To Do

Suppose you are the captain of a spacecraft from another planet that has landed in your community. Write a report giving reasons why the captain would think your community belongs to the modern world.

UNIT REVIEW

A. Check Your Reading

1. On a separate sheet of paper, write the letter of each description. After each letter, write the number of the term that best matches that description.

(a) Area of Earth where growing conditions change least throughout the year.

(b) Planet with large supplies of water and oxygen.

(c) Determined by energy from the Sun's rays and by moisture.

(d) Center of our solar system.

(e) Caused by Earth's revolution around the Sun.

(1) Mars
(2) climates
(3) tropics
(4) Earth
(5) Sun
(6) midlatitudes
(7) seasons

2. Fill in the blanks in the following paragraph by writing the missing term next to the letter of each blank.

The great landmasses of Earth, called __(a)__, contain many different land-forms. __(b)__ are parts of Earth's crust that have been pushed up high above the surrounding land, rising to peaks and sharp ridges. Gently rolling lowlands that are easy to travel across are called __(c)__. If these lowlands have a good climate and soil, they are ideal for supporting __(d)__, or plant life. __(e)__ resources, which include metals and fossil fuels, are found beneath the soil.

B. Think It Over

1. Decide whether each of the following items is a feature of the traditional world (T), the modern world (M), or both (B). Write the initial letter of the type of world(s) to which you think each statement applies.

(a) technology
(b) cutting firewood
(c) making steel
(d) hunting polar bears
(e) fishing
(f) farming
(g) lumber camp
(h) handmade tools
(i) supertanker
(j) computer

2. Choose any two of the items in exercise 1 above, and write their letters on your answer sheet. Then write one paragraph on each item, explaining why it belongs to the modern world, the traditional world, or both.

Further Reading

Powers of Nature. National Geographic Society, 1978. Well-illustrated study of natural forces—such as floods, tornados, droughts—that shape Earth.

Weather Predictions, by Gail Gibbons. Macmillan, 1987. Weather maps are explained to the reader.

The Wonderful World of Maps, by James F. Madden. Hammond, Inc., 1986. An explanation of maps.

2

The United States
and Canada

Chapter 6

Two Giants of the Modern World

It's the end of another busy day. Office workers have filed their last pieces of paper, and factory workers have put down their tools. Cars and buses stream away from downtown areas and industrial parks.

Much of the traffic heads for highways that fan out on all sides. Other traffic squeezes onto a bridge or into a tunnel to cross the river that divides this metropolitan area. Radios in a thousand cars blare out traffic reports, news, and music. Announcers cheerfully urge listeners to fly to far-off vacationlands, or to buy products based on up-to-the-minute technology.

This scene belongs to the modern world — somewhere in North America, north of Mexico. The region made up of the United States and Canada has one of the world's highest living standards. As you can see from the Checklist of Nations on page 581, both Canada and the U.S. have life expectancy and per capita personal income figures that are among the highest in the world. (**Life expectancy** is the average number of years that people live. **Per capita personal income** is the average income of all individuals. In both cases, high figures go with a high standard of living.)

Busy rush-hour scenes can be found in any metropolitan area in the region. But the scene described above is based on a specific place, the home of the region's leading industry. Can you guess where this place might be? Here's a clue: The industry has to do with the way the workers get to and from their jobs. Here's another clue: The metropolitan area includes cities on both sides of an international border.

It is the auto industry that is the biggest in the region. The center of this industry lies along the U.S.-Canadian border, in a metropolitan area that includes the cities of Detroit, Michigan, and Windsor, Ontario. The two cities face each other across the narrow Detroit River.

People from each country cross the Detroit River daily to jobs on the other side. There are no border guards, and the commuters need no passports. Life on one side of the border is much like life on the other side. Houses, schools, and churches look much the same, and stores carry many of the same goods.

But life on the two sides of the river is not exactly alike. For instance, people on opposite sides use different forms of money — U.S. dollars or Canadian dollars, each with a different value. People on opposite sides of the river also follow different sets of laws. And most importantly, people on the two sides are citizens of different nations — the United States and Canada.

Broad highways, tall buildings, and a sparkling skyline are all part of the modern world of North American cities. At left, Kansas City, Missouri, at twilight.

101

Overview of the Region

Both of the nations in the region are giants. In area Canada ranks second in the world (after the Soviet Union). The U.S. ranks fourth (after China). Only Brazil and Australia come anywhere near in size. Most countries in the world are on the scale of a Canadian province or a U.S. state.

Both Canada and the U.S. sprawl across the North American continent from the Pacific to the Atlantic. Both include outlying peninsulas and islands.

Canada is divided into 10 provinces, where most of its people live, and two thinly populated territories. Provinces on the east and west coasts include nearby islands, and one province — Prince Edward Island — *is* an island. The Northwest Territories include hundreds of islands that spread northward through the Arctic Ocean as far as 83°N.

The U.S. contains 48 **contiguous** (joined-together) states, plus two outlying states — Alaska and Hawaii. The U.S. also includes Puerto Rico and the Virgin Islands, both in the Caribbean, and Guam and American Samoa in the Pacific (see the maps on pages 108 and 580).

Although the U.S. and Canada are similar in area, they differ greatly in population. As you will read in the next section, much of the gap is due to the greatly differing environments in the two countries. With some 230 million people, the U.S. is nine times as populous as Canada with 25 million people.

The population difference is not a mere number — it closely affects the development of the two countries. The U.S. has nine times as many taxpayers to support its national programs, nine times as many workers to build up its economy, nine times as many consumers for its business firms to sell to. As a result, the U.S. produces and consumes far more than Canada.

The most northerly parts of North America are bleak and lightly populated areas, which constantly test the survival skills of everything that makes its home there.

Physical Geography

"If this land be not rich, then the whole world is poor." A man named Thomas Morton wrote these words after settling in New England in 1622. Like countless others who came from someplace else, Morton was awed by this new world. "In mine [my] eye," he wrote, "'twas Nature's masterpiece."

Certainly nature has been good to the region. In climate, soils, vegetation, and natural resources, the U.S. and Canada are one of the richest regions on Earth. But not all parts are equally rich. There are icy tundra and barren deserts as well as rich mineral deposits and fertile plains.

If you flew over the region's mainland in an airplane, you would find a vast, flat interior flanked by mountains on both sides. The mountains in the west are younger, and many are high and rugged. Those in the east have had more time to be worn down by erosion. They are lower in elevation and smoother in shape.

Look first at the western mountains, which start in Alaska and stretch beyond the southern border of the United States. In the north, these mountains form a single range, known as the Alaska Range. The highest peaks of North America are found here, notably Mount McKinley (20,257 feet or 6,174 meters).

South and east of Alaska, the western mountains divide into two parallel arms. One arm hugs the Pacific coast and bears different names in different places — the Coast Range (in Canada), the Cascade Range (in Washington and Oregon), and the Sierra Nevada (in California). The other arm is known as the Rocky Mountains. Many western peaks reach 12,000 feet (3,500 meters) or more.

The eastern mountains, called the Appalachians, are much lower. In no place do they top 7,000 feet (2,100 meters). The Appalachian Mountains are made up of a series of ridges running northeast to southwest from eastern Canada to Alabama. From the air, the ridges take on the look of an old-fashioned washboard. But they are high enough to have caused a lot of difficulty to early travelers.

While the region's mountains are impressive, flat lowlands or plains cover a much larger area. The greatest lowland is one that stretches east from the Rockies. In fact, this lowland is so great that the area known as the Great Plains makes up only part of it. You could walk all the way from the Gulf of Mexico to the Arctic Ocean and never rise above 1,000 feet (300 meters).

More plains extend around the southern end of the Appalachians and sweep northward along the east coast, narrowing to a thin strip in New England. These are the Gulf Coastal Plain and the Atlantic Coastal Plain.

A variety of other landforms are also found in the region. The biggest is an area of very old rock known as the Canadian Shield. This rock, which covers half of the land area of Canada, forms a vast horseshoe around Hudson Bay. Because the rock is so old, most of the Shield has been worn down to a low, bumpy plateau.

Water Forms

The U.S. and Canada have some of the largest lakes and river systems in the world. From earliest times, these water forms have had great value, both as highways and as sources of fish. Today rivers are also valued as sources of hydroelectric power.

The most important lakes are the five Great Lakes, which lie on the boundary between the U.S. and Canada. More than half of the water in the Great Lakes is in Lake Superior, which is the largest freshwater lake in the world in area.

If you look at the map on page 108, you will see that the Great Lakes are linked to-

103

Only a few tough plants can survive in parts of the western United States that receive little rainfall. Monument Valley, Utah, is in one of these dry areas.

gether. Water flows from one to another, then drains northeast to the Atlantic Ocean through the St. Lawrence River.

In two places, rapids and waterfalls once blocked traffic between the lakes. Canals were built to bypass those falls, at Sault Ste. Marie (SOO saynt muh-REE) and at Niagara Falls. Now large boats can carry cargoes to all parts of the lakes. Moreover, ships from the Atlantic can also reach the lakes, thanks to the St. Lawrence Seaway, which is a series of canals around rapids in the St. Lawrence River. The Seaway is operated jointly by the U.S. and Canada.

The St. Lawrence is one of three great rivers that drain the interior of the region. A second, which flows north to the Arctic Ocean, is the Mackenzie River. However, the greatest river is the Mississippi, which flows south to the Gulf of Mexico.

Other great rivers, such as the Ohio and the Missouri, pour their waters into the Mississippi. Along with its tributaries, the Mississippi drains more than half the area of the United States, as well as part of western Canada. When the Mississippi floods, it can cause frightful destruction. Most of the time, though, it serves as a valuable highway for trade.

Different Environments

The Gulf of Mexico, where the Mississippi pours its waters, is bordered by beaches where people can sunbathe even in the winter. The Arctic Ocean, where the Mackenzie River ends up, is dotted with ice floes even in midsummer. These extremes are not surprising, given the great difference in latitude between the northern and southern edges of the region.

The U.S. occupies lower latitudes than Canada. The southern tips of Florida and Texas reach almost to the tropics. Even in the north, all of the U.S. (except Alaska and a tiny part of Minnesota) lies south of 49 degrees of latitude.

On the other hand, most of Canada lies north of 49 degrees of latitude. At its northern edge, Canada comes within 600 miles (*1,000 kilometers*) of the North Pole. East of the Great Lakes, Canada projects southward as far as 42°N. This small area is the only part of Canada that is closer to the Equator than to the North Pole.

Thus there is a big difference in the environments of the two countries. Most of the U.S. gets enough of the Sun's energy to allow farming. Most of Canada does not.

However, climate is determined by more than distance from the Equator. Just as important are movements of air masses and the way landforms and water forms affect these air masses. Hot summers are common in southern Canada. You'll find summers cooler in San Francisco (latitude 38°N) than in Winnipeg, Manitoba (50°N).

Most of the region has what is called a **continental climate.** Summers tend to be quite hot, and winters quite cold, because the air over most of the land is a long way from the moderating influence of ocean waters. Since water holds heat longer than land, oceans (and places near them) stay relatively warm in winter. And since water warms up more slowly than land, oceans (and places near them) stay relatively cool in summer. The farther air masses are from oceans, the hotter (in summer) or colder (in winter) they become.

Places that are far from the sea, like Winnipeg, tend to have great contrasts in temperature from one season to the next — and from night to day. Winnipeg's hot summers give way to very cold winters. On a typical January day, the *mean* temperature (halfway between high and low) is −3°F (*−19°C*). No other major city in the region has colder winters.

Besides heat or cold, air masses also carry different amounts of moisture. The U.S. and

---

North America's east coast receives plenty of rainfall. Where the soil is good, farms and mixed forests share the land — as they do in this New England valley.

Canada receive moist air masses from two main directions. Most of the air masses move east off the North Pacific while some move northwest from the Gulf of Mexico or the Atlantic Ocean. As a result, the rainiest parts of the region are coastal areas near the North Pacific and the Gulf of Mexico.

You learned in Chapter 2 that air masses drop their moisture as they rise over mountains. This affects much of the western part of the region. In rising to cross the western ranges, Pacific air masses dry out. Little rain falls between the western ranges or over a broad area east of the Rockies.

On the other hand, moist air masses from the Gulf of Mexico can move long distances northwest across the lowlands of the interior. Moist air from the Atlantic Ocean can bring rain to much of the east. The low-lying Appalachians do not force the air high enough for all its moisture to condense. Thus the eastern half of the region receives plenty of rain.

Plant growth depends not only on heat and moisture but also on the soil. In parts of the Canadian prairies and in places like Iowa, you can dip your hand into soil that is black and crumbly and rich. Elsewhere in the region, you might find far different soils — red or brown, sticky like clay or grainy like sand or gravel.

Some of the best soils are in or near areas that were covered by glaciers during the last Ice Age. The heavy masses of ice crushed the surface rocks into small pieces. Thick layers of such particles, piled up by winds, helped to make Iowa's soil rich.

In some places, however, the glaciers were less helpful. In east central Canada, the moving ice carried the soils away to the south. Today much of the Canadian Shield has poor, thin soils that are almost useless for farming.

In Minnesota the glaciers left mounds of broken rocks that blocked rivers and caused water to collect in thousands of lakes, large and small. Today Minnesotans call their state "The land of 10,000 Lakes."

What Grows Where

You have now read about the main influences on plant growth in the region — latitude, climate, moisture, and soils. It is time to put these together. What grows where in the region — and why?

As you might expect, nothing much of use to humans grows in the far north of Alaska and Canada. This is a land of tundra (a "cold desert"). Reindeer and other hardy creatures graze on the mosses and the small flowering plants that make up the tundra vegetation.

South of the tundra, a broad belt of forest lands stretches across Canada, curving south along the Pacific into northern California. At one time, this belt also covered most of the eastern United States. Today, however, much of the forest has been cleared from southeastern Canada and the eastern United States, and a variety of crops have been planted instead. The forest remains in hilly regions, where farming is difficult.

The forest zone gradually gives way to drier lands where there is not enough moisture for many trees to grow. Early settlers found grasslands west of the Mississippi and across the flatlands of south central Canada. The deep soils of these lands were held in place by the matted roots of tall grasses. With **cultivation** (plowing and planting), the soils produced high yields of crops like wheat in the north and corn farther south.

Farther west settlers found shorter and shorter grass, due to lower rainfall levels. Without extra water, only crops that need little moisture can be grown on the Great Plains. Wheat is such a crop, and it is widely grown here.

Even farther to the west and southwest, desertlike conditions prevail. Sunshine is plentiful, and the growing season is long.

But the area is too dry for most plants. Here farmers must use **irrigation** — that is, they must supply water that is pumped from the ground or brought from somewhere else.

Mineral Resources

The U.S. and Canada are especially rich in mineral resources. The region contains large quantities both of mineral fuels (coal, oil, and natural gas) and of metallic minerals (such as iron and copper).

The mineral fuels are found in places where water has deposited mud and rock on top of dead plants and animals. Over long ages, pressure and heat have converted the dead organisms to fuel. Underground layers of fuel can be found over vast areas of the region, in places that were long ago under the sea. Oil and natural gas are found in the central lowlands and also in offshore waters near the coasts. Coal is found both in the central lowlands and in mountain areas that were once low-lying but have been folded upward by violent movements of Earth's crust (see page 57).

Metallic minerals are found mainly in three areas. One area is in the western

Moderate rainfall, good soil, and level land have created the grassy plains of central North America. Both the U.S. and Canada use the plains for grain crops.

mountains, which contain rich supplies of copper, zinc, silver, gold, and lead. A second is in the Canadian Shield, north of the Great Lakes, where there are vast deposits of iron, copper, and nickel. A third is in parts of the Appalachians, such as around Birmingham, Alabama, where sizable quantities of iron are found.

SECTION REVIEW

1. The two countries in this region are among the biggest in the world in area. What country or countries are bigger than: (a) Canada? (b) the United States?

2. Where is the largest lowland area in the region? How far does it extend from north to south?

3. Parts of inland Canada have very hot summers and very cold winters. Give one reason why this is so.

4. Why does little rain fall in much of the western part of the region?

The United States and Canada

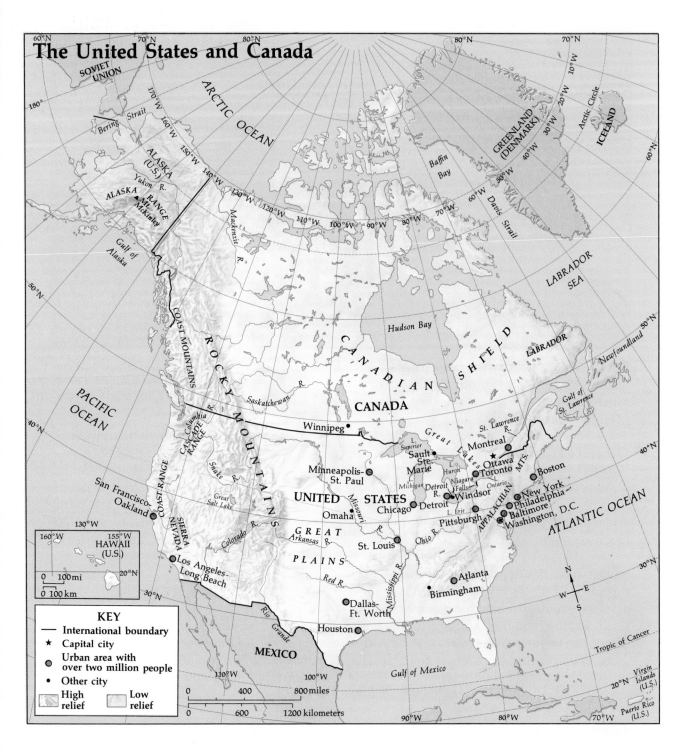

KEY

— International boundary
★ Capital city
● Urban area with over two million people
• Other city
High relief   Low relief

In addition to the 50 states, the U.S. includes Puerto Rico, the Virgin Islands, and several islands in the Pacific Ocean. Canada includes islands in the Arctic Ocean as far as 83°N. Note: The symbol for urban areas (cities and suburbs) with over two million people is combined with the capital star for Washington, D.C.

108

Reading a General-Purpose Map

Suppose you are flying across the U.S. and Canada. From the plane window, you see mountains, hills, lakes, rivers, and other natural features. You also see many signs of human development such as factories, farms, highways, cities, and suburbs.

If you want to get a picture of both natural and human-made features without flying over the land, you can look at a *general-purpose map* such as the one shown on the opposite page. A general-purpose map, you will remember (see Geography Skills 14 on page 94), includes details of both physical and political geography.

The particular details shown can vary widely from one general-purpose map to another. What kinds of information are shown on the map on the opposite page? Answer the following questions on a separate sheet of paper.

1. What two kinds of land surfaces are represented in the map key?
2. According to the key of map symbols, what does the symbol ★ stand for? What does the symbol ◉ stand for?
3. What does a heavy black line stand for on this map?

Study the physical features shown on the map. On your answer sheet, tell whether each of the following is true or false.

4. The Columbia River flows to the east and to the north of the Sierra Nevada.
5. Great Salt Lake is one of the Great Lakes.
6. Areas of low relief border the Gulf of Mexico and Hudson Bay.

The map shows two large countries — the United States and Canada. What political features can you learn from the map? On your answer sheet, tell whether each of the following is true or false.

7. Most of Canada lies north of 50° north latitude.
8. Canada's largest urban areas are close to its southern border.
9. The border between the U.S. and Canada, from the Pacific Ocean to the Great Lakes, is about 1,600 miles long.
10. Montreal, Detroit, Houston, and Winnipeg are all urban areas with populations of more than two million people.

Use the map to compare the physical and political features of the United States and Canada. Tell whether each of the following statements is true or false.

11. The Rocky Mountains extend through both the United States and Canada.
12. Natural systems of lakes and rivers link Detroit, Toronto, St. Louis, and other inland cities to the sea.

Human Geography

Near a creek in Alaska, sharp disks made of chipped stone have been found in the ground. The disks are just like tools used in Siberia more than 10,000 years ago. They make up one of many clues that lead scientists to estimate that humans crossed from Asia into North America 12,000 or more years ago. These humans spread out until they had occupied both North and South America.

The crossing took place so long ago that these first settlers are no longer thought of as immigrants. (**Immigrants** are people who migrate, or move, *into* an area.) Their descendants, the American Indians and Eskimos, are also known as Native Americans.

About 500 years ago, other settlers began moving in. At first they came mainly from Europe and Africa. Later more and more came from Asia and other parts of the world. Thus today Indians and Eskimos make up only a small fraction of the population — under two percent in Canada, and even less in the U.S.

The early European settlers of North America were a mixed group. In the east, they were mainly English, French, Dutch, and Swedish. In the south and southwest, they were Spanish. In the northwest, they were Russian.

After various wars and treaties, much of the eastern **seaboard** (land along the coast) came under British control. Several colonies were set up, and large numbers of English-speaking colonists came to settle them. In 1776, 13 of these colonies broke away from Britain to form the United States. In language, law, and some other ways of life, the U.S. kept its British heritage.

For a long time, Canada was under French control, and many French settlers came. But Britain gained control in the 18th century, and English-speaking settlers then poured in. The Canadian colonies did not join the

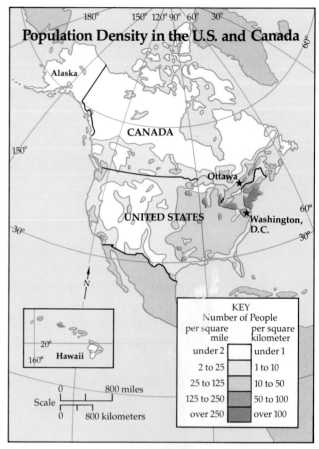

Population Density in the U.S. and Canada

KEY
Number of People

per square mile	per square kilometer
under 2	under 1
2 to 25	1 to 10
25 to 125	10 to 50
125 to 250	50 to 100
over 250	over 100

In this map, the darker the area, the more people are crowded together. Why do you think the least populous part of the 48 states is the inland West? Why do nearly all Canadians live close to the U.S. border?

other American colonies in the American Revolution. In 1867 Britain brought the Canadian colonies together as one nation, with its own parliament and its own prime minister. Although Canada is fully independent, it still recognizes Britain's monarch as its own head of state.

Canada's double heritage — part French, part British — lives on. A majority of today's population speaks English, but a large minority (about 29 percent) speaks French. Both languages are considered "official." French-speaking Canadians are concentrated in the east — mainly in the province of Quebec, where 85 percent of the people are of French descent.

A Variety of People

Many waves of immigrants have arrived in the region over the past 200 years. Thus today's inhabitants represent a rich variety of national and ethnic origins. More than four fifths of the people are at least partly of European descent. Besides countries already mentioned, their ancestors came from places like Germany, Italy, Ireland, Scotland, the Ukraine, Poland, Portugal, and Greece. Many have come from Latin America, and their background may be partly Indian or African as well as European.

People of mainly African descent form a sizable minority. The ancestors of these people came largely from West Africa. Most arrived as slaves. Slavery was concentrated in southern latitudes of the region where crops like cotton and tobacco were grown. These crops required a large labor force in the days before modern farm machinery. There were few slaves in Canada, where cotton cannot grow and tobacco was not grown until later. Today blacks make up about 12 percent of the population in the United States, but less than one percent in Canada.

About 1½ percent of the people in the region are of Asian origin. The first to arrive in large numbers were the Chinese, beginning in the 1850's. Then around 1900, many Japanese came. For many years, U.S. laws barred further Chinese and Japanese immigration, but such discrimination no longer exists. In recent years, Asian immigrants have included many from the Philippines, South Korea, and Vietnam. In the past, a majority of Asian immigrants settled in the areas closest to Asia — Hawaii and the Pacific coast. But today there are also many Asians in large eastern cities.

Immigrants have left their mark in North American life. Turn on the radio and listen to a Jamaican reggae tune or a Polish polka. Stop at a fast-food stand and buy a slice of Italian pizza. Enroll your baby brother in a *kindergarten* (German). Buy a *taco* (Mexican).

Eat an eggroll (Chinese). Almost anywhere you turn, something will remind you of the people's many different origins.

The variety of ethnic origins has led to a variety of religions in the U.S. and Canada. A majority of people in the region are Protestants, belonging to any of a number of separate Protestant churches. A large minority are Roman Catholics. In some areas, there are large communities of Jews and Eastern Orthodox Christians. Two other major world religions, Buddhism (BOOD-iz-uhm) and Islam (is-LAHM), also have followers in some parts of the region.

Two Urban Democracies

In both Canada and the United States, almost three fourths of all people live in large urban areas (with populations of 50,000 or more). Some cities are crowded. But the region as a whole has plenty of open space.

A look at the map on the opposite page will show you how unevenly the population is distributed. Most Canadians live within 100 miles (*160 kilometers*) of the U.S. border, mainly in the southeast — areas where the environment is best for farming and human settlement. The heaviest concentration of U.S. population is in the northeast and around the Great Lakes — areas where the largest number of European immigrants first arrived. Other centers of U.S. population are near the Pacific and Gulf coasts.

Both Canada and the United States have economies based on free enterprise. While governments in both countries play a major role in economic life, basic decisions about producing and selling goods are made by private businesses.

The two nations also share a belief in political freedom. Both have federal systems in which a central government shares power with lower levels of government. The governors of Canada's 10 provinces have somewhat more power than the governors of the U.S.'s 50 states.

Scientists believe that Indians and Eskimos came to America from another part of the world. What part is that?

2. How do the U.S. and Canada differ in their official languages?

3. Which nation, the U.S. or Canada, today has the larger proportion: (a) of Indians and Eskimos? (b) of blacks?

4. Why do most Canadians live within 100 miles (160 kilometers) of the U.S. border?

Economic Geography

It is winter and storms rage over parts of the region. Yet it seems like summer in the grocery store, where there are ripe red tomatoes. In the U.S. and Canada, tomatoes are always in season.

Tomatoes do not appear naturally in the midst of winter storms. In North America as in the rest of the modern world, nature gets an assist from technology.

The tomato you buy in winter may have been grown in a place that was once a desert. It may have been picked by a monster machine that stands two stories high. It may have been trucked all the way across the continent. It may have been sold to you by a corporation that also sells oil or runs hotels.

Agriculture and industry are carried out on a massive scale in the region. A single farm may stretch as far as the eye can see. A single business firm may have branches from coast to coast — and in distant parts of the world as well. Operating on such a scale, both agriculture and industry have enough money to afford new technological advances. In turn, this new technology enables them to cut costs, lower prices, raise profits, and grow bigger.

What makes such large-scale operations possible? Earlier you read about one reason — the size of the population. The U.S. and Canada have millions of tomato eaters (and

car buyers, and tooth brushers). There are lots of customers for businesses to serve.

A second reason is the wealth of these customers. Average incomes are high in both the U.S. and Canada. Most people can afford color TV sets, automobiles, and washing machines. This is not to say that *everyone* in the region is wealthy. Poverty often can be found in the middle of prosperity, and major pockets of poverty dot the land. But a family considered poor in the U.S. and Canada may have many times the income of the *average* family in many other countries.

A third reason is the extensive transportation system. The region is crisscrossed by railroads, public highways, waterways, and pipelines that carry goods to market. Nine of the world's 10 busiest airports are found in the United States. Both products and people can move quickly from place to place.

Putting Resources To Work

The U.S. and Canada have reached their high standard of living by making extensive use of the region's resources. Both agriculture and industry are highly developed.

With plenty of land to choose from, agriculture can be concentrated in areas where climate and other factors are most favorable. Some areas specialize in corn; others, in potatoes; still others, in artichokes or chickens.

Many industries are also grouped in major areas. You have already read that the auto industry tends to center around the Great Lakes. One reason is that the auto industry needs steel, and the U.S. steel industry is located nearby. The steel mills are there because the lakes give easy access to raw materials such as iron (from Minnesota and Labrador) and coal (from the Appalachians). Newer industries, such as electronics, need fewer bulky raw materials and may be more widely scattered.

Wherever farms and factories are located, they may produce either for local markets or for distant ones. Corn from the North Cen-

Modern nations depend on transportation networks. At the port of Montreal, fuels and other resources are transferred from ocean ships to trains and trucks.

tral U.S. may end up in a bowl of corn flakes eaten in Omaha — or in a cow butchered for meat in the Soviet Union. Lumber from Canada's western forests may help build a shopping center in South Dakota — or a private home in Japan.

Exports (goods sold to other nations) play a major role in the economies of both Canada and the U.S. These exports help pay for goods that are imported. (**Imports** are goods bought from other nations.) Many of these imports are resources that are lacking in the region. They include crops such as coffee, which can be grown only in tropical regions, and minerals such as cobalt, which is vital to modern industry.

What's Ahead

In the chapters that follow, you will look more closely at the region's two countries and the way they have used their resources

to build a high standard of living. Chapter 7 deals with the United States; Chapter 8, with Canada.

Both nations belong unmistakably to the modern world and are similar in many other ways. But alongside the similarities are many differences — between the separate countries and among areas within each country. Chapter 9 will explore these similarities and differences.

SECTION REVIEW

1. Give one reason why the auto industry is located in the Great Lakes area.

2. What kind of goods do the U.S. and Canada need to import?

3. Give two reasons why industries in the region can operate on a large scale.

4. Why do many farming areas of the U.S. and Canada specialize in certain products?

YOUR LOCAL GEOGRAPHY

1. Geography Skills 15 tells how to read a general-purpose map. Now suppose you are designing a general-purpose map of your city or county. The scale is five miles to the inch (*three kilometers to the centimeter*). What physical and political features would you consider important to show? For instance, which landforms and water forms would you include? What boundaries would be indicated? What roads and railroads would be included? After deciding on the most important physical and political features of your area, draw a map key with symbols illustrating these features.

2. How does the physical geography of your area fit into the physical geography of North America as a whole? Is your area mountainous? If so, what range do the mountains form part of? Perhaps your community is situated in the plains. Are they interior lowlands, or coastal plains? Is your area typical of the physical geography in your section of North America?

Names on the Map

In the U.S. and Canada, places have been named for a wide variety of reasons. The following paragraphs come from a book on place-names of English-speaking countries. You will find a few examples of the reasons why names were chosen. As you read them, try to think of others.

American place-names have been created by Indians, English sovereigns, Spanish grandees, French Jesuits, seamen, explorers, hunters, trappers, mountain men, businessmen, Congressmen, and a host of ordinary settlers of both sexes; in fact, one might say, by representatives of the entire population.

The first pioneers were often unlettered people who could choose a name without knowing how to spell it. A group who settled in Missouri wanted to call their new home Raleigh, after the town in North Carolina from which they had set out, but they wrote it Rolla. So it remains, and so it has been repeated farther west. In contrast to them were the men of education who could produce classical names such as Phoenix, for a town to be built on the site of a burned-out Indian village, or Akron (Greek, "summit") for the highest point of a projected canal route.

Many place-names were chosen by vote at town meetings and other local gatherings, and names given by the first arrivals were discussed and often rejected as too uncivilized. In this way, Whiskey Hill and Mosquito Gulch, both in California, were changed by their residents to Freedom and Glencoe. Decisions were not always easily reached; and in one place in Kentucky, the matter was so hotly disputed that the only name that could be agreed on was Disputanta. In 1950 the inhabitants of a town with the factual name of Hot Springs were persuaded to change it to "Truth or Consequences"

by the promoters of a radio program of that name. . . .

Much that has been written about the place-names of the United States applies also to those of Canada; and along their 3,000 miles [5,000 kilometers] of frontiers, hundreds of names are held in common. But there are marked differences. The Spanish element, so strong in parts of the United States, is lacking in Canada. On the other hand, French names, which in the States are only a legacy from the colonial period, play a far more dominant part in Canada, where French remains a living language in which new names are still being coined.

As to the antiquity [age] of its naming, Canada has a decided lead over its neighbor. The first recorded name given by Europeans in the United States is Florida, which was given in 1513, when the names Labrador and Newfoundland had already been in use for a decade.

— from *Place-Names of the English-Speaking World,* by C.M. Matthews. New York: Scribner's, 1972.

Ask Yourself . . .

1. A phoenix is a mythical bird that is supposed to burn up every thousand years and then rise again from the ashes. Why was its name chosen for a settlement in Arizona?

2. Can you suggest any geographical reason why Labrador and Newfoundland were named by Europeans earlier than any place in the United States?

3. For what reason did some settlers in Missouri want to call their town Raleigh? Do you think many places in the U.S. have been named for a similar reason? What other reasons can you think of for the choice of U.S. place-names?

CHAPTER REVIEW

A. Words To Remember

Three of the following terms are defined by the numbered phrases below. On another sheet of paper, write each term next to the number of its definition. Then write the two terms that are *not* defined and give a definition for each.

continental climate   immigrant   seaboard
general-purpose map   irrigation

1. Land along the coast.
2. A person who moves into an area.
3. Supplying water for farming in dry areas.

B. Check Your Reading

1. Where are the tallest mountains in the region: in the Alaska Range, the Rocky Mountains, or the Appalachian Mountains?

2. Give two basic reasons why the climate varies widely in different parts of the region.

3. In the past, rapids and waterfalls made it difficult to transport goods across country via the water system of the Great Lakes. How was this problem solved?

4. In what two different ways did glaciers affect the soils found in some parts of the region?

5. Thomas Morton wrote, "If this land be not rich, then the whole world is poor." What did he mean?

C. Think It Over

1. The population of the U.S. is about nine times the size of Canada's. What difference does this make to the economies of the two nations? Explain your answer.

2. The chapter mentions a few of the contributions to American culture made by different ethnic groups. Can you think of any others? Give two examples.

3. You have read that the latitude of a country influences the length of its growing season. Describe one way that Canada's latitude has affected its agriculture.

D. Things To Do

Using the Checklist of Nations on pages 581–587, choose any four nations other than the United States and Canada. Write down the population growth rate and life expectancy figures for each nation. Make a table to compare these figures with those for the United States and Canada. (Geography Skills 10 on page 74 may help you in preparing the table.) Write a paragraph summing up the results of this comparison.

Chapter 7

The United States
Making the Most of Variety

When you look at a slice of pizza, it probably doesn't make you think about geography. All the same, by "studying" a pizza, you can learn something about the way foods are grown and distributed in the United States.

Each part of your pizza is likely to come from a different part of the country. The dough is made of wheat flour and probably comes from the "wheat belt" — say from North Dakota. The sauce is made of tomatoes and probably comes from California, which grows 85 percent of the nation's tomatoes. The cheese probably comes from a dairy state, such as Wisconsin.

In the modern world, no nation relies entirely on its own resources, even for food. Thus if you wanted anchovies on your pizza, the chances are they would be imported. All the same, modern nations try to make the most of their own resources first.

As you can see from your pizza, food resources in the U.S. tend to be **specialized** by area. This means that most farming in an area tends to concentrate on the same kinds of foods. There is the same kind of specialization in mining and industry.

Each area does mostly what it can do best. Wisconsin farmers might grow wheat — and a few do. But most Wisconsin farmers raise corn and hay as feed for dairy cows, and earn their living by selling milk. Then if the farmers need wheat flour, they buy some that has been produced someplace else.

With modern machines and modern transportation, each part of the country can specialize in products best suited to its environment. Thus the environment influences the kind of work that many Americans do.

At the same time, many kinds of work have no direct connection with the environment. Modern transportation and communications make it possible for businesses to locate in a wide variety of places. Moreover, people can move from one area to another in search of different jobs.

In a modern nation like the United States, people and their environment interact in countless ways. The following sections will show the major roles played by geography in Americans' lives.

Physical Geography

Have you watched a weather forecast that used satellite photos to show clouds over the United States? Such photos reveal the general shape of the land, but give you little idea of its size and variety.

The United States is a big place. From coast to coast, the distance is about 2,500 miles (*4,000 kilometers*); from Canada to Mexico, about 1,600 miles (*2,500 kilometers*). And that's just within the contiguous 48 states.

Modern agricultural methods can turn dry and barren land into fertile farms.
At left, sprinklers stretch to the horizon over a cornfield in west Texas.

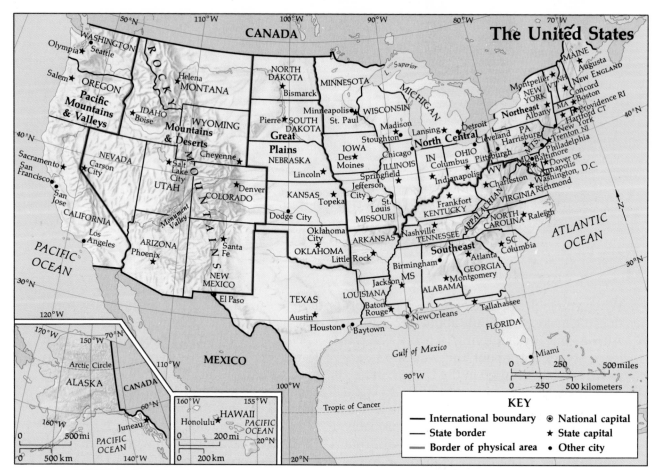

KEY

—— International boundary	⊗ National capital
—— State border	★ State capital
—— Border of physical area	• Other city

From Florida to Hawaii, the distance is more than 5,800 miles (*9,500 kilometers*).

Within these great distances, the environment varies from subtropical to polar climates and from plains to high mountains. To show the main variations, the U.S. can be divided into eight areas.

1. Northeast. This area stretches from Washington, D.C., to Maine. The Appalachians give much of the land a hilly shape. Forests are widespread. Where the forests have been cleared, the hills tend to have rocky, thin soil; but there are also broad valleys that are good for farming. Rainfall is plentiful — often 40 inches (*100 centimeters*) or more a year.

2. Southeast. Much of the area from Virginia to Texas is flat and low-lying, but the Appalachians extend through part of the interior. The area has mild winters and a long growing season, with a subtropical climate in the lowest latitudes. Rainfall is heavy in some places, tapering off toward the west. Some of the land is forested.

The U.S. has eight broadly different environments. The boundaries shown here follow state borders, but in fact there is no sharp break between environments.

3. North Central. This part of the interior flatland stretches from Ohio west to Missouri and north to the Canadian border. Most of the area was once covered by glaciers, so soils tend to be rich. However, parts of the north have thin or rocky soils. Rainfall is usually more than adequate, and summers tend to be hot — good growing conditions for many food crops.

4. Great Plains. The Dakotas, Nebraska, Kansas, Oklahoma, and northern Texas lie in this area. The land tends to be flat. Here it is drier than in the North Central area. The western parts of these states get from 10 to 20 inches (*25 to 50 centimeters*) of rain a year — just enough to grow crops like wheat.

5. Mountains and Deserts. States in this area include all those west of the Great Plains that do not touch the Pacific coast. The land on either side of the Rocky Mountains is largely desert. If you tried to walk across parts of Utah, for example, you might die of thirst before you reached water. With irrigation, however, crops can be grown. In the mountains, rainfall is higher and there are forests below the **timber line** (the highest elevation at which trees can grow).

6. *Pacific Mountains and Valleys.* Another range of rugged mountains is found in the Pacific coast states. Here too the mountains are heavily forested below the timber line. Areas near the coast in the north have mild temperatures and receive more rain than any place in the contiguous states. East of the mountains and farther south, temperatures are higher but rainfall is meager. Much of California gets less than 20 inches (*50 centimeters*) of rain a year, most of it in winter. Irrigation must be used for most crops.

7. *Alaska.* The Pacific mountains extend northward through Canada to run in a curve through Alaska. There are dense forests on the lower slopes of these mountains. Only in central Alaska is there much flat land. Most of the state has poor soil and its polar climate is too cold for farming. Coastal areas in the south have a milder climate but are too rugged for farming.

8. *Hawaii.* This state is an **archipelago** (ar-ki-PEHL-uh-goh; a large group of islands) that sprawls across a vast area of the Pacific. If you flew from one tip to the other, you would cover the same distance as from Chicago to San Francisco. However, the main Hawaiian islands are bunched in the southeastern corner of the archipelago. These islands are volcanic in origin, and contain rugged mountains and sandy beaches. There is not much farmland, but plants such as sugarcane and pineapple are grown. The climate is subtropical.

Many Resources

About one third of U.S. land is wooded. In the west, forests are made up largely of **coniferous** (cone-bearing) trees like Douglas firs and redwoods. These trees may grow as tall as 30-story buildings. In the east, trees are smaller and forests tend to be mixed. Besides coniferous trees, the forests usually include **deciduous** (leaf-dropping) trees like maples and oaks.

Waters in and near the United States support large populations of shellfish and fin fish, from lobsters off the New England coast to tuna off the Pacific coast. The biggest catch in the Gulf of Mexico and the South Atlantic is menhaden (men-HAY-duhn). Don't be surprised if you have never seen this in your neighborhood fish store or supermarket. Menhaden is used to make

A skilled work force is one of the key resources of a modern nation. Below, workers control heavy machinery to build a chemical plant in Puerto Rico.

chicken feed, fertilizer, and other products.

The U.S. is especially rich in mineral resources. Oil is found in many states. The largest supplies have been found along the Gulf coast from Mississippi to Texas, off the northern shore of Alaska, and in parts of the lower Great Plains. California also has oil, as do states in other parts of the country.

A related product, natural gas, is usually found with oil. But sometimes gas is found by itself. The Northeast, which is poor in oil, has been explored for natural gas in recent years, with promising results.

Coal is the mineral fuel that seems most plentiful. The usefulness of coal varies with its age. The oldest type, known as **anthracite** (AN-thruh-siet), has been pressed hard with age. It burns slowly and can be used in furnaces. In the U.S., anthracite coal is found mainly in Pennsylvania. **Bituminous** (bi-TEW-muh-nuhs) **coal** (soft coal), which burns more rapidly and is used widely in industry, is found in the Western Appalachians and in some Rocky Mountain states. **Lignite** (LIG-niet; brown coal) is coal that is not fully developed. It gives off less heat, and is used mainly where other coal is not available.

The main U.S. iron resources are found on the edge of the Canadian Shield, near Lake Superior — in Minnesota, Michigan, and Wisconsin. The best of the iron ores (that is, the cheapest to get at) have already been removed. But lower-grade deposits remain.

Other metal resources include copper, gold, silver, lead, zinc, and uranium. They are found in different parts of the country, from the Rockies to Tennessee.

Still, there are a few metals the United States lacks, or has only in small amounts. For example, the United States lacks many **ferroalloys** (feh-roh-AL-oys). These are minerals that are added to iron to produce high-quality metals such as stainless steel. Chromium, cobalt, and nickel are three ferroalloys that the U.S. has to import.

Human Geography

A good place to glimpse the variety of American people is in a big city like Los Angeles. This is the nation's second most populous city, after New York City.

If you stroll around the downtown streets of Los Angeles near City Hall, you will see many shades of skin colors and hear many accents and languages. For example, you will almost certainly hear Spanish spoken, which is not surprising in a city that has a Spanish name and was once part of Mexico. Most of Los Angeles' Hispanic residents are descended from — or are themselves — recent Mexican immigrants. However, some families have been there longer than any English-speaking Californians.

In downtown Los Angeles, you would also see Chinese Americans, Japanese Americans, and other Asian Americans. Just as many European immigrants settled on the Atlantic coast, where they first arrived, many Asian Americans settled on the Pacific coast.

Like other U.S. cities, Los Angeles has ethnic neighborhoods in which people of one background are concentrated. One neighborhood has mainly Chinese residents. Other neighborhoods are mainly Italian, or Hispanic, or black, or Vietnamese. Still other neighborhoods have people of many groups living side by side.

Americans come in all ages, sizes, shapes, and colors. Every U.S. area has its own ethnic mix; the overall mix of peoples is one of the nation's great strengths.

Chicago in the late 19th century, when industry was expanding and jobs were opening up. Minneapolis became the trading center for the northern Great Plains area, which attracted German farmers partly because the area reminded them of Germany. Cuban Americans settled in Miami because it had a similar climate to nearby Cuba.

Moving Within the Nation

Before slavery ended, small numbers of free blacks could be found in many parts of the U.S. But the great majority of black Americans lived in the rural South. Since World War I, much of the black population has moved to cities in all parts of the U.S. Now some cities (such as Washington, D.C., and Detroit) have more blacks than whites.

Today Americans tend to move into or close to cities because that is where most of the jobs are. For example, Los Angeles has a special attraction for aircraft engineers and actors because its leading industries include aerospace (aircraft, satellites, rockets) and entertainment (TV and movie production).

Until quite recently, the Northeast and North Central areas together have been the center of U.S. industry, and thus the areas where most people tended to move for jobs. But new industries are locating elsewhere. Thus the center of industry has been shifting toward the *Sunbelt* (the belt of states that stretches from coast to coast across the Southeast and Southwest). As its name implies, this area is noted for its milder climate, which makes living easier. In addition, some states attract industry with lower taxes and easier business laws.

As jobs in the U.S. have shifted, so have people. Four Sunbelt states, Florida, Texas, Arizona, and California, gain nearly half a million people each year. At the same time, New York and Rhode Island lose population, while most states in the Northeast and North Central areas grow only slightly.

If you were to leave Los Angeles and visit a rural area, you'd find a less varied population. Overall, rural people have as much ethnic variety as city-dwellers. But people of similar origins tend to cluster together in different areas. For example, suppose you visited the rural area around Stoughton, Wisconsin. You'd find that many people there are of Norwegian descent. One of the town's biggest days each year is May 17 — Norwegian Independence Day.

Each city has its own ethnic mix. For example, Chicago has many Americans of Slavic descent, Minneapolis has many German Americans, and Miami has many Cuban Americans. Many Slavs first settled in

Of course, population shifts are nothing new for the United States. In the 19th century, the movement was largely westward. In the early 20th century, it was out of the South into northern cities. Americans just won't stay put. Each year one family in five moves to a new home.

Wherever people move in the U.S., they will probably find many differences from the place they left. They may find a very different environment, ethnic mix, and choice of jobs.

At the same time, they will certainly find many similarities. One reason is that the U.S. is a nation whose people have many cultural features in common. Another reason is that the U.S. belongs to the modern world, where ideas and goods travel rapidly from place to place.

Thus wherever you live in the U.S., you expect to find public schools that are open to all, paid for by taxes. You expect to find systems of local, state, and national government in which all adult citizens have a voice. You also expect to be able to communicate with people anywhere in the nation by

Three fourths of all Americans live in urban areas with more than 50,000 people. San Francisco, shown here, has an urban area with more than three million.

phone, and to find some easy means of transportation to other places. You will find the same basic goods on sale in stores and supermarkets. You will find many people interested in the same kinds of sports and TV shows. Can you think of other things that would be much the same from one part of the country to another?

SECTION REVIEW

1. Why do many Americans choose to live in or close to an urban area?

2. Give one reason why many German Americans settled in the northern Great Plains.

3. What major shift in population has taken place in the U.S. in recent years?

4. You read in the text that life-styles may vary, depending on where you live in the U.S. However, there are several features of life that are available to just about every resident of the U.S. Describe two.

Economic Geography

If you wanted to prospect (search) for gold, you would have to go where the gold is. Prospecting for minerals is an economic activity that is closely linked to the environment. If you wanted to open a radio station, you would have to go where the listeners are — to a center of population. Radio stations are *not* closely linked to the environment, but they are still influenced by geography.

This section will look first at the parts of the U.S. economy that have close links to the environment. Then it will turn to the parts that are less closely linked.

Farming and Farm Industries

A vast sea of crops stretches from the Appalachians almost to the Rockies. Most of the grain and meat that Americans eat comes from this area. No other area is so completely farmed. It is the agricultural heartland (central, important area) of the United States.

There are two distinct parts of this heartland — the North Central area, and the Great Plains. The North Central area, with more rain, consists largely of a "corn belt," and is the leading area for feed crops that need plenty of moisture, like corn and soybeans.

Much of the corn grown here is fed directly to cattle and hogs. Soybeans are fed to animals after being ground into meal. Thus industries in the North Central area include plants that process pork and beef, and ship meat to other parts of the country.

The northern part of the North Central area belongs to a "dairy belt" that extends eastward across the Northeast. The growing season is too short for corn to develop fully, but the corn can still be used. Together with hay, the "green" corn is fed to dairy cows. Thus industries in this area include milk packing and the processing of milk into butter, yogurt, cheese, and powdered milk.

The Great Plains receive too little moisture for corn, but they can grow wheat. As a result, this area has become the "wheat belt." In dry years, however, the wheat yield can drop far below normal. With low yields, a lot of land is needed to raise enough wheat to make a profit, so farms (and machines) tend to be larger in the Great Plains.

You can tell you're in wheat country by the tall grain elevators that tower over rural towns. These elevators store the wheat grains until they can be moved to market by rail or road. Moving wheat to market is big business. Some wheat goes to cities like Minneapolis to be milled into flour. But much of the wheat goes by railroad to ports to be put on ships and sent abroad.

Fruits, vegetables, and other kinds of crops are grown in other parts of the nation. Each area has some advantage that makes it especially suited to grow what it does. Here are just a few examples:

- Hawaii specializes in tropical crops like pineapples, sugarcane, and orchids.
- California, Florida, and Texas make use of their mild climates to produce citrus fruits and vegetables. Farm workers in these states may be out cultivating lettuce or tomatoes while snow blankets the North.
- Crops like apples, cranberries, and potatoes do best in cool climates and thus are not suited to low latitudes. Look at the map on page 124. Can you find where these crops are grown?

You read earlier that wheat farms tend to be the biggest in area. But every kind of U.S. farm — whether it grows wheat, lettuce, or oranges — is getting a little bigger from year to year. In the 19th century, the average farm covered 160 acres *(65 hectares)*. Today the average farm has some 400 acres *(160 hectares)*. In the Great Plains, farms may exceed 2,000 acres *(800 hectares)*.

One reason for bigger farms is the trend to bigger machines — tractors, harvesters, mowers — that have helped raise U.S. crop

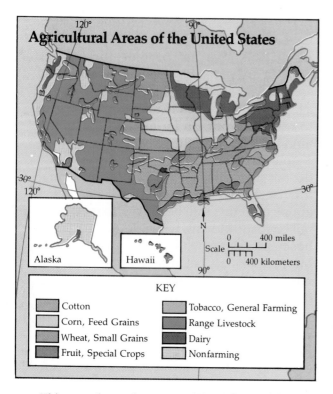

Agricultural Areas of the United States

KEY

- Cotton
- Corn, Feed Grains
- Wheat, Small Grains
- Fruit, Special Crops
- Tobacco, General Farming
- Range Livestock
- Dairy
- Nonfarming

This map shows that mountains can be used for raising livestock, but most crops are grown in lowland areas. Four major farm belts extend across the U.S. east of the Rockies: dairy, corn, wheat, and cotton. Fruits and special crops are grown along the coasts.

yields. Such machines are expensive. To make them pay, farmers must keep them in use. Thus many farmers increase the size of their farms and grow more.

Today people use the term **agribusiness** to describe the complex of giant firms that help to grow, process, and market U.S. farm goods. The term also applies to firms that sell seed, farm machines, and farm chemicals to processors such as meat packers and flour millers, and to distributors such as supermarket chains.

Developing Other Resources

In some parts of the U.S., trees are harvested much like a crop. The timber industry is especially important in the Pacific Northwest and in parts of the South. The wood may end up as rayon blended in the fabric of a shirt, as paper in a book, or as a board in a new house. Factories that make such products tend to be found in timbering areas.

Near seacoasts fishing and fish processing are major industries. Only a small part of the catch is sold fresh. Much is frozen or canned in coastal towns. Some is ground up for animal feed or fertilizer.

Mining and processing are important industries in the areas where minerals are found. To make gasoline, diesel fuel, or motor oil, **crude oil** (oil straight from underground) must be **refined** (treated to remove impurities). Refineries are usually located near a port (such as Baytown, Texas) or in an area with a large population and major industries (such as New Jersey). Natural gas, crude oil, and refined products are transported through pipelines that crisscross the country.

Manufacturing: Where and Why

Industries that manufacture goods are less closely tied to the environment than activities like oil drilling or farming. But geography still plays a part. Imagine you were planning a mill to make steel. Where would you build it — and why?

First, you'd consider *raw materials*. For steel the basic raw material is iron or scrap metal.

Second, you'd need a source of *energy*. Steel mills traditionally use a high-quality type of bituminous coal, called coking coal. New mills often use electricity.

Third, you'd need access to *transportation*. Iron and coal must be brought to the mill, and steel must be moved out. The U.S. has a highly developed network of railroads, highways, waterways, and airways to move goods and people.

Fourth, you'd need *workers* — both skilled and unskilled.

Fifth, you'd need *markets* — places where you could expect to find customers. Most steel is sold to other manufacturers to be made into products like autos, refrigerators, and industrial machines.

Only the first two of these five points are directly related to the environment. Since iron and coal are both bulky and therefore expensive to transport, much American steel is produced in areas with iron mines (such as Birmingham, Alabama) or in coal-producing areas (such as Pittsburgh, Pennsylvania). However, since steel itself is bulky and expensive to ship, it's even more important to consider where the markets are. Thus steel mills tend to be located near other industries. This usually means near metropolitan areas (cities and suburbs).

Metropolitan areas also meet the other two needs. They are centers of transportation, and they are places where many people (workers) live. Thus more than one fourth of all U.S. manufacturing is done in just seven large metropolitan areas — New York, Chicago, Los Angeles, Detroit, Philadelphia, Cleveland, and St. Louis.

You will note that all these cities are in the Northeast, the North Central area, or California. These areas, along with the Gulf coast (Houston, for example) and the Pacific Northwest (Seattle), have the nation's key manufacturing centers.

Let's take a quick tour of the U.S. to see where some major industries are located. As with farm products, most areas have some advantage that leads them to specialize.

New England was the first center of U.S. manufacturing. Not only was it near early markets, but it also had rivers to provide water power. (Before the Industrial Revolution, most factories were built beside rivers and used flowing water to run their machines.) Today New England is handicapped by its lack of modern fuel sources. But it makes use of the brain power available in its many colleges and universities. The area now specializes in high-technology products like aircraft engines and electronics.

The North Central area of the U.S. makes up a large part of the nation's agricultural heartland. (The map on the opposite page shows the area's major crops.) Here corn grown in Iowa is being shipped down the Mississippi River toward a processing center.

Industries in Atlantic coast cities like New York, Philadelphia, and Baltimore often depend on transportation — especially overseas shipping. Ships bring in raw materials such as iron from South America and oil from the Middle East. These materials go into the manufacture of steel and **petrochemicals** (chemicals made from petroleum). The steel is used to make products like ships and railroad cars. Petrochemicals are used to make drugs, pesticides, and other products.

As you have already seen, the North Central area specializes in industries that use steel, such as automobile and machinery manufacturing. The Southeast specializes in *textiles* (fabrics) and clothing. These industries were based not so much on locally grown cotton as on the labor force available and electricity from water power. Today synthetic fabrics (such as nylon and polyester) are based on petrochemicals, produced along the Gulf coast.

More and more modern industries depend mainly on human resources — skilled workers. These industries can be located almost anywhere that large numbers of people live. In California, the most populous state, electronics has become a major industry. The area around San Jose, California, is called "Silicon Valley," after the tiny silicon chips produced there. These chips are vital parts of products like computers and digital watches.

Fuels such as oil, coal, and natural gas are vital to the economy of a modern nation; and the U.S. makes use of nearly all of its known resources. The picture shows part of a natural gas plant in El Paso, Texas.

Manufacturing was once the most important source of jobs in U.S. cities. But with the increasing use of machinery and automation (the automatic control of machinery), blue-collar jobs have been declining. (**Blue-collar jobs** are those that involve manual labor.) More than half of U.S. workers are now in **white-collar jobs** (office jobs).

A large, modern corporation needs a vast staff of office workers. Some corporations prefer to keep their main offices in small towns. But most corporations have headquarters in the largest metropolitan areas, which are handy to transportation and communications networks — and to other businesses.

Large firms that specialize in business services have tended to concentrate in a few major cities. New York City and San Francisco are centers of international banking. New York City and Hartford, Connecticut, are centers of the insurance industry.

Washington, D.C., has its own special "industry" — government. As the nation's capital, Washington is home to many federal agencies — and to Congress. The federal government is the city's biggest employer. Many private firms that do business with the government have offices in Washington.

Trading with the World

Despite its rich resources and massive industries, the U.S. is not self-sufficient. (To be **self-sufficient** is to meet all of one's own needs.) It must depend on other countries for tropical products such as coffee, tea, bananas, and rubber. And it must look abroad for various vital minerals — not only rare metals but also oil. Although the U.S. is rich in oil, it uses so much energy to heat homes and run industries that it must now import one third or more of its oil needs.

Foreign trade is important to the United States. No other country buys and sells more abroad. Foreign trade allows the United States to do four things:

1. *Obtain resources that the U.S. lacks.*

2. *Dispose of surpluses.* (A **surplus** is an amount left over.) For example, U.S. farmers produce far more food than Americans can eat. Up to 40 percent of U.S. farm output (especially grains) is exported.

3. *Buy items that can be made more cheaply or in a different style someplace else.* For example, many of the shoes sold in the U.S. are imported from countries like Italy where labor costs are lower. These shoes also offer Americans a wider choice of styles.

4. *Produce goods more cheaply.* If a manufacturer sets up a factory and produces just one dishwasher, this one item will cost thousands or millions of dollars. But if the factory produces thousands or millions of dishwashers, the cost of setting up the factory can be divided among them all. The **unit cost** (the cost of producing each dishwasher) will then be much lower. By selling worldwide, U.S. manufacturers can take advantage of the lower unit costs that go with large-scale production.

Leading U.S. imports are oil, machinery, autos, metals, and tropical foods. Leading exports are manufactured goods — computers, power generators, chemicals, metal products, and so on. Farm products make up about one fourth of U.S. exports. In recent years, the **balance of trade** (matching of imports and exports) has been negative. In other words, the U.S. has bought more from the world than it has sold.

The U.S. trades with nations in all parts of the world, including the Soviet Union and China. (These Communist nations buy large quantities of U.S. grain in some years.) The U.S. also trades with the **Third World** (developing nations that are not allied with either the U.S. or the Communist powers).

However, the top U.S. trading partners are such non-Communist industrial nations as Japan and Britain — and especially Canada. Chapter 8 will take a closer look at Canada's role in the region and the world.

SECTION REVIEW

1. Why do farms in the Great Plains tend to be larger than in other parts of the country?

2. Why are modern industries such as electronics less dependent on the local environment than heavy industries such as steel?

3. Much of the manufacturing in the U.S. is done in large metropolitan areas. Give two reasons for this.

4. Give two reasons why the U.S. trades with other countries.

YOUR LOCAL GEOGRAPHY

1. What are the main economic activities in your community: farming; mining; fishing; lumbering; industry; business; tourism? From what you have read in the chapter, can you tell which of those activities are closely related to the environment? For example, if there is farming, does your community belong to the corn belt, wheat belt, or dairy belt, or to an area of specialty crops? If there is industry, is it based on local mineral resources?

2. In Geography Skills 16 (page 128), you learn about the information contained in climate graphs. Now try making part of a climate graph yourself. Borrow an almanac from the library and look up "Weather" in the index. Under this heading, you will find an entry for U.S. cities. Choose the city nearest to your community. Note the figures for the city's average monthly temperature. Prepare a line graph (see Geography Skills 11 on page 78), showing average monthly temperatures for an entire year. For the vertical scale, start with the multiple of 10 just below the lowest monthly figures. (For example, if the lowest figure is −15°F, start with −20°F.) For the highest point on the scale, choose the multiple of 10 just above the highest monthly figure. Compare the completed line graph with the temperature graphs in Geography Skills 16. How much does the temperature in your area vary around the year?

Reading a Climate Graph

Near Dodge City, Kansas, wheat is planted in the fall and remains underground as a sprout over the winter. This "winter wheat" is ready for harvesting in late June. Near Bismarck, North Dakota, wheat is planted in the spring and harvested in the fall. The **climate graph** below shows what climate differences account for these different growing seasons. A climate graph gives two kinds of information about a place by combining two kinds of graphs.

Study the section for *average monthly precipitation*. The graph compares two bars — for two cities — for each month. A graph that compares two or more bars is called a **complex bar graph.** Answer the following questions.

1. What scale of measure is on: (a) the left? (b) the right?

2. In which three months of the year do Bismarck and Dodge City have the same amount of precipitation?

Study the section of the graph for *average monthly temperatures*. This graph shows two indicator lines — one for each city. A graph with two or more indicator lines is called a **complex line graph.** Answer the following questions.

3. Which temperature scale is: (a) on the left vertical axis? (b) on the right vertical axis? On which scale is the freezing point of water 32°?

4. In which month(s) of the year is the average temperature below freezing: (a) in Dodge City? (b) in Bismarck?

5. Which factor — temperature or precipitation — probably accounts for the growing of winter wheat in the area of Dodge City? Why?

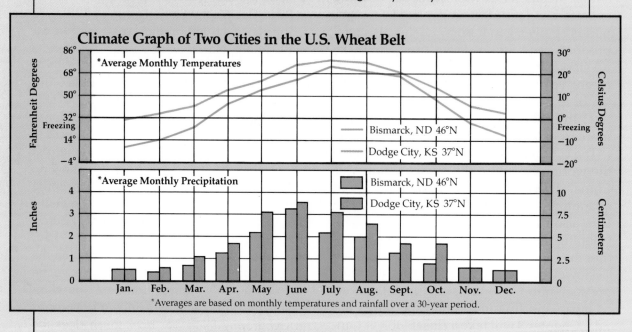

Climate Graph of Two Cities in the U.S. Wheat Belt

*Averages are based on monthly temperatures and rainfall over a 30-year period.

CHAPTER REVIEW

A. Words To Remember

In your own words, define five of the following terms.

agribusiness   balance of trade   deciduous

archipelago   blue-collar job   metropolitan area

B. Check Your Reading

1. What is the agricultural heartland of the U.S.? Name two of its most important products.

2. Give one reason why there are fewer small farms in the U.S. today than in the past.

3. Give two reasons why many businesses and jobs are moving from the Northeast to the Southeast and Southwest.

4. You have read that many areas in the U.S. specialize in growing certain crops. Why is this so? Give two examples.

5. What kinds of jobs (industrial; mining; white-collar; etc.) would be most in demand in Washington, D.C.? Why?

C. Think It Over

1. Give two reasons why New England was once the center of the manufacturing industry. Give one reason why New England is no longer the center of that industry.

2. Would you say that natural resources are distributed widely throughout the different areas of the U.S.? Or are they grouped in just a few areas? Give examples to support your answer.

3. Give examples of two jobs that depend heavily on the environment and two jobs that depend hardly at all on the environment. Name a part of the U.S. where people could work at the two latter jobs but not the first two jobs.

D. Things To Do

The chapter describes several of the reasons why different industries are located in different places. Choose any two of the following: (a) a television station; (b) a mill making wood products from lumber; (c) a factory making steel girders for large buildings. Describe the kind of location that would be best for each of the two businesses, and give reasons. Using the map on page 118, find two specific places where each of the businesses might be located.

Chapter 8

Canada
So Many Resources, So Few People

Fog hangs over the waters of the North Atlantic east of Canada, and icebergs drift along from the Arctic. Yet for centuries, fishing boats have come to these waters from Europe and America in search of rich hauls of fish. Today in addition to the fish, another resource — oil — is being hunted in the North Atlantic.

As you can see from the map on page 132, the easternmost part of Canada is the island of Newfoundland. A broad plateau stretches south and east from Newfoundland not far beneath the surface of the ocean. This underwater plateau is part of Canada's continental shelf. The plateau is known as the Grand Banks, and grand it is — roughly as big as Montana. The fish and oil are found there.

The Grand Banks are rich in fish because this is where cold waters (the Labrador Current) moving south from the Arctic meet warm waters (the Gulf Stream) moving north from the Caribbean (see the map on page 228). The two currents bring with them a great variety of tiny sea plants and animals that are the food of many kinds of fish.

The Grand Banks are rich in oil because layers of dead plants and animals were deposited there ages ago (see page 107 in Chapter 6). Oil is often found beneath a continental shelf.

Oil and fish are only two of Canada's many natural resources. By developing its resources, Canada has become one of the world's leading industrial nations. However, part of this development has been done by companies from outside Canada — mainly from the United States. For instance, the first oil off Newfoundland was struck in 1979 by a U.S. firm.

You read earlier that Canada is even bigger than the United States — in area. You also read that Canada has one ninth as many people. With so few people, Canada has had to call for outside help to raise the huge amounts of money needed to develop its resources. But such dependence on outside help has made many Canadians uneasy. In recent years, Canada has taken steps to cut back the role of outsiders in the Canadian economy.

In this chapter, you will learn more about Canada's resources, and about the people these resources help support.

Physical Geography

From an oil rig over the Grand Banks, a boat can take you to Newfoundland in 10 or 12 hours. Then by car and ferry, you can head west on the Trans-Canada Highway. There's no telling how long it will take you to cover the 5,000 miles *(8,000 kilometers)* of

As far as the eye can see, grain waves and billows in Saskatchewan fields.
Canada is larger in area than the U.S., but has one ninth the people.

131

Canada

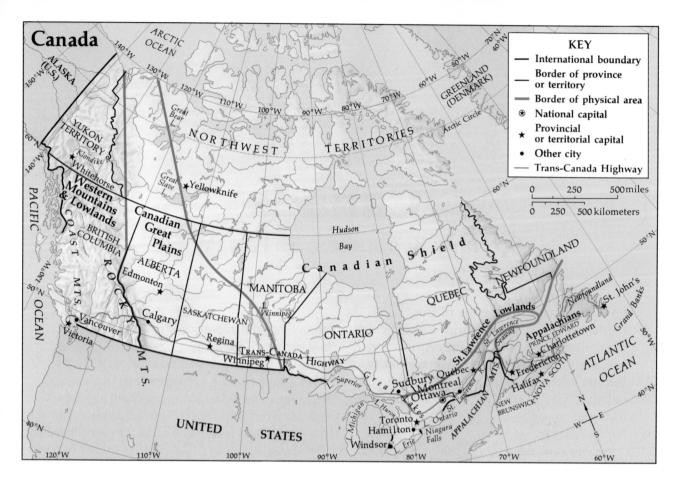

Follow the route of the Trans-Canada Highway from St. John's, Newfoundland, to Vancouver, British Columbia. Does it pass through all the physical areas of Canada? Why wasn't it built farther north?

mostly two-lane highway, finished only in 1965. But at last, you will reach Victoria, British Columbia, on the edge of the Pacific. There you will be farther from where you started than you are from Tokyo, Japan. Canada is a *big* country.

You may wonder why most of the Trans-Canada Highway has only two lanes instead of four or six, as on a U.S. interstate. Well, in most places, the highway is just not busy enough. For much of its length, the road passes through sparsely settled lands.

To learn where these lands are, and why they are sparsely settled, imagine you are making that east-west journey across Canada. You will pass through each of the country's five main geographic divisions.

1. Appalachians. The Appalachian Mountains that run through the eastern U.S. extend into Canada. They make eastern Canada hilly and poorly suited to farming, so settlements are scarce. This area includes the four **maritime** (coastal) provinces — Newfoundland, Prince Edward Island,

Nova Scotia, and New Brunswick — plus eastern Quebec. The coastline is deeply indented (cut like the teeth of a saw). Much of the area is forested. Farming is limited to the few flat places and to river valleys. Only cool-weather crops like hay, barley, and apples can grow.

2. St. Lawrence Lowlands. This is a low-lying area stretching from Lake Erie northeast through the valley of the St. Lawrence River. The southern part is a peninsula jutting into the Great Lakes. The area resembles the North Central U.S. and contains more than half of Canada's people. Southern Ontario and southern Quebec make up this area. Only in the far south is its growing season long enough for corn to mature.

3. Canadian Shield. Almost half of Canada's land lies in this rocky, U-shaped area that curves around Hudson Bay (see the map opposite). The southeast edge of the Shield drops steeply to the St. Lawrence Lowlands, and rivers flow out of the Shield strongly enough to generate electricity.

As you read earlier (page 106), soils on the Shield are thin, so farming is difficult and settlements are few. Northern parts of the Shield are too cold for crops, in any case. Much of the land is forested, except in the far north, where forests give way to tundra.

4. *Canadian Great Plains.* This flatland area begins at the western edge of the Shield and is an extension of the U.S. Great Plains. The area is shaped like a wedge, narrowing in the north as it approaches the Arctic Ocean. Manitoba, Saskatchewan, and Alberta lie mainly in this area and are called "the prairie provinces." (**Prairie** means "grassy plain.")

Although tundra and forests cover the northern parts, the southern parts have a rich soil and are heavily farmed. As you drive through, you will see broad fields of wheat, oats, and hay. There is generally enough rain for farming except in the southwest, where irrigation must be used.

5. *Western Mountains and Lowlands.* Past the city of Calgary, Alberta, the Trans-Canada Highway climbs high into the Rocky Mountains, and you will see breathtaking scenery that includes thick forests and snowcapped mountain peaks. Much of this area lies in Canada's westernmost province, British Columbia, and in the Yukon territory to the north. Only in the valleys and basins between the Rocky Mountains and the Coast Range is there much level land for farming. Rainfall is heavy along the coast, which has a **marine** climate (mild and wet), but tends to be sparse inland. However, rivers from the mountains provide water for irrigation and hydroelectric power.

Natural Resources

There are forests in each of the five areas. In fact, about 40 percent of Canada is forested — more than four times as much as is farmed.

As in the United States, western forests tend to be made up of evergreens, and eastern forests tend to be mixed. Canadians have chosen an eastern tree, the maple, as their national symbol. The Canadian flag displays a large red maple leaf.

Canada's forests are home to many species of animals, from tiny wrens and chipmunks to bighorn sheep and caribou. Wildlife was once so abundant in the maritime provinces that forests there were called "the moose farm."

Canada's waters provide many resources. Fish can be found in many places besides the Grand Banks. Cod and lobster are plentiful along the Atlantic coast, and salmon along the Pacific coast.

Canada also has many inland lakes. Four of the Great Lakes lie half in Canada and half in the United States; the fifth, Lake Michigan, is entirely in the U.S. A string of lakes at the western edge of the Canadian Shield includes three of the world's 15 largest lakes (Winnipeg, Great Bear, and Great Slave). In addition, Hudson Bay is a vast inland sea up to 600 miles *(1,000 kilometers)* across. All of these bodies of water contain fish. Hudson Bay even has whales, although far fewer than in earlier times. Canada's lakes and Hudson Bay freeze over for four to eight months every year.

Thanks to one of Canada's mineral resources, a small river in the frozen north once became famous around the world. A major gold strike near the Klondike River in 1896 drew waves of prospectors to the Yukon territory. Today minerals such as iron, nickel, and uranium are even more vital to industry than gold. So are coal and oil. A continuing search keeps turning up new deposits of such resources in Canada — as with the Grand Banks oil strike of 1979.

There are major oil and gas fields in the Canadian Great Plains. Oil is also found in another form, embedded in sandy soil. Some of the richest deposits of oil sands in the world lie in northern Alberta. Coal is found in the West and also in Nova Scotia.

The Canadian Shield contains many valuable metal ores. For example, rich sources of iron occur in Labrador and eastern Quebec; and nickel, cobalt, and copper are found around Sudbury, Ontario. (You will recall from Chapter 7 that nickel and cobalt are largely lacking in the United States.) Other metals are found outside the Shield.

Unfortunately for Canada, many sources of minerals are far to the north, away from settled areas, which means the minerals are hard to get at and costly to develop. But potential riches are there.

SECTION REVIEW

1. What area of Canada contains more than half of the country's population?

2. What are the major resources of the Canadian Great Plains?

3. Why does most of the Trans-Canada Highway have only two lanes?

4. Why are there such rich fishing grounds on the Great Banks?

Human Geography

You have stopped to buy food on your drive across Canada. You pick up a can of peas. On one side, the label says "peas." On the other, it says "petit pois" (puh-TEE pwah). By law food labels and many other printed notices in Canada must be in two languages — English and French.

The reason is simple: Many Canadians speak English — but not French. Others speak French — but not English. By requiring that labels be printed in both languages, Canada is trying to ease the problems of living in a **bilingual** (bie-LING-gwuhl; two-language) society.

English is the dominant language in most of Canada. In all provinces but Quebec, most of the people speak English. Quebec was the center of French settlement in colonial times, and most of its people are descendants of French settlers. Today nearly two out of three people in Quebec speak only French. Communities of French-speaking people also live in other provinces, especially in the maritime provinces.

French Canadians are bound together not only by language but also by religion. Like their French ancestors, most are Roman Catholics. Most English-speaking Canadians are Protestants.

In recent years, a strong separatist movement has grown up in Quebec. (**Separatists** are people who want their area to form a separate nation — in this case, a French-speaking nation centered in Quebec.) Pro-separatist leaders were elected to run the government of Quebec Province in the 1970's and 1980's.

From Many Lands

English-speaking Canadians do not all trace their ancestry to Britain. Almost one in three has roots elsewhere.

Different parts of Canada contain people of distinctly different origins. In the eastern province of Nova Scotia, for example, many people have Irish or Scottish ancestors. (*Nova Scotia* means "New Scotland.") In the prairie provinces, many people have roots in Germany, Poland, and the Ukraine. In British Columbia, there are many people with roots in Asian countries.

Canada also has many people whose roots reach south to the United States. After the American Revolution, large numbers of *Loyalists* (people who remained loyal to Britain) moved to what is now Ontario. In the late 19th century, many Americans carved out homesteads on the Canadian prairies. In recent years, there has been some two-way migration across the border.

Like the U.S., Canada has people who are *not* immigrants. There are less than 400,000

The red maple leaf flag of Canada flies alongside the blue and white flag of Quebec Province in the background of this street scene. The city is Montreal, the urban center of French-speaking Canada.

Indians and Eskimos in Canada — about half as many as in the U.S. Because of Canada's smaller population, however, Indians and Eskimos make up a larger proportion — nearly two percent.

Also like the U.S., Canada still receives large numbers of immigrants every year — an average of nearly 150,000, or 0.6 percent of its population. If the U.S. received the same percentage, it would have nearly 1.5 million immigrants every year.

City Life

In Canada, as in the United States, about three fourths of the people live in urban areas. The largest cities are Montreal, Quebec, and Toronto, Ontario. Each city has about three million people in its metropolitan area.

Most of the people in Montreal speak French. Montreal boasts one of the world's most modern subways, with trains that roll silently on rubber wheels. Montrealers are glad to stay underground during their city's bitterly cold winters. A vast underground shopping mall allows people to buy what they need without stepping outdoors.

Montreal sits on a large island in the middle of the St. Lawrence River. The city is at the foot of Mount Royal (also on the island), from which Montreal takes its name.

Toronto is on the shore of Lake Ontario, across from Niagara Falls. While Montreal is the urban center of French Canada, Toronto is the urban center of English Canada. Toronto is also the capital of the province of Ontario. But Ottawa, a smaller city in Ontario, is the national capital of Canada. Toronto has become the home of many of

Canada's recent immigrants — including Poles, Hungarians, Vietnamese, and Cubans. It is a **cosmopolitan** (international) city.

These two big cities, so different in many ways, are only 300 miles (*500 kilometers*) apart. They are both within the St. Lawrence Lowlands, which, as you read earlier, contain more than half of all Canadians.

Rural Life

Do the words *rural life* make you think of farming? Most often rural life in Canada involves some other occupation. Only one rural Canadian in four lives on a farm.

Of the other rural Canadians, some live in forest areas and work in the woods. Some live in remote mining camps. A few live in the far north and trap fur-bearing animals.

Most Canadian farm families live either in the St. Lawrence Lowlands or the prairie provinces. The shape of a farm varies greatly from place to place.

In Quebec early French settlers laid out long, narrow farms that stretched back from the St. Lawrence River. Houses were built close together along the river, which served as the chief "highway." This pattern was repeated elsewhere in Quebec. Today in western Quebec, farmhouses stand shoulder to shoulder along the main roads.

In the prairies, by contrast, farmlands look/like those of the North Central U.S. Fields are closer to squares, and houses are spaced far apart.

Canadians and the U.S.

Three out of four Canadians live within 100 miles (*160 kilometers*) of the U.S. border. If these Canadians want to shop in U.S. stores, they can hop in their cars and do so. If they want to watch U.S. television, they can aim their antennas south.

It's not surprising that U.S. and Canadian ways of life have much in common. Two major-league baseball teams are located in Canadian cities. Other Canadian teams play in hockey and soccer leagues with U.S. teams. Business links between the two countries add to the similarities. On a drive through Canada, you'd see familiar names on many gas stations, fast-food restaurants, and hotels.

Canada resembles the United States socially. For example, large numbers of women work outside the home. Most young people finish high school, and many go on to college. People in the two countries lead the world in use of the telephone (with Canada ahead of the U.S.). Politically the two countries also seem to have much in common. They are both democracies, and they are linked by a defense treaty.

Yet Canada differs from the United States in many ways. For example, Canada's national government does many tasks that the U.S. government does not do. Among other things, the Canadian government operates a nationwide broadcasting system, a national oil company, an airline, and a major railroad.

Individuals have some different rights and obligations under Canadian and U.S. laws. In Canada there is no such thing as a constitutional "right to bear arms." Since 1892 a Canadian national law has required the registration of private handguns. In Canada basic health care is free, and families receive a monthly payment from the government for each child under the age of 18.

SECTION REVIEW

1. Why did some Americans decide to move to Canada after the American Revolution?

2. What two major cultural features do French Canadians share?

3. Only one rural Canadian in four works at farming. Give two examples of the kind of work that other rural Canadians do.

4. Give two examples of similarities between ways of life in Canada and the U.S. Give two examples of differences.

Economic Geography

As you approach Winnipeg, Manitoba, on your east-west drive across Canada, you find a change in the Trans-Canada Highway. It widens first to four and then to six lanes — but not just because there is more local traffic around the city. At Winnipeg the Trans-Canada Highway is the *only* car, truck, and bus route linking points east and west. Moreover, the only two cross-country railroad lines in Canada both pass through Winnipeg.

If you look at the map on page 132, you can see why. A series of lakes extends some 300 miles (*500 kilometers*) north from Winnipeg, blocking any east-west land routes. And north of those lakes, the population density is less than one person per square mile (*2½ square kilometers*). Any useful cross-country routes have to pass through the narrow but more densely settled strip of land to the south.

Transportation is a bigger problem in Canada than in the contiguous states of the U.S. Canada's prairie lands, which are good for farming, are cut off from the east by 1,300 miles (*2,100 kilometers*) of forested land on the Canadian Shield. This land is not good for farming and is still lightly settled. Just north of Lakes Superior and Huron, you can travel for hours without seeing more than an occasional cluster of houses.

Development of the prairies only made sense if settlers could find markets for what they grew. In 1885 a cross-country railroad bridged the gap, and the prairies prospered. Modern highways have bridged the same gap only in the last few decades. Until the Trans-Canada Highway was completed, Canadian drivers often dipped south and used U.S. highways for their long east-west journeys.

In recent years, Canada has built new roads and railroads to reach mines being opened in the far north. The new highways are dubbed "roads to resources." Frozen ground and a forbidding environment are just two of many obstacles that must be overcome. In many places, the only practical method of transportation is the airplane.

Harvesting Natural Resources

Canada's farmers face two challenges. First, they must choose crops that fit their environment. Second, they must choose crops for which they can find markets.

Three fourths of Canada's crop-growing farmland is in the prairie provinces. There the climate is suited to wheat. The "wheat belt" spreads from Manitoba to Alberta.

Canada's other key farming area is in southern Ontario, near major population centers. Where the climate is mildest, crops like corn, tobacco, and fruit are grown. Elsewhere hay and oats are grown for livestock

Logging is a major Canadian industry. British Columbia, where loggers are shown felling trees, is a major producer of lumber and plywood used in the U.S.

137

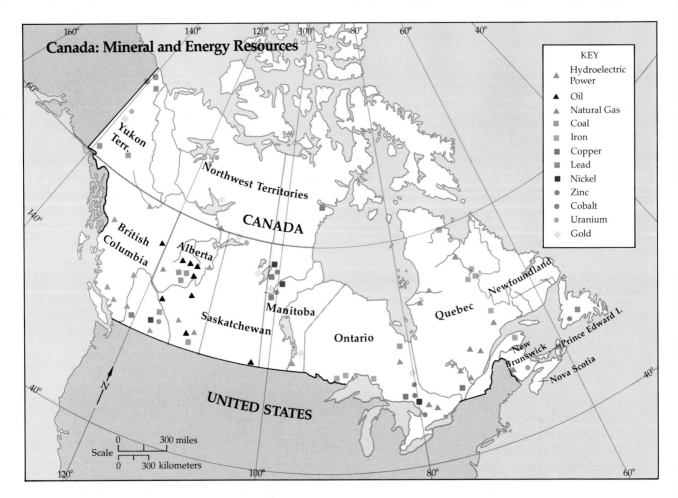

Canada: Mineral and Energy Resources

KEY
- ▲ Hydroelectric Power
- ▲ Oil
- ▲ Natural Gas
- ■ Coal
- ■ Iron
- ■ Copper
- ■ Lead
- ■ Nickel
- ● Zinc
- ● Cobalt
- ● Uranium
- ◆ Gold

Yukon Terr.

Northwest Territories

CANADA

British Columbia

Alberta

Saskatchewan

Manitoba

Ontario

Quebec

Newfoundland

New Brunswick

Prince Edward I.

Nova Scotia

UNITED STATES

N

Scale
0 — 300 miles
0 — 300 kilometers

— dairy cows nearest to cities, beef cattle in more remote areas.

The next time you pick up a newspaper, take a close look at the paper it's printed on. Chances are the paper comes from Canada. Such paper is called **newsprint**, one of the many products of Canada's vast forests.

Tens of thousands of Canadians work in the woods. They live in scattered logging towns. To reach lumber mills, logs are sometimes floated down rivers.

Especially in Quebec and Ontario, many of these logs go to pulp and paper mills. Besides newsprint such mills turn out products like paperboard and turpentine.

Some logs go instead to sawmills. British Columbia is a major producer of lumber, plywood, and similar products. About one fourth of the lumber used in the United States comes from British Columbia.

Both the east and west coasts have major fisheries. Because markets are usually distant, most fish are processed in some way —

Compare this map with the population density map on page 110. Overall, are Canada's mineral resources near the most populous areas of the country? Do you think it is easy for Canada to develop its resources?

frozen, canned, or dried. Frozen haddock from Canada is often seen in U.S. stores.

Indians in northern Alberta are among the thousands of Canadians who trap animals for their furs. With a snowmobile, a trapper can cover many miles in a day and check many traps. Muskrats and beavers are among the animals trapped. In some areas, silver foxes and mink are raised on "fur ranches."

Mining

Oil is Canada's most valuable mineral resource. Yet Canada has become a major oil producer only since 1947, when large finds were made in Alberta. This province now accounts for some 85 percent of Canada's oil output. Pipelines carry oil and gas to nearby parts of Canada and the United States. Because Canada's population is so widely scattered, it would be costly to pipe oil from Alberta to the eastern provinces. Thus Canada sells oil to the U.S. in the west and buys it from the U.S. in the east.

Almost three fourths of Canada's electricity is generated by the nation's rivers. Hydroelectric power is a major energy resource in Canada, which produces more than any other nation. Some of this electricity is sold to the United States. Some is used to run industries that require lots of cheap power. For instance, Canadian firms import bauxite (a mineral that Canada lacks) and use electricity to turn it into aluminum. The metal is then exported to Europe.

Canada is the third largest metal producer in the world (after the United States and the Soviet Union), and exports more metals than any other nation.

Most of Canada's metal mining takes place north of the Great Lakes in Ontario, and in the far north. Large quantities of iron from Labrador are shipped through the St. Lawrence Seaway to Canadian and U.S. ports on the Great Lakes. Nickel, zinc, and other metals move by railroad and ship to markets all over the world. Near many mines are processing plants that extract the metals from the ores. This reduces the bulk that has to be transported.

Industry

Where would you expect to find most of Canada's manufacturing industries?

You read in Chapter 7 that industries tend to locate close to centers of population (for workers and markets), sources of raw materials, and transportation routes. The part of Canada that best meets all of these needs is the St. Lawrence Lowlands.

As you read earlier, most of Canada's population lives in this area. There is good water transportation on the Great Lakes and the St. Lawrence Seaway, and good road and rail transportation along the shores. There are also many mineral resources nearby.

Canada's auto industry is located in Windsor, Ontario, and nearby cities, not far from the U.S. auto industry. The auto firms tend to be Canadian **subsidiaries** (branches) of U.S. firms. Three fourths of the autos and auto parts made in Canada are exported to the United States. Your family may own an "American" car made in Canada.

The Canadian shores of Lake Ontario are known as the "golden horseshoe" because so many industries border the lake. The area includes the cities of Toronto and Hamilton. Here there are steel mills and factories that turn out airplanes and other metal goods.

Similar factories are found around Montreal, which was long the **head of navigation** (the farthest inland point that oceangoing ships can reach) on the St. Lawrence River. Because ships could reach Montreal from the Atlantic Ocean, the city became a center of production for export. Today Montreal is no longer the head of navigation. Since 1959 the St. Lawrence Seaway has opened the Great Lakes to ocean shipping.

Trading Across the Border

On your journey across Canada, suppose you wanted to buy something of the same kind you're used to in the United States. It might be your favorite toothpaste or a record album. You would find that it costs quite a lot more in Canada than it does back home.

Like many other countries, Canada sets **tariffs** (taxes on imported goods) on many manufactured items. The aim is to make imports more expensive and thus encourage Canadians to buy Canadian-made goods.

Canadian manufacturers say tariffs are needed because of Canada's small population. The fewer the customers in a country, the more it costs to produce goods there. Thus Canadian firms must charge more.

Because of Canada's tariffs, U.S.-made goods are more expensive in Canada. As a result, many U.S. firms have set up Canadian subsidiaries to produce goods within Canada. Your favorite toothpaste and the record album would most likely have come from such subsidiaries.

In setting up subsidiaries, U.S. companies have invested money in Canada's economy. (To **invest** money is to put it to use so that it will bring in more money.) U.S. companies are also involved in developing Canada's natural resources—especially in mining and oil production. In the late 1980's, roughly 70 percent of Canada's oil industry was controlled by U.S. firms.

Canada's government has tried to hold down such outside control of Canadian business. For instance, the government has put limits on outside investment in banking and in newspaper publishing. It has also set a goal of 50 percent Canadian control in the oil industry by 1990.

Canada must perform a delicate balancing act with its economy. It cannot set tariffs too high, or other nations will refuse to trade with Canada. It cannot act too strongly against foreign investors, or they will be scared away. Canada depends heavily on the outside world. Its prosperity is largely due to foreign trade and foreign investment.

Roughly one fourth of all that Canada produces is sold to people in other countries. (The U.S. exports only about one ninth of its production.) Without those foreign markets, Canada's economy would suffer.

Canada's leading exports are motor vehicles, minerals, forest products, and wheat. In return Canada mainly buys manufactured items such as computers, TV sets, and machinery. It also buys food like bananas that do not grow in Canada.

Some 70 percent of all of Canada's trade is with the United States. As you can see, the two neighbors are linked not only by geography but also by strong economic ties. The next chapter will consider some of the issues that involve both Canada and the U.S.

SECTION REVIEW

1. What are Canada's two main energy resources?

2. What are the key factors that influenced most Canadian manufacturing industries to locate in the St. Lawrence Lowlands?

3. What is a tariff? Give one reason why Canada has placed tariffs on imported goods.

4. Describe two of the economic ties that bind the U.S. and Canada.

YOUR LOCAL GEOGRAPHY

1. Geography Skills 17 (next page) tells how the time of day varies around the world. Earth rotates through 15 degrees of longitude in one hour, or one degree in four minutes. Thus by the position of the Sun, each place on a different longitude has a different time. However, official time zones group many of these different times together. For example, Mountain Time is based on longitude 105°W, which is seven hours earlier than the Prime Meridian (0°) — since 7 times 15 equals 105. At Coeur d'Alene, Idaho, which is more than 116°W, Sun time is nearly 45 minutes earlier than official time. Use an atlas or a large-scale map to find the longitude of your community. Then work out how many hours and minutes your Sun time differs from 0°. (Divide the number of longitude degrees by 15 for the hours. Multiply any remainder by four for the minutes.) How does your Sun time compare with your official time?

2. One of the ways in which Canada differs from the U.S. is in using the metric system. The U.S. is the only major nation that still uses its own measures. Make a list of three items each in your home, your school, and your community that are measured in feet, quarts, pounds, or other customary measures. Would any problems arise if these were changed to the metric system? If so, what problems? Do you think the U.S. should or should not change to the metric system? Give reasons for your answer.

Reading a Time-Zone Map

People in Victoria, British Columbia, and in Los Angeles, California, are finishing breakfast at 8 A.M. At the same moment, people in Halifax, Nova Scotia, and in New York City are getting ready for midday lunch. The reason for this time difference is Earth's rotation.

You know that every 24 hours Earth turns around (rotates) a full 360 degrees of longitude (see Geography Skills 5 on page 35). This means that Earth moves 15 degrees from west to east every hour. At any given moment, the Sun shines most directly along only one meridian of longitude. There it is noontime. At the same moment, it is a different time of day along every other meridian.

In the past, travel and communications were so slow that people did not have to think about time differences. Today people jet from one continent to another in hours, or phone the other side of the world in an instant. It's important to know what the local time will be when they — or their calls — arrive.

How Time Changes as Earth Rotates

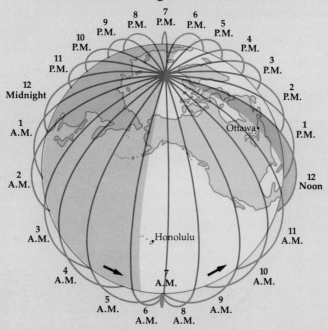

The drawing above shows times around Earth when it is noontime near Ottawa, along the meridian for 75°W. (The letters A.M. stand for Latin words meaning "before noon"; the letters P.M. stand for "after noon.") An hour later, after Earth has rotated 15 degrees to the east, it will be noontime 15 degrees west of Ottawa along the meridian for 90°W.

(Turn page.)

To make it easier to measure time around the world, people have divided the globe into 24 **time zones** — one for each hour of the day. The zones are centered on meridians of longitude 15 degrees apart. The map below shows the time zones for North America. Notice how the boundaries between time zones are sometimes in zigzag lines. Often it is more convenient to have time-zone boundaries follow national or local boundaries. Study the time zones on the map, and answer the following questions on a separate sheet of paper.

1. The far northwest coast of North America is in what time zone? Newfoundland, off Canada's eastern coast, is in what time zone? Including these two zones, how many time zones are there altogether across North America?

2. Which time zone is centered on the meridian for: (a) 75°W? (b) 60°W?

3. Which meridian(s) on the map run(s) through the: (a) Alaska-Hawaii Time Zone? (b) Yukon Time Zone? (c) Pacific Time Zone?

4. What U.S. cities on the map are in the same time zone as:
(a) Edmonton? (b) Ottawa? (c) Regina? (d) Victoria?

If you know what time it is in one city or time zone, you can use the time-zone map to figure out times in other zones. Notice the clock faces on this map, for example. They show what time it is in each zone on the map when it is 12 noon in Ottawa. Answer these questions.

5. When it is 12 noon in Ottawa, what time is it in: (a) New York? (b) New Orleans? (c) El Paso?

6. When it is 12 noon in Anchorage, what time is it in: (a) Yakutat?
(b) Whitehorse? (c) Edmonton? (d) Regina?

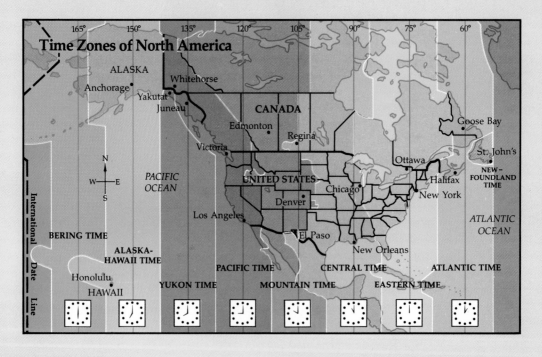

CHAPTER REVIEW

A. Words To Remember

From the following list, select the term that best completes each sentence below. Then write the two terms that are *not* defined and give a definition for each.

head of navigation   subsidiary   time zone
maritime   tariff

1. The ＿＿＿ on a river is the farthest inland point that oceangoing ships can reach.

2. A firm that is a branch of a larger foreign firm is called a ＿＿＿ .

3. A ＿＿＿ is a tax on imported goods.

B. Check Your Reading

1. Why are products labeled in both French and English in Canada?

2. Why has Canada had to have outside help in developing its economy?

3. Describe two ways that the Canadian government has tried to limit outside control in its economy.

4. Why did Canadians wait until the late 19th century to develop the prairie provinces?

5. Most of Canada's mineral wealth is located far from settled areas. What problems has this caused?

C. Think It Over

1. In what ways is the makeup of Canada's population similar to that of the U.S.? In what ways is it different?

2. Why does the Canadian government set tariffs? What problems are involved in setting tariffs?

3. "The construction of the Trans-Canada Highway was an important step in uniting Canada." Do you agree? Give reasons for your answer.

D. Things To Do

During one week, look through newspapers and magazines for news items and photographs of Canada. Choose one news item and one photo. Write two paragraphs on each, describing (a) what features of the news item or photo suggest that it refers to Canada rather than the U.S., and (b) what features of the news item or photo are similar to the U.S.

Chapter 9

How Similar?
How Different?
The Region in Perspective

If you walked through rural Quebec on a frosty winter weekend, you'd hear the laughter of young people on ice skates, chasing a puck across a frozen pond. Ice hockey is the winter sport of Canada.

Around Dallas, Texas, you'd have trouble playing ice hockey outdoors, even in the dead of winter. It gets cold in Dallas, but rarely cold enough for ponds to freeze. So young people there are more likely to be playing football on a winter weekend.

The U.S. and Canada form such a big region that there are bound to be many differences from one place to another.

Some differences — like the winter sports just described — are based on physical geography. Some are based on culture. But underlying the differences are some basic similarities. You'll note that the hockey sticks and footballs came out on *weekends*, because most young people throughout the region go to school on weekdays.

Both the differences and the similarities are important to understanding the region. This chapter examines both.

More Different Than Alike?

If you grow up in Quebec, you live with ice several months of the year. You're not surprised by a cold spell that makes your breath form droplets of ice in front of your face. But if you grow up in Dallas, you become used to sunshine and warmth. You shiver at the thought of a "blue norther" blowing across the plains from Canada. Temperatures might then drop to about 15°F (–9°C) — a mild winter day for Quebec.

Quebec and Texas have very different climates. Moreover, there are differences even *within* Quebec and Texas. Northern Quebec has tundra, while southern Quebec has forests and fields. East Texas is cotton country, while the drier and cooler panhandle of northwest Texas is wheat country.

The environment accounts for other differences too. It may even influence what you eat for breakfast. If you fancy grits and molasses, chances are you live in the southern U.S. Grits are made from corn; molasses, from sugarcane — both southern products. If you live in Vermont or Quebec or Ontario, you may prefer buckwheat cakes and maple syrup. Ever since Indian times, people in these regions have been tapping maple trees and boiling the sap down into syrup or maple sugar.

Just as foods often vary from place to place, so do economic activities. If you want to be an oil-field roughneck, you won't look for a job in Quebec. You'll look in Alaska, or West Texas, or Alberta. If you want to be a sawyer (a worker who runs the saw at a saw-

One of many links between the U.S. and Canada is Lake Superior. Like this giant ore freighter, ships of both nations cross the lake even amid the ice of winter.

mill), you won't look for a job on the Great Plains. You'll look in British Columbia, or Georgia, or Maine.

You read earlier that most of Quebec's people speak French, while most other people in the region speak English. Language is one of the most obvious cultural differences between places, but there are many others. For example, country people are more likely than city people to drop in unannounced to visit a neighbor. In rural areas, old ways are still strong. To take another example, Canadians are more likely than Americans to stand up when they hear "God Save the Queen," which is the British national anthem. In Canada it's the custom — since Britain's monarch is also Canada's monarch.

While geographic and cultural differences may at times be great, most of them cause little trouble. But a few differences do raise serious problems.

The "Other" Canada

Passing through Quebec on your drive across Canada, you would have seen many homes with a flag flying outside. In most cases, the flag would not display the red maple leaf of Canada but the *fleur-de-lis* (flur-duh-LEES; an ancient French flower design) of Quebec. As you read earlier (page 134), many French Canadians feel that they belong to Quebec more than they do to Canada. French-Canadian separatism has become a serious issue.

Separatists argue that French Canadians do not receive equal treatment in Canada. Even in Quebec, the argument goes, top positions in business are held by English Canadians. To get ahead, a French Canadian must learn English. Only in an independent Quebec, the separatists say, can French-speaking Canadians be full citizens with control over their own economic and political futures.

Many French Canadians strongly oppose separatism. They argue that it would not make economic nor political sense. A united Canada, they say, is a major world nation. If Canada were split, neither Quebec nor what was left of Canada would count for much. Most English Canadians also take this position.

Canadians are aware that the issue of separatism once split the United States and touched off a civil war. Few people think Canadian separatism will come to that. But no one denies that the dispute may have serious effects on Canada in the years to come.

SECTION REVIEW

1. Give one example of the way in which differences in physical geography affect the kinds of sports played in different parts of the region.

2. What is the main difference in physical geography between Quebec and Texas?

3. Give an example of a cultural difference that you might find between different areas *within* either Quebec or Texas.

4. What does the *fleur de lis* represent? What group of people would be most likely to display it?

Canada and the U.S.

Disputes can easily arise not only within a country but also between neighboring countries. Canada and the U.S. have a history of friendship — but some touchy problems have arisen just the same.

Both countries trade extensively with each other. However, about 70 percent of Canada's exports go to the U.S., and the same percentage of its imports come from the U.S. Only 20 percent of the United States' trade is with Canada. Thus Canada relies much more on U.S. trade than the U.S. relies on Canadian trade. U.S. government decisions on economic matters can have a considerable effect on Canada — which makes many Canadians uneasy.

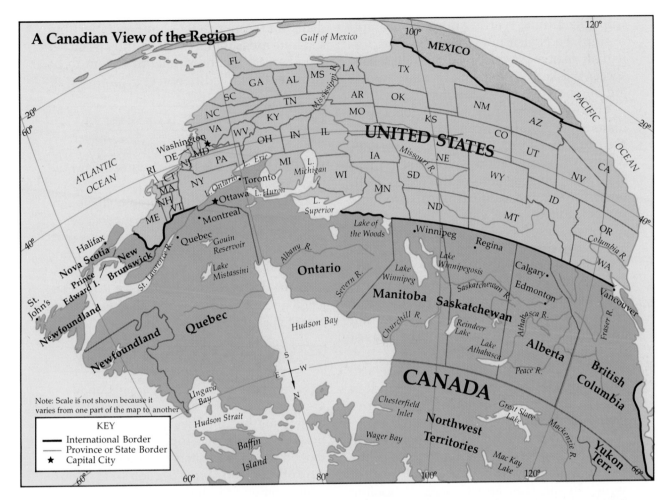

A Canadian View of the Region

Note: Scale is not shown because it varies from one part of the map to another

KEY
— International Border
— Province or State Border
★ Capital City

In recent years, a key issue has been the role of U.S. firms in Canada's economy. Canada's leaders have tried to curb this role (see page 140). U.S. leaders have protested.

Another issue has been the **pollution** (dirt from industrial or household wastes) that moves from one country to the other. Canada and the U.S. have taken joint action to clean up water pollution in the Great Lakes. But Canadians are still concerned that polluted air from the U.S. may be harming their environment. They are most concerned about *acid rain* — chemicals that fall from the air into rivers and lakes and make the water acid.

Such differences often arise between nations. After all, each side looks out for its own needs or interests first. Yet each side re-

A map does not have to show north at the top. This map gives the effect of an aerial view south from northern Canada. There is no distance scale. You can see why if you compare the shape and size of U.S. states here with those in the map on page 118.

alizes that there are interests that both share. The many similarities between Canada and the United States strengthen these common interests — and are more important than any differences.

More Alike Than Different?

Sports are one example among many of links between Canada and the U.S. Ice hockey is not *just* a Canadian sport; many U.S. cities have professional teams, and some of those teams have won the Stanley

Cup (the annual North American championship). Football is not *just* a U.S. sport; Canadian teams play football too. Baseball, soccer, basketball, and many other sports are common to the two nations.

Although geography can divide Canadians and Americans, it can also bring them together. Canada and the U.S. are both North American nations, which sets them apart from other regions of the world. Both must import the same tropical products (rubber, coffee, cacao, etc).

In many cases, geography brings together areas on opposite sides of the U.S.-Canadian border. As you have seen, the Great Plains states of the U.S. and the prairie provinces of Canada grow many of the same crops and share much of the same terrain and climate. Can you think of other areas of the two countries that are alike?

Culturally Canada and the U.S. are also much alike. Each is part of the modern world. Most people in the two countries share a fairly high standard of living.

The fact that you can read this book is also a sign of "modernity." In many parts of the world, a majority of people cannot read or write. In the U.S. and Canada, schooling and literacy are almost universal — that is, they are found almost everywhere.

Stop and think a minute about the kind of life you lead.

What is the farthest you have been away from home? The next valley? The state capital? Another nation?

Do you know where the ingredients of your breakfast this morning came from? Your own garden? Another state? Another country?

Where were your shoes made? What electric appliances have you used since yesterday? What machines do your parents use in the course of a week?

A Canadian's answers to such questions would be much the same as an American's.

Similarities like this tend to bring Cana-

dians and Americans — and people from separate areas within each country — closer together. So far the similarities seem to be stronger than the differences.

SECTION REVIEW

1. Give one example of an issue on which the U.S. and Canada have disagreed. How did they disagree?

2. How does physical geography create similarities between the U.S. and Canada?

3. Name an area where you would expect to find many similarities between the U.S. and Canada. Explain your choice.

4. The U.S. and Canada are alike in both belonging to the modern world. Give two examples of this.

YOUR LOCAL GEOGRAPHY

1. Suppose some visitors from Canada or another part of the U.S. are coming to your community by car. They are on a combined business and pleasure trip. What kinds of information would the visitors find useful on a road map of your area? Draw a rough sketch map of your community, including the information you consider most important. Use Geography Skills 18 (next page) as a guide in choosing symbols.

2. The U.S. and Canada are considered to be a cultural region. Many people and places throughout the region share features in common, and both nations belong to the modern world. Now look at your community as if it were a small cultural region. First look at the environment. What physical features does the community have in common? Are there any important exceptions? Next look at cultural features — language, ethnic groups, religion, leisure activities, etc. Are there any features that *all* people have in common? Are there any that a majority have in common? What major exceptions are there? Summarize the profile of your community in two columns under the headings "Majority" and "Other."

Reading a Highway Map

Suppose you are planning a trip in the Pacific Northwest. From Seattle, Washington, you plan to drive across the Canadian border and continue to the port city of Prince Rupert, British Columbia, before reaching the state of Alaska. To plan such a trip, you would need a **highway map** such as the one below.

Highway maps have many of the basic features of other maps. In addition, they give information of special use to drivers trying to find their way through unfamiliar places. One such feature is an alphabetical index of place-names with the map coordinates of each place.

(Turn page.)

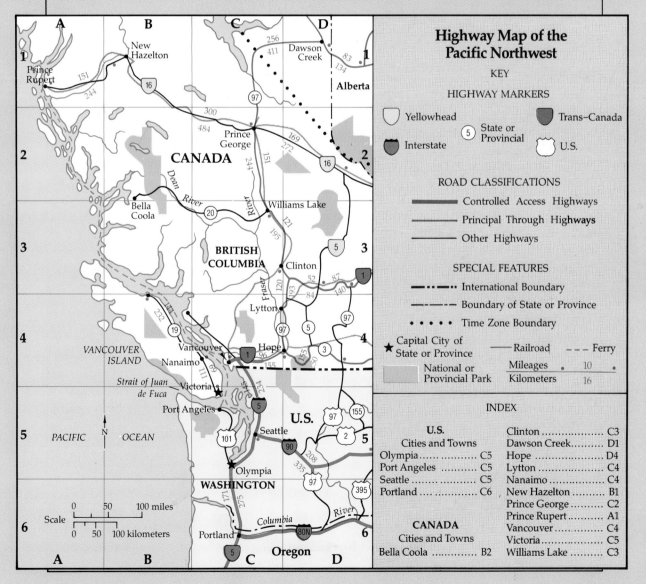

Highway Map of the Pacific Northwest

KEY

HIGHWAY MARKERS

Yellowhead · State or Provincial · Trans–Canada · Interstate · U.S.

ROAD CLASSIFICATIONS

Controlled Access Highways
Principal Through Highways
Other Highways

SPECIAL FEATURES

International Boundary
Boundary of State or Province
Time Zone Boundary
★ Capital City of State or Province
Railroad
Ferry
National or Provincial Park
Mileages 10
Kilometers 16

INDEX

Use the sample index and map to answer the following questions on a separate sheet of paper.

1. What are the map coordinates of: (a) Seattle? (b) Vancouver? (c) Hope? (d) Prince Rupert?
2. Use the coordinates to find Seattle and Prince Rupert on the map. Prince Rupert is located in what general direction from Seattle?

Highway maps give especially detailed information about roads. Notice also that each highway route on the map has special markers with numbers. Study the colored lines, markers, and other symbols in the map key. Then answer the following questions.

3. To drive from Seattle to Prince Rupert, you would travel on highway routes numbered 5, 1, 97, and 16. In what direction would you travel along each route?
4. Is Route 5 between Seattle and Vancouver a "controlled access" or "principal through" highway?
5. Is Route 97 between Hope and Prince George an interstate, a provincial, or the Trans-Canada Highway?
6. If you do not want to go from Vancouver to Prince Rupert by car, what other two forms of transportation could you take?
7. If you take a side trip from Prince George to Dawson Creek, how should you reset your watch? Why?

Many highway maps have colored numbers along major routes to indicate distances. On this map, red numbers tell the number of miles between red dots; blue numbers tell the number of kilometers. Notice the red dots near Seattle and Vancouver. The red number printed along Route 5 shows that the actual distance between the two red dots is 145 miles. What is the distance in kilometers? Use the colored numbers or the scale line to answer the following questions.

8. What is the distance in miles between: (a) Vancouver and Hope? (b) Hope and Williams Lake? (c) Vancouver and Prince Rupert?
9. Using the scale line, what is the railroad distance between Vancouver and Clinton? How much longer is the highway route?
10. Again using the scale line, what is the distance between Vancouver and Prince Rupert by ferry? How much longer is the highway route?

CHAPTER REVIEW

A. Words To Remember

In your own words, define the following terms.

common interests	pollution	universal
map coordinates	separatism	

B. Check Your Reading

1. What is represented by: (a) the maple leaf flag? (b) the *fleur de lis*?

2. Give an example of the way in which differences in physical geography affect the kinds of foods eaten in different parts of the region.

3. Would someone looking for a job in a sawmill be likely to find one in: (a) British Columbia? (b) Georgia? (c) Kansas? Why, or why not?

4. What do separatists mean when they argue that French Canadians do not receive equal treatment?

5. What is the main argument of Canadians who oppose separatism?

C. Think It Over

1. Why do you think Canada might be concerned with air pollution in the U.S.?

2. Why do you think it is important for the U.S. and Canada to maintain good relations?

3. The chapter includes these questions: "Where were your shoes made? What electric appliances have you used since yesterday? What machines do your parents use in the course of a week?" Do you think that your answers to these questions would be much the same as those of a Canadian student? Why, or why not?

D. Things To Do

Find a road atlas of the U.S. and Canada in your school or public library. Plan an automobile tour of about 500 miles (*800 kilometers*), starting anywhere on the U.S. side of the U.S.-Canada border. The tour should cross into Canada and return to your starting point by a different route. List the major points on your route so that another person could easily follow it. Then using the maps and information in the text, briefly describe the landform(s) and climate(s) you would pass through on the tour. Are they different or similar on both sides of the border?

A. Check Your Reading

1. On a separate sheet of paper, write the letter of each description. After each letter, write the number of the area that best matches that description.

(a) This province is on the Pacific Coast.
(b) The land tends to be very flat and dry, and wheat is grown there.
(c) This area is a rich fishing ground.
(d) This area is an archipelago.
(e) This area contains more than half of Canada's population.

(1) British Columbia
(2) St. Lawrence Lowlands
(3) Alberta
(4) Canadian Shield
(5) Alaska
(6) Hawaii
(7) Great Plains
(8) Grand Banks

2. Fill in the blanks in the following paragraph by writing the missing term on your answer sheet.

Here are some of the ways in which the U.S. and Canada are similar. In __(a)__ Canada is the second biggest country in the world and the U.S. is the fourth biggest country. In vegetation about one third of the U.S. and 40 percent of Canada are covered with __(b)__. Large parts of both countries have a __(c)__ climate, becoming quite hot in summer and quite cold in winter. Three fourths of the people in each country live in __(d)__ areas. Both countries jointly operate the __(e)__ Seaway.

B. Think It Over

1. The following statements may be true of either or both the United States (U) and Canada (C). Write the initial letter of the nation(s) to which you think each statement applies.

(a) It is rich in mineral and agricultural resources.
(b) Its climate ranges from subtropical to polar.
(c) It has few land transportation routes between east and west.
(d) Almost one fourth of the products it produces are sold to other countries.
(e) It has several large minorities.

2. Choose any of the statements in exercise 1 above, and write its letter on your answer sheet. Then write a short paragraph that shows how the facts in the statement have affected the development of either the U.S. or Canada.

Further Reading

Canada, by Linda Ferguson. Scribner's, 1979. A survey of Canada's natural environment, history, and culture.

Our National Parks: America's Spectacular Wilderness Heritage. Reader's Digest Assn., 1985. Each park's chapter illustrates the diversity of U.S. environments.

Travels with Charley in Search of America, by John Steinbeck. Penguin, 1980. Steinbeck travels with his dog Charley across the U.S. and back.

3

Latin America and the Caribbean

Chapter 10

Our
Southern Neighbors

What do the following people have in common?

- José (hoh-SAY) works in the new oil fields of Reforma (reh-FOR-muh), Mexico. He is of mixed Indian and Spanish descent. Though he speaks Spanish, he is proud of his country's Indian heritage. He lives in a town that was once a sleepy village. Now thousands of people have come here from all over Mexico to work in the oil fields.

- Maria is an opera singer in Rio de Janeiro (REE-oh deh zhuh-NAYR-oh), Brazil. She is of Portuguese descent, and she speaks Portuguese. She lives in a modern apartment house beside one of the finest beaches in the world, the Copacabana (koh-puh-kuh-BAH-nuh).

- Topa (TOH-puh) is a sheepherder in the Andes (AN-deez) Mountains of Peru. He is of Inca (IN-kuh) Indian descent and speaks the ancient language of his people. Topa lives with his family in a one-room hut. He uses a llama (LAH-muh) to carry his supplies. This camel-like animal is slow and cranky — but it also provides Topa's family with wool to make blankets and coats.

- Juan (WAHN) works on a sugar plantation in the Dominican Republic. He is of African and Spanish descent. He speaks Spanish and a little English, which he learned in school. At harvest time, Juan cuts

about 2½ tons of sugarcane a day. By November his job is finished until the next planting season. Juan makes money between seasons by raising pigs and chickens.

All of these people lead very different lives. But they have one thing in common: They all live in a vast region to the south of the United States known as Latin America. It stretches from the Rio Grande, the river that separates Texas from Mexico, to Tierra del Fuego (tee-EHR-uh del FWEH-goh), the southernmost tip of South America. If you traveled this distance by car, you would cover about 7,000 miles (*11,000 kilometers*).

Overview of the Region

Altogether about 360 million people live in Latin America. As you have seen from the profiles that began this chapter, the people come from many different cultures — mostly European, Indian, and African. Two European countries, Spain and Portugal, have had the most obvious influence on the region.

Most Latin Americans use the languages of Spain and Portugal. Spanish is spoken in most of the region, while Portuguese is spoken in Brazil. Both languages come from the Latin language of ancient Rome. That's why the region is called Latin America.

Look at the map on page 159, and you will

Influences of many cultures meet and mix in Latin America. On a square in São Paulo, Brazil, modern buildings tower over a display of traditional African-style crafts.

see that Latin America is made up of three main areas. Mexico and Central America occupy the strip of land that begins south of the United States. This strip leads to the biggest of the three areas, the continent of South America. The third area in the region consists of the islands in the Caribbean (kar-uh-BEE-uhn or kuh-RIB-ee-uhn) Sea.

If you look at the Checklist of Nations on page 581, you will see that the countries of Latin America and the Caribbean vary greatly in area. As you might expect, the smallest countries are found among the Caribbean islands. The smallest island that is an independent nation, Grenada, could fit inside the city limits of Mobile, Alabama. The region's two biggest nations in area are found in the landmass of South America. Brazil is the biggest, covering one third of the continent. Argentina is a distant second.

Brazil also has the biggest population in the region. The second most populous nation is Mexico. To a large extent, population depends on environment. For example, places that are mountainous or very dry cannot provide much food and are therefore likely to have small populations. On the other hand, lowlands or plateaus that receive adequate rain are likely to have large populations.

Physical Geography

Suppose the space visitors you met in Unit 1 flew over the United States and came to the mountain range we call the Rockies. They might wonder how far the range extends and decide to follow it southward. What would they find?

The map on page 159 shows how the great mountain ranges of the western U.S. continue southward into Mexico. There they form two chains with a high plateau in between. As Mexico narrows toward the south, the two mountain chains squeeze together into one. This single chain runs through Central America into South Amer-

ica, where the chain links up with the Andes Mountains.

The Andes Mountains extend to the southernmost tip of South America and form the world's longest mountain chain. They are narrower than the Rocky Mountains, but much higher. Topa, the sheepherder you met at the beginning of this chapter, lives 2½ miles (*four kilometers*) above sea level — or 2½ times higher than Denver, Colorado. Yet his home is in a valley. The Andes are so high that snow covers their peaks, even at the Equator. The highest peak, Mount Aconcagua (ah-kuhn-KAHG-wuh), in Argentina, reaches 22,835 feet (*6,960 meters*). It is the highest mountain in the Western Hemisphere.

The Andes run along the west coast of South America. In this way, they are similar to the Rockies, which run through the western part of the United States. If you compare the whole of South America with the whole of the United States, you can find other land features in common (see the maps on pages 574 and 575).

On the eastern side of the United States, there are mountains that are lower and less rugged than the Rockies — the Appalachians. Similarly, South America has a chain of lower mountains in the east — the Guiana and Brazilian highlands. Great plains (lowlands) and river systems stretch across the central areas between the mountains of both continents. The main river systems that lie between the mountains in the United States are the Mississippi and the Missouri. In South America, the Amazon (AM-uh-zohn) and the Rio de la Plata (REE-oh deh-luh PLAH-tuh) systems divide the higher lands on either side.

The Amazon River is the second longest river in the world, after the Nile in Africa. In some places, the Amazon is as much as six miles (*10 kilometers*) wide and 300 feet (*90 meters*) deep. It sends more water into the ocean than any other river in the world.

Small ocean liners can travel along it as far as 2,300 miles (*3,700 kilometers*) inland.

The Amazon flows through the world's largest tropical **rain forest** (a tree-covered area that remains hot and wet throughout the year). The forest stays green throughout the year because its trees continually grow new leaves as they shed old ones. The rain forest of the Amazon covers an area more than five times the size of Texas.

Another major area in the South American lowlands is the Argentine **pampa** (PAHM-puh). This is a grassy plain on which huge herds of cattle are raised. Many cattle ranches here cover 100,000 acres (*40,000 hectares*) or more.

The islands of the Caribbean form a chain 2,000 miles (*3,200 kilometers*) long that loops from the tip of Florida to the coast of Venezuela. Most of the islands are the tops of mountains that rise from the floor of the Caribbean Sea. Thus they form another link between the U.S. and South America.

Different Climates

Because much of Latin America lies within the tropics, you might expect every place to be hot. Most of the region is indeed warm all year round, but very high temperatures are rare.

In the mountain and plateau areas, elevation keeps the climate mild or even cool. Quito (KEE-toh), Ecuador, is on the Equator, at an elevation of nearly two miles (*three kilometers*). The highest temperature ever recorded in Quito is 86°F (*30°C*). Usually temperatures reach only about 70°F (*20°C*).

Even the lowland areas of Latin America rarely get as hot as the corn belt of the United States during a summer heat wave. In the Amazon rain forest, normal temperatures range from 70° to 90°F (*20° to 32°C*). The highest temperatures in Latin America

The Amazon carries more water than any other river in the world. This aerial view shows how wide it has become 1,000 miles (*1,600 kilometers*) from the ocean.

are found in the dry area of northern Argentina.

The warm water seas around much of South America readily give up moisture. They provide two thirds of the continent with moderate to heavy rainfall. The rainiest place in Latin America is Quibdó (keeb-DOH), Colombia, where an average of 415 inches (*1,060 centimeters*) of rain falls every year. The rainiest place in the continental United States is the Olympic Peninsula in Washington, where the annual rainfall may reach about 150 inches (*375 centimeters*).

The waters of the Caribbean Sea are particularly warm. From August to October, 60 to 70 inches (*150 to 175 centimeters*) of rain usually fall on the Caribbean islands and on Central America. Many of the islands have thick forests and lush tropical plants, as do the coastal lowlands of Central America. In these areas, food is grown all year round.

During the rainy season, **hurricanes** may sweep across the Caribbean. These powerful storms create winds of up to 150 miles (*240 kilometers*) an hour, with heavy rains. The winds and rains whip up huge waves that flood the land.

The driest places in Latin America are the desert coasts of Peru and northern Chile. Along here the Pacific Ocean is chilled by a current rising from the deeps. Cold water evaporates very slowly — too slowly to form rain clouds over these coasts. In the Atacama (ah-tuh-KAH-muh) Desert of northern Chile, there are places where no rain has fallen for several decades.

The Search for Resources

At one time, Europeans came to Latin America in hope of finding huge deposits of gold. There was — and still is — gold in parts of the region, as well as silver and diamonds. But there is not enough to bring wealth to all the people of Latin America.

All kinds of minerals have been found in different parts of Latin America. Chile is fa-mous for its copper mines in the Atacama Desert. Bolivia is famous for its tin mines high in the Andes. Other minerals range from asphalt in Trinidad-Tobago to zinc in Argentina. As with gold, these minerals have not made nations rich.

Venezuela has benefited most from its minerals. It is the second largest producer of oil in the Western Hemisphere, after the United States. With money from that oil, the government of Venezuela has built new roads, schools, and parks. New industries have been developed, and many people have left the farms to work in the city.

Mexico seems to be headed for the same kind of wealth. Rich deposits of oil were discovered in Mexico in the 1970's. You will read more about this in Chapter 11.

Brazil is the region's big question mark. Many different minerals have been discovered there, and it seems certain that many more will be discovered in the country's vast interior. Some of these deposits may be huge. What Brazil would like to find is large deposits of oil.

Some nations that lack minerals have other natural resources they can put to use. The Caribbean islands have fine beaches, beautiful scenery, and clear waters full of tropical fish. These resources are the base of the Caribbean's major industry — tourism.

SECTION REVIEW

1. What is a tropical rain forest? Where can one be found in Latin America?

2. What is the Argentine pampa? What is the main economic activity there?

3. What resource has benefited Venezuela? What other Latin American country has found rich deposits of the same resource?

4. What general effect do the waters around Latin America have on its climate? Why is the climate different on the coasts of Peru and northern Chile?

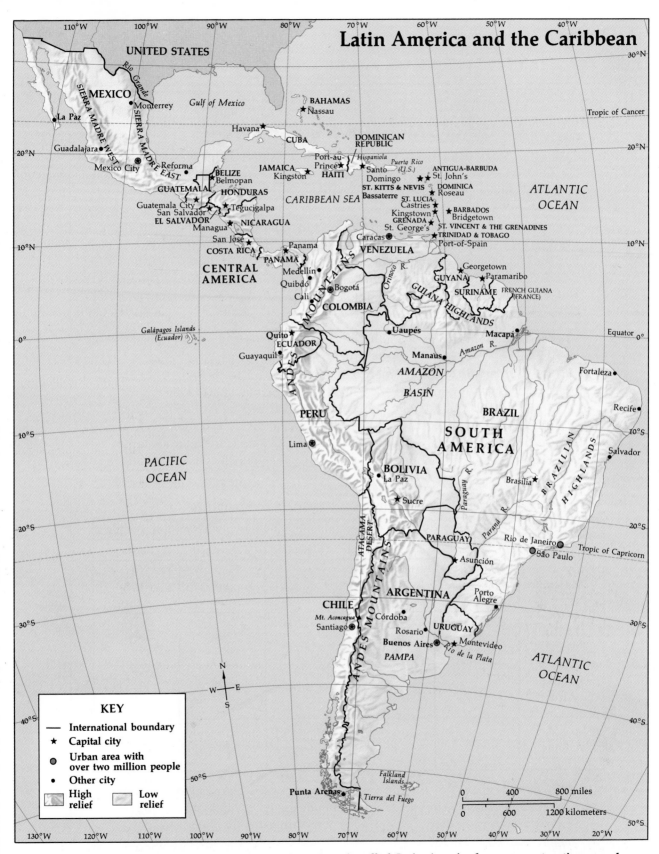

Latin America and the Caribbean

UNITED STATES

Tropic of Cancer

MEXICO
- La Paz
- Monterrey
- Guadalajara
- Reforma
- Mexico City

SIERRA MADRE WEST
SIERRA MADRE EAST
Rio Grande

Gulf of Mexico

BAHAMAS
- Nassau

- Havana
CUBA

JAMAICA
- Kingston

DOMINICAN REPUBLIC
Hispaniola
Port-au-Prince
HAITI
Santo Domingo
Puerto Rico (U.S.)

ANTIGUA-BARBUDA
St. John's
ST. KITTS & NEVIS
Basseterre
DOMINICA
Roseau
ST. LUCIA
Castries
BARBADOS
Bridgetown
Kingstown
GRENADA
St. George's
ST. VINCENT & THE GRENADINES
TRINIDAD & TOBAGO
Port-of-Spain

ATLANTIC OCEAN

CARIBBEAN SEA

BELIZE
Belmopan
GUATEMALA
Guatemala City
HONDURAS
Tegucigalpa
San Salvador
EL SALVADOR
Managua
NICARAGUA
San José
COSTA RICA
PANAMA
Panama

CENTRAL AMERICA

- Medellín
- Quibdó
- Cali
- Bogotá
COLOMBIA

VENEZUELA
- Caracas
Orinoco R.
GUYANA
Georgetown
SURINAME
Paramaribo
FRENCH GUIANA (FRANCE)

GUIANA HIGHLANDS

Galápagos Islands (Ecuador)

- Quito
ECUADOR
- Guayaquil

Equator

- Uaupés
- Manaus
Amazon R.
- Macapá

AMAZON BASIN

- Fortaleza

ANDES MOUNTAINS

PERU
- Lima

BRAZIL

SOUTH AMERICA

BRAZILIAN HIGHLANDS

- Recife

BOLIVIA
- La Paz
- Sucre

Paraguay R.

- Brasília

- Salvador

ATACAMA DESERT

Paraná R.

PARAGUAY
- Asunción

- Rio de Janeiro
- São Paulo

Tropic of Capricorn

CHILE
Mt. Aconcagua
- Santiago
- Córdoba
- Rosario

ANDES MOUNTAINS

ARGENTINA

- Porto Alegre

URUGUAY
- Buenos Aires
- Montevideo
Rio de la Plata

PAMPA

PACIFIC OCEAN

ATLANTIC OCEAN

Falkland Islands

- Punta Arenas
Tierra del Fuego

KEY
— International boundary
★ Capital city
◉ Urban area with over two million people
• Other city
High relief
Low relief

| 0 | 400 | 800 miles |
| 0 | 600 | 1200 kilometers |

N
W — E
S

The region has three main areas: Mexico with Central America; South America; and the Caribbean islands. It is called Latin America because most nations speak languages that come from the Latin of ancient Rome.

159

Using a Map and Tables Together

Suppose you traveled the length of South America from north to south. Then you traveled across the continent's central bulge from west to east. Both ways you would notice great differences in the temperatures of places at different latitudes and at different elevations.

You will often find temperature information about places in tables similar to the ones in this lesson. You can also use a map, such as the one on page 159, to find the latitudes and elevations of places.

Look first at the map. Study the symbols in the map key and on the map itself. Then complete each of the following statements by writing the best choice on a separate sheet of paper.

1. The northernmost part of South America has a latitude of about (12°N; 0°; 12°S).
2. The southern tip of South America is near (55°N; 55°S; 70°S).
3. Most of South America is in the (Antarctic; temperate; tropic) zone.
4. The areas of highest relief in South America are near the continent's (north; west; east) coast.
5. The city of Quito, on the Equator, is located in the (Amazon Basin; Guiana Highlands; Andes Mountains).

Table A
Differences in Latitude and Average Temperature

City	Latitude	Elevation		Average Yearly Temperature	
		Feet	Meters	Fahrenheit	Celsius
Manaus	3° S	87	26	80°	27°
La Paz (MEXICO)	24° N	85	26	75°	24°
Buenos Aires	35° S	82	25	60°	16°
Punta Arenas	53° S	89	27	45°	7°

You already know that latitude affects climate. Table A shows what happens to temperatures as you travel to higher latitudes south or north of the Equator. Study the column headings across the top of the table and the cities listed down the left side. Then use Table A to answer the questions at the top of page 161.

6. What four cities are compared in Table A? What three kinds of information are compared for these cities?
7. Elevations are given in what two units of measurement? Temperatures are given in what two scales?
8. What is the latitude of: (a) Manaus? (b) La Paz? (c) Buenos Aires? (d) Punta Arenas? What is the difference in degrees of latitude between the lowest and highest latitudes?
9. What is the difference in elevation between the cities at the highest and lowest levels?
10. Which city has the highest average yearly temperature? Which city has the lowest average yearly temperature?
11. What effect does latitude generally have on temperatures as you go farther from the Equator?

Table B
Differences in Elevation and Average Temperature

City	Latitude	Elevation		Average Yearly Temperature	
		Feet	Meters	Fahrenheit	Celsius
Quito	0°	9,243	2,815	55°	13°
Uaupés	0°	278	85	77°	25°
Macapá	0°	36	11	80°	27°

Table B compares facts for three other cities shown on the map of South America. Study the title and other information in Table B. Answer the following questions on your sheet of paper.

12. What is the subject of Table B? What three cities are being compared?
13. What is the latitude of each city? Which city has the highest elevation? Which city has the lowest elevation?
14. Which city has the highest average yearly temperature? Which city has the lowest average yearly temperature?
15. What effect does elevation seem to have on temperature?

Human Geography

You are traveling along a country road in Latin America. You meet a Central American, a Guatemalan, a woman with Indian ancestors, a woman with Spanish ancestors, a woman who speaks Spanish, a woman who speaks an Indian language, a Roman Catholic, and a follower of an Indian religion. How many people have you met?

The answer could be: just one. Today many different backgrounds and cultures can be found throughout Latin America, and many people share in two or more of them.

At one time, the region was inhabited only by American Indians. It is believed that at least 13 million Indians lived here before 1500 A.D. Some of them had highly advanced civilizations.

In Central America, for example, the Maya (MIE-uh) built great cities, palaces, and pyramids. They created a system of picture writing and a calendar more accurate than the European calendar. The Maya developed efficient farming methods and carried on a lively trade with their neighbors.

The Aztecs (AZ-tehks) lived in what is today southern Mexico. They too developed agriculture and erected large buildings. They built a great city called Tenochtitlán (teh-nohch-teet-LAHN). It was bigger than any city in Europe at the time.

In the Andes Mountains, the Inca Indians ruled a great empire that spread through present-day Ecuador, Peru, and Bolivia. The Incas built roads and bridges that linked every part of their mountain empire. They lived in cities built with stones that were fitted together without mortar. The job was so well done that the walls of many Inca buildings still stand today.

The Incas also developed advanced farming methods. They used dead plants as fertilizers to make the soil richer. They dug long ditches to bring in water for the soil. They built terraces on the mountainsides to make them suitable for growing crops.

Then in 1492, Christopher Columbus made the first of his four voyages across the Atlantic. Spanish and Portuguese *conquistadores* (kohn-kees-tuh-DOR-ehs; conquerors) arrived soon afterward. They destroyed the Indian civilizations and brought Latin America under European rule.

As the Europeans conquered Latin America, large numbers of Indians died. Some were killed by the invading soldiers. Many more died of diseases that the Europeans brought with them.

A Mix of Cultures

Over many years, those Indians who survived adopted much of the culture of their rulers, including the Spanish or Portuguese language and the Roman Catholic religion. In addition, both Indians and Europeans borrowed each other's resources.

The Europeans soon developed a taste for the New World's corn, sweet potatoes, tomatoes, and squash. They also adopted Indian words such as *mahiz* (maize), *xocoatl* (chocolate), and *tomatl* (tomato). In return, the Europeans brought plants such as bananas, citrus fruits, and rice, which today are among major food crops in Latin America.

Other new sources of food came from the Europeans' animals — cattle, pigs, and chickens. The Spanish also brought horses and mules, which became the chief means of transportation in Latin America.

The mixing of European and Indian cultures in Latin America was furthered by intermarriage between people of the two groups. The first Spanish and Portuguese settlers in the New World were men. Many of them married Indian women. The children of Spanish fathers and Indian mothers were called **mestizos** (mehs-TEE-sohs). Today the word *mestizo* is used to describe any Latin American of mixed Indian and Spanish descent.

The Europeans brought large numbers of blacks from Africa to work as slaves on the plantations and in the mines of Latin America and the Caribbean. Like the Indians and Europeans, the blacks kept many of their old customs. But they also adopted much of the Latin culture of Spain and Portugal. Latin Americans who are of mixed African and European descent are known as **mulattos** (moo-LAT-ohs).

The ethnic pattern varies from one country to another. In Mexico most people are mestizos. In Guatemala, just south of Mexico, about half the people are Indians. In Costa Rica, also in Central America, the great majority of people are of Spanish descent, while about one third of the people in neighboring Panama are blacks.

Outside Central America, most of the region's Indians live in the steep valleys and high plateaus of the central Andes Mountains. This area was the heart of the Inca Indian empire. Indians make up about half the populations of Ecuador, Peru, and Bolivia.

In these countries, the Indian and Spanish cultures exist almost separately. The Indians continue to live in isolated villages in the highlands. Many speak only the languages of their tribes. Most people of Spanish descent live along the coast and in the cities.

Wherever the Europeans first landed, they tended to settle near the coast. They did so partly because the coastal lowlands were fine for crops, and partly because forests and mountains blocked the interior.

In South America today, the majority of

Many people of Indian descent live in the highlands of the Andes. Here Peruvians walk to market in a mountain valley 2½ miles *(four kilometers)* above sea level.

Europeans are descendants of immigrants who arrived only within the past 100 years. Many of these people chose to settle where the environment was most like the "old country." Thus South Americans of European descent are numerous in the cooler areas south of the tropics. They make up as much as 90 percent of the populations of Argentina and Uruguay, and many also live in southern Brazil and central Chile.

In addition to Indians, Europeans, and Africans, many of the people of Latin America today trace their origins to Asia. Hundreds of thousands of laborers from India and China were brought to the Caribbean in the 19th century. Many of their descendants still live there today. Since World War II, more and more Asians have been migrating to parts of South America, especially to Brazil. The largest number of these Asian immigrants comes from Japan.

SECTION REVIEW

1. What effect did the arrival of Europeans in Latin America have on the Maya, Aztec, and Inca civilizations?

2. Why did the Europeans bring large numbers of blacks from Africa to Latin America?

3. Name two resources the Europeans "borrowed" from the Indians. Name two the Indians "borrowed" from the Europeans.

4. When the first Europeans came to Latin America, where did they tend to settle? Give one reason why.

Economic Geography

In downtown Tegucigalpa (teh-goo-see-GAHL-puh), capital city of Honduras, you can see high-rise hotels, lawyers' and dentists' offices, stores selling TV sets, ads in flickering lights, and traffic jams at rush hours. You can also hear jets roaring overhead on their way to or from the airport.

Now take a bus out of the city on the modern road to the east. In a few minutes, you will see farmers working on tiny plots of land, some only a little bigger than your classroom. The homes are small huts built by the farmers themselves.

Like most farmers in Latin America, these people are raising the same foods with the same methods used by their ancestors many hundreds of years ago. They use a pointed stick to punch holes in the earth. Then they plant corn, beans, or squash by hand. They use a homemade hoe to loosen the soil around the plants.

Fertilizers, insecticides, and machinery play no part in this farming, and the small plots grow barely enough to feed the farmers' own families. This kind of farming is known as **subsistence farming**.

Other farms belong to the modern world. These are large in area and usually specialize in one crop. They are known as **plantations**.

On plantations crops are produced not to be eaten by the farmer but to be sold, mostly in other countries. Because these farms are large and make money, their owners can afford to use machinery, fertilizers, and other modern techniques.

The farm work on plantations is done by paid employees. Many of these workers are **tenant farmers**. In return for their work on the plantation crops, they are allowed to grow their own food on small plots of land.

The tropics, with their long growing season, provide the setting for the region's plantations. These plantations are found mainly in the Caribbean islands, along the Caribbean coast of Central America, and around the northern coast of South America. Sugarcane is the main crop in the Caribbean islands and in northeast Brazil. Bananas are grown largely in Ecuador and Central America. Coffee is the main crop in the highlands of Colombia and Brazil.

South of the tropics, in southern Brazil, Uruguay, and Argentina, there are large modern farms that specialize in crops such as corn and wheat. Large numbers of ani-

Coffee grows best in highland areas of the tropics. Above, a worker picks coffee beans on a plantation in the mountain chain that extends through Costa Rica.

mals are also raised for food. Grass grows naturally on the plains that sweep down through southern Brazil, Uruguay, and the Argentine pampa. It takes a lot of grass to feed a herd of cattle—but these plains are vast. Millions of cattle are raised on the huge ranches. Many of them end up as beef in supermarkets around the world.

Toward the Modern World

Before World War II, Latin America had very little industry. The region exported food and raw materials, chiefly metals. It imported most manufactured products. In recent times, however, industry has grown rapidly in Argentina, Brazil, Chile, Colombia, Mexico, and Venezuela. These countries now manufacture their own cars, trucks, refrigerators, TV sets, radios, and other products.

New factories generally are located in and around the large cities. In recent years, millions of poor farmers have looked for work in the cities. This has caused serious overcrowding. Yet the cities still act as a magnet for the poor.

165

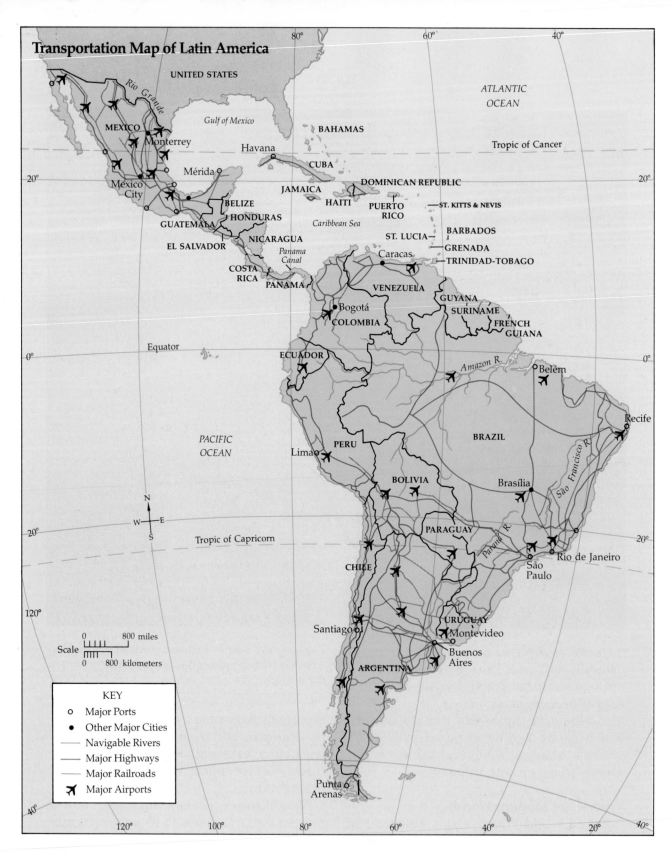

Transportation Map of Latin America

UNITED STATES

ATLANTIC OCEAN

Rio Grande

Gulf of Mexico

BAHAMAS

Tropic of Cancer

MEXICO

Monterrey

Havana

CUBA

Mérida

DOMINICAN REPUBLIC

JAMAICA

HAITI

ST. KITTS & NEVIS

Mexico City

BELIZE

PUERTO RICO

GUATEMALA

HONDURAS

Caribbean Sea

ST. LUCIA

BARBADOS

EL SALVADOR

NICARAGUA

GRENADA

Panama Canal

Caracas

TRINIDAD-TOBAGO

COSTA RICA

PANAMA

VENEZUELA

GUYANA

SURINAME

Bogotá

FRENCH GUIANA

COLOMBIA

Equator

ECUADOR

Amazon R.

Belém

Recife

PACIFIC OCEAN

PERU

Lima

BRAZIL

São Francisco R.

BOLIVIA

Brasília

PARAGUAY

Paraná R.

Tropic of Capricorn

CHILE

Rio de Janeiro

São Paulo

URUGUAY

Santiago

Montevideo

Buenos Aires

ARGENTINA

Scale

0   800 miles

0   800 kilometers

Punta Arenas

KEY
- ○ Major Ports
- ● Other Major Cities
- — Navigable Rivers
- — Major Highways
- — Major Railroads
- ✈ Major Airports

You can see that land routes in the region are most complete in Mexico, Argentina, and in areas along the coasts. Suppose you wanted to travel from Mexico City to Rio in Brazil. Could you go by major highways? By railroad? What routes could you take via a Mexican port? What other way could you make the journey?

Transportation is a problem in much of the region, especially in the vast spaces of South America. Many modern roads have been built in recent years, but most of them are near the coast and serve the large cities. In rural areas, oxcarts still travel over dirt roads. And in vast inland areas covered by forests, there are no roads at all.

Latin Americans rely heavily on air travel to overcome land barriers such as the Andes Mountains and tropical rain forests. More than 50 airlines now crisscross the region. Some carry freight and food to distant areas that cannot be reached by roads. Planes help to bring some of the advantages — and problems — of the modern world to areas that were cut off from rapid change.

What's Ahead

In this chapter, you studied the three main areas of Latin America. You saw how people of many backgrounds have created a culture with much in common. You also saw how they use their environments to support themselves.

The next three chapters take a closer look at the major areas of Latin America. Each chapter focuses on one or two countries in the area.

Chapter 11 deals with Mexico, which is today one of the fastest developing nations in Latin America. Mexico is the closest Latin American neighbor of the U.S.

Chapter 12 deals with Brazil, the Latin American giant, with a vast area that makes it the fifth largest country in the world.

Chapter 13 focuses on Haiti and the Dominican Republic, the two contrasting Caribbean nations that share the island of Hispaniola (his-puhn-YOH-luh). Haiti has a black population that speaks a language based on French. The Dominican Republic has a mixed population that speaks Spanish.

Chapter 14 takes a fresh look at the region as a whole to see what its prospects are for the future.

SECTION REVIEW

1. What is subsistence farming?

2. Name three ways in which plantations differ from traditional farms.

3. Name three Latin American countries in which industry has grown rapidly.

4. In what latitudes are most of Latin America's plantations located? Why?

YOUR LOCAL GEOGRAPHY

1. You read that many farmers in Latin America build their own houses. Suppose your family had to build its own house out of local resources. What materials could you use? Are there trees for lumber, or some kind of stone or clay for making bricks? What could you use to make the roof?

In planning the house, you would want to make it as comfortable as possible in your local climate. Would you need it to be warm in winter, or cool in summer, or both? Would the house have to shelter you from heavy rain or snow? Make a list of your local requirements, and decide how you would design the house to meet them.

2. Tourism is a major industry in the Caribbean and some other parts of Latin America. Do many tourists visit your community? If so, what are the main attractions — natural resources, or things made by humans, or both? Is there a tourist season, or do people visit all year round? Do many local people rely on tourists for their livelihood? Does your local government encourage tourism? If so, how? Could more be done to encourage tourism? Should more be done? Give reasons.

If few or no tourists visit your community, what is the reason? Are there any natural or human-made features which *could* attract tourists? Plan a tourist brochure of your community. Write a brief description of places that might be of interest to visitors. Suggest illustrations that should be included in the brochure.

Tropical Surprises

Zoologist William Beebe spent many months at a time in different parts of tropical South America. Here he describes a journey of about five miles (eight kilometers) from the Caribbean coast of Venezuela south into the forest. As you read, note whether these tropical environments are different from what you expected.

With the seashore only a few yards behind, Venezuela fades from our northern eyes and in its place are Arizona and New Mexico. Here and there the tall fingers of cacti point up to the cloudless sky.

On through the desert to the south, our car sends up clouds of dust. Then around a sharp turn, our car climbs out of sight of the sunny plains. A rush of cool air excites us and dense foliage (mass of leaves) shuts out the sun.

What is a tropical jungle like, as seen by an ordinary human on a walk? You can answer this yourself if you climb up the trail to the north. A bomb bursts under your feet and a hump-backed, bright reddish animal leaps away like a rabbit. You have seen a tropical aguti, your first jungle rodent. Startled, you stand still, and your ankles take fire. You have stopped full in the path of a legion of army ants. This is one of the supposedly terrible dangers of the jungle, but a single step takes you out of their line of march, a few minutes clears the last ant from your person, and the inconvenience is over. (In our northern woods, death would result if you stuck your head into a hornet's nest.) As with almost all tropical terrors, there is not the slightest danger. For example, to be bitten by a vampire bat is no worse than losing a teaspoonful of blood. Afterward you can always be the center of conversation by murmuring casually, "The last time I was bitten by a vampire. . . ."

As you walk on, it is the silent, terrible warfare of the plant world which is most impressive. A great tree, two centuries old, is being strangled by a huge climbing vine. Its leaves reach out above those of the tree and intercept the sunlight.

From close behind comes the snarl of a jaguar, and then three jaguars begin roaring at once. Other creatures join in; a wind rises, although around you the leaves hang motionless.

You have been assured that no jaguar will ever attack a human being, so you slowly stalk the uproar. In the distance, a big red, long-tailed monkey is silhouetted against the sky. He is joined by three others, and through your glasses you watch his mouth open. The great male howler reproduces jaguar, wind, chorus of other beasts, all with the aid of a "sounding board" of bone in his throat.

— From *High Jungle*, by William Beebe. New York: Duell, Sloan and Pearce, 1949.

Ask Yourself . . .

1. What two environments are described? In what ways would their climates differ? What do you think accounts for the sudden change from one environment to the other?

2. The author says that the dangers of the tropical jungle are greatly exaggerated. What examples does he give? Why might visitors imagine that the jungle is dangerous?

3. This description was published in 1949. Do you think it would still be accurate today? Why, or why not?

CHAPTER REVIEW

A. Words To Remember

Three of the following terms are defined by the numbered phrases below. On another sheet of paper, write each term next to the number of its definition. Then write the two terms that are *not* defined and give a definition for each.

mestizos   rain forest   tenant farmers
mulattos   subsistence farming

1. Persons of mixed African and European descent.

2. A tree-covered area that remains hot and wet throughout the year.

3. Paid employees who work on the plantation crops and in return are allowed to grow their own food on small plots of land.

B. Check Your Reading

1. Latin America can be divided into three parts: Mexico and Central America, South America, and the Caribbean. Which part has the biggest land area? Which contains the wettest, driest, and hottest places in the region?

2. On about what latitude is Quito, Ecuador, located? How does it differ in temperature from other places on the same latitude?

3. Give two reasons why the first Europeans settled near the coast of South America rather than farther inland.

4. The ancient civilizations of the Maya and Inca Indians were highly developed. Name two of their accomplishments.

5. Outside Central America, where do most Indians live today in Latin America? Why do you think they survived in this environment while many died or lost their homelands in others?

C. Think It Over

1. New factories now being built in Latin America are usually located in or around big cities. Give two reasons for this.

2. In your opinion, in which area of Latin America (Mexico and Central America, South America, or the Caribbean) is air travel most important? Why?

3. Many Europeans who came to South America within the past 100 years chose to settle where the environment was most like their homeland. Give two reasons why they might have made this choice.

D. Things To Do

Imagine that you are taking an automobile trip all the way from the northern part of Mexico to the southern tip of South America. You are keeping a diary of your journey. Write one day's entry from that diary, describing any one area you might pass through. As in Geography Skills 19, include a map and a table. The map should show the area's location within Latin America. The table should give latitude and approximate elevation (low; high; in between), and should include any two places from Tables A and B (pages 160 and 161). How would the area you chose compare with those places in temperature?

Chapter 11

Mexico

A Land of Promise and Challenge

Until 1980 Ramón (rah-MOHN) and Inés Torres (ee-NEHS TOR-ehs) lived in a village in northern Mexico. Like many other rural Mexican farmers, they had no land of their own. Instead Ramón worked on a large farm that belonged to a wealthy family.

Ramón's wages were so low that they barely provided food and clothing for the family. Ramón, Inés, and their three children lived in a one-room adobe (uh-DOH-bee) hut. (**Adobe** is clay mixed with straw. It is shaped when wet and then dried in the sun.) The hut had no windows, no running water, and almost no furniture. The entire family slept on straw mats on the floor.

Finally Ramón and Inés decided that they had had enough. Pemex, Mexico's government-owned petroleum company, was hiring people to work in the oil fields. It was paying good wages, so the Torres family moved to the city of Tampico (tahm-PEE-koh) on the coast of the Gulf of Mexico. Tampico is the main center of Mexico's booming oil industry. Ramón Torres was hired as a laborer on a construction crew. He was soon helping to build an oil refinery.

Today the Torres family lives in a small apartment in a high-rise building. Ramón and Inés were able to buy furniture, a TV set, and carpets on credit. Each member of the family sleeps in a bed. The kitchen has a modern gas stove and an electric refrigerator. The family even owns a used car, which Ramón uses to drive to and from his job in the oil fields.

Of course, they do find drawbacks in their new life. They are still not used to the noise and bustle of the city. Because there are so many workers with money to spend, prices in Tampico are high. Road construction has not caught up with the increase in traffic, so Ramón spends much of his driving time in traffic jams.

The Torres family is only one of many Mexican families who have jumped from the traditional world into the modern world. Since the 1940's, Mexico has made great efforts to develop its rich mineral resources. Deposits of coal and iron have gone into the making of steel, automobiles, electric appliances, and many other products.

In the 1970's, huge deposits of oil and natural gas were discovered on the Gulf Coast. The Mexican government estimates that there may be enough oil to fill 200 billion barrels — more than all the known oil left in the United States. Oil towns are now booming all along the Gulf Coast.

The wealth from new industries has helped change Mexico. The government has spent huge sums of money on new housing, schools, hospitals, and highways. In the

Mexico's booming oil industry has opened up new jobs and brought more wealth to the nation. Modern refineries are being built near oil fields along the Gulf Coast.

cities, modern high-rise buildings, department stores, and supermarkets are spreading everywhere.

Unlike the Torres family, many Mexicans have not yet shared in the nation's new wealth. About two Mexicans out of five still live in rural areas and farm the land. Most of these farmers work small plots that barely give them a living. In this chapter, you will learn more about the contrasts of Mexico and about the people who face its promise and its challenge.

Physical Geography

One of the first Europeans to describe Mexico was Hernán Cortés (ehr-NAHN kor-TEHS), a Spanish soldier. In 1521 an army led by Cortés defeated the great American Indian empire of the Aztecs. When Cortés returned to Spain, the Spanish king asked him what the country was like.

At first Cortés was at a loss for words. Then he picked up a piece of paper, rolled it into a ball, and crumpled it. The paper became a mass of peaks and valleys.

"This, Your Majesty," he said, "is a map of New Spain [Mexico]."

Cortés' "map" described Mexico well. More than two thirds of the country is covered with towering mountains, high plateaus, and deep canyons. In these regions, the distance between two villages may be no more than four or five miles (*seven kilometers*). But because the mountains are so steep, a mule path linking the villages may twist and wind for 20 miles (*30 kilometers*).

The chief mountain ranges in Mexico share the same name: Sierra Madre (see-EHR-uh MAH-dreh). One branch of the Sierra Madre extends along the west coast, while another extends along the east coast. Both the Sierra Madre West and the Sierra Madre East are linked to the Rocky Mountains in the United States. As you see on the map on page 173, the western and eastern ranges come together just south of Mexico

City to form a huge V-shape. From there the mountains run as a single range into Central America.

These mountain ranges divide Mexico into two very different kinds of environments. A high plateau covers the inside of the huge V-shape. On the outside, lowlands run along the Pacific and Gulf coasts.

Plateau and Lowlands

One day in 1943, a farmer was plowing his cornfield when he felt the ground shake and become hot. The next morning he saw a volcano rising above the surface of the cornfield! The volcano poured lava and ashes over his farm and soon covered nearby forests and fields. It even buried a town. Named Paricutín (pah-ree-koo-TEEN), the volcano grew to 1,700 feet (*500 meters*) before it stopped erupting in 1952.

Paricutín is the youngest of the volcanoes that mark the southern end of the Central Plateau. As you know, a *plateau* is a highland area with a surface that generally is flat. However, Mexico's Central Plateau is far from level. The volcanoes in the south reach heights of 12,000 to 18,000 feet (*3,500 to 5,500 meters*).

These volcanoes are a valuable natural resource. When volcanic rocks are eroded by wind and rain, they crumble into a fine soil. This soil is ideal for farming. What's more, the mountains help to bring rain to the area. In spring and summer, the prevailing winds carry moisture from the Atlantic Ocean. Forced upward by the mountains, the winds drop rain on the southern Central Plateau almost every day from June to September.

The southern half of the Central Plateau is the heart of Mexico. With its rich soil and favorable climate, it provides a home for about half of all Mexicans. About 15 million people live in and around Mexico City, the nation's capital and largest city.

The northern half of the Central Plateau has a rich soil and enough warmth to grow

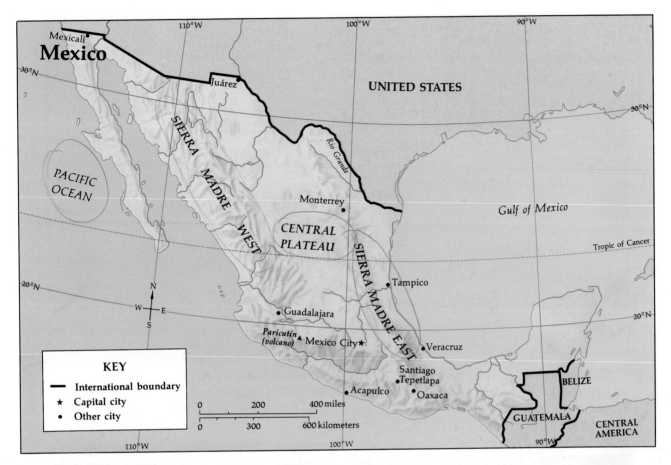

Mexico

as many crops as the south. What it does *not* have is enough water. Here the prevailing winds come from the landmass of North America. These winds are so dry that they *take* moisture from the land.

The Pacific Coast of Mexico is much like the Central Plateau — dry in the north, rainy in summer in the south. The driest parts of Mexico are found on the northern Pacific Coast. The warm, dry winters of the southern Pacific Coast help to make the area a magnet for tourists. The famous seaside resort of Acapulco (ah-kuh-POOL-koh) has curving white beaches, palm trees, and brilliant flowers.

On the Gulf Coast too, the southern half is wetter than the northern half. But the northern coast gets more rain than the rest of northern Mexico, and the southern coast is the wettest part of the country. This area of

Much of Mexico's population lives in the Central Plateau around Mexico City. This area is not too dry (like most of the north) or too wet (like parts of the south).

the Gulf Coast is a tropical rain forest, where plants grow thickly. Underground, the southern Gulf Coast is rich in mineral resources. Large deposits of oil and natural gas were first tapped here in the 1970's.

Elevation and Climate

Suppose you wanted to visit the hottest place in Mexico. How far south of the U.S. border would you have to travel?

The surprising answer: no distance at all. The hottest place in Mexico is the town of Mexicali (mehk-see-KAH-lee), just across the border from California. During July, Mexicali's hottest month, the temperature can reach 110°F (43°C).

Mexicali is about as far from the Equator as you can get in Mexico. Why aren't there hotter places farther south? After all, the closer the Sun's rays are to the Equator, the more heat they bring to Earth. For example, Oaxaca (wuh-HAH-kuh) is 1,200 miles (*2,000 kilometers*) closer to the Equator than Mexicali. Yet summer temperatures in Oaxaca range around a pleasant 70°F (*20°C*).

The answer is that elevation keeps Oaxaca cooler than Mexicali. As you read in Unit 1 (see page 57), temperatures drop with increases in elevation. Mexicali is below sea level. By contrast, Oaxaca is about a mile (*1¹/₂ kilometers*) above sea level, in the mountains. This difference in elevation has much more effect on temperature than the difference in latitude.

Mexicans say that their country has five directions — east, west, north, south, and up-and-down. Because elevation plays such a big role in the climate, Mexicans divide the country into three zones, like a layer cake.

The warm zone takes in the land from sea level to about 3,500 feet (*roughly 1,000 meters*). Tropical crops such as sugar, rice,

Volcanic soil and a favorable climate produce fertile farmland in the southern half of Mexico's Central Plateau. A small farm can supply a family's crop needs.

cacao, and bananas are grown here. The temperate zone is the land from above 3,500 feet to about 7,000 feet (*roughly 1,000 to 2,000 meters*). In this zone, coffee is grown, as well as such basic Mexican foods as corn, beans, squash, and chili peppers. The cold zone lies above 7,000 feet (*2,000 meters*). Some food still can be grown here, while pine and oak trees provide lumber.

SECTION REVIEW

1. What are the two major landforms found in Mexico and Central America?

2. In what area of Mexico is Mexico City located? What features of this area helped to make it a large population center?

3. What are the driest and wettest areas of Mexico?

4. In most parts of Mexico, what has the most effect on the temperatures: latitude, season, or elevation? Give an example.

Human Geography

The usually quiet village square has come alive with music, dancing, and feasting. It is *fiesta* (fee-EHS-tuh; holiday) time. Today people are honoring the patron saint of their village. This is the Roman Catholic saint who they believe watches over them.

The noise is almost deafening. Church bells ring, firecrackers explode, and whistles blow. A merry-go-round clangs gaily. A band plays popular folk songs. In this band, the flute and drum are like those used by the Indians to make music before the Spanish arrived. The guitar and the trumpets were introduced by the Spanish.

The mixing of Indian and Spanish instruments in the band reflects the Mexican people themselves. About three out of five Mexicans are mestizos. This mixing of Indian and Spanish people produced a wholly new culture — Mexican. But in some parts of Mexico, there are Indians who still live much as their ancestors did.

The great majority of Mexicans speak Spanish, the official language of the country. Most Indians of Mexico speak Spanish as well as one of about 80 Indian languages that still survive. (There were once almost 200.) But more than a million speak only an Indian tongue.

Mexicans are proud of their Indian heritage. It has inspired many of the country's greatest artists, writers, and composers. One of the most honored figures in Mexican history is Cuauhtémoc (kwow-TEH-mohk), the last Aztec emperor, who led a rebellion against the Spanish and was put to death. There are statues of him in many cities and towns. Statues of Cortés are rare.

Mexicans have inherited strong feelings of family loyalty from both their Indian and their Spanish ancestors. Families are usually large, with parents, children, and grandparents often living under the same roof.

The population of Mexico has been growing more rapidly than that of most other countries. Between 1957 and 1980, the number of people increased from 31 million to 69 million. If the population keeps growing at this rate, there will be about 130 million Mexicans by the year 2000. By that year, Mexico City may be the world's biggest city, with almost 30 million people!

Mexico's Ethnic Makeup

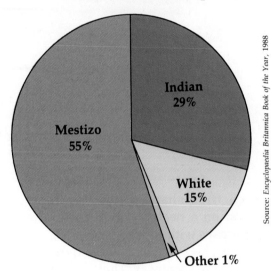

Source: Encyclopaedia Britannica Book of the Year, 1988

Mestizo 55%
Indian 29%
White 15%
Other 1%

There are many ethnic differences among Latin American nations. Mexico is unusual in having a majority of mestizos — people of Indian and Spanish descent.

Because the population is growing so fast, Mexico has a large percentage of young people. Some 46 percent of all Mexicans are under 15 years of age. In the United States, only 22 percent of the people are under 15.

Adapting to the Land

Traditional ways of life in Mexico can often be traced back to Indian ways before the Spanish conquest. Many of the old ways have survived because they were well suited to the environment.

In much of Mexico, the soil and climate are suitable for two highly nutritious food crops — corn and beans. For thousands of years, the Indians made corn into dough,

Different traditions blend in Mexican life today. In Mexico City, the Palace of Fine Arts shows the influence of Spanish architecture. It houses a dance company based on Indian traditions. Beside it, cars and telephones show the influence of the modern world.

which they shaped into flat cakes and cooked on a heated slab of rock. They ate these cakes with cooked beans.

When the Spanish arrived, they began to eat corn the same way. They called the cakes *tortillas* (tor-TEE-yahs). The Spanish also brought new foods — beef, chicken, sausage, cheese, olives, onions, and lettuce. Since then tortillas have been eaten with many of these foods.

When a tortilla is rolled around other foods and covered with a hot sauce, it becomes an *enchilada* (ehn-chee-LAH-duh). When it is fried crisp and topped with these ingredients, it becomes a *tostado* (tohs-TAH-doh). When it is fried crisp and folded in half around them, it is called a *taco* (TAH-koh).

There's another blend of cultures in sports that Mexicans enjoy. As in nearly all of Latin America, the most popular sport is soccer. At the same time, Mexicans share an enthusiasm for baseball with their neighbors in the north, in the U.S.

Mexicans who live far from big-city stadiums often follow sports on the radio. To-day the radio is the most popular source of news and entertainment. TV sets are too expensive for most families to afford. But in poor city neighborhoods, and even in some country villages, one family may buy a TV set on credit, and will pay for the set by charging neighbors who want to view the programs.

Some of the most popular shows on Mexican TV are imported from the United States. But the influence is not all one way. For a long time, Mexico has been the biggest movie-producing nation in Latin America. Its movies are shown all over the Spanish-speaking world — including Hispanic communities in the U.S.

To show how all these facts about the Mexican people fit together, the following two sections give you a closer look at daily life.

Village Life

Santiago Tepetlapa (sahn-tee-AH-goh teh-peht-LAH-puh) is a small village not far from Oaxaca, at the southern tip of the Central Plateau. On the main street, which is paved with stones, pigs and chickens wander around. There is no danger that they will be run over by a car because there isn't a car in the entire village.

Near the main street is the home of Miguel (mee-GEHL) and Pilar Flores (pee-LAR FLOR-ehs) and their three children: Esther, 11; Abel, 7; and Rubén, 3. Their house is a solid one-room adobe building. Nearby is a hut that Pilar Flores uses as a kitchen.

The kitchen has a gas stove that Pilar Flores uses only on special occasions, like a birthday or a holiday. To save money, she cooks most meals over a wood fire. The kitchen also has a radio, a small electric clock, and one light bulb. The family is proud of these things, for until a few years ago, the village had no electricity.

Like most Mexican villagers, the Flores family relies heavily on local resources for its daily needs. The family has several hens, one rooster, and two pigs. The animals wander around freely, for there is neither barn nor fence. Miguel Flores owns a piece of land about a mile (*about 1 1/2 kilometers*) from the house. There he grows corn for his family. However, he also works on a large farm that grows tomatoes for the markets of Mexico City. This work brings in pesos (PEH-sohs; Mexican money) for cash needs.

Esther goes to school in Santiago. At home her main chore is to help her mother prepare tortillas. These are an important part of every meal. Esther also helps wash the family's clothes by hand, and presses them with an iron heated over the fire. Another of her duties is to sweep the house with a broom made of twigs.

Miguel and Pilar Flores hope that one day their children may have a better life. Esther would like to become a lawyer. This is no longer an impossible goal for village children. Many Mexican universities are almost free — for those who have good grades.

City Life

Jorge Fernández (HOR-heh fehr-NAHN-dehs) is 14 years old and lives in Mexico City. His everyday life includes some activities that would be the same for teenagers in many other parts of the modern world. Others apply only to Mexico. As you read this section, can you tell which are which?

Jorge goes to a military school called Héroes de Chapultepec (EHR-oh-ehs deh chuh-POOL-teh-pek). Its name honors a group of 14- to 16-year-old boys who were national heroes in 1847 when Mexico and the United States were at war. The boys were students at a military school in a castle in Chapultepec Park, in Mexico City. When U.S. soldiers surrounded the castle, the boys tried to defend it.

All but one of them were killed in the battle. The one who lived did not want the Mexican flag to fall into foreign hands. So he wrapped himself in the flag and dived from the castle wall, killing himself in the fall.

Jorge's family and friends are proud of Mexico's history and culture. They often visit the museums in Mexico City and study the arts and crafts of the Indians. Jorge has also visited many of the temples and cities built by ancient Indian civilizations.

Jorge's daily schedule starts about 7:30 when he has breakfast with his sisters Isabel and Dolores, who also go to school. Classes run straight through until one or two in the afternoon, with no study periods. Jorge studies chemistry, physics, math, English, Spanish literature, civics, and modern history. Chemistry interests him the most, and he hopes to become a chemical engineer.

Jorge usually doesn't get his lunch until he comes home from school at 3:30. Then he has soup, meat with potatoes, beans, salad, lemon or orange soda, fruit, and pastry. In the evening, he does his homework and often watches television.

Visitors to Mexico City sometimes ask Jorge what it's like to live high up where the air is thin. But Jorge has never noticed any difference. He is more concerned that the air is getting too *thick* — with smog. At one time, the city's high elevation gave it pure air all year round. But the spread of industry and traffic have brought air pollution.

SECTION REVIEW

1. Mexican culture is a combination of what other two cultures?

2. What two highly nutritious food crops have long been grown in Mexico? Name at least one Mexican dish.

3. How does the rate of population growth in Mexico compare with that of other countries? What age group is very large as a result?

4. Esther Flores and Jorge Fernández lead very different lives. Name two cultural features they share.

Economic Geography

In the last section, you met Miguel Flores, who lives in a village near Oaxaca (see page 177). He makes his living directly from the land — but on two very different kinds of farms. One is his own small plot, where he raises corn with simple hand tools. The other is a large, modern farm that uses machinery and fertilizers. These farms are typical of the two worlds of Mexican farming today.

Although all of Mexico's farms together produce large amounts of food, most of the country's land surface is not farmed at all. Either it is too mountainous or it lacks rain. In recent decades, huge irrigation projects have greatly increased the areas that can be farmed. Even so, only about one eighth of Mexico's total land surface is suitable for growing crops.

For hundreds of years after the Spanish conquest, the great majority of Mexicans had no land of their own. They worked on large estates, called **haciendas** (ah-see-EHN-duhs), that were owned by wealthy families of Spanish descent. Most of the farm workers were always in debt to the landowners and could not leave the estates.

In 1910 Mexico's poor farmers began a revolution. One of their main demands was for land. In 1917 a new, democratic government came to power. This government began to break up the large estates and divide them among the landless farmers.

Today privately owned farms are limited to 500 acres (*200 hectares*). Vast areas of land have been "shared out" since 1917. Most land was given to communities of 20 or more families, rather than to individuals. Under this system, the land is owned by the community, and each family is assigned a plot. Farmers can leave their plots to their children, but they cannot sell their lands or rent them. Community-owned lands in Mexico are known as **ejidos** (eh-HEE-dohs).

Most *ejido* farmers use traditional methods of growing food. They plow the land with oxen, and plant and harvest by hand. Often the soil is worn out, or there is not enough rain or irrigation. As a result, the small plot may not produce enough food to support the family.

In recent decades, more and more poor farmers have left the land and gone to live in the cities. The biggest magnet is Mexico City. About 400,000 people from rural villages arrive there each year to look for jobs.

All the same, about two out of five Mexicans still make their living from the land. Like Miguel Flores, some of them work on large farms that use modern technology. The biggest area of modern farming is the north Pacific Coast. In this desert area, farms rely on irrigation for their water. The results are impressive. Half of Mexico's farm products come from the north Pacific Coast.

The crops grown in this area range from

wheat to oranges, from cotton to cauliflowers. They are sold all over Mexico and exported to the U.S.

Booming Industry

In the city of Juárez (WAR-ehs), close to the U.S. border, factory workers are busy assembling everything from cowboy boots to TV sets. Since 1976 the number of factories in Juárez has almost doubled. The number of factory workers has jumped from 13,000 to 40,000 and keeps growing.

Booming industrial cities like Juárez can be found in all parts of Mexico. In the past 40 years, industrial growth has raised the standard of living of millions of Mexicans. Today about one Mexican out of three is a member of the prosperous middle classes.

Most manufacturing in Mexico centers in and around Mexico City. Hundreds of factories produce cars, electric appliances, chemicals, and many other products. The area is not rich in mineral resources, however. What brings the factories here? One answer is the large population, which provides plenty of both skilled and unskilled workers. In addition, this area is at the heart of Mexico's transportation system.

Mexico's most valuable mineral resources today are oil and natural gas. Mexico now exports oil to the U.S. and other countries. Mexico still ranks behind Venezuela as an oil exporter, but could catch up by the late 1980's. The United States probably will become Mexico's biggest customer.

Mexico's exports consist mainly of raw materials — such as oil and silver — and farm products. The goods Mexico manufactures are used mostly within the nation itself. At present it does not manufacture enough to meet all of its own needs. It has to import goods such as machinery and cars.

In most years, the value of Mexico's imports is greater than the value of its exports. How does Mexico make up the difference? The answer is tourism, which Mexicans call "the industry without chimneys." Hundreds of thousands of people from the United States alone visit Mexico every year.

In this century, Mexico has had two revolutions. The first, begun in 1910, led to a democratic government and land for poor farmers. The second has been an industrial revolution that still continues.

SECTION REVIEW

1. Give two reasons why most of Mexico's land surface is not farmed.

2. Give two reasons why most manufacturing in Mexico centers around Mexico City.

3. What are *ejidos*? How do they differ from *haciendas*?

4. In what main ways do farms on the north Pacific Coast differ from most others in Mexico? How successful are they?

YOUR LOCAL GEOGRAPHY

1. Elevation has such a great effect on life in Mexico that Mexicans call their country's fifth direction up-and-down. But elevation plays at least some part in the geography of any place.

If you don't already know it, look up the elevation of the place where you live. (If you don't have a map that gives elevations, ask at the library.) If you live at a high elevation, is the weather mild or cool in summer? If you live in a low-lying area, is flooding a problem? Are there nearby places at a very different elevation? If so, do these differences create any transportation problems? Write a paragraph on the role elevation plays in your life.

2. Like Héroes de Chapultepec School in Mexico City, there are places and monuments dedicated to important people in most cities and towns throughout the United States. Check the names of a few schools, libraries, and streets or parks in your neighborhood. See if you can discover for whom they were named, and why. Then choose one name and report to the class on his or her role in your community.

Locating Places on a Large-Scale Map

Imagine you are a tourist in the heart of Mexico City. You want to visit the central plaza, or the *Zócalo* (ZOH-kuh-loh). To find the landmarks of any city, it helps to have a large-scale map. A large-scale map is a close-up view of an area. For example, the map below shows the Mexico City metropolitan area, which includes the city itself and its *suburbs* (the communities surrounding the city). Study the map key. Answer these questions on a separate sheet.

1. What does the heavy unbroken line represent?

2. What do shaded areas represent on this map?

3. What symbol is used to represent: (a) a point of interest?   (b) a suburb?

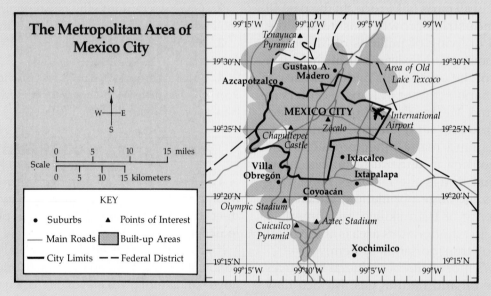

This close-up view of the Mexico City area does not cover a full degree of either latitude or longitude. So lines of latitude and longitude are marked off in parts of a degree called **minutes**. There are 60 minutes in one degree. An accent mark (') after a number indicates minutes, in the same way a small circle (°) indicates degrees.

For example, the suburb of Coyoacán (koy-wah-KAHN) is located at a latitude 19 degrees north of the Equator, plus about 20 minutes. This is written 19°20'N. Find Coyoacán along the 19°20'N line of the map. What is the town's longitude, in degrees and minutes? Write the number on a separate sheet of paper. Then on the same sheet, answer the following question.

4. What point of interest would you find nearest each of these four locations?   (a) 19°25'N, 99°11'W;   (b) 19°18'N, 99°9'W;   (c) 19°32'N, 99°11'W;   (d) 19°26'N, 99°8'W.

A. Words To Remember

In your own words, define each of the following terms.

adobe   minutes (of latitude or longitude)
ejidos   plateau
haciendas

B. Check Your Reading

1. Name two major ways in which the volcanic mountains of Mexico's Central Plateau help farmers in the area.

2. In which of the following areas do about half of all Mexicans live: the Pacific Coast, the southern Central Plateau, or the northern Gulf Coast? What attracts them to that area?

3. Is the hottest part of Mexico in the northern, central, or southern part of the country? Why are temperatures higher in that area?

4. How did the dry area of Mexico's north Pacific Coast become a highly productive farming area?

5. Mexico's exports fall into two main categories. What are they? What type of goods or products does Mexico import?

C. Think It Over

1. Which culture — Indian or Spanish — do you think has a greater influence on Mexican life today? Give reasons for your answer.

2. Name two resources besides raw materials that a city would probably need in order to become an industrial center.

3. Why is the discovery of oil and natural gas so important to Mexico? Give two reasons.

D. Things To Do

One way to learn more about the different environments in Mexico is to study the animals that live there. For example, there are alligators and crocodiles in the coastal lowlands; rattlesnakes and scorpions in the desert lands of the northern Pacific Coast; and armadillos, anteaters, and jaguars on the southern Gulf Coast.

Look up one of these animals in an encyclopedia and find out both how it meets its needs in its own particular environment and how it can be used as a resource by the people of Mexico. (To give just one example, many rural Mexicans keep nonpoisonous snakes as house pets because these snakes eat mice and bothersome insects.)

Chapter 12

Brazil

The Search for Hidden Treasure

It looks like a city of the future. It is Brasília (brah-SEEL-yuh), the super-modern new capital of Brazil.

Seen from the air, Brasília looks like a huge bird or an airplane. The body is formed by a wide boulevard with government buildings of marble and glass on either side. The wings are formed by the apartment houses where most of the people live. Each block of four buildings has its own shops, theaters, playgrounds, and a school.

Brasília was built inland, in a part of the country that was a wilderness, like the Old West of the United States. Work on the city began in 1956. Four years later, the government of Brazil moved from Rio de Janeiro, known simply as Rio, to the new capital.

Building Brasília was a tremendous task. Only dirt roads led to the site, 690 miles (*970 kilometers*) northwest of Rio. At first workers and supplies had to be flown in from distant cities. Later a modern highway was built to bring in supplies on trucks. About 60,000 people worked around the clock to get the city ready by 1960.

Today Brasília is a major business center as well as the seat of government. It is also the center of a network of roads that reach out to all parts of the country (see map on page 166). Farms and towns have sprung up along the new highways. Settlers are moving to distant parts of Brazil that until now had few, if any, people.

Brazil's young capital city now has a population of more than one and a half million. It is a symbol of the pioneering spirit that is opening the vast interior of Latin America's largest country.

Physical Geography

Brazil is big. It spreads over almost half of South America. It has a common border with every country on the continent except Chile and Ecuador. In the entire world, only the Soviet Union, Canada, China, and the United States are larger.

Most of Brazil lies in the tropic zone. Except for the southern tip of the country, which extends into the temperate zone, Brazil's climate is hot and generally humid. Because of the heat and heavy rainfall, crops can be grown all year round in most of Brazil.

Like the U.S., Brazil is large enough to have a wide variety of environments. A narrow strip of fertile lowlands runs along the Atlantic Coast. Except in the north, a chain of mountains rises sharply behind these lowlands. In some places, mountains come to the ocean, which is what happens at Rio. The city owes its picture-postcard views to sheer peaks that hang over sandy beaches.

Brasília, the new capital city of Brazil, has plenty of open space — it was built in the nation's undeveloped interior. These structures are the home of the Brazilian congress.

Brazil

KEY

— International boundary

★ Capital city

• Other city

Most of Brazil lies west of the mountains, and can be divided into two main areas, south and north. Most of the southern half of the country is covered by a series of plateaus. Though much larger in area than the plateaus of Mexico, Brazilian plateaus are not nearly so high. They range from about 600 to 3,000 feet (*about 200 to 1,000 meters*) above sea level.

These highlands are a huge patchwork of grassland and forest. Large areas near the

In both area and population, Brazil is by far the biggest country in the region. It touches all of the other countries in South America except Chile and Ecuador.

coast have been cleared for farming. Farther inland the forest consists of tangles of low trees and bushes that are hard to clear. This region is known as the **cerrado** (seh-RAH-doh; closed land).

Some big rivers run through the southern half of the country. The Paraná (par-uh-

NAH) is 2,500 miles (*4,000 kilometers*) long; the São Francisco (sowm frahn-SEES-koh), 2,000 miles (*3,200 kilometers*). But these rivers are overshadowed by the network of rivers in the north — the Amazon Basin.

The Amazon River is 4,000 miles (*6,400 kilometers*) long, slightly shorter than the world's longest river, the Nile in Africa. But in nearly every other way, the Amazon is far bigger. The Nile has few important tributaries. The Amazon has dozens. Look on the map opposite at the Theodore Roosevelt River, named for the U.S. President who helped to explore it. This river runs into a bigger river, which runs into an even bigger river, which runs into the Amazon. Yet the Theodore Roosevelt River alone is longer than either the Hudson or the Sacramento rivers!

The land area drained by the Amazon and its branches is twice as large as that drained by the Mississippi. The Amazon Basin is covered by the world's largest rain forest. Trees in this forest produce about as much oxygen as all the rest of the world's trees put together.

Many Resources

Almost 500 years ago, Portuguese explorers discovered a new kind of tree in the Amazon rain forest. This tree produced a dye that reminded them of the color of a bonfire. So they named the tree *brasa* (BRAH-suh), the Portuguese word for burning coals. The whole area became known as the land of the brasa tree — or simply Brazil.

Later many other resources were discovered in the Amazon forest. There were useful products such as rubber, resin, nuts, and waxes. An even wider range of riches was found in other parts of the country.

Much of the plateau land south of the Amazon Basin has rich soil and a moist, tropical climate. Many different food crops can flourish here.

Beneath the soil, Brazil is also rich in re-

Brazil's Portuguese heritage can still be seen in the elegant buildings of this former gold-mining town.

sources. In the highlands between Rio and Brasília, there is a large state named Minas Gerais (MEE-nuhs zheh-RIES), which means "general mines." Minas Gerais contains huge deposits of iron — perhaps one fourth of the world's supply. It is rich in aluminum, manganese, and other useful minerals. Gold and diamonds have also been found there.

Today prospectors travel by helicopter over the cerrado and by canoe up remote branches of the Amazon. They are trying to find what further resources may lie hidden in Brazil's vast interior.

SECTION REVIEW

1. How does Brazil rank among world nations in area? How much of the South American continent does it cover?

2. What is the cerrado? Where is it?

3. What does *Minas Gerais* mean? Name two resources found there.

4. Would you say that the Amazon River is bigger or smaller than the Nile? Explain your answer.

Human Geography

Around 500 years ago, the land that is today Brazil was inhabited by about one million Indians of different groups. Unlike the Aztecs, Maya, and Incas you read about earlier, these Indians lived very simply. Some hunted and gathered food in the wild. Others were subsistence farmers (people who grow just enough food for their own needs).

In 1500 a Portuguese explorer named Pedro Alvares Cabral (PEH-droh AHL-vuh-rehs kuh-BRAHL) landed on the northeast coast of Brazil and claimed the territory for his king. Soon other Portuguese began to arrive in the country.

The first European settlers built towns and cities along the east coast. Later, when settlers moved inland, they built towns along the rivers. But most European life centered on the coast — especially on the hot, moist, fertile coast in the northeast.

Here the Portuguese set up plantations to grow such tropical lowland crops as sugarcane and bananas. At first the Portuguese used Indians as slaves to work on the plantations. But the Indians soon died of diseases and overwork. So the Portuguese used African slaves. The west coast of Africa is only 1,600 miles (*2,500 kilometers*) from northeast Brazil — shorter than the distance up the Amazon to the border with Peru.

After Brazil became an independent nation in 1822, many Europeans began to migrate there. Thousands of Germans, Italians, Poles, and Russians settled in southern Brazil, where the climate was temperate.

People of European descent live in all parts of Brazil today. In fact, they make up about 53 percent of all Brazilians. The main European groups living in Brazil today, starting with the largest, are Italian, Portuguese, Spanish, German, and Slavic.

In recent decades, large numbers of Japanese have also settled in Brazil. About a half-million Japanese live in and around the city of São Paulo. A few have developed plantations along the Amazon River.

Today Brazil has more people than all other countries of South America combined. In 1980 the total number of Brazilians was estimated at about 125 million. The population is growing rapidly. By the late 1980's, it will probably exceed 150 million.

Brazil's Ethnic Makeup

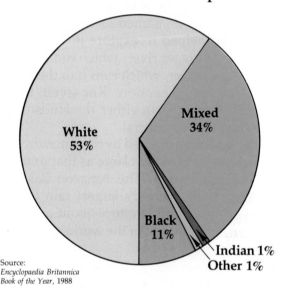

Source:
Encyclopaedia Britannica Book of the Year, 1988

The "Other" group in this pie graph includes many recent immigrants from Japan and other Asian countries.

A United Nation

Despite many differences in racial and national backgrounds, there is a strong feeling among Brazilians that they form a single people and a united nation. What accounts for this sense of unity?

Much is due to the early Portuguese settlers of the country. They gave the nation their Portuguese language, which today is spoken by the vast majority of Brazilians. They also gave Brazil their religion, Roman Catholicism. More than 90 percent of all Brazilians today are Catholics. In fact, Brazil has a larger Catholic population than any other country in the world.

Brazilians are united by other traditions and interests. Almost all Brazilians celebrate the four days and nights before Lent with exciting carnivals. All businesses close, and the streets are filled with people parading, singing, and dancing in colorful costumes. The Carnival in Rio attracts visitors from all over the world.

As in nearly all of Latin America, there are big differences between the ways of life in Brazil's cities and in its countryside. Almost two out of three Brazilians live in towns and cities. The rest live on small farms, plantations, ranches, or in villages.

On the following pages, you will see first how people live on a typical plantation. Then you will look at life in a modern city.

On a Plantation

The plantation village of Fazenda das Moças (fah-ZEHN-duh duhs MOH-sahs) is on the northeast coast of Brazil in the state of Bahia (bah-EE-ah). The plantation covers about 700 acres (*300 hectares*). Half is planted with sugarcane. The rest is used for grazing farm animals, and for the workers to grow their own food.

The plantation is owned by Donha Sinha (DOHN-yuh SEEN-yuh), a widow. It is home not only to her but also to the workers and their families. About 200 men, women, and children live on the plantation. Donha Sinha knows all of them by name.

The buildings of the plantation form a village. Two houses are separate from the rest. One, made of **stucco** (plaster wall covering) with a tile roof, is occupied by Donha Sinha. The other, which is smaller, is occupied by her foreman and his family. These are the only houses that have electricity, radios, and telephones. The workers and their families live in a long, one-story adobe building that is divided into 45 apartments.

The workers are paid according to the tasks they perform — weeding, and planting or cutting cane. They each have a small garden where they grow crops such as manioc and corn. (**Manioc** is an edible root that is eaten in much of Latin America.) The workers may also own an ox or a donkey, which they are allowed to graze on plantation land.

The people at Fazenda das Moças have close ties to each other. Most of them have spent their entire lives on this plantation. In recent years, the population of the village has dropped sharply. This is true of many plantations and farms in the northeast of Brazil. In the dry interior, where some years bring no rain, small farmers may face starvation if they stay. Even on the fertile coast, there are too many people trying to grow food by inefficient methods. Since most Brazilians have large families, the problem increases from year to year. The result is that many people leave the land, hoping to find a better life in the cities.

In a City

In Brazil nearly all of the big cities lie along the coast. Migration from the countryside squeezes more and more people into a tiny part of the country's area. Inland there are vast regions that could support a much larger population. This is why the Brazilian government built its new capital, Brasília, in the interior of the country.

Sergio (SEHR-jee-oh) and Maria Tourinho (mah-REE-uh toh-REEN-yoh) live with their five children in a four-room apartment in Brasília. Sergio is a member of Brazil's Congress.

The Tourinhos come from Salvador (sahl-vuh-DOR) on the Atlantic Coast. Salvador is an old and beautiful city. Brasília is new and exciting. It has not only the offices of the Brazilian government but also the embassies of other countries. Like Washington, D.C., it attracts visitors from around the nation and the world.

The Tourinho boys are too young to remember much about their lives in Salvador.

The two girls, Sylvia (SEEL-vee-uh), 15, and Aneta (ah-NEH-tuh), 14, remember Salvador and miss their friends.

The girls like to swim and play tennis. The apartment buildings in Brasília are planned so that sports centers are nearby.

About twice a month, the Tourinho family has dinner in one of Brasília's many restaurants. They often order *feijoada* (fayzh-WAH-duh), which is a very popular dish in Brazil. It brings together several of the country's major meat and vegetable products. The meats include tongue, spareribs, sausages, beef, and bacon. They are cooked with black beans and sprinkled with onions, tomatoes, and peppers.

In Brasília, Sylvia and Aneta have a close-up view of developing Brazil. They are interested in careers that will make them part of that development. Sylvia plans to be an economist. Aneta has two interests — she wants to be either a diplomat or an electrical engineer.

SECTION REVIEW

1. What people inhabited Brazil 500 years ago? Name three major ethnic groups who live there today.

2. What features of Brazil's northeast coast attracted settlers?

3. Why has the population of villages such as Fazenda das Moças dropped sharply?

4. Why would Sylvia and Aneta Tourinho consider Brasília a new and exciting city?

Economic Geography

Power shovels roar and clatter on the outskirts of Poços de Caldas (POH-sohs deh KAHL-duhs), a small city in the southern highlands. The shovels are scooping more than 1,000 tons (*1,200 metric tons*) of **bauxite** (BAWK-siet; aluminum ore) out of the hills every day. The ore is taken in trucks to a huge refinery nearby where it is made into bars of aluminum.

In 1980 Brazil produced four times as much aluminum as in 1970. Today the country is just beginning to mine some of the vast bauxite deposits that have been found in the Amazon Basin.

In recent decades, Brazil's economy has grown so fast that it is now the tenth largest in the world. The country is racing to develop all of its varied resources — crops, livestock, minerals, water power, and many kinds of industry. This is a change from the past, when Brazilians pinned their hopes on just one or two leading products at a time.

It is a change for the better, since the old way involved a lot of risk. Demand for a product could suddenly drop, pushing down prices. Or another country might start selling the product more cheaply and thus take over much of the trade. As you will see, Brazil often ran into these kinds of trouble.

A Search for Riches

The first Portuguese who arrived in Brazil in the 1500's were hoping to find gold. The best treasure they could find was the brasa or brazilwood tree and its red dye. But felling and shipping the brazilwood trees was a long and exhausting job. The Brazilian colony did not get rich on it.

Northeast Brazil seemed ideal for growing sugar. It had (and still has) a hot, moist climate, together with fertile soil. It did not have any natural sugarcane, but around 1530 the Portuguese brought some from islands off the coast of Africa.

For almost 200 years, the plantations on the northeast coast of Brazil supplied most of the world's sugar. Later, however, they faced competition from new plantations in the West Indies. Brazil's sugar production fell. As you saw at Donha Sinha's plantation, sugarcane is still widely grown today in northeast Brazil. But it is only fourth or fifth among the country's farm exports.

Gold and diamonds were discovered in the 1690's, in what is now the state of Minas

Gerais. Prospectors and settlers came rushing to the area. Gold and diamonds are still mined there, but they are no longer as important as they once were.

In the 18th century, the coffee plant was brought from Africa to Brazil and another boom followed. Between 1850 and 1950, Brazil produced as much as three quarters of the world's coffee supply. Even today Brazil is the largest supplier of coffee in the world.

Brazil's next economic boom took place in the Amazon Basin. There in the tropical rain forest, rubber trees grew wild. Demand for tires, first for carriages and later for cars, sent the price of rubber soaring late in the 19th century.

But the rubber boom in Brazil did not last long. An Englishman had smuggled some rubber tree seeds out of the country in 1876. Soon after, the English started rubber plantations in some of their tropical Asian colonies. These plantations produced rubber more cheaply than the wild-growing trees of the Amazon. Brazil's rubber production shrank.

Fitting the Pieces Together

In the middle of the 19th century, São Paulo was a small and rather quiet city. It then became the center of the coffee trade and started to grow — fast. Before long it became Brazil's biggest city. Today with the coffee boom long past, São Paulo has 12 million people and is still growing fast. Why?

The wealth that São Paulo earned from coffee was put into industry. There were plenty of workers to run the new factories. Immigrants from Europe and Japan were already arriving because of the coffee boom. Water power from the mountains along the coast generated electricity to run the new factories. And the neighboring state of Minas Gerais could supply key mineral resources. Thus an industrial boom began.

Coffee plantations still flourish in the highlands near São Paulo. But today other

São Paulo is one of the world's most industrialized and fastest-growing cities. Here, high-rise apartments tower over a modern sports stadium. The fans are watching a game of soccer, which is Latin America's favorite sport.

crops are important too. With the help of Japanese experts, Brazilian farmers began growing soybeans, which are used for livestock feed and processed foods.

São Paulo has kept booming because it developed many resources, not just coffee. The whole country now plans to vary its economy in much the same way.

On the Frontier

Brazil's number-one aim today is to develop farming. There are still enormous areas in the interior of Brazil without a single village, road, or farm. In fact, the country's farm products come from just 11 percent of its total area.

A large part of the remaining 89 percent of the land consists of the Amazon Basin. This area has great growing potential, but it is not easy to farm. Trees have to be cleared. The soil is not very fertile. The heavy rains quickly wash away the nutrients that plants need. Developers also find it hard to attract enough workers to the area, since it is a long way from the population centers of São Paulo and Rio.

There is also widespread concern about the effects of clearing the Amazon forest. Development has pushed many Indians of the Amazon from their forest homelands. Only about 100,000 Indians survive today. In addition, any large-scale clearing of the rain forest could affect the world's climate. The trees of the Amazon Basin not only produce oxygen but also act as a huge reservoir for moisture.

Today the Brazilian government is focusing more on the other part of its wild interior — the cerrado, which begins south of the Amazon rain forest.

Few Indians live in the cerrado. And opening the land to farming will not affect the world's climate. On the other hand, much of the soil is not very fertile. Still, food crops can be grown more easily here than in the Amazon.

Brasília is still at the center of developing Brazil. In Brasília you can climb into a car and set out on a 1,300-mile (2,100-kilometer) journey along the highway to Belém, near the mouth of the Amazon.

On the way, you will pass through undeveloped cerrado and the world's biggest rain forest. Monkeys may shout at you from treetops, and you may have to brake to avoid an armadillo. And somewhere under the earth around you, there may be vast mineral riches that someday will be discovered.

SECTION REVIEW

1. Name three crops that were important to Brazil in the past. Which two of these are still grown in large quantities today?

2. Name two disadvantages that Brazil faced in relying on one major product at a time.

3. Give two reasons for São Paulo's fast industrial development.

4. Why is the Brazilian government eager to develop the country's interior?

YOUR LOCAL GEOGRAPHY

1. In this chapter, you read how some Brazilian cities came into existence. Brasília was built to be the nation's new capital. São Paulo was founded to meet the business needs of the coffee plantations.

Find out why your own community started. What was it that attracted people to your area? Did natural resources or mineral wealth have anything to do with it? When your community was still new, how did most of its people support themselves (by farming, raising livestock, coal mining, manufacturing, etc.)? What kinds of jobs do most people work at today?

2. When Pedro Alvares Cabral landed on the coast of Brazil in 1500, he knew nothing about the territory he was claiming for his king. He had no idea whether it was mostly mountainous or flat, had many rivers or only a few, had deserts or rain forests, and so on. A map of Brazil could have told him all these things. But no such map existed.

Draw a simple freehand map of your own community that would show visitors what kind of environment they could expect to find there. Use shading to show high land and add x's for barriers such as cliffs and mountains. Also mark in water forms (oceans, rivers, lakes, streams, etc.). Then show what the surface is like by labeling areas "forest," "grassland," "farmland," "desert," "residential" (homes), "commercial" (offices, stores), or "industrial."

Understanding a River System

The Amazon is the world's mightiest river. It carries more water to the sea than the Mississippi River, Africa's Nile River, and China's Yangtze River put together — enough to color the ocean brown for a full 200 miles (*320 kilometers*) out from the coast!

Despite its great size, the Amazon River has features in common with most other rivers. These features are shown in the map on page 192. Like the Mississippi, for example, the Amazon River receives much of its water from tributaries. A **tributary** is a stream that flows into a larger stream. Some tributaries of the Amazon are themselves large rivers, with tributaries of their own. Such a network of rivers makes up a **river system**.

The Amazon, like every river, also has a beginning — a **source**. At its source, the Amazon is known as the Marañón (mar-uhn-YOHN). Usually the source of a river is made up of small streams coming together in a highland or mountain area. These streams and the beginning of the river are also known as the **headwaters**. Since water flows downhill, all rivers flow **downstream** from areas of high elevation to areas of lower elevation.

The separation between two river systems is called a **divide**. For example, the Andes Mountains create a divide between sources that flow downstream into the Amazon River and those that flow downstream into the Pacific Ocean. Below the mountains and highlands where a river system begins is a **basin** (lower expanse of land), through which the river and its tributaries flow. Sometimes the entire region drained by a river system is called a **drainage basin**.

The end of a river's course, where it empties into another body of water, is called its **mouth**. Find on the map where the Amazon empties into the Atlantic Ocean. The many islands there are part of the Amazon's delta. A **delta** is a broad, usually triangle-shaped area of islands or low-lying ground at the mouth of some rivers. A delta is formed from soil that has been carried downstream by the flow of the river.

Refer to the map of the Amazon River System on page 192. Complete each of the following statements by writing the best choice on a separate sheet of paper.

1. The (source; basin; mouth) of the Amazon River is located at 0°, 50°W.
2. The city of Manaus is located in the (basin; mouth; delta) of the Amazon River System.
3. Marajó Island is part of the Amazon (Basin; source; delta).
4. The city of (Belém; Brasília; Iquitos) is near a divide between two river systems.

(Turn page.)

Refer to the map of the Amazon River System and its key. Answer the following questions.

5. The latitude/longitude figures following each river below tell the location of its source. In which highlands or mountain range — and at what elevation — is each source located? (a) Marañón (Amazon) River (11°S, 77°W); (b) Branco River (3°N, 64°W); (c) Tocantins River (17°S, 48°W).

6. The Branco River flows downstream from its source — first east and then south. In which directions do these rivers flow? (a) Ucayali River (11°S, 73°W); (b) Purus River (11°S, 72°W).

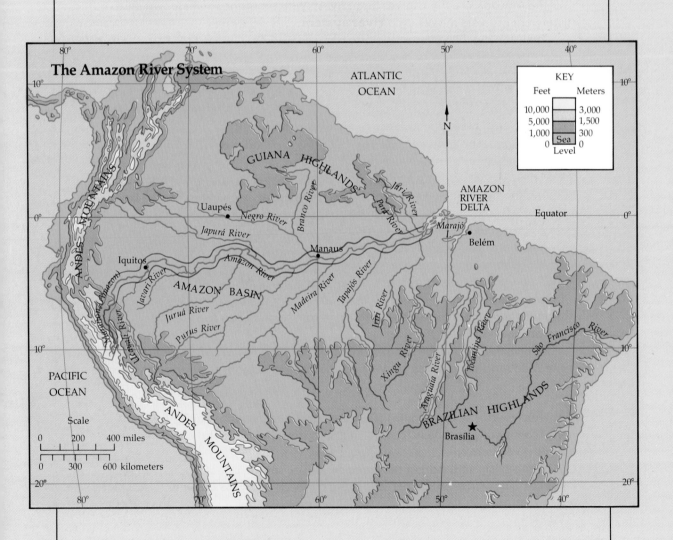

The Amazon River System

KEY

Feet		Meters
10,000		3,000
5,000		1,500
1,000		300
0	Sea Level	0

ATLANTIC OCEAN

GUIANA HIGHLANDS

Jari River

AMAZON RIVER DELTA

Equator

ANDES MOUNTAINS

Uaupés

Negro River

Branco River

Pará River

Marajó I.

Belém

Japurá River

Manaus

Iquitos

Amazon River

Javari River

Madeira River

Tapajós River

Iriri River

São Francisco River

AMAZON BASIN

Marañón (Amazon) River

Ucayali River

Juruá River

Purus River

Xingu River

Araguaia River

Tocantins River

BRAZILIAN HIGHLANDS

PACIFIC OCEAN

Scale

0   200   400 miles

0   300   600 kilometers

ANDES MOUNTAINS

Brasília

A. Words To Remember

From the following list, select the term or terms that best complete each sentence below.

Amazon Basin	colonies	settlers
bauxite	manioc	stucco
cerrado	plantations	subsistence farmers

1. _____ grow just enough food to meet their own needs.

2. Brazil's inland area that is covered by low trees and bushes is known as the _____.

3. _____ is an ore that contains aluminum.

4. _____ are farms that are large in area and usually specialize in one crop.

5. _____ is an edible root.

B. Check Your Reading

1. Most of Brazil lies in what zone? Describe its overall climate.

2. What language is spoken in Brazil? Why?

3. Why did the Brazilian government build a new capital city?

4. The Amazon Basin has great growing potential. Give two reasons why it is not widely farmed.

5. Brazil's number-one aim today is to develop farming. In what area outside the Amazon Basin does it plan to do this? Name two obstacles it will have to overcome.

C. Think It Over

1. Give two reasons why nearly all of Brazil's large cities are located along the coast.

2. Why is Brazil now trying to develop a variety of crops and minerals rather than rely on one main resource?

3. Give two examples of cultural differences among Brazilians. State some of the factors that helped Brazilians to become a united people.

D. Things To Do

The Amazon River and its rain forest together make up one of the most extraordinary regions on Earth. This chapter gives only some of its dramatic facts and figures. Use encyclopedias, travel books, and other library sources to collect further information about the region. Make at least one note under each of the following headings: size of the river; size of the forest; plant life; animal life; Indians; difficulty of travel.

Now imagine you are preparing a TV program about the Amazon. Using the information you have collected, write two paragraphs that might be read to introduce the program. Make them as dramatic as you can.

Chapter 13

Haiti and the Dominican Republic
The Two Faces of Hispaniola

Let's suppose you have never heard of the Caribbean island called Hispaniola. You know nothing about Haiti or the Dominican Republic, the two nations that share the island. You are now given the following clues. What picture do you get of the two nations?

■ In Haiti four people out of five live in rural areas. In the Dominican Republic, about half the people live in cities.

■ Haiti has about 560 people on each square mile (*212 on each square kilometer*) of its land. The Dominican Republic has about 285 people on each square mile (*110 on each square kilometer*).

■ One out of five people in Haiti can read and write. Two out of three people in the Dominican Republic can read and write.

■ The average income of a Haitian is less than half of the average income of a Dominican.

■ Haiti has one doctor for every 11,000 people. The Dominican Republic has one doctor for every 1,800 people.

■ People born in Haiti have a life expectancy of 51 years. (The life expectancy of a nation is the average number of years that its people live.) People born in the Dominican Republic have a life expectancy of 60 years.

These clues tell you that people in the Dominican Republic generally live better than their neighbors in Haiti. Dominicans are less crowded, have more industry, are better educated, earn more money, have better health care, and live longer.

What the clues *don't* tell you is the reason for these differences. Why is one nation on the island of Hispaniola more fortunate than the other? The answer to this question is closely linked to the island's environment.

Physical Geography

Hispaniola is the second largest island in the Caribbean. Only Cuba is bigger. Hispaniola, like all of the Caribbean islands, is within the tropics.

The eastern two thirds of Hispaniola is occupied by the Dominican Republic. It is almost twice as large as Haiti, which occupies the western part of the island. Haiti is about the same size as the state of Maryland.

You read earlier that the Caribbean islands are the tops of mountains that rise from the floor of the sea. Hispaniola itself has four rugged mountain ranges that cut across the island from east to west.

While the entire island is mountainous, Haiti is much more mountainous than the Dominican Republic. About three quarters of Haiti's surface is covered by steep mountains. *Haiti*, the name given to the country by the Indians who once lived there, means "The Land of the Mountains."

Haiti and the Dominican Republic share an island smaller than South Carolina but with nearly three times the population. This crowded street is in Port-au-Prince, Haiti's capital.

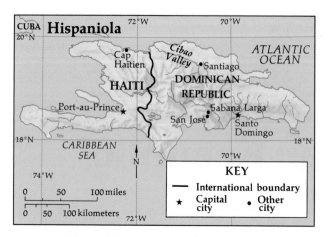

Hispaniola is the second largest Caribbean island. It lies between Cuba (the largest island) and Puerto Rico.

Hispaniola's mountains affect the island in a number of ways. In the Caribbean, moist air from the Atlantic Ocean is blown inland by the **trade winds** (the name given to the prevailing winds of the tropics). In the days of sailing ships, traders relied on these winds to cross the oceans. North of the Equator, the trade winds generally travel from the northeast to the southwest. As they move across Hispaniola, they meet the mountains and give up their rain on most parts of the Dominican Republic.

By the time the trade winds reach the western part of the island, they have already lost most of their moisture. As a result, much of Haiti gets only 20 to 25 inches (*50 to 65 centimeters*) of rain a year. By contrast, rainfall in the Dominican Republic averages about 60 inches (*155 centimeters*) a year.

In cooler parts of the world, such as most of the U.S., the amount of rain that falls on Haiti would be enough for growing crops. But in a tropical area such as Haiti, water evaporates more quickly, before it can benefit the crops. Many farms need irrigation.

Resources of the Land

The mountains do more than cause rain to fall unequally on Haiti and the Dominican Republic. They also limit the amount of land that can be planted with crops. The soil on the mountains is thin and poor.

The best places for farming in Hispaniola are in the level valleys that lie between the mountain ranges. The most fertile region is the Cibao (see-BOW) valley in the Dominican Republic. In some places, the Cibao valley's rich topsoil is several feet thick.

Haiti is the most crowded nation in the Western Hemisphere. Its farmers must grow food anywhere they can, whether the land is suitable or not. To grow crops, Haitian farmers have cut down most of the country's trees, even on the mountainsides. The roots of trees help to hold loose topsoil in place. On mountainsides without trees, rain washes away the topsoil and carries it out to sea. As the mountains in Haiti become barer each year, the soil becomes less fertile.

In the Dominican Republic, land for farming is less scarce, so people don't have to cut down the trees. The trees trap rainfall, and prevent the soil from eroding. The water is released gradually into streams and rivers.

In times of little rainfall, there is still enough water in Dominican rivers to irrigate farmlands. But in Haiti, the rivers often run dry, and irrigation becomes impossible.

Other than its farmlands, Hispaniola has very few natural resources. There is no iron ore, coal, or oil. The Dominican Republic does have a rich deposit of bauxite in the southwest, and some gold and nickel.

SECTION REVIEW

1. Compare Haiti to the Dominican Republic. Which is larger in area? Which is more mountainous? Which gets more rain?

2. What are the trade winds? How did they get their name?

3. Give two reasons why the Dominican Republic is better for farming than Haiti.

4. Is either Haiti or the Dominican Republic well supplied with mineral resources for industry? Explain.

Human Geography

Hippolyte (ee-poh-LEET) and Jeanne Valmé (zhahn vahl-MEH) live in a village about 25 miles (*40 kilometers*) from Port-au-Prince (por-toh-PRINS), the capital of Haiti. Like most of Haiti's six million people, the Valmés live on a small farm.

Their house is about 5,000 feet (*1,500 meters*) up in the mountains. Hippolyte built the house himself, with cement blocks and a roof of corrugated iron. There are two rooms on the road level, with a porch in between. Because the land is so steep, a third room downstairs in the back of the house cannot be seen from the road.

If you drop anything outside the house, it's likely to roll down the mountainside. On this steep slope, the Valmés have about half an acre (*one fifth of a hectare*) of land. Hippolyte grows potatoes, cabbages, carrots, bananas, and beans. The weather is warm enough for him to plant three times a year — if he has money for seed.

The oldest son is Noileus (noy-leh-OOS), who is 14. He helps his father plant and weed. The land is very rocky, so Hippolyte and Noileus always have a job picking out stones. They use the stones to make walls at convenient places down the mountain.

The Valmés also have a cow and a goat.

Haiti is famous for painters who produce colorful scenes of daily life. Like other Caribbean nations, Haiti relies on tourism for much of its income. Many tourists buy paintings such as this farming scene.

Few Haitian farmers keep large animals like cows and horses because there is so little spare land for grazing. The Valmés' cow and goat have to wander around and find food wherever they can. Noileus gives the animals water that he carries in buckets from a stream about a mile down the mountain.

Noileus is the only one in the family who has gone to school. The school is about a mile away. There are 30 or 40 students, but none of them attends every day. Like Noileus they have to help their parents.

Every week or two, Jeanne walks to a market, which is at a crossroad about a half hour away. She carries a basket on her head with any food the family has to sell.

It takes her nearly all day to sell the food crops. She uses the money to buy soap, flour, matches, bread, and sometimes clothes. She may buy a shirt or cotton pants for her husband or son.

Noileus does not want to work as a small farmer like his father. He would like to learn to be a mechanic at a garage in Port-au-Prince.

Haitian Ways

A small number of Haitians speak pure French and wear the latest European styles in clothing. They work in the government, in business, or in professions such as the law and medicine. But most people in Haiti still follow many customs that have their roots in Africa.

One example is religion. Officially the great majority of Haitians are Roman Catholics, but many of them also follow beliefs and practices brought from West Africa.

As in West Africa, women play an important role in Haitian farming. Their most important task is to carry farm products to a market and sell them. The women can carry up to 50 pounds (*more than 20 kilograms*) in baskets that they balance on their heads.

Although every town and crossroad has a market, the women may walk as far as 30 miles (*50 kilometers*) to get a better price. The women of Haiti often spend half of each week on the road and in the marketplace.

In some parts of Haiti, farmers grow African grains like sorghum (SOR-guhm) and millet. Other African foods include vegetables like okra, plantain (cooking banana), and yams.

As you read earlier, the poor soil and dry climate of Haiti make farming difficult. Most farmers produce just about enough food to feed one family. Thousands of Haitians spend part of the year earning a living in Cuba or the Dominican Republic, where they work on sugar plantations.

Dominican Ways

The people of the Dominican Republic are strongly Hispanic in their outlook and culture. Roman Catholicism is the official religion of the country, and Spanish is the official language. The cathedral of Santo Domingo (SAHN-toh doh-MEEN-goh), the nation's capital city, is the oldest in the Western Hemisphere, and contains the tomb of Christopher Columbus.

In one way, however, Dominicans have been influenced by the United States. In most Latin American countries, the favorite sport is soccer. In the Dominican Republic, however, the favorite sport is baseball. Most small towns throughout the country have their own baseball team.

Let's look now at everyday life in a village in the Dominican Republic. Amanda Castillo (ah-MAHN-duh kahs-TEE-yoh) lives in the village of Sabana Larga (sah-BAH-nuh LAR-guh). She is 15 and lives with her widowed mother Argentina (ar-hehn-TEE-nuh) and two brothers. Her uncle lives next door. He farms an acre (*about one third of a hectare*) of land that he and Argentina own. The two families share the food he grows.

Most Dominican farmers live in a home made of palm branches with a roof of palm leaves. Usually it has two rooms, with floors

The Roman Catholic religion is a strong cultural influence in the Dominican Republic. Many village children attend Catholic schools, as can be seen by the school uniforms that these young Dominicans are wearing.

of earth. But Amanda's home is better than most. It is made of cement blocks covered with plaster and has cement floors. There are three rooms. Like most village houses, there is a separate building, like a shed, for the kitchen. The family usually eats in the kitchen, so the three rooms are for sleeping.

Amanda starts her day by making breakfast — usually bread and *plátano* (PLAH-tuh-noh; plantain), with strong, sugary coffee.

After breakfast Amanda straightens up the house and gets ready for school. Classes start at eight and end at noon. Then it's time for lunch — usually a meal of beans, rice, and some coffee. Sometimes the Castillos have *sancocho* (sahn-KOH-choh), a meat and vegetable stew.

In the afternoon, Amanda helps her mother wash and iron clothes. They have electric lights and the iron is electric, but there are no other appliances. They don't have running water. They pay a girl to bring fresh water from a tap down the street. She carries it on her head.

Amanda will graduate from the village school in a year. After that she hopes to go to the high school in San José (sahn hoh-SEH), a small town about five miles (*eight kilometers*) away. The students in Sabana Larga

199

have to take a jeep there every day. As for her future, she would like to work in San José, but there are not many jobs available.

Two Different Cultures

In reading about the Valmés of Haiti and the Castillos of the Dominican Republic, you came across two different kinds of names. Hippolyte and Jeanne are names of French origin, and Port-au-Prince is made up of French words meaning "Prince's Port." By contrast, Amanda and Argentina are names of Spanish origin, and San José is Spanish for "Saint Joseph."

If you traveled to Haiti, you would find that practically all the people have French names. They speak a language called Creole (KREE-ohl) that is a mixture of French and some African tongues. In the Dominican Republic, as you read earlier, the great majority of people have Spanish names and speak Spanish.

Many of the cultural differences between Haitians and Dominicans can be traced to Hispaniola's early settlement. When Christopher Columbus discovered the island on his first voyage to the New World in 1492, he claimed it for the king of Spain. (*Hispaniola* means "Little Spain.") At that time, the island was home to perhaps 300,000 Indians.

Soon thousands of Spanish settlers began to arrive in Hispaniola. They conquered the Indians, and forced them to work on plantations. The Indians were unable to survive

---

The nations of Hispaniola have different historical backgrounds. As a result, Haiti and the Dominican Republic each have a distinct ethnic mix. In both nations, however, the Indians who once lived on the island have left hardly any trace in today's population.

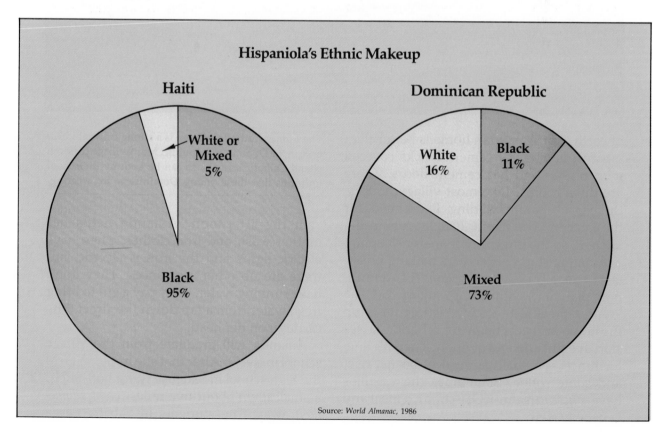

Source: *World Almanac*, 1986

the harsh treatment they received as slaves. Within 50 years, few Indians were left. The Spanish replaced them with slaves brought from Africa.

Spain lost interest in Hispaniola after gold was discovered in Mexico and Peru. Meanwhile French settlers began taking over the western part of the island. Following a war in Europe, Spain gave up western Hispaniola to France in 1697.

Over the years, the French imported one million slaves from Africa to work on their coffee and sugar plantations. Before long the slaves outnumbered the French 10 to one in the French colony. These slaves rebelled in 1791 and eventually drove the French out. In 1804 the former French colony became the free and independent nation of Haiti.

Later the Haitians conquered the Spanish colony in eastern Hispaniola. In 1844 the Dominicans rebelled against Haitian rule and won *their* independence. Ever since, the boundary line between Haiti and the Dominican Republic has remained unchanged.

SECTION REVIEW

1. When and how did Haiti become an independent nation?

2. Is the influence of African culture still important in Haiti today? Give two reasons for your answer.

3. What European country most influenced the culture of the Dominican Republic? Give two examples of this influence.

4. Name two things that are similar in the everyday lives of Noileus Valmé and Amanda Castillo.

Economic Geography

Today there are few plantations in Haiti. The great majority of farmers work small plots of land. As a result, much of Haiti's farm produce is eaten by the families that grow it. Some is sold in village markets to be

eaten by other people in the area. Only a small amount finds its way to more distant places, such as Port-au-Prince or foreign countries.

Coffee is Haiti's chief **cash crop** (a crop grown for sale). When Haiti was a French colony, coffee was grown on plantations. During the Haitians' struggle for independence, the plantations were destroyed and coffee seeds were scattered by the winds. The seeds took root on the mountainsides and grew wild. Most Haitian coffee today comes from this natural growth.

Haiti's other cash crops include sugar, which is grown in irrigated lowland areas, and sisal (SIE-suhl), a plant used to make rope. Both of these crops are raised mainly on plantations owned by U.S. companies.

In the Dominican Republic, there is more available farmland, and plantations are more numerous. Plows and tractors are used on these large farms. Crops are moved to markets by trucks and railroad cars. Sugar is the leading cash crop, making up more than half of the country's exports.

Cacao and coffee are next in importance. Unlike sugarcane, they are raised on small plots by farmers who use simple tools and methods. Cacao trees produce beans from which chocolate is made (see the illustration on page 437).

Help from Industry

In Haiti, Noileus Valmé hopes to leave the farm to work as a mechanic in the city. In the Dominican Republic, Amanda Castillo would like to leave her rural village — if she could find a job in the nearest town.

Their futures may depend on changes now taking place in their two countries. Both Haiti and the Dominican Republic see their best hope in developing more industry.

Unlike Brazil, or even Mexico, Hispaniola has no large tracts of undeveloped land that could be opened to farming. The land already farmed *could* be made more productive. But to do this, small plots worked by traditional methods would have to be changed into large, modern farms and plantations. New jobs and homes might have to be found for the small farmers and their families.

Both Haiti and the Dominican Republic have tried to provide new jobs by attracting industry from other, more developed countries. Therefore, foreign companies that build factories in certain parts of the Dominican Republic do not have to pay taxes for a number of years. These areas are known as "free zones." One of the free zones is in Santiago (sahn-tee-AH-goh), the Dominican Republic's second largest city, which is located in the Cibao valley.

For example, a U.S. tobacco company ships tobacco leaves from the U.S. to a big shed it has built in Santiago. There 300 women sort the leaves according to size and color. The women receive the minimum wage of 55 cents an hour. It is cheaper for the company to ship the tobacco to Santiago for sorting than to have the job done in the U.S., where wages are much higher.

Thousands of jobs have been created for Dominicans in the free zones. This is a big step for a nation in which 20 to 30 percent of the work force is unemployed.

Haiti is also attracting foreign companies by not taxing them for 10 to 20 years. In addition, about 40,000 Haitians are now employed in what are called "transformation industries." In a transformation industry, products are assembled from parts that are shipped from other countries. The finished products are then shipped back. For example, a U.S. company sends wire, plastic, and electronic parts to its factory in Haiti. There Haitian workers assemble the materials into switches that will be used in telephones in the United States.

Some Haitians are hopeful that transformation industries will help transform Haiti itself. Jobs provided by industry could ease the problem of too many people farming too little land. As you read earlier, Haiti has one of the poorest environments in Latin America. So if Haiti can break out of poverty, there is hope not only for the Dominican Republic but also for the region's other developing nations.

SECTION REVIEW

1. What is a cash crop? Which one is most important to Haiti? Which is most important to the Dominican Republic?

2. What are free zones? What are their purpose?

3. What are transformation industries? How might they play a vital role in Haiti's future?

4. How might the creation of new jobs in industry help increase farm production in Hispaniola?

YOUR LOCAL GEOGRAPHY

1. Haiti has a population density of 560 people per square mile. If you live in a city, this figure may not seem very high. For example, the District of Columbia has a population density of more than 11,000 people per square mile. But the figure for Haiti is an average for the whole country. The average for the whole of the United States is 64 people per square mile.

From *The World Almanac* or other reference source, find the population density of your state. It will be somewhere from the high figure of D.C. to less than one person per square mile in Alaska. How do you think your own community compares with your state figure? Would your community's density be more, less, or about the same? Give reasons for your answer.

2. Suppose Amanda Castillo of the small Dominican village is coming to live in your community for a while. Write a short letter telling her what differences to expect in environment (including climate) and in everyday life.

Reading a Topographic Map

The side view, or profile drawing, of Pico de Macaya (below) shows how Hispaniola might look to someone on a ship off Haiti's southwest coast. In Geography Skills 8 (page 54), you studied a profile drawing of Hawaii that used bands of color to show elevation. In the drawing of Pico de Macaya, however, numbered lines are drawn around the mountain at intervals to show how many feet above sea level the land rises in different places. The top view, above the profile drawing, shows how this mountain with numbered lines around it might look if you viewed it from above.

The same top view of Pico de Macaya is also shown on the topographic map on page 204. A **topographic map** is a large-scale map of an area, showing physical and cultural features in detail. A topographic map also uses contour lines to show the shape and elevation of the land. A **contour line** joins all points of a map that have the same elevation on Earth. Each line is numbered to show how high places are in feet above sea level.

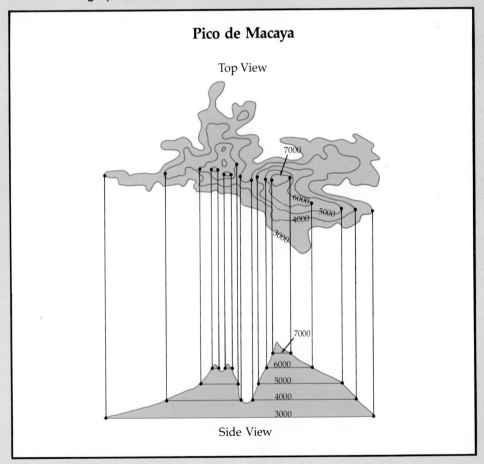

Pico de Macaya

Top View

Side View

(Turn page.)

Find Port-à-Piment on the map at about 18°15'N, 74°6'W. Now look toward Pico de Macaya. Notice that the first contour line connects places with an elevation of 1,000 feet. What elevation is represented by the second contour line? Using the map, decide whether each of the following statements is true or false. Write your answers on a separate sheet of paper.

1. Pico de Macaya is more than 7,000 feet above sea level.
2. The town of Marceline (located at about 18°21'N, 73°52'W) is between 2,000 and 3,000 feet above sea level.
3. Anse d'Hainault (18°28'N, 74°27'W) is at a higher elevation than Marceline.

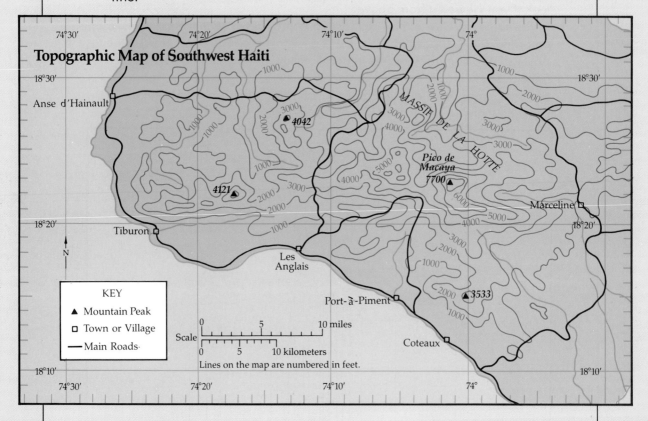

Besides showing elevation, contour lines also show relief (see Geography Skills 8 on page 55). Where contour lines are close together, they indicate high relief. Where contour lines are farther apart, there is low relief. On your answer sheet, tell whether each of the following statements is true or false.

4. South of Marceline is an area of very high relief.
5. To the northeast of Les Anglais, on the southern coast, the land rises fairly steeply — from sea level to over 4,000 feet in less than 10 miles.

CHAPTER REVIEW

A. Words To Remember

Three of the following terms are defined by the numbered phrases below. On another sheet of paper, write each word next to the number of its definition. Then write the two words that are *not* defined and give a definition for each.

cash crop   life expectancy   transformation industry
elevation   trade winds

1. Height above sea level.
2. The average number of years that people live.
3. Food grown for sale.

B. Check Your Reading

1. The mountains of Hispaniola cause Haiti and the Dominican Republic to receive an unequal amount of rain. How does this happen?
2. In what ways do the forests of the Dominican Republic help to keep its land suitable for farming?
3. The amount of rainfall that is adequate for farming in the U.S. is too little in Hispaniola. Why?
4. Why do few Haitian farmers keep cows or horses?
5. Both Haiti and the Dominican Republic want to attract foreign industry. What special benefit do they offer to companies that agree to build factories there?

C. Think It Over

1. What are some of the ways in which the influence of Spain and France can be seen in Hispaniola today?
2. Why do both Haiti and the Dominican Republic see more hope for their people in industry than in farming? Do you agree? Why, or why not?
3. In what ways would you expect other parts of the Caribbean to resemble Hispaniola? Explain your answer.

D. Things To Do

Imagine that you are a citizen of Haiti or the Dominican Republic who is trying to persuade a foreign company to build a factory in your country. Make a list of all the reasons you can think of for the company to build a factory. Consider the location, climate, work force, and government programs such as free zones. Then write a letter to the company's president, outlining all the benefits to be gained by building in Hispaniola.

Chapter 14

More and More People on the Land
The Region in Perspective

The family of Gilberto (geel-BEHR-toh) and Vitória Archango (vee-TOR-ee-uh ar-CHAHN-goh) live on a hill that overlooks the city of Rio de Janeiro, Brazil. In the distance, they can see all of Rio's wonders — high-rise buildings, wide avenues with fine shops, and miles of beautiful beaches. At night the city glows with lights.

The hillside community where the Archango family lives contrasts sharply with the central city. The Archangos' community is a *favela* (fah-VEH-luh; slum) inhabited by about 35,000 people. The homes in which they live are shacks that they have built themselves. These shacks are made of junk materials — scraps of wood and iron, cardboard, and flattened cans.

None of the shacks has running water. A fireplug at the bottom of the hill is the main source of water. Women spend several hours a day carrying water up the hill in cans. There are no sewer pipes, and no garbage collection. All of the shacks are crowded close together.

A mud path leads to the Archango family's home. Their shack is built mainly of boards with a roof of corrugated iron. There are two rooms. One serves as a kitchen. The other contains a double bed and two double bunks, and serves as both bedroom and living room. Eight people live here — Gilberto, Vitória, their four sons and daughters, a niece, and an uncle.

The family pays no rent. Like the other people who live in this favela, they are squatters. They have simply taken over an empty lot and built their shack on it.

Gilberto works at any kind of job he can find — construction laborer, car washer, or street sweeper. His wages are low, but three to four times as much as he earned working on a large farm. Vitória takes in washing and earns about 35 dollars a month.

Life in the favela is hard. Sickness and crime are common. But the Archangos will never go back to the rural area they left. That area, in the northeast, is often struck by long droughts (dry periods). When this happens, the land turns barren, and many people starve. More and more people are leaving the region. Thousands of them arrive every day in Rio, São Paulo, and other cities to look for work.

Gilberto and Vitória feel that they are much better off living in a favela than in the village they left. At least their children can go to school and learn a trade. The entire family can look forward to an occasional movie, a soccer game, or the excitement of the Carnival.

Providing enough housing to keep pace with rapid population growth is a problem facing many Latin American nations, as is seen in this sprawling Bolivian city.

The Archangos wait patiently for the city to build a new housing project so the people of the favela can have decent homes. They also hope that Brazil's growing industries will one day provide Gilberto with a regular job. Already many people have moved out of the favela into better neighborhoods. But as soon as they move out, other poor people from the country move in.

Rising Population

In Rio alone, about 650,000 people live in shack towns like the one you have just read about. And almost every other big city of Latin America has its share of slums. Most of these slums have spread rapidly since the end of World War II. Why has this happened?

In no other large part of the world is the number of people increasing so fast as in Latin America. In recent decades, the population has been growing at an average rate of almost three percent a year. This means that the number of people is doubling about every 25 years.

Dramatic population increases in Latin America are due mainly to greatly improved health services. **Infant mortality** (the percentage of children who die at birth) is much lower than in the past. Most people today live longer than people in the past.

The population of Latin America's cities is growing at an even faster rate — about five or six percent each year. The reason is that millions of poor farm families like the Archangos are moving to the cities. As you read earlier, Mexico City is expected to be the biggest city in the world by the end of this century, with nearly 30 million people.

The cities of Latin America cannot keep up with all the new people who arrive each year. There is not enough housing, electricity, and water to go around. Nor are there enough jobs. Until the government can provide more services, and industry can provide more jobs, the new arrivals live as best they can in slums.

SECTION REVIEW

1. What is a favela? State two of the problems faced by the inhabitants of favelas in Latin America.

2. What is infant mortality? Is infant mortality in Latin America higher or lower now than it was in the past? Why?

3. As you have read, the population of Latin American cities is increasing dramatically. State two needs that result from this increase.

4. Give two reasons why families living in rural areas of Latin America might decide to move to a favela.

Meeting the Demand

Each *month* there are about 700,000 additional people in Latin America — enough to fill a new city the size of Phoenix, Arizona. These people must be fed, clothed, and housed. At some later time, they also will have to be educated and employed.

What are the countries of Latin America doing to meet these growing needs? Here are some of the solutions they are trying.

Producing More Food

Most of Latin America is tropical or subtropical. Its year-round warmth gives it tremendous growing potential. Thus some areas that are now barren could be farmed. Other areas where one crop is grown each year could be made to yield two or three.

In some places, water is needed to make the most of the growing potential. As you read in Chapter 11, Mexico has built large irrigation projects in the northwest, an area that gets very little rainfall. Canals and pipes carry water from rivers to farmlands along the coast. As a result, this area now provides half of Mexico's food.

In Brazil the interior has been largely undeveloped because most people did not want to leave the coast. Those who did want to leave could find little transportation. Today the government is encouraging farmers to settle in the empty interior.

Some experts believe that Brazil's undeveloped cerrado could become one of the major food-producing regions of the world. They predict that it could grow enough soybeans, rice, corn, and wheat to feed 150 million people — more than the present population of Brazil.

As you have seen, many Latin American farmers like the Flores in Mexico (page 177) and the Valmés in Haiti (page 197) work very small plots. They use simple tools like the hoe and the machete, and know very little about the use of fertilizers, insecticides, and other modern techniques.

Many Latin American countries have begun programs to teach modern methods to farmers with small plots. It is hoped that these farmers will be able to produce a bigger surplus of crops for sale.

At the same time, there are some *large* farms that are run by inefficient methods. Most of these belong to old landowning families. Such estates have been broken up in Mexico and some other countries. But they still exist in many parts of the region.

Meanwhile, there is a limit to the land available for the increasing rural population. When parents die, their farm is often divided among the children. More and more people try to live on smaller and smaller plots. Sooner or later, the plots will not produce enough food.

Thousands of farm people do what the Archangos did — move to a city in search of jobs. Creating such jobs is the second way in which Latin America is trying to cope with its rising population.

Expanding Industry

New industries do more than offer jobs for factory workers in cities. They also create a need for more engineers, accountants, lawyers, teachers, clerks, storekeepers, barbers, bus drivers, and so on.

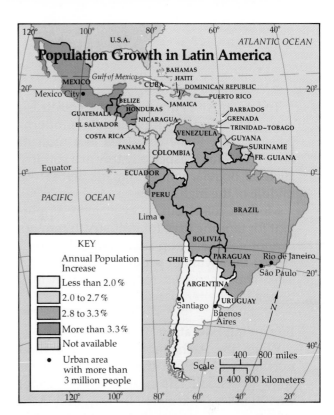

Yearly population growth in the U.S. is about one percent. Thus most nations in Latin America and the Caribbean are growing more than twice as fast as the U.S.

Some Latin American countries are already being transformed by the rapid development of industry. Countries rich in minerals can use these resources to launch their new industries. In the past, for example, Mexico exported most of its iron ore to other countries. Today it uses that ore in its own steel and manufacturing industries. Mexico is creating jobs at home instead of abroad.

Some countries that are poor in minerals are encouraging foreign companies to build factories there. These transformation industries bring only part of their work to the developing country. They are only half a step toward **industrialization** (building industries). But nations such as Haiti see this as being better than no step at all.

Improving Transportation

Brazil discovered its rich deposits of iron ore a long time ago. Yet for many years, hardly any of it was mined. There weren't enough roads or railroads to carry miners into and the ore out of the vast interior.

209

Today most Latin American countries, along with Brazil, are building modern, paved roads. Some countries face difficult obstacles. This is especially true of a country such as Bolivia, which spreads across the middle of the high Andes.

Modern roads can be even more important for farming than for industry. If farmers cannot get their crop to a market before it spoils, they will not produce extra food for sale. Thus increased food production may depend on better roads.

Improving Education

When the Archangos lived in the rural northeast of Brazil, their children never went to school. The nearest grade school was 20 miles (*30 kilometers*) away, and there was no transportation. In any case, the children were needed to help work on the farm.

Now, in the Rio favela, there is a grade school within walking distance. Classes are overcrowded, and the children have to catch up with the years they missed. But at least they have the chance for an education.

As you read in Chapter 11, Mexico has made a big effort to improve education in its rural areas. Other Latin American countries such as Bolivia, Cuba, and Venezuela have launched similar programs.

However, major problems remain. In most countries, only a small percentage of students goes on to secondary school. Higher education usually has been limited to the well-to-do. Brazil has tried to offer new opportunities at the higher level by creating technical and agricultural colleges. But in Brazil, as in most countries in the region, the most urgent task is to provide elementary schooling for all.

Latin American countries today are growing more food, providing more jobs, and building more links between their traditional and modern worlds than at any time in the past. With their booming populations, they have to do even better in the future.

SECTION REVIEW

1. Describe two ways in which the small farm plots of Latin America could be made more productive.

2. What is industrialization? How does it help to meet the needs of an expanding population?

3. New industries create more jobs than those in the factories themselves. Give two examples of such jobs, and explain why they are created.

4. Why is it more difficult to provide schooling in the rural areas of Latin America than in the cities?

YOUR LOCAL GEOGRAPHY

1. Various land barriers have made it hard for some Latin American countries to build new roads. For example, Bolivia has to contend with the Andes Mountains. Are there land or water forms in your community that make travel difficult? If so, what are they? Are there other kinds of natural barriers, such as forests?

At your local library, try to find pictures or descriptions of your community around the year 1900. See if you can find what means of transportation existed at that time. Is travel easier now than it was then? If so, how has transportation been improved? In your opinion, what remains to be done?

2. Look through a local telephone directory to find the types of schools that exist in your community. (Public schools will probably be listed with the departments of your local government.) Are there elementary and secondary schools? What about trade schools? If you decide to become an auto mechanic or a broadcaster, will you have to travel outside your community to get training? How far away is the nearest college or university? If you want to attend that school when you get older, what means of transportation will you have to use to get there?

Comparing and Contrasting Information

In the coastal areas of Brazil and in Haiti, many people live in crowded conditions. Despite this similarity, Brazil has a large land area and a large population, while Haiti has a much smaller land area and population. Noting similarities and differences between people and places is an important skill for understanding and thinking critically. When you note similarities between things, you *compare* them. When you point out differences, you *contrast* them.

Comparing and contrasting are also useful skills for understanding and thinking about graph information. What similarities and differences can you find between the two graphs in this lesson? Here is a list of true statements about Graph A (shown below) or Graph B (on page 212). For each statement, decide whether the *other* graph is "the same" or "different." Write your answer for each statement on a separate sheet of paper.

1. Graph A gives facts about population in Latin America.
2. Graph B is a bar graph.
3. Graph A shows the population totals of six countries.
4. Information in Graph B is for the year 1986.
5. Information is given in Graph A for the countries of Brazil, Mexico, Argentina, Guatemala, Haiti, and El Salvador.

Graph A **Estimated Population Totals in Selected Latin American Countries: 1986**

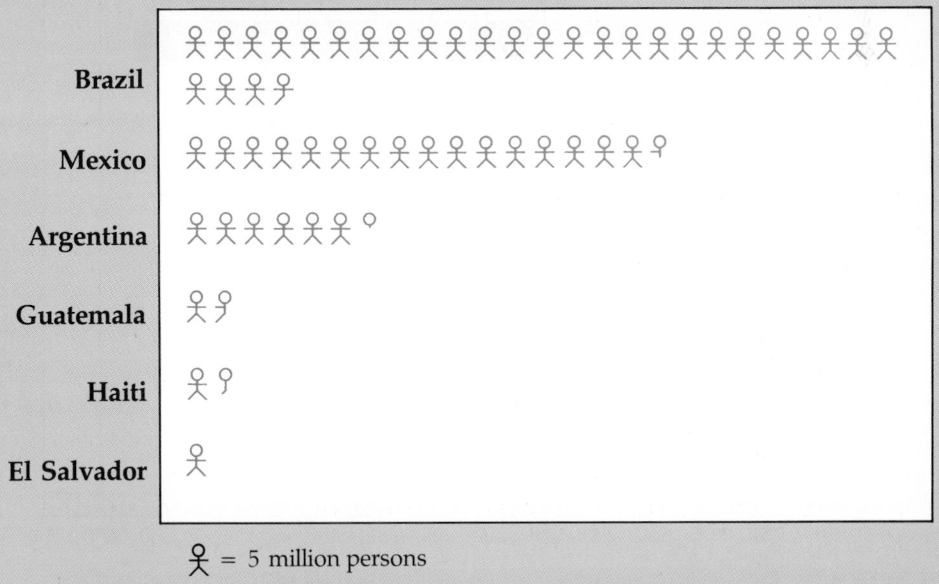

Brazil	𝑅𝑅𝑅𝑅𝑅𝑅𝑅𝑅𝑅𝑅𝑅𝑅𝑅𝑅𝑅𝑅𝑅𝑅𝑅𝑅𝑅𝑅𝑅𝑅𝑅𝑅 𝑅𝑅𝑅𝑅
Mexico	𝑅𝑅𝑅𝑅𝑅𝑅𝑅𝑅𝑅𝑅𝑅𝑅𝑅𝑅𝑅𝑅𝑅
Argentina	𝑅𝑅𝑅𝑅𝑅𝑅
Guatemala	𝑅
Haiti	𝑅
El Salvador	𝑅

𝑅 = 5 million persons

Source: *World Almanac,* 1986

(Turn page.)

It is sometimes useful to compare and contrast information *within* the same graph. Use Graph A or Graph B to answer the following questions on your sheet of paper.

6. In which country (or countries) are there: (a) fewer than 10 million persons? (b) more than 50 million persons?

7. In which country (or countries) is the average population density: (a) more than 400 persons per square mile? (b) less than 50 persons per square mile?

You can use both graphs together to compare and contrast the total population and population density of one or more countries. Answer each of the following questions on your sheet of paper.

8. Which Latin American country in 1986 had the smallest total population but the highest population density?

9. Which country had the largest total population but the second lowest population density?

10. Which country had almost the same population density as Brazil but a total population less than one quarter the size of Brazil's?

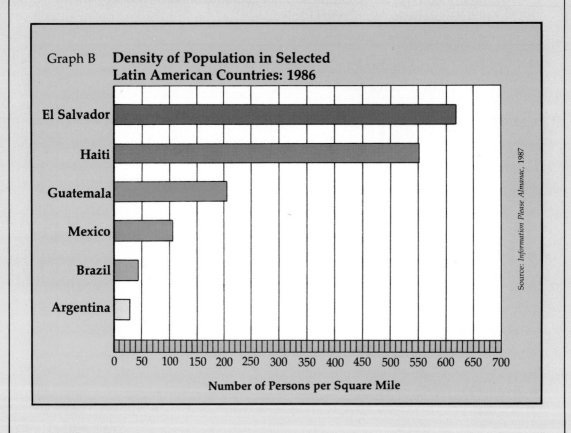

Graph B **Density of Population in Selected Latin American Countries: 1986**

Number of Persons per Square Mile

Source: *Information Please Almanac, 1987*

CHAPTER REVIEW

A. Words To Remember

In your own words, define each of the following terms.

drought   industrialization   irrigation project
favela   infant mortality

B. Check Your Reading

1. Why do more and more Brazilians move to the cities?

2. What is the main cause of rapid population growth in Latin America? What are some of the problems it causes?

3. Describe three ways in which food production in Latin America might be increased.

4. Why is modern transportation necessary for industrialization? How can it help increase food production?

5. What is the most urgent task that most Latin American countries face in improving the education of their people?

C. Think It Over

1. The population of Latin America's big cities is increasing dramatically. This chapter mentions major needs created by that increase — more food, housing, jobs, and schools. Name four other needs that must be met if city-dwellers are to have an adequate life.

2. Suppose a country has (a) a farming area, (b) a mining area, (c) an inland city with growing industry, and (d) a seaport. It wants to build new roads linking at least some of these areas. Choose two areas that you think would be most important to link with roads, and explain why.

3. Although a large population can cause many problems for a developing nation, it can also be an asset. List some ways in which this might hold true for Latin America.

D. Things To Do

Imagine that you are a Latin American farmer. You have three children, aged 6, 10, and 14, and you also provide for your own parents. In the last few years, several of your neighbors have left the village to find work in the city. Now you are trying to decide if this is the best thing for your family to do too. Most of your former neighbors live in the favela, but at least one of them has been able to move from the slum to an apartment.

Make two columns on a piece of paper. In one of them, list all the benefits that city life could offer your family. In the other column, list all the good things about your life now in the village. Then compare the columns. What would you do? Why?

UNIT REVIEW

A. Check Your Reading

1. On your answer sheet, write the letter of each description. After each letter, write the number of the city that best matches that description.

(a) Latin America's newest capital city.

(b) The hottest place in Mexico.

(c) It grew with the coffee trade and then turned to industry.

(d) French and Creole are spoken here.

(e) By the year 2000, this may be the world's largest city.

(1) Sabana Larga
(2) Brasília
(3) São Paulo
(4) Mexico City
(5) Mexicali
(6) Port-au-Prince
(7) Tegucigalpa

2. Fill in the blanks in the following paragraph by writing the missing term on your answer sheet.

Most of Latin America lies in the __(a)__ zone. It has a wide variety of environments, including the world's longest mountain chain, the __(b)__. In this region it is common for farmers to work on small plots of land, growing just enough to feed their families. This is known as __(c)__ farming. Brazil is now encouraging people to farm its __(d)__, the undeveloped lands of its interior. Mexico has used __(e)__ to make food crops grow in the desert area of its northwestern coast.

B. Think It Over

1. The following statements may be true of any one or more of Mexico, Brazil, Haiti, and the Dominican Republic. Write the initial letter of the nation(s) (M, B, H, D) to which you think each statement applies.

(a) It is located entirely in the Northern Hemisphere.

(b) Most of the people speak Spanish.

(c) Many people are fully or partly of African descent.

(d) The land is overcrowded with people.

(e) It has large areas of undeveloped land that could be used for farming.

2. Choose any one of the five statements in exercise 1 above and write its letter on your answer sheet. Then write a paragraph giving reasons to support your choice of nation(s).

Further Reading

A First Geography of the West Indies, by F.C. Evans. Cambridge, 1974. The land, history and culture of the Caribbean islands.

Mexico: Crucible of the Americas, by Lila Perl. Morrow, 1978. Surveys the geography, history, economy, and cultures of Mexico.

Brazil, by Wilbur Cross. Children's, 1984. Describes culture, geography, and government of the nation.

4
Western Europe

Chapter 15

Where the Modern World Began

It is Saturday morning, and you are shopping with your parents in a large supermarket. Your mother stops first at the dairy foods section to look for cheese. There is Edam cheese from the Netherlands, Danish blue cheese from Denmark, and Brie cheese from France. Finally she picks the Danish blue cheese, which she will mix into salads.

In the next aisle are various kinds of canned fish. Among them are sardines from Norway, anchovies from Portugal, and kippers from Iceland.

"Which would you like?" your father asks. You're not familiar with anchovies or kippers, so you choose the sardines.

"We're almost out of olive oil," your father says, and plucks a bottle from the shelf. A look at the label tells you that the oil comes from Italy.

"While we're here," your mother says, "we might as well get some olives too." She selects a jar of olives, which comes from Spain.

Before the supermarket visit is over, you have bought a number of other products. Among them are cookies from Britain and chocolate bars from Switzerland.

You and your parents carry the packages to the family car, which was made by a West German company. On your way home, your father fills the gasoline tank at a station that is owned by a Dutch and British company.

"What time is it?" your mother asks. You look at your watch and tell her it is noon. Your watch was made in Switzerland.

Clearly people in the United States buy a great many products that come from Western Europe. People in Western Europe also buy many products that come from our country. If you visited cities like London, Paris, or Amsterdam, you'd have no problem finding American-style fast foods. You would also see a good many U.S. cars in the streets. Our country and Western Europe are closely linked both by trade and by a network of communications, including TV, radio, and newspapers.

Overview of the Region

The term *Western Europe* today describes more than a geographical area. It describes a group of nations that are quite different from the Communist nations of Eastern Europe, which you will study in the next unit. The nations of Western Europe have democratic governments and close ties to the United States.

The region can be divided into five parts. In each of these five areas, there are similarities in the physical environment, in the ways people live, and in the ways they use their resources. Here are the five areas,

This modern bridge in Austria is part of a network of modern highways that crisscross Western Europe. The farming village below it seems to belong to another time.

listed roughly from north to south, with the major nations in each:

- Scandinavia (Norway, Sweden, Finland, Denmark, and Iceland);
- Britain and Ireland;
- Central Western Europe (France, Belgium, the Netherlands, Luxembourg, and West Germany);
- Alpine Europe (Austria and Switzerland);
- Mediterranean Europe (Spain, Portugal, Italy, and Greece).

Look at the Checklist of Nations on pages 582–583 and compare the areas and populations of countries in the region. How many are smaller in area than Rhode Island (1,214 square miles or *3,144 square kilometers*)? How many are larger in area than California (158,693 square miles or *411,015 square kilometers*)? How many have fewer than one million people? How many have more than 50 million people?

Altogether these nations cover an area that is about one third the size of the United States. Within this area live more than 350 million people, about one-and-a-half times as many as in the U.S. This makes Western Europe one of the most densely populated regions in the world.

Physical Geography

A look at the map opposite will show you that Europe is a big peninsula thrusting westward toward the Atlantic Ocean.

If you look at the map more closely, you will see several smaller peninsulas jutting out from the larger one into the surrounding waters. In the north is the Scandinavian peninsula, consisting of Norway and Sweden. Just south of this peninsula, and almost touching it, is a much smaller peninsula occupied by Denmark. In the south, three peninsulas extend from the main body of land. The largest of them, Iberia (ie-BEER-ee-uh), consists of Spain and Portugal. It lies between the Atlantic Ocean and the Mediterranean Sea. The other two southern peninsulas, Italy and Greece, are almost surrounded by the Mediterranean Sea.

Western Europe also has many islands that lie off its coasts. The biggest islands are Britain, Ireland, and Iceland. But there are many smaller islands. Norway has more than 10,000 islands, and Greece has 2,000.

In no other large part of the world do the sea and the land intermix as much as they do in Western Europe. The surrounding waters have long served the area as routes for settlement and trade in distant parts of the world. At the same time, these waters have favored the area with a climate that is generally mild. As you know, water cools and warms more slowly than land. The result is that oceans tend to warm nearby land areas in the winter and cool them in the summer.

The waters that lie off the shores of Western Europe are unusually warm. This is due to the effect of ocean currents that start in the tropical Caribbean Sea and gradually flow northeast toward Europe.

These currents, which are known as the North Atlantic Drift, even keep parts of northern Scandinavia fairly warm in the winter. For example, the Norwegian port of Narvik (NAR-veek) lies within the Arctic Circle. Yet because of the North Atlantic Drift, the port is ice-free all year round.

Most of Western Europe's mountain ranges run from west to east, and none are as high as the Rockies or the Andes. Thus there are few barriers to the winds that blow inland from the Atlantic Ocean. These winds bring warm, moist air from the ocean. Western Europe gets an average of 20 to 40 inches *(50 to 100 centimeters)* of rainfall a year — about the same as Nebraska and Missouri receive, and ample for growing crops.

Oceans and seas have a strong influence on the climate of Western Europe. Use the scale to mark off 300 miles on a piece of paper. Then, using the paper, try to find a place in the region that is 300 miles from a sea. Most places in Western Europe are much closer.

Western Europe

KEY
— International boundary
★ Capital city
⊙ Urban area with over two million people
● Other city

ICELAND
★ Reykjavik

ATLANTIC OCEAN

Arctic Circle

Narvik ●

SCANDINAVIAN PENINSULA

FINLAND

NORWAY

SWEDEN

Oslo ★

★ Helsinki

Stockholm ●

NORTH SEA

DENMARK
Copenhagen ●

Dublin ★
IRELAND
IRISH SEA
UNITED KINGDOM (BRITAIN)
Manchester ⊙
Birmingham ⊙
London ⊙

Hamburg ⊙

NETHERLANDS
Amsterdam ★
The Hague ★
Rotterdam ⊙
Ruhr R.
Rhine-Ruhr cities ⊙
West Berlin ⊙
EAST GERMANY

Brussels ★
BELGIUM
Bonn ●

Luxembourg ⊙
LUXEMBOURG

WEST GERMANY

Rhine R.

Paris ⊙

FRANCE

MASSIF CENTRAL

Mont Blanc ●

Rhône R.

SWITZERLAND
Bern ★
LIECHTENSTEIN
ALPS

Vienna ★
AUSTRIA

EASTERN EUROPE

BLACK SEA

Danube R.

Milan ⊙

SAN MARINO

MONACO

APENNINES

ITALY

PORTUGAL
PYRENEES
SPAIN
Madrid ⊕
ANDORRA
Barcelona ⊙
Corsica (France)
VATICAN CITY
Rome ⊕
Naples ⊙

Lisbon ★
IBERIAN PENINSULA
Sardinia (Italy)

TURKEY

GREECE
Athens ★

Sicily (Italy)

Crete (Greece)

NORTH AFRICA

MALTA
Valletta ★

MEDITERRANEAN SEA

CYPRUS
Nicosia ★

0   200   400 miles
0   300   600 kilometers

219

Nature has been generous to Western Europe in two other ways:

■ A large part of the region consists of a level or gently rolling plain. The plain has good soils, and is one of the richest farming areas in the world. (You will read more about this lowland plain later.)

■ Western Europe has many navigable rivers. (A river is said to be **navigable** when ships can use it.) Many rivers are linked by canals, and carry an enormous amount of traffic. The most important of these rivers, and also the longest, is the Rhine. Its 800-mile *(1,300-kilometer)* course passes through the industrial centers of West Germany and the Netherlands.

Within this broad picture of the region, there are also many differences — especially among the five areas listed on page 218. On the following pages, you will take a brief tour of each of these areas.

Scandinavia

If you lived in the far north of Norway, Sweden, or Finland, you might have a problem sleeping between May and July. At that time of the year, the Sun never sets north of the Arctic Circle, and daylight is continuous. This part of Scandinavia is called "The Land of the Midnight Sun." From November to January, you would have a different problem. At that time, the area is tilted away from the Sun. Then the Sun never rises, and darkness is continuous.

Scandinavia is the northernmost part of Europe and occupies about the same latitudes as Alaska. Thousands of years ago, during the Ice Age, all of Scandinavia was covered with glaciers. These slowly moving fields of ice acted like giant bulldozers. They scooped up millions of tons of rocks and earth and carried them southward. Along the coast of Norway, the glaciers carved deep U-shaped valleys. When the glaciers melted, the sea flooded these valleys, turning them into steep-sided inlets known as fjords (fi-YORDZ). Some fjords extend as much as 100 miles *(150 kilometers)* inland and form natural harbors for ships.

Norway, Sweden, and Finland are all heavily forested. Finland's forests cover three quarters of the land and are by far the country's most valuable natural resource.

Except in Denmark, only a very small part of the land of Scandinavian countries can be used to grow food. At these high latitudes, the growing season is short. In addition, a chain of steep mountains covers most of Norway and western Sweden.

One natural resource that all Scandinavian countries have in common is the sea. The waters around Scandinavia are filled with herring, cod, mackerel, and haddock. On the island of Iceland, which lies just south of the Arctic Circle, survival depends on fishing. About three quarters of the island is barren. More than 90 percent of the people live on the coasts of Iceland and earn their living by fishing and canning.

Sweden and Norway have important mineral resources. Sweden has large amounts of iron ore and some other metals. In the 1970's, scientists discovered large oil and natural gas fields in Norway's North Sea waters.

Norway and Sweden have one other valuable resource in abundance — water power. Streams rushing down from the mountains are used to generate electricity for industry.

Britain and Ireland

On a perfectly clear, sunny morning, many people in London, England, may carry an umbrella as they leave home. If you think these British are an odd bunch, you will soon learn better. By afternoon there may have been one or two brief rain showers — and the umbrella has come in handy.

As an island country, Britain is strongly affected by the waters that surround it. The weather is often cloudy, damp, or rainy. The rainfall usually comes in long, gentle drizzles, rather than heavy downpours. Yet there is plenty of it—an average of 41 inches *(105 centimeters)* a year.

Across the narrow Irish Sea to the west lies the island of Ireland, which is closer to the Atlantic and thus even wetter than Britain. Almost all of Ireland gets 40 to 80 inches *(100 to 200 centimeters)* of rain a year. The rain drenches the ground and makes the grass very green. For this reason, Ireland is called "The Emerald Isle." However, the damp air makes it difficult to grow food. Without periods of dryness, most grain crops will rot. So the land is used mainly to graze livestock.

As you will see in Chapter 16, Britain has large deposits of coal and iron that helped it become the world's first industrial nation. In recent years, huge reserves of natural gas and oil have been discovered in British coastal waters.

Ireland has large deposits of **peat** (decayed plants that can be used as fuel and fertilizer). However, Ireland has few other resources that can support industry.

Central Western Europe

You are on a train that is traveling from Amsterdam, the Netherlands, to Brussels, Belgium. You see endless fields, every bit of which is planted with food crops. There are scarcely any trees. Vegetables are growing beside the railroad tracks. The land seems as flat as a pool table and just as green.

Belgium and the Netherlands are called the "Low Countries" with good reason. Most of the land lies very near sea level. In the Netherlands, 40 percent of the land actually lies *below* sea level. This land was once covered by the sea, but the Dutch (as the people of the Netherlands are known) have worked hard and long to drain it. To drain an area, the Dutch first build a **dike** (earth barrier) around it. The water is then pumped into canals that flow into the North Sea.

Because the land is below sea level, pumping goes on constantly. In the past,

The sea is a key resource for much of Western Europe. Here, ships bustle through a Norwegian *fjord* (inlet).

windmills provided the power for this job. Today, however, the Dutch use diesel engines. The Dutch need this reclaimed land badly, since the Netherlands is one of the most densely populated countries in the world.

The level lowlands of Belgium and the Netherlands extend westward across France and eastward into Germany (see the map on page 219). In addition to good soils, this great central plain also contains large deposits of coal and iron ore. It is the most heavily industrialized area of Western Europe, and has many big cities.

France has not only lowlands but also highlands and mountains. The highest mountain in Western Europe, Mont Blanc (mawn BLAHNG), lies in the French Alps near the Italian border. Mont Blanc towers well above 15,000 feet (4,800 meters).

Alpine Europe

If you traveled through Switzerland and Austria, you would be in the heart of the Alps, the highest mountains in Western Europe. This mighty ice-capped range extends about 750 miles (1,200 kilometers) from west to east. Its mountains spill over not only into France but also into parts of Italy, West Germany, and Yugoslavia. However, the Alps are centered in Switzerland and Austria, covering 60 percent of Switzerland and 75 percent of Austria.

The mountains have made most of the land unsuitable for growing food crops. In Austria farming is carried out in valleys and on mountain slopes that face south and get plenty of sun. In Switzerland most farming is carried out on a large plateau that extends across the middle of the country between two mountain ranges.

Neither of these countries has a seacoast, but both are served by important rivers. The Rhine, which starts in the Alps, flows through Switzerland on its way to the North Sea. The Danube, which starts in southwest Germany, flows through Austria on its way to the Black Sea (see the map on page 219).

Perhaps the greatest natural resource of these countries is the scenery of the Alps. Tourism is a major industry in Austria and Switzerland, and adds hundreds of millions of dollars a year to their incomes.

Mediterranean Europe

During the month of January in Rome, Italy, the temperature in the afternoon will probably rise to 54°F (12°C). Yet Rome is on the same latitude as Des Moines, Iowa, where January temperatures are likely to remain below freezing all day. What makes the difference?

The countries of southern Europe — Italy, Greece, Spain, and Portugal, together with the island nations of Malta and Cyprus — are all warmed by the Mediterranean Sea. The Mediterranean is surrounded almost entirely by land. (Mediterranean is a Latin name that means "in the middle of land.")

During the summer, the Mediterranean stores heat from the sun, and its surface temperature may reach 80°F (27°C). Even during the winter, the Mediterranean remains warm and helps produce mild temperatures in southern Europe.

The Mediterranean countries have only two seasons — winter and summer. Winters are mild and rainy, while summers are hot and dry. In the summer, the Sun's direct rays fall north of the Equator, reaching the Tropic of Cancer. As a result, Mediterranean Europe gets the hot, dry climate of North Africa. In the winter, the Sun's direct rays fall south of the Equator. Then the Atlantic climate of northern Europe spreads southward, bringing rain and cooler temperatures to the Mediterranean.

Not all parts of Mediterranean Europe enjoy mild winters. Most of central Spain is on a high plateau that can be bitterly cold during the winter. Heavy snowfalls are frequent in the high mountains of Italy and Greece.

Except for Spain, the Mediterranean countries are poor in mineral resources. Above all, they lack fossil fuels — coal, oil, and natural gas. Later in this unit, you will read how this problem is being solved.

SECTION REVIEW

1. Is Western Europe more densely populated than the U.S., less densely populated, or about the same in population density?

2. Which of the five areas of Western Europe extends farthest south? Does this area include the Land of the Midnight Sun?

3. What effect does the Atlantic Ocean have on the climate of Western Europe in winter?

4. Which two of the five areas are the most mountainous? What major effects do the mountains have on life in these areas?

Human Geography

If you traveled in Switzerland, you would find that railroad schedules and official notices are printed in three languages. A majority of the Swiss people speak a form of German. In western Switzerland, however, the people speak French; in the southeast, they speak Italian. In addition, a smaller number of Swiss people speak a fourth language, Romansh, which is similar to Latin.

Switzerland is a small country, and the fact that it has four different languages says a great deal about Western Europe generally. This is a region where peoples belong to many different cultures and speak many different languages (see the map on this page). Yet these peoples have had to find ways of sharing their small part of Earth's land.

The mix of cultures in Western Europe developed over a long period of time. For example, people throughout the region once belonged to the Roman Catholic Church. Then more than 900 years ago, the Eastern Orthodox Church broke with Rome. Today most of the people in Greece and Cyprus are members of the Eastern Orthodox Church.

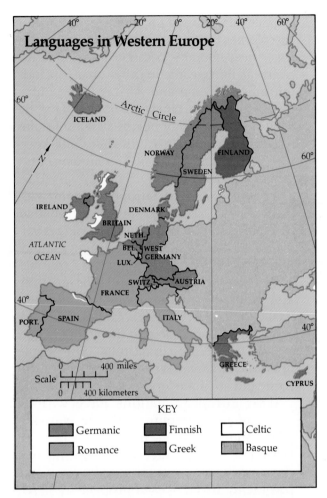

This map shows only major language *groups*. The number of individual languages in the region is far greater. For example, the Romance languages (those that come from Latin) include French, Spanish, Portuguese, and Italian. The Germanic group of languages includes English, German, Dutch, Danish, Swedish, Norwegian, and Icelandic.

In the 1500's, other groups broke away from the Roman Catholic Church to form various Protestant churches. Today, roughly speaking, most people in the southern half of the region are Roman Catholics, while most people in the northern half are Protestants.

Around 1500, Europeans began to make use of their surrounding waters to travel by ship to other parts of the world. Their main aim was to bring back resources (see page 225). At the same time, Europeans often

223

claimed other lands for themselves and set up colonies there. In areas such as the Americas, these colonies became homelands for overseas Europeans. In other areas, including much of Asia and Africa, fewer Europeans settled, and they did not displace the local peoples. Today nearly all colonies are independent, but many of them still keep the Europeans' languages, religions, and other cultural features.

In some parts of Europe, people with similar cultural backgrounds are divided between different countries. The biggest division is found among the German people. After World War II, Germany was split between the West and the Soviet **bloc** (a tightly knit group of nations). Today the people of East Germany live under a Communist government and are linked to the Soviet Union and Eastern Europe (see Unit 5). The people of West Germany have a democratic government and are linked to the West.

West European nations have various forms of government. Some are **constitutional monarchies** — that is, the head of state is a king or queen who has no real power. Others are **republics**, headed by a president. In all West European nations, however, government is in the hands of officials who are elected by the people.

In many countries, several key industries and services have been **nationalized** (taken over and run by the government). For example, railroads and telephone systems are government-run in nearly all West European countries. Most countries also have social security and health-care systems run by the government. However, most industries and services remain in private hands.

Growth and Change

A large majority of West Europeans live in cities. In Britain 85 percent of the people live in urban centers. The metropolitan area (city and suburbs) of London alone contains more than 10 percent of Britain's population.

The cities of Western Europe have changed in many ways in recent decades. The rapid growth of industry has drawn many people from farm areas, making the cities crowded. Some city people are moving to suburbs.

Rising standards of living have made it possible for more and more people to own automobiles. Today there is hardly a city in Western Europe that does not have a severe traffic problem. In many cities, you can still see churches and other buildings that are hundreds of years old. More and more, however, you will see modern high-rise buildings, large supermarkets, and fast-food restaurants.

Life is changing rapidly even in the rural areas of Western Europe. In some southern countries like Greece and Portugal, many farmers still raise food in traditional ways. They use the same simple equipment and methods that their ancestors did. But most farmers in Western Europe use modern machinery, chemical fertilizers, and insecticides to increase food production. TV sets, refrigerators, and washing machines have become as common in farmhouses as they are in city homes.

SECTION REVIEW

1. One of the two main groups of languages spoken in Western Europe is the Germanic. What is the other? Do people in Switzerland speak languages from both groups or only one?

2. Why does East Germany belong to a different cultural region from West Germany? What does West Germany's government have in common with the rest of Western Europe?

3. Give two examples of differences between governments in Western Europe. What do all West European governments have in common?

4. Name two ways in which the use of modern technology has changed life in Western Europe.

Economic Geography

If you wanted to visit the biggest cities in Western Europe, Rotterdam (ROT-ur-dam) in the Netherlands would be low on your list. It is not even the biggest city in the Netherlands. With about two thirds of a million people, it is far smaller than London, Paris, Rome, or Madrid. It is certainly not in the same league as New York City, the biggest city in the U.S. Since New York City is also the busiest port in the U.S., you would not expect the port of Rotterdam to be a match for New York. Yet Rotterdam handles more than twice as much cargo as New York.

Rotterdam is a busy port because of its location on the Rhine River near the North Sea. The Rhine flows through Switzerland, France, and West Germany as well as the Netherlands. These countries, like the others in the region, rely heavily on trade.

As you read earlier, European ships began bringing back resources from other parts of the world in the 1500's. For a long time, however, they brought back mainly luxury goods, such as silk and spices. The big change in trade came with the Industrial Revolution that began in the late 1700's.

The Industrial Revolution was based on two of West Europe's resources — coal for power, and iron for machinery and construction. Even today the biggest industrial areas in the region are those where coal and iron were first found.

The Industrial Revolution created a demand for worldwide resources. The machines in the new factories could produce more goods faster than in the past. Thus

Western Europe's lowlands are good for crop growing and dairy farming. In these densely populated areas, agriculture and industry are often found side by side.

more and more raw materials were needed. New inventions and discoveries found uses for other materials — such as oil to run engines and rubber to make tires. Often these resources were found only, or mainly, in other regions of the world. Many of these resources were brought in at low cost from European colonies.

Later in this unit you will read how the need to trade resources is bringing the nations of Western Europe closer together. The rest of this chapter will focus on the differences in resources between nations.

Scandinavia

Most Scandinavian countries have harsh environments. Large areas are cold and mountainous, with poor soils and few mineral resources. Yet the people of Scandinavia generally have found ways to live well.

In Norway, for example, only about four percent of the land is suitable for farming. So the Norwegians rely on the sea for much of their food. Norway's fishing industry is one of the largest in the world. Some Norwegian fishing fleets sail all over the world.

Norway, Sweden, and Finland make use of their large timber resources. They export

Much of Alpine Europe is too mountainous to farm, but crops are grown on the lower slopes of the Alps. Here, a Swiss woman prepares the soil for planting.

wood and wood products such as pulp and paper, furniture and matches.

A major problem for most Scandinavian countries is fuel for industry. The mountain streams of Norway and Sweden provide hydroelectric power, while Iceland taps the steam heat of its **geysers** (GIE-zurs; spouts of water heated by molten rock underground). However, only Norway has sources of oil — from under the North Sea.

Britain and Ireland

Large deposits of coal and iron helped to make Britain the leading nation in the Industrial Revolution. Today the iron and coal are being used up, but oil and natural gas from the North Sea are helping to fill the gap.

Britain is a crowded island with only a limited amount of good land for farming. Thus the nation must import more than half the food that its population needs. The Republic of Ireland has little industry, but raises large numbers of farm animals for both meat and dairy products.

Central Western Europe

One of the greatest industrial areas of the world is located in the Ruhr River Valley of West Germany. This area specializes in the products of heavy industries, including steel, chemicals, and automobiles.

The Ruhr Valley became an industrial center because it has both iron ore and the largest deposits of high-quality coal in all of Europe. It has remained important because the Ruhr River flows into the Rhine, which makes it easy for the area to ship its products to other ports of the region and the world.

France has less coal than West Germany and is not so heavily industrialized. However, France leads all other countries of Western Europe in agricultural production.

Mediterranean and Alpine Europe

Much of the land in these areas is mountainous or hilly, with soils that are too poor for growing crops. Many of these highlands are not wasted, however. They are used as pasture lands for goats and sheep.

In Mediterranean Europe, farmers have to cope with the hot, dry summer. They have found various ways to solve this problem. Where rivers and streams are available, the land is irrigated. In other areas, grain crops are planted in the fall. They soak up moisture from the winter rains and are ready for harvesting in early summer.

Water — or the lack of it — affects industry too. In northern Italy, streams rushing down from the Alps generate enough electricity to power industry. In most of Mediterranean Europe, however, there are no fossil fuels or water power. As a result, most people work at farming.

What's Ahead

In the following chapters, you will visit three countries in different parts of the region — Britain (Chapter 16), France (Chapter 17), and Italy (Chapter 18).

These three chapters reveal much of the diversity of the region. At the same time, they point to some of the interests and concerns that all West European countries have in common. Chapter 19 surveys the region to see how much unity can be found within its diversity.

SECTION REVIEW

1. Norway has one of the largest fishing industries in the world. Why does it rely on fishing more than on farming?

2. Why did big industrial centers develop in Britain and in the Ruhr Valley?

3. What part did overseas colonies play in the development of industry in many West European countries?

4. Why would the five different areas of Western Europe want to trade with one another?

YOUR LOCAL GEOGRAPHY

1. In this chapter, you read about many items which the United States imports from West European countries. Many such products can no doubt be found in your community. See how many you can find, and make a list as you go along. In your own home, you can start looking in the refrigerator and kitchen cabinets. Then check out your appliances, clocks, watches, jewelry, motor vehicles, etc. The catalog of a department store in your area will probably also list many imports from Western Europe.

Prepare to answer questions such as: Are most of the imports items that are not produced in the U.S.? Why might people in your community choose them? What products made in the U.S. do you think are exported to Western Europe?

2. The one major influence on the climate of Western Europe is its surrounding water. Is there one outstanding influence on the climate in your part of the U.S.? If so, what is it? If not, why not?

Understanding Arrow Symbols on a Map

Britain is farther north in latitude than North America's Great Lakes. Why, then, do lemon trees grow in southwestern Britain, just as they do in southern Florida? To find the answer, look at the ocean.

Ocean waters do not stay in one place. They flow continually in regular paths known as **currents**, which move over vast areas. The map below shows the ocean currents in the North Atlantic. The **arrow symbols** indicate motion, pointing out the directions in which the currents flow.

The waters of some ocean currents are warm; others are cold. Use the map and map key to answer the following questions on a separate sheet of paper.

1. Are the following currents warm or cold?   (a) Labrador Current;   (b) Caribbean Current;   (c) Gulf Stream;   (d) Canary Current;   (e) North Equatorial Current;   (f) East Greenland Current;   (g) North Atlantic Drift.

Remember that arrow symbols point out the directions in which the ocean's currents move. Now study the map to answer these questions.

2. Which current flows mostly westward across the Atlantic?
3. Which current moves in a northeast direction from the coast of Florida?
4. Which currents begin north of the Arctic Circle and flow south or southwest?

The temperature of an ocean current affects the climate of nearby land. Notice which land areas on the map are near warm or cold currents.

5. What effect do you think ocean currents have on the climate of: (a) the northwest coast of Africa?   (b) southern Florida?   (c) southern Britain?
6. Why can lemon trees grow in southwestern Britain, just as in Florida?

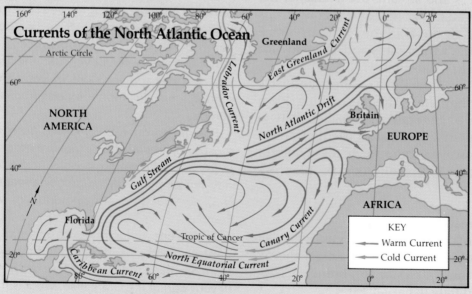

CHAPTER REVIEW

A. Words To Remember

From the following list, select the term that completes each of the sentences below. On a separate sheet of paper, write your answer next to the number of each sentence.

Alps	navigable	plain
fjord	North Atlantic Drift	Rhine River
Mont Blanc	peninsula	

1. _____ rivers can be used by ships.

2. The great mountain range that covers 60 percent of Switzerland and 75 percent of Austria is called the _____.

3. A _____ is a body of land that juts out into water.

4. The warm ocean currents that come from the tropical Caribbean Sea and flow northeast toward Europe are referred to as the _____.

5. A _____ is a steep-sided inlet along the coast of Norway.

B. Check Your Reading

1. What is Western Europe's longest river? Does it flow through Central Western Europe, Alpine Europe, or both?

2. Which of these areas is most suitable for farming: Belgium and the Netherlands, Alpine Europe, or Scandinavia? Why?

3. What two conditions have helped to make the Ruhr River Valley of West Germany one of the greatest industrial regions of the world?

4. What are nationalized industries? Name one that is common in Western Europe.

5. Why does Mediterranean Europe have less industry than other parts of the region?

C. Think It Over

1. Where would you find the rainiest parts of Western Europe — in the east, west, north, or south? Why?

2. Why does Central Western Europe contain the most densely populated parts of the region?

3. Would you expect to find more cultural differences if you crossed Western Europe from east to west or from north to south? Why?

D. Things To Do

Plan a two-week vacation to Western Europe. Pick the area you would most like to visit and make a day-by-day schedule. Decide how long you would spend in each country and what you would like to do or see there. Also decide what clothing you would need and what types of souvenirs you would want to bring back. Then draw a map showing the route you would follow from your departure to your arrival back home.

Chapter 16

Britain

At Home with the Sea

At the British port of London, a band played songs of the sea and fireboats sprayed water high into the air. On a summer day in 1979, 12 young British explorers were setting out on a difficult journey around the world. They were planning to circle the globe by way of the North and South poles. Such a journey had never been tried before. Their trip by steamship, with help from trucks, snowmobiles, and skis, would cover 52,000 miles (*84,000 kilometers*) and take about three years.

British people have been sailing from their island nation to explore the world for hundreds of years. Britain became a major sea power in the 1500's, after Europe's discovery of the New World and an ocean route to India. Until then the Mediterranean Sea had been the center of European trade. Any trade with Asia moved slowly along land routes from the eastern end of the Mediterranean.

With the new discoveries, the Atlantic Ocean became the main avenue for ships and trade. Britain, which had been a rather isolated island off the northwest tip of Europe, suddenly found itself at the very center of the new trade routes.

By the late 1500's, daring English sea captains were sailing all over the world in search of wealth and adventure. They were followed by settlers who created British colonies from the Americas to Australia. Eventually Britain ruled an empire that covered one fourth of the world's land surface.

Ships and colonies played an important part in the Industrial Revolution, which began in Britain in the 18th century. The new factories needed more and more raw materials from far-off places. Britain became the first nation to use resources from all around the world.

Two destructive World Wars in the 20th century greatly weakened Britain, as well as other European powers. Moreover, in the years following World War II, Britain gave up all but a few small outposts of its empire. Yet even today, the sea continues to shape the character of Britain, and the nation's fortunes still depend largely on ocean-going trade.

Physical Geography

The island of Britain, which includes England, Scotland, and Wales, has a long, jagged coastline that extends more than 2,500 miles (*4,000 kilometers*). The coastline is broken by many deep bays and inlets. Some of Britain's best harbors, like those in London, Liverpool, and Bristol, are actually located inland on rivers that open to the sea. Because the island is small (about the same size

Britain's surrounding waters have shaped its history and way of life. Today offshore oil is fueling Britain's economy. Here, an oil rig is towed from Scotland out to the North Sea.

as the state of Oregon), no part of Britain is more than 75 miles (*120 kilometers*) from open waters.

As you read earlier (page 220), the surrounding seas have influenced Britain's climate. Winters in Britain are quite mild, thanks to the warming influence of the ocean current known as the North Atlantic Drift. During the summers, the ocean breezes keep the island comfortably cool. Meanwhile the surrounding waters also produce a considerable amount of rainfall. The heaviest rains fall on the western side of the island, where they average about 60 inches (*150 centimeters*) a year. Southeastern England, which is the sunniest and driest part of the island, gets about 25 inches (*65 centimeters*) of rainfall a year.

The Shape of the Land

If you traveled in Britain, you would probably be surprised by the great variety in the country's landscape. During a journey of only a few hours, you may pass through bleak **moors** (low hills with poor soil where only tough plants like heather can grow), narrow mountain valleys covered with grass, and gently rolling fields planted with wheat. Yet even in these rural areas, industrial cities, mines, and mills are nearby.

Britain is made up of highland and lowland areas. Much of England is a lowland plain with its most fertile soils in the southeast. This area is actually an extension of the great level lowlands that cross the northern mainland of Europe (see page 219).

Britain's highland areas lie mainly in the west and north. In the west is the peninsula of Wales. Southern Wales is a plateau cut deeply with green valleys. It has large deposits of coal that have been mined for more than 150 years. Northern Wales is covered with steep mountains and is thinly settled.

A chain of high hills called the Pennines runs along the western side of northern England. The largest deposits of coal in En-

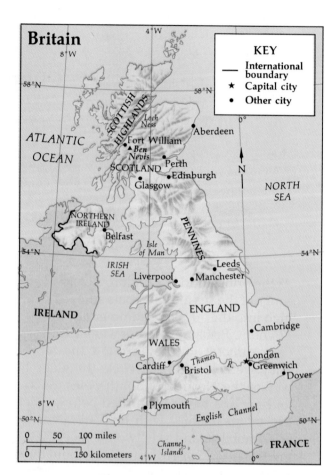

In latitude Britain extends as far north as the Alaska panhandle, but warm ocean currents give Britain a mild climate. Longitude around Earth is measured from the Prime Meridian at Greenwich (GRIN-ij).

gland are located on either side of the Pennines, and the largest iron-ore deposits lie nearby.

Britain's largest highland area covers the northern half of Scotland. The Scottish Highlands contain some of the most beautiful scenery in Britain, and also its highest peak — Ben Nevis, which is 4,406 feet (*1,343 meters*) high. The soils in this area are poor, however, and only a few people live here.

Changing Resources

For many years, Britain's chief natural resources were coal and iron ore. Today, however, many of the older coal fields in England and southern Wales have been largely used up. Until recently Britain had to import increasing amounts of both coal and oil.

During the 1960's, large natural gas fields were discovered off Britain's North Sea coast, and natural gas is now Britain's second most important mineral resource. In the 1970's, large oil deposits were also discovered off the North Sea coast. Britain hopes that before long its North Sea oil deposits will supply most of its petroleum needs.

Britain still has large amounts of iron ore, but most of it is of low quality. High-grade iron ore needed for making steel has to be imported. Britain once had plenty of copper, tin, and lead, but these have been largely used up. Except for coal, natural gas, oil, and low-grade iron ore, Britain today has very few natural resources. Most of the raw materials Britain needs for manufacturing are imported from other countries.

SECTION REVIEW

1. What mineral resources were discovered off Britain's seacoast in the 1960's and the 1970's?

2. Which part of Britain has the driest climate? Why?

3. Where does Britain get most of the raw materials it needs for manufacturing?

4. Where would you expect to find most crop farming — in England, Wales, or Scotland? Why?

In the west and the north of Britain, wild hills sweep up from the densely populated lowlands. The hilly area shown below is in Wales, in the west. The ancient castle is one of hundreds that still remain all over the country. These ruins are 600 years old.

Human Geography

Imagine that you are taking a bus tour of Britain. In England your guide greets you at the start of the day with a cheery, "Good morning!" When you get to Wales, your new guide also speaks English but first greets you by saying, "Bore da!" (BAW-ruh DAH). The meaning is the same (good morning), but the language is Welsh, which bears little resemblance to English. When you arrive in the Scottish Highlands, a new guide greets you by saying, "Maduin mhath!" (MAH-tin VAH). Again the meaning is the same, but this time the language is Scottish Gaelic (GAY-lik).

A common mistake that many visitors make when they visit Britain is to call everyone who lives there English. The people of Scotland or Wales have traditions of their own, and like to be called Scots or Welsh.

How did these ethnic differences come about? Ancient Britain was a favorite target of invaders from the mainland of Europe. The invaders were attracted to the island by its mild climate, its fertile soils in the southeast, and its trees for huts and firewood.

Among the earlier invaders were the Celts (keltz), who arrived a few hundred years B.C. (Before Christ) and conquered both Britain and Ireland. Later invaders included Romans, Anglo-Saxons from Germany, Vikings from Scandinavia, and finally Normans from France.

As the newcomers seized the lowlands, the Celtic tribes retreated to the distant highland areas of Britain. There they kept their own customs and languages. Their strongholds were northern Scotland, Wales, and the southwestern peninsula of England.

Varied Traditions

Today the people of these areas are proud of their special cultural traditions. About a quarter of the people of Wales speak Welsh, as well as English. The Welsh language

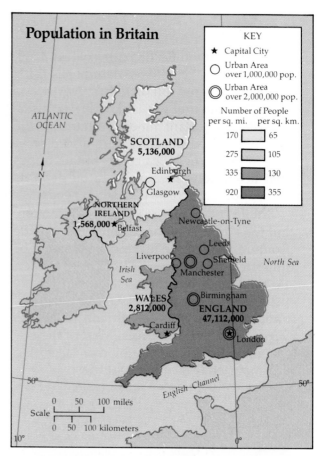

Population densities around London and Britain's other leading industrial cities are among the highest in the world. But there are some parts of Britain that remain thinly populated. This is especially true of the rugged highlands of Wales and northern Scotland.

comes from the ancient Celtic language. Some newspapers are printed partly or entirely in Welsh. Radio and TV programs are broadcast in both Welsh and English.

In the Scottish Highlands, about 75,000 people speak another Celtic language, Gaelic, in addition to English. In a few remote areas, the people speak only Gaelic.

Scottish Gaelic is similar to Irish Gaelic, the official language of the Republic of Ireland. At one time, all of Ireland was ruled by Britain, but most of it broke away under its own government after World War I. While the people of Britain are mainly Protestants, the Irish are mainly Catholics.

However, Northern Ireland has a large Protestant population, and remains a part of

Britain. (The entire nation is officially known as the United Kingdom of Great Britain and Northern Ireland.) Northern Ireland also has a large Catholic population, and conflict between Catholics and Protestants has erupted in violence since the 1960's.

Life in the highland areas of Britain contrasts sharply with that in the lowland areas. In the following sections, you will see first how a family lives in the Scottish Highlands. Then you will see how another family lives in England.

In the Highlands

People who live in the western part of the Scottish Highlands are surrounded by craggy mountains; deep valleys; and long, narrow lakes. The soil is too poor for growing many crops. Most of the people are tenant farmers who raise cattle. Ian Henderson runs a trucking business, but he also keeps some cows. His son Alexander, 14, takes care of the cows as his main chore.

The family often goes to Highland gatherings where the people sing, dance, and play bagpipes. Alexander's father takes part in bagpipe contests, and Alexander himself is learning to play the pipes.

Alexander attends high school in Fort William, the town nearest to his home. Classes include Gaelic as well as English. Gaelic is easy for Alexander because many of the people where he lives speak it.

After school Alexander often plays soccer with his friends — if it's a nice day. Parts of the western Highlands receive up to 150 inches (*nearly 400 centimeters*) of rain a year, and no one likes to get soaked.

In the Lowlands

The hometown of 13-year-old Ann Crutchley makes a sharp contrast with that of Alexander Henderson. Cambridge, England, lies 50 miles (*80 kilometers*) north of London in the flattest part of the nation. Some parts of the nearby countryside are below sea level. With their flat fields, canals, and dikes, these parts look very much like the Low Countries of mainland Europe.

Although Cambridge is not a large city, it is both busy and famous. Here you will find one of the oldest and best-known schools of higher education in the world — Cambridge University. Ann's father is a professor there.

Ann's mother is a marriage counselor, but also works afternoons in a shop that sells antiques. Women in Britain work in almost every kind of job and profession. In the 1970's, for the first time, a woman (Margaret Thatcher) became prime minister (head of the nation's government).

Ann attends a Roman Catholic school. Although the majority of British people are Protestants, more than two million are Catholics. Ann's classes include French, which is the traditional second language in British schools. France is Britain's closest neighbor on the mainland. Today, however, more and more schools offer a choice of second languages, including German, Spanish, and Russian.

The British and the World

From the time that Britain became a great sea power until the end of World War II, millions of British people emigrated to distant lands. They settled in the original 13 American colonies, Canada, Australia, New Zealand, South Africa, and other parts of the vast British empire.

Many of these countries still show the imprint of British culture. Place-names are one example.

Ann Crutchley's hometown, which gave its name to Cambridge, Massachusetts, is just one of countless British place-names that can be found along the eastern seaboard of the U.S. Other British names dot the globe from London, Ontario, in Canada to Perth in Australia.

British settlers also spread the English language. Though more people speak Chinese

than English, most of them are in China. English is the most widely spoken language around the world.

Other traditions spread by British settlers included the ideas of liberty, representative government, and the right to a fair trial. Wherever representative government took root, Britain's Parliament (legislature) served as the model. Even after the American colonies became independent, their new Constitution was influenced by British traditions of freedom and justice.

Since the breakup of the British empire after World War II, fewer British people have been leaving their island. Instead many people from British Commonwealth nations have been entering Britain. The Commonwealth is an association of nations that were once part of the old British empire.

The Commonwealth immigrants are chiefly blacks from the West Indies and Asians from India and Pakistan (PAK-is-tan). They come in hopes of finding jobs and better lives in Britain. These new groups now make up about four percent of the Brit-

Traditions such as representative government spread from Britain to much of the world. Here, the Houses of Parliament tower over the Thames River in London. The flying flag shows that Parliament is in session.

ish population and they have settled in the larger industrial cities.

There has been conflict between some of the old and new British groups. Leaders of the latter groups complain of discrimination, which is against the law. Meanwhile more and more of the "immigrants" belong to new generations born in Britain.

SECTION REVIEW

1. How does the population density of Britain compare with that of the United States?

2. Name one British language other than English. Where is it spoken?

3. You read about the everyday life of a Scottish boy and an English girl. How do the environments where they live differ?

4. Name two of the legacies that British settlers have given to the world.

Economic Geography

Let's suppose you are driving north from London on the first motorway (multilane highway) built in Britain. It runs 180 miles (*300 kilometers*) to the industrial city of Leeds. Traffic whizzes past fields of crops and cattle, and villages that date back hundreds of years.

If you look carefully at the crop fields, you will notice that they cover nearly all of the land area and that machines do most of the work. Since the end of World War II, British farming has become highly mechanized, and food production has risen steadily. Although less than four percent of the British people work on farms, they produce about half the food the nation consumes. Only about one third of the land in Britain is suitable for raising crops, so the country must still import large quantities of food.

In the lowland areas, most of the land is used for growing food crops. Britain's main crops are barley, oats, sugar beets, and wheat. In the highland areas, where the soils are poor, farmers use most of their land to raise sheep and other livestock. The sheep provide both wool and meat.

The sea also supplies Britain with much of its food. The country's fishing industry is one of the largest in the world.

The Rise of Industry

When you leave the motorway to enter downtown Leeds, you find yourself in a different world. Your route loops and winds along one-way streets where 19th-century warehouses are scattered among more modern buildings. Here you still see traces of the first great boom in industry that arose out of the Industrial Revolution.

For hundreds of years before that revolution, Britain had had a thriving wool industry. The wool came from much the same highland areas where sheep are farmed today, such as the Pennines of northern England. The wool was woven on hand looms, usually in village homes.

In the late 18th century, the steam engine was improved to the point where it could run machinery — including looms. Weaving was now done more and more in factories, where one engine could drive many looms.

The steam engines were powered by coal. As you saw earlier, there are large coal fields just to the east and west of the Pennines. Thus wool factories sprang up alongside the Pennines, close to their sources of raw material and power. One of the biggest of these manufacturing centers was Leeds.

Leeds is on the eastern side of the Pennines. A textile (cloth) industry also sprang up on the western side, in and around Manchester. Other industries and cities developed in Britain, especially around the coal fields of northern and central England. In the early 19th century, Britain became the world's leading industrial nation.

Industry Today

Britain is still a great industrial nation, although it has been surpassed in this century by the United States, the Soviet Union, Japan, and West Germany. Britain's leading industries are steel and products made from it. Among the products are ships, cars, trucks, and machinery.

Britain now relies much less on coal to power its industries. It has turned to other sources of energy, including nuclear power plants. A long strike of coal miners in 1984 had no effect on industry or production because much of the country's electricity is generated by nuclear stations. Some of Britain's coal is even exported.

Britain's economy depends on trade. It must import most of the raw materials it uses in manufacturing. To pay for these imports, it must sell things to other countries.

Among the things it sells are services such as banking, shipping, and insurance. For example, Lloyd's of London is an international grouping of insurance companies. It arranges insurance for risks that would be too costly for any one company alone — such as Japanese supertankers and American bridges. Lloyd's has also insured the Apollo moon buggy, racehorses, and movie stars' and basketball players' legs.

Britain must also sell its manufactured goods to other countries. But since the end of World War II, newer industrial nations have been winning a larger share of the world's markets. In most years, Britain buys more than it sells.

The British government has tried many ways to balance the country's imports and exports. One way has been to raise taxes sharply, so that families like the Hendersons and the Crutchleys cannot buy as much as they did in the past. In many British homes, the traditional Sunday dinner of roast beef is giving way to chicken, which is cheaper.

A Time of Change

Since World War II, Britain has gone through many changes. Nearly all of its former colonies have become independent nations, and Britain now has closer ties with its neighbors in Europe. (You will learn more about those ties in Chapter 19.) People from Africa and Asia, who once made up only a handful of Britain's population, now number more than two million.

High-rise buildings tower over the old skylines of London and other big cities. Busy highways slice through the British countryside. Along these highways, gasoline is sold by the liter. Temperatures are given on the radio in degrees Celsius. Britain, which gave such traditional measures as feet and inches to the world, has now switched to the metric system.

Britain is no longer the heart of the world's largest empire but a small island once again. Yet it is still a major industrial and trading nation, with people who have always known how to adapt to change.

1. Why must the British import food?

2. Britain's earliest industries included the manufacture of woolen goods. From what areas did the wool come? What great event in history changed the way such goods were manufactured?

3. Britain today has more difficulty exporting manufactured goods than in the past. Why is this a matter of concern to the British?

4. "Britain is a small island once again." What does this mean?

YOUR LOCAL GEOGRAPHY

1. The chapter describes how Britain's economy boomed in the 18th and 19th centuries and has declined since World War II. What is the trend in the economy of your community today? Make a table listing the following headings and kinds of information: Population (Is it increasing, decreasing, or remaining about the same?); Construction (Are new factories, office buildings, and homes being built?); Transportation (Have there been any recent changes in highways, railroads, airports, etc.?); Jobs Available (Have any businesses started up or moved into the community, or closed down or moved out?); Other (any other changes affecting the economy, such as the discovery of a resource, a natural disaster, etc.). A librarian can suggest sources of detailed information. After completing the information for each heading, write a brief report on the general economic state of your community.

2. In the past, Britain relied mainly on coal as its energy source. Today Britain also relies on natural gas, oil, and nuclear power plants.

Do you know what energy sources are in use in your community? You might first investigate how your own home and school are heated. Then check out nearby industries, apartment buildings, or stores. Also check the means of transportation (buses, trains, planes, boats, cars, etc.) in your area.

Reading Temperature Maps

Imagine you are in London, England, at the start of a tour of Western Europe. You want to know what clothes you should take with you. Will it be much colder in northern Scotland or in Scandinavia than it is in London? Will it be much warmer in France or West Germany? Would the answers be the same in both winter and summer?

You can find all of these answers on temperature maps. The two maps in this lesson show average temperatures at sea level throughout Europe. One map is for January, the coldest month; the other one is for July, the warmest month.

The wavy lines on these maps are called **isotherms**. These lines connect all points on a map that have the same average temperatures. Isotherms will

(Turn page.)

Temperatures in Western Europe

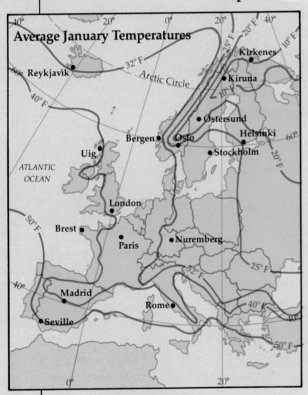

KEY											
Fahrenheit	80°	75°	70°	60°	55°	50°	40°	32°	25°	20°	10°
Celsius	27°	24°	21°	16°	13°	10°	4°	0°	−4°	−7°	−12°

remind you of contour lines (Geography Skills 22, page 203), which connect all points on a map with the same elevation.

The isotherms on these maps are numbered in Fahrenheit degrees. On the January map, find the isotherm for 40°F in southern England and the isotherm for 50°F beginning in the Atlantic Ocean. Areas between these two lines have average January temperatures between 40° and 50°F.

Notice the key, which matches the isotherm temperatures to Celsius degrees. Now use the January map. Answer the following questions on a separate sheet of paper.

1. What is the average January temperature in Fahrenheit degrees for each of these cities? (a) Madrid, Spain; (b) Oslo, Norway; (c) Seville, Spain.

2. The average January temperature for each of the following cities falls between what two isotherm temperatures shown on the map? (a) Helsinki, Finland; (b) Bergen, Norway; (c) London, England.

3. How many isotherms are on the map between Bergen, Norway, and Kiruna, Sweden? About what is the difference in average temperatures between the two cities?

4. Which city has the cooler January temperature on average — London, England, or Uig, Scotland?

5. Compare average January temperatures for the cities of Brest, Paris, and Nuremberg. Which city is warmest? Which is coolest?

Look now at the July map to answer the following questions.

6. Which city has the cooler July temperature on average — London, England, or Uig, Scotland?

7. What are the average July temperatures in Brest, Paris, and Nuremberg?

Now compare the two maps. You can see that the isotherms form different patterns in winter and summer. In winter, for example, isotherms tend to run from north to south. In summer they tend to run from east to west. What conclusions can you draw? Refer to both maps to answer the following questions.

8. In which month of the year, January or July, are temperatures in Europe determined mostly by latitude?

9. In which month, January or July, are temperatures in Europe most affected by distance from the Atlantic Ocean? Do temperatures near the ocean in this month tend to be warmer or cooler than those farther inland at the same latitudes?

CHAPTER REVIEW

A. Words To Remember

Three of the following terms are defined by the numbered phrases below. On another sheet of paper, write each term next to the number of its definition. Then write the two terms that are *not* defined and give a definition for each.

British Commonwealth   North Atlantic Drift   textile industry
moors   Parliament

1. Low hills with poor soil.
2. A group of factories that make cloth.
3. An association of nations that were once part of the British empire.

B. Check Your Reading

1. Name one group of recent Commonwealth immigrants to Britain.
2. In the lowland regions of Britain, most of the land is used for growing crops. How are the highlands used?
3. What is Britain's traditional energy resource? Why has Britain been developing other energy sources?
4. What happened in the late 18th century that started the growth of factories in Britain?
5. What recent change in Britain affected things such as weather announcements and gas pumps?

C. Think It Over

1. England has more farmland and a bigger population than either Scotland or Wales. How does the environment help explain these differences?
2. What effect have the surrounding waters had on Britain's history?
3. What kinds of things do you think tourists from other countries would visit Britain to see?

D. Things To Do

People's ways of life and beliefs are often preserved in folk songs (songs made up and handed down by ordinary people). At your library, look for records of folk songs from Wales, Scotland, Ireland, or England, or for books of folk song lyrics. What do these songs tell you about the people of that area?

Chapter 17

France

A Fortunate Land

The people of France have always been deeply attached to their own country, the areas in which they lived, and their homes. They were — and still are — very proud of being French.

France has long seacoasts, high mountains, forests, colorful apple orchards, and miles and miles of green farmlands. It has charming towns and villages with pleasant places to enjoy a picnic. The cities have large areas set aside for gardens and parks.

The French people are proud also of their history and arts. The country has magnificent cathedrals, castles, and palaces that are hundreds of years old. It has museums with the works of the world's greatest artists, including many who are French.

France has long been a magnet for visitors from abroad. More than six million people visit France each year, especially to see its beautiful capital city, Paris. But until the 1960's, few French people were interested in traveling to other countries.

This attitude has been changing rapidly in recent decades. The French people have become much more curious about the countries beyond their borders.

One reason for this change may be France's membership in the European Common Market (which you will read more about in Chapter 19). The Common Market is an economic grouping of Western European nations that was formed in 1957 to promote trade. Today people of each member nation can visit the other member nations without a passport.

For whatever reasons, the French are getting to know their neighbors better. In this chapter, you will get to know more about the French and their land.

Physical Geography

France is the largest nation in Western Europe, with an area more than twice the size of Britain. Compared to the United States, however, France is small. Mainland France and its island of Corsica (KOR-si-kuh) in the Mediterranean Sea together are not quite as big as Texas.

If you look at the map of France on page 244, you will see that the country is shaped like a six-sided box. Most French boundaries are natural ones. On three sides, France faces bodies of water — the English Channel on the north, the Atlantic Ocean on the west, and the Mediterranean on the south.

On two other sides, France's borders are formed by mountain chains. The Pyrenees (pir-uh-NEEZ) Mountains extend along France's boundary with Spain. The Alps separate France from Switzerland and Italy. On only one side, facing Belgium and Ger-

France's fine palaces are among its many tourist attractions. Built for a king in the 1500's, Chambord (shahm-BOR) castle rises out of the fertile lowlands of northern France.

France has natural borders on nearly all sides. If you check, you will find a sea, a bay, a channel, two mountain ranges, and a river. What section of the country's border does not have any natural barrier?

France is located on the same latitude as southern Canada, but France's climate is much more temperate. In Paris in February, the temperature at midday is likely to be about 42°F (*6°C*). In Marseille (mar-SAY) on the Mediterranean coast during January, the temperature might reach 53°F (*12°C*).

The Shape of the Land

The terrain of France is even more varied than the climate. Most of northern France is a fertile plain, which is part of the great level lowland that extends across Belgium, the Netherlands, Germany, and Poland. The rich soils of this area produce a great variety of crops.

The plain continues south along much of the Atlantic and Mediterranean coasts. These areas also contain many rich farmlands and vineyards.

As you read earlier, mountains lie along France's borders with Spain, Italy, and Switzerland. In addition, France has a large highland area all to itself — the *Massif Central* (mah-SEEF sahn-TRAHL), which covers much of south-central France. Its wild and rugged scenery attracts many tourists every year, but the soils are generally too poor and stony for extensive farming.

Rivers and Other Resources

Many rivers rush down from the Massif Central to join into a few bigger rivers that wind across France's lowlands. There some of them run close to other rivers that started in the Pyrenees and the Alps.

These rivers are navigable over long distances, and many are linked by canals. As a result, they form a vast network of highways for carrying goods. If you got on a barge at Dunkirk on the English Channel, you could travel all the way to the port of Marseille on the Mediterranean Sea by rivers and canals.

France's most important river is the Rhône (rohn). It enters France from Lake Geneva

many, does France lack a natural boundary. Even here the Rhine River separates a small part of France from Germany.

France has three different kinds of climates. In western France along the Atlantic coast, winters are mild, summers are comfortably cool, and rainfall is plentiful. This marine climate results from westerly winds blowing across the North Atlantic Drift.

As you travel away from the Atlantic coast, the influence of the North Atlantic Drift decreases. Eastern France has colder winters, warmer summers, and less rainfall.

In the south along the Mediterranean coast, the lowlands have hot, dry summers and mild winters with some rainfall. However, no part of France gets fewer than 20 inches (*50 centimeters*) of rain a year.

on the border with Switzerland, then turns south, and empties into the Mediterranean Sea. The valley formed by the Rhône is the main avenue for traffic between the Mediterranean and the northern part of France.

The rivers of France are a useful resource not only for transportation but also for power. Water power is plentiful in the mountains, the Massif Central, and the fast-flowing Rhône River. Many dams have been built in these areas to generate electricity for industry.

Power also comes from France's mineral resources. Large deposits of natural gas were discovered near the Pyrenees in 1951. France has some oil deposits, but not nearly enough to meet the nation's needs.

Perhaps the most valuable of France's many natural resources is the land. Fertile soils and a temperate climate have made France the leading producer of food in Europe.

SECTION REVIEW

1. How does France compare in area with the other nations of Western Europe? How does it compare with the U.S.?

2. France has natural borders on nearly all sides. What are some of these natural borders?

3. Where would you find the coldest winters in France: in the west, the east, or the south? Explain your answer.

4. In what ways are France's rivers useful?

Since the end of World War II, many French farm families have moved into the cities. These new apartment buildings are in the city of Limoges (lee-MOHZH).

Human Geography

In the northwest part of France, a large, rocky peninsula called Brittany juts out between the English Channel and the Atlantic Ocean. The people who live on this peninsula are known as Bretons. Do "Brittany" and "Bretons" sound like familiar names to you? They should. In the last chapter, you read about the people of Britain, who are often called Britons.

In fact, the people who live in Brittany are descendants of early Britons. As you read in Chapter 16, when Anglo-Saxons and other invaders stormed into Britain, many of the Celtic Britons fled to distant parts of the country.

Some left the country altogether. They crossed the English Channel and settled in the northwestern peninsula of France. Today many of the people in Brittany still speak a Celtic language that is similar to Welsh.

The French language stems from Latin, the language of the ancient Romans. The great Roman commander Julius Caesar conquered France in the first century B.C. The Romans called the country Gaul. The people of Gaul soon adopted the language and customs of the Romans. Over many years, they gradually changed the language.

Germanic tribes later conquered Gaul and added new words to the language. The most powerful of these tribes, the Franks, gave the country a new name — France. But the Romans had ruled the country for so long that their influence remained strong. Today France is still considered a Latin country, and French is a Romance language.

As in Brittany, however, there are parts of France where people speak other languages in addition to French. Along the western Pyrenees, for example, there are about 120,000 French Basques (bahsks). Very little is known about the origins of these people, or of the Spanish Basques who live on the other side of the Pyrenees. The language spoken by the Basques is unlike any other language in the world today. On the island of Corsica in the Mediterranean, the people speak a form of Italian as well as French.

Despite these differences, the people of France are proud of their historic traditions and their common language. A club called the Académie Française (frahn-SEHZ) tries to "protect" the French language from foreign influences. Its members include some of France's greatest writers and scholars.

City and Country

France has about 54 million people. About three quarters of them live in cities and towns of at least 2,000 persons. Nine million live in and around the capital city, Paris, which is also the center of France's transportation network, education system, and art world. Other cities in France are much smaller. The next largest cities of France, Marseille and Lyon (lee-AWN), have just over one million people each.

France is one of the less crowded nations of Western Europe. It has an average of 256 people for each square mile (*99 people for each square kilometer*) of land. Britain, as you have seen, has 596 people for each square mile of land (*230 people for each square kilometer*). The Netherlands, Belgium, and West Germany are even more crowded than Britain.

Outside Paris and the other large cities is the France of small villages and farms. Since the end of World War II, the number of people in this "other" France has been declining. Many farm families have moved to large industrial centers to find jobs that offer steady pay and regular vacations.

Benjamin Lesage (luh-SAHZH), 16, is a member of this "other" France — with one important difference. Benjamin and his family originally lived in Paris. But now their home is a 400-year-old stone farmhouse near Montignac (mawn-teen-YAHK) in southwestern France.

A leading art center, Paris attracts both tourists and artists from all parts of the world. Above, in an old city square, tourists watch two artists at work.

Benjamin prefers living on a farm because it is quieter and calmer than the city, and there are many things for him to do. Of course, much of his time is spent attending school in Montignac. To get there, he first rides almost two miles (*three kilometers*) on a motorbike from the family farm to a crossroads. There he catches the school bus. The bus ride into Montignac takes about an hour.

Benjamin is in the ninth grade at the junior high school. He studies English as his second language. His courses are basically the same as those taken by ninth-graders in Paris, Brittany, or any other part of the country. France has a **centralized** education system in which the basic courses for each grade are decided by the national government in Paris.

At home Benjamin and his older brother have plenty of chores. Benjamin feeds the chickens and gathers the eggs. The chickens in his part of the country are in great demand because they are fed only with grain. In most parts of France, fish meal is added to the feed; but grain alone makes the chickens plumper and tastier.

In the summer and autumn, Benjamin looks for wild mushrooms, which seem to spring up almost overnight. His mother uses a lot of mushrooms in her cooking. If there are more mushrooms than she needs, Benjamin sells them in the local market.

The Lesages eat well, but they do not have any of the elaborate dishes served in expensive city restaurants. The big meal of the day is lunch, which usually starts with soup. The main course may be beef, veal, or pork, served with a salad and cheese. Dessert is a

247

piece of fruit. Benjamin's favorite dish is *bif-teck frites* (BEEF-tehk FREET), which is steak served with french-fried potatoes.

Benjamin has only one ambition. After he's finished going to school, he wants to become a farmer. For him there is no better life.

In a Southern City

If you traveled to the southeastern corner of France along the Mediterranean coast, you would find a much different way of life. This area is known as the French Riviera (riv-ee-EHR-uh).

Here the summers are warm and sunny, and the sea looks as blue as the sky. The long, sandy beaches are lined with palm trees and other subtropical plants. There are colorful flowers everywhere. The Riviera is famous throughout the world as a resort area, and attracts many thousands of vacationers every year.

Not all the people you would meet here are visitors, however. Many French people live in towns along the Riviera all year round. For example, the town of Cannes (kahn) is home to 14-year-old Annick Ferré (ah-NEEK feh-REH), whose father is an accountant. Annick lives in a six-room house that is painted yellow. From her bedroom window, she can see the gentle waves of the Mediterranean Sea breaking on the shore.

From April through October, the Ferrés enjoy swimming in the Mediterranean. In the winter, they ski in the Alps, which are a short drive to the north. In the springtime, they can ski in the mountains in the morning, and then drive back to Cannes for an afternoon swim.

Even in summer there is snow on the Alps that form France's southeastern border. This ski resort is not far from the sunny beaches of the Mediterranean Sea.

Annick attends classes at the Lycée Van Gogh (lee-SEH vahn GOH). The school is named after the famous painter Vincent Van Gogh, who lived near Cannes for a time. Like Benjamin Lesage, Annick studies English as a second language. Unlike Benjamin, she and her class often swim or take sailing lessons when the weather is warm.

SECTION REVIEW

1. What city is the center of France's transportation network and education system? How does it compare in size with France's other cities?

2. Is France more densely populated than, less densely populated than, or about the same in population density as the other countries of Central Western Europe?

3. What is the purpose of the Académie Française?

4. Who do you think would have more opportunity to practice speaking English — Benjamin Lesage or Annick Ferré? Why?

Economic Geography

The peninsula of Brittany is a rugged land whose people have always depended heavily on the sea for a living. Fishing off the coast of Brittany was — and still is — dangerous. The waters are rough, and the offshore tides are powerful.

In 1961 a group of French engineers decided they might put those rough waters to good use. Water power from fast-running streams had long been used to generate electricity. Why not use the strong tides off Brittany to do the same thing? It was a new idea, but it was worth trying.

The engineers began building a dam near the mouth of the Rance (rahns) River, which flows into the English Channel. The dam was completed in four years. The idea worked. As the ocean tides rushed into the river mouth, they turned giant turbines (propellor-like blades) at the dam, producing an enormous amount of electrical energy.

Today the Rance Tidewater Dam supplies all of Brittany, and other areas as well, with electricity. Since then the Soviet Union, the United States, and Canada have also built tidewater dams. But it was the French who led the way.

Since the end of World War II, France has become one of the world's leaders in both industry and engineering. In addition to building the Rance Tidewater Dam, France has had a major role in developing supersonic jet airliners, high-speed electric railroads, and powerful antennas and satellites to transmit TV broadcasts all over Europe.

From Steel to Silk

Huge deposits of coal and iron ore are found in an area that runs from the northwest of West Germany across Belgium into northeastern France. Heavy industries have sprung up in this area in all three countries, with an emphasis on steel and other metal products. Factories in northeastern France produce heavy machinery, railroad equipment, and automobiles. There are also many linen and cotton mills.

However, France's greatest center of manufacturing lies in and around Paris. Although this area contains few mineral resources, Paris is the hub of a vast transportation network of motor roads, railroads, and rivers. Raw materials reach Paris easily, and much of its electric power comes from streams in the Massif Central and the Alps, and from the Rance Tidewater Dam.

The Paris area produces automobiles, airplanes, electronic equipment, and chemicals. The city is most famous, however, for its luxury industries, including fashions, furs, jewelry, cosmetics, and perfumes.

Tourism is an important industry in many parts of France, and adds many millions of dollars to the country's economy. The tourist industry is based on a wide variety of the country's resources. These include the landforms and climate, and also various cultural features, such as art, architecture, and food.

Riches of the Land

As you read earlier, France leads all of Western Europe in the production of food. Since the end of World War II, farming has become highly mechanized. French farmers now have about one million tractors, compared to 30,000 before the war.

French farms have also become much larger in recent years, because many farmers who had small, inefficient plots sold them to their neighbors and moved to the cities. With fewer farmers, but with more machines and larger farms, France produces more food than ever before.

Wheat is France's leading crop. It is grown on the best soils, which are found in the fertile plains of the north. Altogether about one third of the land is used for growing food crops. About one quarter of the land is used to graze meat and dairy animals. Much of the milk produced on dairy farms is used to make butter and cheese. French farmers earn more money from livestock products than they do from food crops.

Southern and Mediterranean France produce many of the same crops as Italy, Spain, and Greece. Vineyards for making wine thrive in the warmth of southern France. Citrus fruits, olives, and cork oaks grow well along the Mediterranean coast.

Because France is rich in natural resources, and produces an abundance of food and manufactured products, it carries on a lively trade with other nations. Only the United States, Japan, and West Germany export more goods. France's chief exports are machinery, motor vehicles, chemicals, luxury goods, and food. Its chief imports are oil, rubber, and raw materials — silk, wool, and cotton — for its textile industries.

In all the world, few nations have been as fortunate as France in the richness and variety of its resources.

SECTION REVIEW

1. France's greatest center of manufacturing lies in and around what major city? Why did manufacturing develop there?

2. Why does France produce more food than any other country in Western Europe?

3. Would you expect to find wheat fields alongside olive trees in France? Why, or why not?

4. Which of these French industries has to import most of its raw materials: iron and steel, textiles, or dairy products?

YOUR LOCAL GEOGRAPHY

1. On five of its six sides France has natural boundaries formed by either water or mountain chains. What are the boundaries of your community like? (Your local Chamber of Commerce should be able to provide you with a map of your town, county, city, or state.) Are any of the boundaries formed by natural features? If so, what are they? Do these natural boundaries still separate areas where people live and work today? Are there other kinds of boundaries? Do any of them follow human-made features such as railroads or highways? Do any of them follow straight lines that cut across both natural and human-made features? In general, does the shape of your community depend on the environment?

2. Nearly all major highways and railroads in France radiate outward from Paris like the spokes of a wheel. Paris has traditionally played a dominant role in the nation. Look in an atlas for a map showing major transportation routes in the state where you live. Is there one dominant center of these routes? If so, is the center the state capital? Or are there two or more centers of transportation? If you wanted to travel to different parts of your state by major highways, would you have to make any big detours? Overall, would you say that transportation in your state is highly centralized?

The Faces of Europe

The 19th-century painter Vincent Van Gogh was born in the Netherlands, where he painted the first picture below. Later he lived in the south of France, where he painted the second picture. What do the paintings tell you about the different environments of Northern and Mediterranean Europe?

In 1882 Van Gogh had a studio on the outskirts of The Hague, a city in the Netherlands. This drawing shows the view from his window.

Van Gogh painted this scene in 1890, near the mouth of the Rhône River in southern France. The trees are olive trees.

Ask Yourself . . .

1. How would you describe the difference in the shape of the land in each area?

2. In which area would you expect to find the: (a) most sunshine? (b) most rain? Why?

3. In which area do you think farming would be easier? Why?

4. Which area would you expect to find more densely populated today? Why?

Reading a Divided-Bar Graph

In 1975 the French government took a **census** (count of the population) to find out how men and women in the French labor force were employed. The results are shown in the **divided-bar graph** below.

Like a simple bar graph (page 63), a divided-bar graph compares total amounts of things by using bars of different lengths. However, a divided-bar graph also divides each bar into **segments** (smaller parts) to compare the smaller amounts that make up each total.

On this graph, for example, each bar represents the total number of workers in a particular grouping of jobs. The top bar groups together jobs from agriculture, hunting, and fishing. Besides farmers, what other kinds of workers are represented by this bar? Each bar is also divided into two segments to show how these job groupings were divided between male and female workers.

The marks between numbers along the horizontal axis can help you measure what each bar and its segments represent. Notice the vertical lines rising from the numbers. You can also use these as guides to figure amounts.

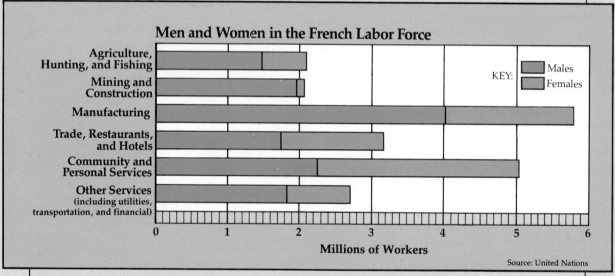

Men and Women in the French Labor Force

KEY:
Males
Females

Agriculture, Hunting, and Fishing
Mining and Construction
Manufacturing
Trade, Restaurants, and Hotels
Community and Personal Services
Other Services (including utilities, transportation, and financial)

Millions of Workers

Source: United Nations

Now answer each of the following questions on a separate sheet of paper.

1. According to the key, what group of workers is represented by the right segment of each bar?
2. In "Manufacturing," there were about how many: (a) male workers?   (b) female workers?   (c) workers altogether?
3. Which job category had the: (a) largest number of workers?   (b) smallest number of workers?
4. Which category had the largest number of: (a) male workers?   (b) female workers?

A. Words To Remember

In your own words, define each of the following terms.

census   marine climate   tidewater dam
centralized   Romance language

B. Check Your Reading

1. If you were in Paris and wanted to see mountains, would you travel north, west, or south?

2. Where in France is the Riviera located? What is the Riviera famous for?

3. Which of these is *not* a part of present-day France: Brittany, Gaul, or Corsica? What *does* it refer to?

4. How has the rural population of France changed since World War II? Why?

5. One third of France's land is used for growing food crops, but French farmers earn more money from another kind of product. What is it?

C. Think it Over

1. Why does France have a wider range of climates than other West European nations?

2. Do you think each of the following areas would be more densely or less densely populated than the average for all of France?   (a) the Riviera; (b) the area around Paris;   (c) the Massif Central.   Explain your answers.

3. Why do you think it is said that France's most valuable natural resource is land?

D. Things To Do

Make a minicrossword puzzle about France. Read through this chapter and look for terms having to do with the geography and culture of France (cities; crops; resources; rivers; etc.). Pick a term that is rather long, like *Académie Française*, to begin. Write the term on a piece of graph paper, with one letter to a space. Then build on the term, always using words that have something to do with France. For example, you could use Corsica, writing it downward so that the second "c" shares the "c" in Académie.

When you have connected about 15 terms, write clues for each term. Outline the areas of the graph paper you are using. Copy the outline on another piece of graph paper, this time with no words filled in. Give it to one of your classmates, together with the clues, and see if he or she can figure it out.

Chapter 18

Italy
A Place in the Sun

A team of Italian scientists stands in a church cellar, looking at one of its walls. The wall has been photographed under infrared light and probed by electronic scanners. The chemical makeup of the wall's surface has been analyzed by the most modern equipment. In all these tests, the scientists have been careful not to cause the slightest damage.

Now with the test results in, the scientists discuss the next step. They must decide the best ways to restore and preserve the surface of the wall. Why is this surface so important? On it the great Italian artist Leonardo da Vinci (leh-oh-NAR-doh duh VEEN-chee) painted a masterpiece called "The Last Supper." Leonardo created the painting 500 years ago. During that time, both natural and human-made conditions have eaten away at the painted wall.

The church that houses the painting is in Milan (mi-LAN), which is in one of the rainiest areas of Italy. The church cellar was usually damp. Molds and other tiny plants grew on the wall and damaged the paint.

In recent times, Milan has become Italy's biggest industrial city. Its air has been polluted by factory smoke and auto fumes, and these too have damaged the painting.

Millions of tourists visit Italy every year. Many of them come to see the country's art treasures — including "The Last Supper." The breath and body heat of this continual stream of visitors have also been harmful to the painting.

Work on restoring "The Last Supper" began in the late 1970's. Air conditioning is being installed to keep out dampness and pollution. The picture will look — and remain — as good as new.

Italy is a historic land with a long tradition of civilization and culture. In ancient times, Rome was the center of a great empire that spread around the Mediterranean Sea and beyond. Later Rome became the spiritual capital of the Western world — the center of the Roman Catholic Church. In the 14th and 15th centuries, a period of great energy and artistic creation began in Italy. This period became known as the Renaissance (REHN-uh-sahns), meaning "rebirth." Later the Renaissance spread to other parts of Europe.

Today Italians still have the greatest admiration for their past achievements. But Italy does not live in the past. It has become one of the most modern nations in Western Europe. Without advanced technology, Italian scientists would not have been able to restore a 15th-century masterpiece like "The Last Supper."

Milan, Italy's leading industrial city, also reflects the nation's religious and artistic heritage. Here, citizens and tourists stroll in front of Milan's cathedral.

255

Physical Geography

If you glance at the map of Italy on page 257, you will see that it looks like a boot that is about to kick a ball. The "ball" is the island of Sicily (SIS-i-lee), which lies close to the "toe" of the Italian peninsula. Italy also includes the large island of Sardinia (sar-DIN-ee-uh), which lies about 250 miles (*400 kilometers*) to the west of the peninsula.

Except for northern Italy, which is part of the European mainland, the country is surrounded by seas. The east coast of the peninsula faces the Adriatic (ay-dree-AT-ik) Sea. The western and southern coasts face the Mediterranean.

Rugged mountains and hills cover more than three fourths of the land. The country's northern border is formed by the Alps, which extend in a great semicircle from France on the west to Yugoslavia on the east. Although the Alps are very high, they have never cut Italy off from the rest of Europe. Since ancient times, the Mediterranean Sea has provided a highway for travel. Today the Alps are easier to cross than they were in ancient times. Railroad and motor vehicle tunnels have been cut through the mountains, and modern roads climb the passes.

Italy's other great chain of mountains, the Apennines (AP-uh-nienz), forms the "leg" inside the "boot." Starting in the northwest, they twist and wind along the center of Italy all the way to the "toe." Then they rise again on Sicily. Although these mountains generally are not very high, they are quite rugged and cover about two thirds of the peninsula.

The southern Apennines contain the three largest volcanoes in Europe, all active. Mt. Vesuvius (vuh-SOO-vee-uhs) towers over the city of Naples (NAY-puhlz); Mt. Etna (EHT-nuh) rises 11,000 feet (*3,350 meters*) in Sicily; and Stromboli (STROM-buh-lee) sits on an island between Sicily and the peninsula. In ancient Roman times, an eruption of Mt. Vesuvius completely buried two large cities, Pompeii (pom-PAY) and Herculaneum (hur-kew-LAY-nee-uhm).

As you learned in Unit 1, earthquakes often take place in areas with volcanoes, since they are both caused by pressures in Earth's crust. Thus the southern Apennines are also an area of frequent earthquakes. In November 1980, southern Italy was struck by its worst earthquake in 65 years. The quake killed at least 3,000 people and left more than 300,000 others homeless.

Italy's largest plain is the Po River Valley in the north. This valley lies between the Alps and the Apennines, and covers about 15 percent of the total land area of the country.

Because the mountains slope so sharply in Italy, many of the country's rivers flood during the rainy season. Occasionally the rivers overflow the dikes that have been built along their banks, and cause heavy damage.

A Mediterranean Climate

The word that is used most often to describe the climate of Italy is *sunny*. Most of Italy has long summers with almost cloudless skies and warm temperatures. The temperature in July averages about 75°F (*24°C*) throughout the country.

Winters in Italy generally are short and mild. As you learned in Chapter 15, the main reason for these mild winters is the Mediterranean Sea. This stores heat in the summer and remains warm all year round.

Because of Italy's physical length, however, the climate does vary slightly from north to south. Temperatures in the north are cooler than in the south, especially in the winters. High in the Alps, of course, the winters are quite cold. But these great mountains act as a weather shield for the rest of Italy. They prevent cold air masses from the north of Europe from entering the country. As a result, January temperatures even in a northern city like Milan will usually reach 40°F (*4°C*) by midday.

For tourists the climate of Italy is almost

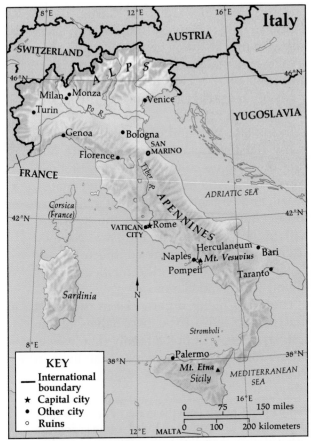

Italy

0   75   150 miles
0   100   200 kilometers

Italy's shape is often compared to a boot kicking a ball. The varied history of this area has given it two other nations — San Marino, Italian in culture, and Vatican City, center of the Roman Catholic Church.

Italy has very few mineral resources. It especially lacks coal and iron ore. Some deposits of natural gas and oil were discovered in the Po Valley after World War II, but they are not nearly large enough to meet Italy's growing energy needs.

The only minerals that Italy has in large enough quantities to export are mercury, sulfur, and marble. Italy supplies about one third of all the mercury used in the world.

Marble is a fine stone used in building and sculpture. It comes in a wide variety of patterns and colors. Its surface can be polished to a shine. Marble from quarries near Rome went into temples and palaces built 2,000 years ago. For example, the inside of a temple called the Pantheon glows with ancient marbles of various colors.

In ancient times, the Apennine Mountains were heavily covered with forests. Over the centuries, trees were cut down to burn as fuel or to clear the land for grazing animals. Today large parts of the Apennines are barren of trees and grass. Since there are no roots to absorb rainfall and hold the topsoil together, water rushes down the bare slopes during the rainy season and washes away the topsoil.

This erosion of the soil has added to the problems of farming. In recent years, the Italian government has started a number of programs to plant trees and restore the forests.

perfect. For farmers, however, the climate can create serious problems. The chief problem is the lack of rainfall during the summer months — the usual season for growing crops. As in other Mediterranean countries, most rainfall in Italy occurs during the winter months. The long, hot summers are generally dry, especially in the south.

Natural Resources

Italy's most valuable natural resource is the rich soil of the Po River Valley. Another important resource is the water that flows down from the Alps. This water is used both to irrigate the Po farmlands and to generate electricity for industry.

SECTION REVIEW

1. A mountain range and a sea form the northern and southern borders of Italy. What are their names, and which is where?

2. In what two ways does the climate of southern Italy differ from that of the north? What causes the difference?

3. In what two ways can pressures in Earth's crust cause damage in parts of Italy?

4. Name two major resources of the Po Valley.

Human Geography

For a long time, the mountains of Italy were a barrier to travel and kept people in different areas apart from one another. As a result, many areas developed their own local dialects. (A **dialect** is a regional language that differs from the official language of a nation.) In Italy today, there are hundreds of dialects. Quite often the people of one area of Italy cannot understand the dialect spoken by the people of another area.

In recent decades, the mountain barriers have been largely overcome. Modern highways and railroads have made traveling easy. Moreover, radio and television programs are broadcast to the entire nation in standard Italian. Thus more and more Italians can now speak the standard language as well as their local dialect.

Italy once was divided by more than its mountains. Until 1870 the country was separated into a number of small states. Some of these states were ruled by foreign powers. The Italians had to struggle to join together as a nation.

They were willing to struggle because they felt they were one people. They shared a culture that went back to ancient Rome and a heritage that included the work of Renaissance artists such as Leonardo da Vinci. Italians also shared many beliefs and customs. The beliefs that united them most were — and still are — the faith of the Roman Catholic Church and the importance of family life.

North and South

It is often said that Italy is really two countries — the north and the south. The north is modern and industrial with many large cities. Its Po River Valley also contains the country's richest farmlands. Some 40 percent of Italy's people live in the Po Valley.

The south is more traditional and rural. Most of its people live in small towns and farming villages. The only large city in southern Italy is Naples. In the south, earning a living is much more difficult. The area has few natural resources that can support industry. And as you have seen, farming is often a struggle because of the lack of rainfall and level land.

In the late 19th and early 20th centuries, many Italians looked for a better life outside their home country. Millions of them crossed the Atlantic to the United States, Brazil, and Argentina. The great majority of those emigrants came from the southern half of Italy. As Italian industry boomed after World War II, many southern Italians began to migrate to northern cities such as Milan, Turin, and Genoa.

There are many differences in the ways that people live in the two halves of the country. Let's visit first a family that lives in the south and then a family that lives in Rome.

Compare this graph with the graph of France's labor force on page 252. Although the two graphs are very different, they show that both nations have modern economies. In Italy as in France, far more people work in manufacturing and in services than in agriculture.

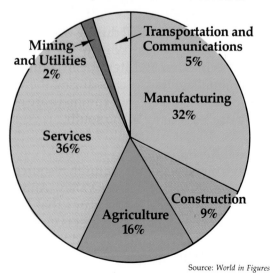

Italy at Work
Distribution of Jobs in the Italian Labor Force

Mining and Utilities 2%

Transportation and Communications 5%

Manufacturing 32%

Services 36%

Construction 9%

Agriculture 16%

Source: *World in Figures*

Life on a Southern Farm

The Lanzetta (lahn-ZEH-tuh) family rents a farm of eight acres (*about three hectares*) near the town of Cassino (kuh-SEE-noh). The farm is high in the Apennines, about halfway between Rome and Naples. Giuseppe (joo-SEH-peh) and Gabriella (gah-bree-EH-luh) Lanzetta have six children.

Anna, who is 14, attends a nearby school with her brother Vittorio, 13, and her sister Elena, 10. The three oldest boys have finished going to school and now work full-time on the farm. Anna, Vittorio, and Elena help on weekends and during the summer.

The Lanzettas live in a two-story stone house on the side of a hill. An outside staircase leads to the second story, which has three bedrooms and a kitchen. The ground floor is used to shelter farm animals.

In the summer, the entire family gets up at sunrise, about 5 A.M., and goes to bed at sunset, about 8 P.M. Anna feeds the smaller animals, which include ducks, chickens, geese, rabbits, and pigs. Some of the animals are butchered to provide the family with meat. The meat is salted to preserve it. Other animals are sold alive in the market at Cassino. The Lanzettas' landlord gets half of everything the family raises as his rent.

Near the house, grapevines are grown on wooden supports built over the paths. When the vines are full, they shade the paths. In late September, the grapes are picked and crushed. Then the juice from the grapes is placed in barrels where it ferments (goes through a natural chemical change) into wine.

Almost everything the family eats is raised on the farm. In the summers, breakfast usually consists of eggs, milk, and bread. A kind of "coffee" is also made from toasted wheat. In the winters, breakfast is likely to be beans baked with pork fat. Lunch consists of soup, rice, or spaghetti. Gabriella Lanzetta makes the spaghetti herself from wheat flour mixed with water. For

Old and new can be found side by side in Italy, as in most of Western Europe. Above, a modern highway runs past a small farm in the fertile Po River Valley.

dinner the family usually has a salad, vegetables, and either ham or fowl (chicken, duck, or goose).

Anna and Vittorio would like to live in a large city when they are old enough, and are preparing themselves by learning new trades. Vittorio is studying automobile mechanics in school. Anna works after school for a seamstress who is teaching her to become a dressmaker.

Life in the Capital City

Rome, the capital and largest city of Italy, is visited every year by millions of tourists.

They come to see its beautiful churches and palaces, and the ruins of the ancient Roman civilization. Rome was the center of that civilization, which flourished 2,000 years ago.

The most magnificent church in Rome is St. Peter's, which contains some of the greatest paintings of Italian Renaissance artists like Michelangelo (mie-kuhl-AN-juh-loh) and Raphael (RAF-iel). St. Peter's is a part of Vatican City, the world center of the Roman Catholic Church. Although Vatican City lies entirely within Rome, it is an independent state.

Rome has many buildings that are hundreds of years old, but it also has modern high-rise apartment houses. Most new apartment houses are on the outskirts of Rome. In one of them lives Claudia Simonetti (CLOWD-yuh see-moh-NEH-tee), who is 16.

Claudia is an only child who lives with her parents in a five-room apartment. The apartment has modern conveniences, such as an electric stove, a refrigerator, a dishwasher and a vacuum cleaner. Italians call these appliances "electric servants."

Claudia's father works for a large export company and speaks many languages. Claudia is also learning to speak several languages — English, French, and German. She goes to a special high school that prepares students to become travel agents. Tourism is one of Italy's most important businesses, and many travel agents are needed. Claudia's school is in the center of Rome. She goes there by subway.

Southern Italy has only one major city, Naples. In the background is Mt. Vesuvius, a still-active volcano.

Classes end at two o'clock. Then Claudia goes home for lunch — the big meal of the day. The first course is almost always *pasta*, the staple food of all Italy. The main ingredient is a dough made from wheat flour. Pasta comes in many different shapes and sizes — such as spaghetti, linguine (lin-GWEE-nee), ravioli (rav-ee-OH-lee), or lasagne (lah-ZAN-yuh). Pasta is generally served with a tomato sauce.

On a cold day, the family may have minestrone (min-uh-STROH-nee) — a mixed vegetable soup. The main dish usually consists of meat, but may be roast chicken or fish. There is always a salad — usually prepared with olive oil and vinegar — and a dessert.

After lunch Claudia plays tennis with her friends in a nearby park. In the winter, they often go skiing in the mountains.

SECTION REVIEW

1. What feature of the environment helped to keep Italians separate from one another in the past? Why is this feature much less of a barrier today?

2. In recent decades, what kind of migration took place *within* Italy? Why?

3. What Italian city was the center of a civilization 2,000 years ago? What role does this city play in Italy today?

4. Why is farming more difficult in southern Italy than in the north?

Economic Geography

The racing car driven by Mario Andretti roared across the finish line in first place. No other car was even close. The crowd at Monza (MOHN-zuh), Italy, cheered wildly. Andretti was born and raised in Italy, though his family moved to the United States when Mario was 15. He has won more great races than any other driver in the sport. Very often the cars he drives are designed and made in Italy.

In recent times, Italian racing cars have become famous throughout the world. So have many other products of Italian industry, such as automobiles, motor scooters, typewriters, and sewing machines. The skill and creative design of such products have helped Italy become a world leader in the manufacture of high-quality machines.

Since the end of World War II, Italian industry has grown enormously. Production of manufactured goods has gone up more than 400 percent. Most of Italy's industrial centers are in the northwest, within a triangle formed by the cities of Milan, Turin, and Genoa. Here water power from the Alps provides an abundance of electricity to power industry. An excellent system of modern motor roads and railroads speeds the delivery of raw materials.

The Milan-Turin-Genoa triangle forms one of the world's most advanced industrial areas. Milan is at the heart of Italy's largest urban area and leading manufacturing center. Its factories produce steel and a wide variety of machinery, electronic equipment, and textiles. Turin is the home of one of the biggest automobile companies in the world. It also produces locomotives, airplanes, and engines for ships. Genoa has large iron, steel, and chemical industries, but is even more important as Italy's leading seaport.

As you saw earlier, the south has much less industry than the north, and many southern Italians migrated to the north to find work. Most industry in the south is centered on the port city of Naples.

In recent years, the government of Italy has been helping the south to develop more industry. For example, a giant steel mill built at Taranto (tuh-RAHN-toh) now produces more than 10 million tons of steel a year.

Water and Food

Some years when there is less rain than usual, the Lanzettas get hardly any wheat from their fields. The area where they live is

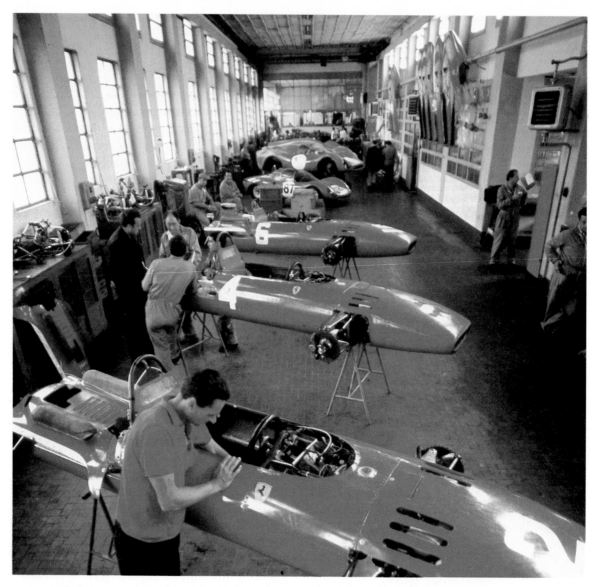

A skilled work force is one reason why Italy is famous for well-designed machinery. Above, racing cars are being assembled by hand at a small factory.

not good for crop farming. The three best farming areas in Italy can all rely on a regular supply of moisture.

As you saw earlier, by far the richest farming area is the Po Valley, which receives the country's heaviest rainfall and is crisscrossed by rivers from the Alps. The Po Valley produces half the country's wheat; three fourths of its corn, sugar beets, and rye; and nine tenths of its rice.

On a smaller plain around Naples, the soil has been enriched by volcanic lava. Streams that flow down from the Apennines provide ample water. Almost every inch of this plain is planted with tomatoes and other vegetables, citrus trees, and some grains.

The coast of northern Sicily gets enough

rainfall to grow both citrus fruits and vegetables. In fact, nine tenths of Italy's lemon crop and almost three fourths of its orange crop are produced here.

As in Greece, Spain, and other Mediterranean countries, the drier and more mountainous areas of Italy are heavily planted with olive and fig trees. These trees have long roots that can reach deep underground for water, so they can survive the summer dryness much better than small plants. Grapevines also do well in hot, sunny areas without much rain. About four million acres

(1½ *million hectares*) of land in Italy are planted with grapevines.

Some land cannot support any food crops, because it is either too steep or too dry. These areas are used to graze sheep, goats, pigs, donkeys, and cattle.

Italy and the World

At grocery stores in Sweden, West Germany, and other parts of northern Europe, many of the fruits and vegetables on display may have been imported from Italy. Since World War II, Italy has tried to export as much as possible to pay for the fuels and iron ore it needs to import for its expanding industries.

Italy was one of the first nations to join the European Common Market (which you will read about in Chapter 19). Membership in the Common Market helped the nation's economy greatly. For the first time, Italy was able to buy coal from other member nations without having to pay import taxes on it. This made it possible for Italy to buy more coal and produce more steel than ever before. Today Italy's most valuable exports are products made from steel — cars, trucks, tractors, motor scooters, and machinery.

Italy is now one of the world's top 10 nations in foreign trade. Yet in most years, it imports more than it exports. How does Italy balance its economy? It makes up the difference with the money it earns from tourism. About 11 million tourists visit Italy every year—more than any other country in Europe. The sunny climate and art treasures such as Leonardo's "The Last Supper" earn more money for Italy than many factories.

SECTION REVIEW

1. What is important about the triangle formed by Milan, Turin, and Genoa?

2. In southern Italy, has industry increased, decreased, or remained about the same in recent years? Why?

3. Much land in Italy is too steep and dry for growing crops. How do farmers use it?

4. What major advantage did Italy gain from joining the European Common Market?

YOUR LOCAL GEOGRAPHY

1. The Po Valley in northern Italy has plenty of water while other parts of the country, especially in the south, often have water shortages. Is the water supply in your community more like that of the Po Valley or of southern Italy?

To find the answer, you may need information from a source in your local government. Look in your phone book for the local government (county or city) listings. (You will find them under the name of the county or city, either in the White Pages or in a special government section, the Blue Pages.) The various departments of the local government are listed alphabetically under the main heading. Look down the list for "Water Department" or "Water Supply" (the name may vary). If you cannot find a listing under "Water," look for an "Information" listing or other general department, which can tell you where to inquire about water. One student from your class may be chosen to telephone or write to the water department for the following information: Where does your community's water come from? Is the water supply regular and adequate? If not, what measures have been taken or are being planned to improve the supply?

2. Italy is a long distance from north to south, compared to its width, and its environment changes from one end to the other. As a result, Italy is divided into two major areas — the cooler, wetter north and the hotter, drier south. Is the state where you live divided into broad areas in a similar way? For example, is there a mountainous and a lowland area, a drier and a wetter area, a rich and a poor soil area, etc.? Is there a simple division between north and south or east and west? Summarize the ways in which people commonly divide your state, and give environmental reasons.

Reading a Band Graph

Millions of foreigners visit Italy. Do most of them come from nearby countries or from far-off places? Is there a different "mix" from year to year?

Answers to such questions can be shown on a **band graph** like the one below. The graph may remind you of a line graph (Geography Skills 11, page 78). Like a line graph, a band graph uses indicator lines to show how amounts of something change over time. However, a band graph also uses bands of various widths to show how *different parts* of something change.

In this graph, the top indicator line shows how the total number of people entering Italy from abroad changed over several years. The total is divided into two groups represented by two bands. Visitors from the countries of the European Common Market (Britain, France, West Germany, Belgium, the Netherlands, Luxembourg, Denmark, Ireland, Italy) are represented by the top band. Visitors from the rest of the world are represented by the bottom band. (The total could be divided into more bands to show more groups.)

From left to right across the graph, each band gets wider or narrower, depending on whether the number in that group increased or decreased. You can compare changes in both groups by comparing the widths of the bands.

Answer each of the following questions on a separate sheet of paper.

1. About how many visitors came to Italy from "all other countries":
(a) in 1975? (b) in 1980? By how many did this group increase?

2. About how many visitors came from Common Market countries:
(a) in 1975? (b) in 1980? By how many did this group increase?

3. From 1975 to 1980, did both groups increase, decrease, or stay the same?

4. From 1975 to 1980, did visitors from Common Market countries make up about one quarter, one half, or three quarters of the total number?

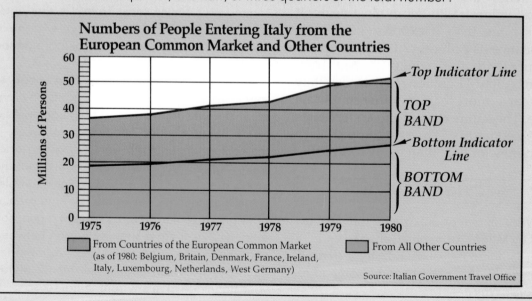

Numbers of People Entering Italy from the European Common Market and Other Countries

From Countries of the European Common Market (as of 1980: Belgium, Britain, Denmark, France, Ireland, Italy, Luxembourg, Netherlands, West Germany)

From All Other Countries

Source: Italian Government Travel Office

A. Words To Remember

From the following list, select the term that best completes each of the sentences below. On a separate sheet of paper, write your answer next to the number of each sentence.

balance   irrigate   mercury
dialect   lava   Renaissance
Herculaneum   marble

1. In addition to standard Italian, many people in Italy speak a local _____.

2. Leonardo da Vinci and Michelangelo were two of the outstanding artists of the _____.

3. Because Italy imports more goods than it exports, it has to find some way to _____ its economy.

4. The beautiful stone called _____ is one of the few mineral resources of Italy.

5. Volcanic _____ has helped to produce fertile soil around Naples.

B. Check Your Reading

1. Give two reasons why "The Last Supper" needed to be restored.

2. What is the southernmost part of Italy, and what kind of landform is it?

3. Both farming and industry flourish in the Po River Valley. What do farming and industry both need that the valley offers as a natural resource?

4. Name two crops that are grown in Italy's drier and more mountainous areas. Why are they able to grow there?

5. Italy generally imports more than it exports. How does it make up the difference?

C. Think It Over

1. Why have large sections of the Apennines become barren of trees and grass? What problem does this cause? How is it being tackled?

2. How are the barriers created by Italy's many dialects being overcome today?

3. Name two ways in which the north of Italy differs from the south.

D. Things To Do

Draw a large outline map of Italy. (Don't forget Sardinia and Sicily!) Then make up map symbols for Italy's main crops, resources, and products. Write these symbols in a map key at the bottom of your outline map. Using the information in this chapter, draw the symbols in the appropriate parts of Italy. Add a scale of miles and kilometers, and give the map a suitable title.

DOUANE
DOGANA

Chapter 19

A United Europe?
The Region in Perspective

About every 75 years, people can see a brilliant comet lighting up the night sky. It is called Halley's (HAY-liz) Comet. In 1704 an English astronomer named Edmund Halley realized that the comet belonged to our solar system. Halley and later astronomers studied the comet through telescopes. They could not get a close look at it, because it was still millions of miles from Earth each time it swung past the Sun.

However, Halley's Comet appeared again in 1986, and this time it was studied at a close distance. A spacecraft was launched that met Halley's Comet in order to collect scientific data and to take pictures of it.

This space mission was a project of the European Space Agency, which was formed in 1972 by 11 nations of Western Europe. Together these nations developed both the rocket that launched the mission to Halley's Comet and the spacecraft itself.

The European Space Agency is only one example of the increasing cooperation among the nations of Western Europe. A movement toward unity began soon after the end of World War II. At that time, much of Europe was in ruins. Cities, factories, railroads, and shipyards had been bombed out. How could the nations of Western Europe rebuild themselves?

The U.S. helped with billions of dollars in aid. Yet even with that aid, it would have been hard for the European nations to rebuild themselves separately. As you have seen, these nations are very small in area compared to the United States. The largest of them, France, is smaller than Texas.

The Western European nations were (and still are) heavily dependent on each other for food, raw materials, and manufactured goods. Yet trade among these countries was always held back by tariffs (import taxes). Each nation placed taxes on the goods of other nations to protect its own industries from competition. Tariffs raised the prices of imports, whether they were dairy products from Denmark, coal from Belgium, or steel from Germany.

If the nations of Western Europe could pool their economic resources, their combined wealth would be great. If they did away with import taxes, goods would be able to move from one country to another as freely as they do between the states of the U.S. Prices would come down, and trade would grow. Western Europe would become a huge market of more than 350 million people, spurring industry and providing more jobs.

Western European leaders began to discuss these ideas more and more after World

---

Barriers to trade and travel have been lifted between West European nations. Here, a West German drives from France to Italy past a "Customs" sign in French and Italian.

Modern buildings in Athens contrast with ruins dating back over 2,000 years. Ancient Greek civilization is a common heritage of Western Europe today.

War II. Some leaders looked beyond economic cooperation. They dreamed of a day when all barriers in Western Europe would fall, and the nations would be politically united.

Europe Together and Apart

The idea of European unity goes back a long way into the past. Its roots can be traced to Greece some 2,500 years ago. The ancient Greeks shaped a civilization that was to leave its mark on all of Europe.

In government the Greeks developed the political system still known by the Greek word **democracy,** meaning "rule by the people." They also set high standards in art, architecture, and literature. They laid the foundations of mathematics, sciences, and other branches of learning — including geography. It was a Greek, for example, who first measured the size of planet Earth. He did it by comparing the angle of the Sun at two places in different latitudes.

Greek civilization had a strong influence on the ancient Romans, who spread their ideas to large parts of Europe. The Roman Empire, which grew steadily until around 200 A.D., was a first attempt at European unity. At its height, the Roman Empire included present-day Italy, Greece, France, Spain, Switzerland, Austria, Belgium, the Netherlands, and parts of Britain and Germany, as well as other areas around the Mediterranean. However, communication at that time was slow, and the empire was too big and diverse to stay together.

There was one important Roman gift that Europeans were to share for many hundreds of years, even when they were at war with one another. This gift was the Latin language. Educated people in all parts of Europe spoke Latin until about 400 years ago.

Latin was also the language of the Roman Catholic Church. Until the Protestants broke away to form their own churches in the 1500's, Roman Catholicism was part of the civilization that nearly all Europeans had in common.

In spite of a common heritage, Europe splintered into hundreds of states. Many of these states later joined together to form larger nations, such as Italy and Germany. Yet even today, Western Europe is divided into 17 nations, plus a number of tiny states such as Andorra and Monaco.

SECTION REVIEW

1. How did World War II lead to cooperation among the nations of Western Europe?

2. What was the main barrier to trade between West European nations before the move toward unity? Why did that barrier exist?

3. How did ancient Greece prepare the way for today's move toward European unity?

4. What cultural feature did ancient Rome leave in nearly all parts of Western Europe? What happened to this feature?

Lowering the Barriers

The first important step toward modern European unity was taken in 1952, when six nations formed a group called the European Coal and Steel Community. The members were France, West Germany, Italy, Belgium, the Netherlands, and Luxembourg. Their goal was to do away gradually with the tariffs that held back their trade in coal, iron ore, and steel.

In 1957 the six members formed the European Economic Community (EEC), which is better known as the Common Market. It did away with *all* tariffs on trade among the member nations, and also removed many barriers to the free movement of workers, services, and money among them. This meant, for example, that French perfumes could be sold in any Common Market coun-

try without the payment of import taxes. Italian workers could get jobs in West Germany, France, and the Netherlands, where their labor was needed. Businesspeople in Belgium could start a company in any of the other countries as equals.

The Common Market has been a great success. Trade and industry have grown, and the member nations have become more prosperous. Britain, Ireland, and Denmark joined the Common Market in 1973. Greece became a member in 1981, and Spain and Portugal joined in 1986. The Common Market is the largest group formed to unite Western Europe's economies.

Other nations formed a separate group called the European Free Trade Association in 1960. This group includes Norway, Sweden, Iceland, Finland, Austria, Switzerland, and Portugal. The association did away with most tariffs on the goods of its members. In the next 10 years, their trade almost doubled, and living standards improved in each country.

West European nations work together in other areas besides trade. For example, an organization called the European Atomic Energy Community (Euratom) is part of the Common Market. In Euratom the Common Market nations cooperate to develop nuclear power for industry.

Members of both the Common Market and the European Free Trade Association cooperate with each other in the following two fields of work:

■ *Nuclear Research.* The European Organization for Nuclear Research is the world's largest center for the study of particles smaller than atoms. The organization was formed in 1954 by 12 nations — Austria, Belgium, Britain, Denmark, France, Greece, Italy, the Netherlands, Norway, Sweden, Switzerland, and West Germany.

■ *The Study of Space.* The European Space Agency was formed in 1972 to develop space equipment for peaceful purposes. The

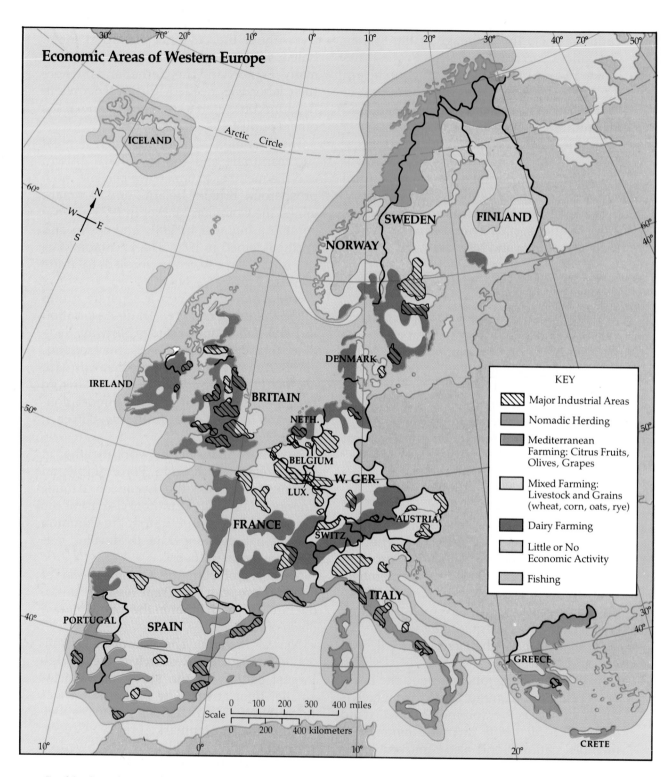

Economic Areas of Western Europe

KEY

Major Industrial Areas
Nomadic Herding
Mediterranean Farming: Citrus Fruits, Olives, Grapes
Mixed Farming: Livestock and Grains (wheat, corn, oats, rye)
Dairy Farming
Little or No Economic Activity
Fishing

ICELAND

Arctic Circle

NORWAY SWEDEN FINLAND

DENMARK

IRELAND BRITAIN

NETH.

BELGIUM

LUX. W. GER.

FRANCE SWITZ. AUSTRIA

ITALY

PORTUGAL SPAIN

GREECE

CRETE

Scale
0 100 200 300 400 miles
0 200 400 kilometers

In this densely populated region, all land that can be used *is* used. Thus industrial and farming areas cluster together. Only scattered mountain areas and northern Scandinavia have little economic activity.

agency's biggest project was to launch the spacecraft that took pictures of Halley's Comet. The members are Belgium, Britain, Denmark, France, Ireland, Italy, the Netherlands, Spain, Sweden, Switzerland, and West Germany.

Flowing through several West European nations, the Rhine River is part of a major transportation network for both trade and tourism. Above, sight-seeing boats from different nations carry tourists along the Rhine.

How Much Unity?

Do these moves toward closer cooperation mean that some day there will be a United States of Europe? This is the goal that many members of these groups are working toward. They believe that removing trade barriers and working together on scientific projects are steps toward political unity.

No one can say whether this unity will ever be achieved — and if so, how long it will take. In the meantime, however, there are many signs that some of the divisions between the nations of Western Europe are fading.

People in Western Europe still think of themselves as members of particular nations, but are becoming more aware of themselves also as Europeans. They have discovered that the differences between the peoples of Britain, France, Italy, and other European nations are not so great after all.

TV news programs beamed by satellites to all parts of Western Europe remind the people of how close they have become. License plates stamped "Europa!" may be seen on cars from Rotterdam to Rome. Road signs in Switzerland say, "You are now in Switzerland. But remember that this country is part of Europe."

Europeans have changed many of the customs that once separated them. Britain switched to the metric system of measurements used by the other nations of Europe. Sweden switched to driving on the right side of the road, instead of the left. (In Western Europe, only Britain, Ireland, and Malta still drive on the left side.)

Western Europe has one of the best transportation systems in the world. Airlines, highways, railroads, rivers, and canals crisscross the entire region. While most of these routes were developed separately by each country, they are being adapted more and more to form a united system. For example, Trans-Europa Express trains link the main cities in nine Western European countries. These trains do not have to stop at borders for immigration and customs inspections.

In recent years, travel for pleasure has grown tremendously, and is helping to bring Europeans closer together. Germans fly to Spanish and Greek islands in the Mediterranean to enjoy the climate. People from Mediterranean countries visit northern cities like London, Paris, Amsterdam, and Stockholm.

This year Annick Ferré, the girl from Cannes, France, goes skiing in the French Alps. Next year, perhaps, she may ski in the Italian Alps.

Right now Alexander Henderson helps care for his father's cows in the Scottish Highlands. Next year, perhaps, he may be waiting in line to see "The Last Supper" in Milan.

SECTION REVIEW

1. What was the importance of the European Coal and Steel Community?

2. The European Economic Community (EEC) is usually known by another name. What is this name? Is the EEC's membership fixed or growing?

3. Name one way in which members of the EEC cooperate with other West European nations.

4. If you visited Western Europe, you might find various signs of European cooperation. Give two examples.

YOUR LOCAL GEOGRAPHY

1. To understand why the nations of Western Europe have made such efforts to cooperate, you might "bring" their region to the U.S. Find maps of Western Europe and of the U.S. that are both on the same scale, or as close to the same scale as you can find. (If possible, these maps should also have the same projection.)

On a piece of tracing paper, copy the outlines and national boundaries of Western Europe and mark in the following capital cities: Athens (Greece), London, Madrid (Spain), Paris, Rome, Stockholm (Sweden), and Vienna (Austria). Then place the tracing paper over the map of the U.S. so that one of the West European capital cities is on the location of your community. (Choose a capital city that will allow most of the European land area to fit within the North American land area.)

Now on a separate sheet of paper, write down the names of the other capital cities and, beside them, the town or city nearest their location on the U.S. map. Measure the distance from "your" capital city to each of the others. Do two or more of these cities fall within the same state?

(If you live in Hawaii, of course, no part of Western Europe will reach another state. You may wish to try the exercise with any place you know on the mainland.)

2. Does the United States cooperate with its Canadian and Mexican neighbors in any ways similar to those described for Western Europe? Start your research by finding answers to these questions: Do U.S. citizens need a passport to travel to Canada or Mexico? Do many Canadians and Americans make frequent business and vacation trips to each other's country? Can you make phone calls to Canada and parts of Mexico in the the same way as to other parts of the U.S.? Do any sport leagues include teams in both Canada and the United States? Are there international labor unions that include both U.S. and Canadian unions?

Evaluating and Synthesizing Information

Imagine that you have just arrived in Amsterdam. You have a ticket that you can use to travel from one country to another by special trains. You also have the map shown below. Does the map show all the information you would need to get around Western Europe by train?

Before using any kind of information, you often must decide whether the information is useful for what you want to do. This skill is called **evaluating** (judging) information. Sometimes when you evaluate a map, you may discover that the map shows only part of the information you need. Then you have to **synthesize** (combine) the map with information from somewhere else.

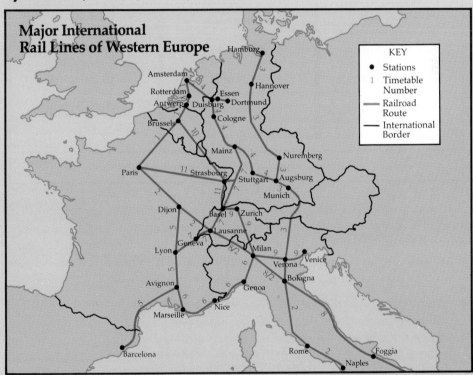

Major International Rail Lines of Western Europe

KEY
- Stations
1 Timetable Number
— Railroad Route
— International Border

Study the key and the map. Then follow the directions below.

1. Each of the following phrases describes something you might need to know to travel around Europe. On a separate sheet of paper, tell whether the map would be "useful" or "not useful" for finding each kind of information.
   (a) Cities where stations are located.   (b) Travel time between cities.
   (c) International boundaries you would cross on each train route.
   (d) Mountains and rivers along each train route.
   (e) Which timetable you need for a route.
   (f) Where the train stops on the route from Amsterdam to Paris.

(Turn page.)

2. Find the train route on the map between Amsterdam and Paris. Then answer the following questions.

(a) In which three other cities does the train stop on this route?

(b) Does the train begin its journey at Amsterdam or somewhere else?

(c) Does the train end its journey at Paris or continue somewhere else?

(d) What is the timetable number for this route?

		Timetable 1		
		AMSTERDAM–PARIS		
arrives	departs		arrives	departs
	5:55	**Amsterdam**	**12:30**	
6:54	**6:55**	Rotterdam	11:31	11:33
7:55	**7:56**	Antwerp	10:29	10:30
8:26	**8:27**	**Brussels**	10:00	10:01
11:11		**Paris**		7:19
	A.M. (before noon) (10:00)	**P. M.**(afternoon) **(10:00)**		

When you travel by train, you must often synthesize map information with a timetable. To find out when the train leaves Amsterdam for Paris, for example, you could read Timetable 1, which appears above. Like other kinds of tables, a **timetable** lists information in columns and rows. It also has arrows to indicate direction. Use Timetable 1 to answer the following questions.

3. What kind of information is listed down the middle of the timetable? What kind of information is listed on either side?

4. Notice the arrow at the left of the listed cities. The arrow points down the list, toward Paris. This means that times on the left of the timetable are for trains going *from* Amsterdam *to* Paris. What does the arrow at the right mean?

5. Study the key below the timetable. What do the times in light numbers mean?

6. The left side of the timetable shows the times when the train arrives and departs (leaves) each city. At what time does the train: (a) leave Amsterdam? (b) arrive in Rotterdam? (c) leave Rotterdam? (d) arrive in Paris? How long does the trip take altogether?

7. Study the right side of the timetable. If you want to make the return trip, what time does the train: (a) leave Paris? (b) arrive in Amsterdam?

8. Look at the map again. Which timetable(s) would you need if you wanted to travel from Paris to: (a) Strasbourg? (b) Barcelona?

CHAPTER REVIEW

A. Words To Remember

From the following list, select the term that best completes each of the sentences below. On a separate sheet of paper, write your answer next to the number of each sentence.

cooperation
democracy
European Economic
    Community

evaluate
heritage
living standards

political unity
synthesize
tariffs

1. For many years, trade among the nations of Western Europe was held back by _____ that raised the prices of all imports.

2. The ancient Greeks developed the political system known as _____, which means "rule by the people."

3. Some Europeans hope that all barriers between West European nations will be lowered and the region will achieve _____.

4. You _____ information by deciding whether it is useful for you.

5. You _____ information by combining it from different sources.

B. Check Your Reading

1. After World War II, many Western European leaders began to discuss practical ways of achieving what goal?

2. Why did nations set tariffs in the first place?

3. How would Western Europe become "a huge market of more than 350 million people"? Has it become this yet? Explain.

4. Name two important cultural features that West European nations do *not* all have in common.

5. Besides doing away with tariffs, the EEC removed other economic barriers between member nations. Give one example.

C. Think It Over

1. Does the environment of Western Europe do more to help or hinder cooperation among its nations?

2. What part do modern transportation and communications play in the move toward European unity?

3. In your opinion, is European political unity a realistic idea? Explain your answer.

D. Things To Do

Pick one of the characters mentioned in this unit and imagine that you are this person. You have been selected by your nation to be a delegate at a conference that will discuss the prospects for further cooperation among the nations of Western Europe. Prepare two suggestions for further ways in which these nations could work together. Use the map on page 219 and the Checklist of Nations on pages 582–583 as sources for making suggestions.

UNIT REVIEW

A. Check Your Reading

1. On a separate sheet of paper, write the letter of each description. After each letter, write the number of the nation that best matches that description.

(a) Birthplace of the Industrial Revolution
(b) The leading food producer in Europe.
(c) The Alps form its northern border.
(d) Home of the Ruhr River Valley.
(e) Its growing season is too short for much farming, but it has iron ore and water power.

(1) Spain
(2) Italy
(3) Sweden
(4) West Germany
(5) Britain
(6) France

2. Fill in the blanks in the following paragraph by writing the missing terms on your answer sheet.

Western Europe is warmed by ocean currents known jointly as the __(a)__ . Much of the land in Scandinavia is covered by __(b)__, a valuable natural resource. Rainfall is so heavy in Ireland that much of the land cannot be used for __(c)__ . The mountains of Alpine Europe make most of the land there unsuitable for growing food, but they help support its major industry, __(d)__ . Much of Mediterranean Europe is mountainous. Since many of its trees have been cut down, much land is now bare and __(e)__ of the soil has resulted.

B. Think It Over

1. Some of the items listed below are connected with the movement toward European unity. These items may be causes, effects, or both. Other items may *not* be connected with the movement toward European unity. On your answer sheet, write the letter of each item followed by C (cause), E (effect), B (both cause and effect), or N (neither cause nor effect).

(a) Increased trade.
(b) Heritage of ancient Greece.
(c) Language differences.
(d) Increased tourism.
(e) Trans-Europa Express.

(f) Discovery of North Sea oil.
(g) Lowering of tariffs.
(h) Britain's switch to metric system.
(i) Breakup of European empires.

2. Choose one of the items in Exercise 1 above, and write its letter on your answer sheet. Then write a paragraph giving reasons to support your answer.

Further Reading

France. Silver Burdett, 1985. A survey of France.

Italy: The Land and Its People, by Michael Leach. Silver Burdett, 1976. Describes the geography, culture, history, and economic life of the country.

The Crack in the Teacup: Britain in the Twentieth Century, by Marina Warner. Houghton Mifflin, 1979. Discusses Britain's economic problems and the end of its colonial era.

5

The Soviet Union and Eastern Europe

Chapter 20

Across
Two Continents

Imagine you are in eastern Siberia, at the eastern edge of the Soviet Union. It is a spring morning, and the sun is just rising. At this same time, night's shadows lie across most of the Soviet Union; and in Eastern Europe, the sun has not yet set on "yesterday."

A look at the clock can give a clearer idea of the distances involved. Say it is 6 A.M. at the eastern tip of Siberia, and people are beginning to wake up. At this moment, it is 8 P.M. (the evening before) in Moscow, the Soviet capital. In Warsaw, Poland, in Eastern Europe, some people are sitting down to supper at 6 P.M. (See Geography Skills 17 on page 141 for more about time zones.)

Compare these times to the U.S. From Maine to California, the U.S. has four time zones. Add Alaska and Hawaii, and the U.S. has seven time zones. By contrast there are 11 time zones in the Soviet Union alone.

In area the Soviet Union is the largest nation in the world, bigger than the U.S. and Canada combined. With Eastern Europe, the region stretches from longitude 10°E (East Germany) to 169°W (Bering Strait) — more than halfway around the world.

Overview of the Region

As you can see from the map on page 280, three sides of this vast region have clear natural boundaries. On the north and east lie bodies of water. The Baltic Sea marks the northern boundary of Eastern Europe; the Arctic Ocean, the northern boundary of Soviet Asia; and the Pacific Ocean, the eastern boundary of the Soviet Union.

The southern boundary also follows natural landmarks. Towering mountains cut off the Soviet Union from the rest of Asia. Farther west the region's southern border is formed by three seas and a mountain range. The seas (east to west) are the Caspian, the Black, and the Mediterranean. The mountains are the Caucasus, between the Caspian and Black seas.

On the western rim of the region, natural features are less help in setting boundaries, since waterways and mountain ranges are smaller. In this area, wars and human decisions have played a great role in deciding where one nation ends and another begins. Eastern Europe has been a meeting place of cultures throughout history, and its nations' borders have shifted often with the tides of battle.

Today the Communist system of government binds the nations of Eastern Europe to the Soviet Union. Until the end of World War II (1939–1945), the Soviet Union was the only Communist-ruled nation in the world. In fighting against Nazi Germany, the Soviet army occupied the eastern half of Europe,

Spanning two continents, the Soviet Union and Eastern Europe contain many cultures.
Europeans and people of Moslem background mingle in this market in Soviet Asia.

while U.S. and other Western forces moved in from the west. With the backing of the Soviet army, Communists came to power in eight East European nations (see the Checklist of Nations on page 583).

Under communism the government controls the nation's economy, and can make decisions affecting all areas of people's lives — including where they live and the kind of work they do. In practice most Communist governments have had to back off from complete control because of popular opposition. Even so, private businesses remain few and small, and major industries are owned by the state.

Most of the East European countries are dominated by the Soviet Union. For East Europeans, the Soviet Union is a giant on their doorstep. The Soviet Union is huge not only in area but also in population, with 270 mil-

lion people. Only two countries (China and India) have more people. If you add up the populations of the other countries in the region (see the Checklist of Nations on page 583), you will find that they total only half the Soviet figure.

However, the eight East European countries cover an area only one eighteenth as large as the area of the Soviet Union, which means that Eastern Europe is more densely populated. In other words, many people live in a small area in Eastern Europe, while there is far more elbow room in the Soviet Union. A closer look at the environment will show why this is so.

Oceans, seas, and mountains provide natural borders for most of this vast region. The major exception is in Eastern Europe. Most of these nations were occupied by the Soviet army at the end of World War II.

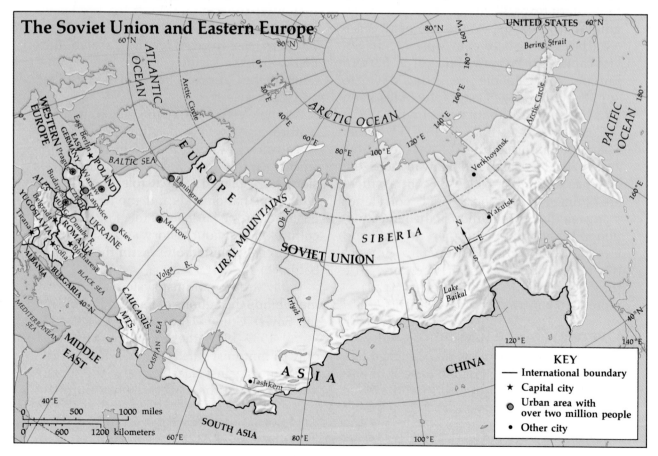

The Soviet Union and Eastern Europe

KEY
— International boundary
★ Capital city
◉ Urban area with over two million people
• Other city

Physical Geography

Maria Thalvieser (tahl-VEE-sur) lives on a government-run farm near Budapest (BOO-duh-pehst), the capital of Hungary. Her mother works with flowers grown on the farm, while her father looks after the livestock. Thunderstorms are common in June and July, bringing rain for the farm's crops. Hungary gets 20 to 25 inches (*50 to 65 centimeters*) of rain each year, about as much as Amarillo, Texas. This is less rain than falls on most farmlands in the U.S.

Only rare areas of Eastern Europe and the Soviet Union get as much as 30 inches (*75 centimeters*) of rain, the amount that falls in Des Moines, Iowa. In general, the farther east you go, the less rain you find. The driest areas are in the southeast.

Amin Rakhim (AH-meen rah-KEEM) lives in the Soviet city of Tashkent (tahsh-KEHNT), near the Chinese border. Some places in this area get less than four inches (*10 centimeters*) of rain a year, the same as the desert city of Las Vegas, Nevada. If you take a stroll through Amin's hometown, you will see irrigation ditches next to the sidewalks. The ditches are needed to carry water to the trees that shade the streets.

There is a simple explanation for this pattern of rainfall. North of the Equator, winds carrying moisture travel mostly from west to east, which means they approach the landmass of Europe and Asia from the Atlantic Ocean. The farther the winds get from the ocean, the less moisture they are likely to be carrying.

Thus the great size of the Soviet Union is something of a disadvantage. Rainfall is moderate in the west, but scanty in most of the far-flung regions. People tend to cluster in areas with the greatest rainfall, where growing food crops is easier. Three out of every four Soviet citizens live in the west — the European part of the country.

Temperature is affected also by oceans and landmasses. Oceans lose heat much more slowly than does land. As summer turns to winter, ocean temperatures tend to remain warmer than land temperatures, and winds carry this warmth inland. The farther the winds travel from the ocean, however, the colder they become.

Look at the map on page 282, which shows average temperatures in January. Can you tell why Moscow is colder than Poland? The coldest part of the Soviet Union is Siberia, and a place in Siberia called Verkhoyansk (vehr-koy-YAHNSK) is said to be the coldest town in the world. Its average temperature in January is −58°F (−50°C).

Of course, distance from the sea is only one reason for cold winters. Distance from the Equator is another. Most of Eastern Europe and the Soviet Union lies north of the latitude of Boston, and even a "southern" city like Tashkent is roughly as far north as Pittsburgh, Pennsylvania. If you look again at the map on page 282, however, you will see that winter temperatures depend more on distance from the Atlantic than on distance from the Equator.

Plains and Mountains

The region is dominated by a vast plain that crosses from west to east. The plain is narrow in the northwest, in East Germany and Poland. To the east, it widens rapidly until it spans almost the entire north-south width of the Soviet Union.

The plain is broken only by a low range of mountains called the Urals, which mark the border between Europe and Asia. East of the Urals, the plain resumes, stretching all the way across Siberia to the Pacific.

To find more mountains, you would have to go south of this plain. You have already read about the mountains along the southern edge of Soviet Asia (see page 279). Eastern Europe has several ranges of mountains that are extensions of the Alps found in Western Europe.

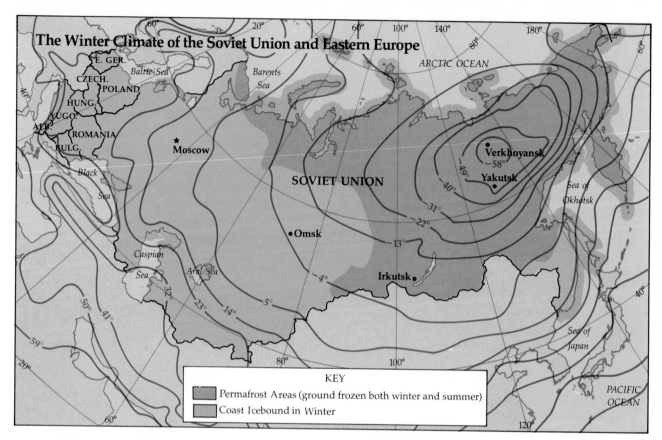

The Winter Climate of the Soviet Union and Eastern Europe

KEY
- Permafrost Areas (ground frozen both winter and summer)
- Coast Icebound in Winter

Only parts of the vast plain are suitable for farming, since growing conditions vary greatly within the region. The areas of the plain most suitable for crops are those that get the most rain and the least cold. As you read earlier, the region is generally wetter in the west and drier in the east. It is also generally warmer in the south, becoming colder in the north. Thus the best growing conditions are found mainly in the west, above all in the southwest.

West of the Urals, the area around the Baltic Sea is the northern limit of farming. Here the growing season is short, and summer days are cool. Poland and parts of East Germany also belong to this Baltic area. Cool-weather crops such as rye, potatoes, and hay grow well here.

Farther south summers are hot and sunny, and growing conditions are much like those in the U.S. Middle West. Corn and wheat are common crops in Hungary, Romania, and the Ukraine (yoo-KRAYN; a southwestern area of the Soviet Union).

East of the Urals, much of the region is too cold for farming. Throughout vast areas of

This map shows isotherms for January, with temperatures in degrees Fahrenheit. It also shows where the coast is icebound for at least part of the year, and where the ground is permanently frozen. The mildest climates are in the southwest while the coldest are in the northeast.

Soviet Asia, the ground remains frozen all year long. During the summer, the sun may warm a shallow part of the soil, but cold weather returns before deeper layers can thaw. This permanently frozen ground is called **permafrost**, and it covers 45 percent of all land area in the Soviet Union.

It is not easy to live in the permafrost zone. Since water on the surface cannot sink to lower ground levels, much of the land becomes marshy in summer. This marshy ground is home to hordes of mosquitoes. There are other problems too. In Yakutsk (yah-KOOTSK), a Siberian city, high-rise buildings have to be placed on stilts, six feet (*two meters*) off the ground. Otherwise heat from the buildings would melt the ground, causing foundations to shift.

In the northern permafrost region, near the Arctic Ocean, only tundra plants can survive. Beginning around the Arctic Circle, there is a wide belt of cool, forested land called **taiga** (TIE-guh). Most of the trees are coniferous (evergreen), such as spruce and pine. Farther south there are more deciduous trees (whose leaves fall each year).

Around 55°N the forest gives way to a dry grassland called **steppe**. This in turn gives way to deserts and mountains in Soviet Central Asia.

Natural Highways

It is much easier to travel through the European part of the region in the west, than it is through the Asian half in the east. One reason is that the western part has more natural highways — such as rivers.

Two great rivers flow through Eastern Europe. One is the Danube, which gave its name to a famous waltz ("Blue Danube").

The other is the Volga, also famed in song ("Volga Boatman").

Find the Danube on the map on page 280 and follow its course. You will see that it touches all of the countries in the region except East Germany, Poland, and Albania. Its length is 1,750 miles (*2,815 kilometers*).

The Volga River is even longer — 2,290 miles (*3,685 kilometers*), making it the longest river in Europe. It lies entirely within the Soviet Union and flows into the Caspian Sea.

Eastern Europe has some other large rivers. Along with the Danube and the Volga, they serve as highways for travel and trade. Most of the major cities of Eastern Europe

With old buildings, people in Western dress, and modern street lamps, this city might be anywhere in Europe. However, a West European city would have more private cars. This is Prague, Czechoslovakia.

and the Soviet Union lie on or near them.

Soviet Asia has great rivers too, most of them longer than the Volga. For example, the Ob and Irtysh (eer-TEESH) rivers form a system bigger than that of the Mississippi and Missouri rivers. However, these eastern Soviet rivers are not very helpful for travel and trade, mainly because they are frozen for much of the year. In addition, most of them flow northward into the Arctic Ocean, far from population centers.

Rich in Resources

Taken as a whole, Eastern Europe and the Soviet Union is rich in natural resources. Most of these resources are found in the Soviet Union. The countries of Eastern Europe have built up industry over recent decades, but they have had to import many raw materials from the Soviet Union to do so.

The Soviet Union has more than half of the world's known reserves of coal, oil, and manganese (used in making steel). It has natural gas, iron, aluminum, gold, and other important minerals. It has rich fishing waters, and many of its rivers can be harnessed for hydroelectric power.

By contrast the eight nations of Eastern Europe are poor in resources, especially fuels. Romania, which has the most oil, produces about as much oil as the island nation of Trinidad-Tobago. Poland has deposits of good-quality coal.

SECTION REVIEW

1. The Urals mark the border between Europe and what other continent?

2. What is permafrost? Name one problem it causes.

3. How does the Soviet Union's environment help explain why three out of every four Soviet citizens live in the west?

4. A great distance from the sea and a great distance from the Equator combine to produce what kind of climate?

Human Geography

If you lived in Poland and wanted some bread, you would ask in Polish for *chleb* (klehb). In the Soviet Union, the Russian word for bread is pronounced exactly the same. You would also hear the same word for bread in Yugoslavia, Bulgaria, and Czechoslovakia. This is a clue to an important fact: Many of the people of the region are related to one another.

These people are called Slavs. At some point in the past, their ancestors migrated to Europe from southern Asia. Later, beginning about the third century A.D., different groups of these Slavs spread to the areas where their descendants now live.

Today's Slavs are divided into about a dozen groups: Poles (mainly in Poland); Czechs and Slovaks (mainly in Czechoslovakia); Serbs, Croats (KROH-ats), Slovenes, and Macedonians (mainly in Yugoslavia); Bulgars (mainly in Bulgaria); Russians, Byelorussians (bee-EHL-oh-ruhsh-uhns), and Ukrainians (mainly in the Soviet Union).

Many given and family names are similar among the various Slavic peoples. Slavic names have something else in common. You will notice that the same family name has different endings for men and women. In Chapter 22, for example, you will meet Ewa Lubowska (EH-vuh loo-BOHV-skuh) of Poland. Her husband is Jerzy Lubowski (YEH-zhee loo-BOHV-skee).

All Slavic languages *sound* pretty much alike, but the words do not all *look* alike. Some Slavs write with the Latin alphabet (the one we use), while others have the Cyrillic (suh-RIL-ik) alphabet. (Here, in the Cyrillic alphabet, is the Russian word for *bread*: ХЛЕБ .) In Yugoslavia, Serbs and Croats speak exactly the same language, but the Serbs use the Cyrillic alphabet and the Croats use the Latin.

This difference in alphabets comes from the influence of two branches of Christian-

Contrasts are found not only in the vast Soviet Union but also in the much smaller countries of Eastern Europe. In this town in Yugoslavia, modern buildings loom over farm women at a traditional food market.

ity. People in areas where the Roman Catholic Church was strong learned to use the Latin alphabet. People in areas where the Eastern Orthodox Church was strong learned to use the Cyrillic alphabet.

A Range of Peoples

Besides the Slavs, many other peoples live in Eastern Europe. They include Germans in East Germany, Magyars in Hungary, Albanians in Albania, and Romanians in Romania.

All the countries of the region contain **minority peoples** (people who belong to different ethnic groups from most of the nation's population). For example, some Germans live just outside East Germany in Poland and Czechoslovakia, while Bulgars live on both sides of the Bulgarian-Yugoslav border. However, the largest number of minority groups are found in the Soviet Union.

People often use the terms *Russia* and *Russians* when talking about the entire Soviet Union. But the Russians are only one of the many dozens of ethnic groups in the Soviet Union. They are the most numerous, how-

ever, making up half of all Soviet citizens. Some Soviet peoples (such as the Ukrainians) are Slavs like the Russians. Some (such as the Latvians, Lithuanians, and Estonians who live near the Baltic Sea) are not Slavs but share the traditions of Western Europe. Others (such as the Georgians and Armenians who live near the Caucasus Mountains) are not Slavs but have their own Christian traditions. Others (such as the Kazakhs [kah-ZAHKS] and Tadzhiks [tah-JEEKS] who live in Soviet Central Asia) have Moslem traditions and bear a physical resemblance to the peoples of East Asia.

At one time, millions of Jews lived in the Soviet Union and Eastern Europe. Harsh treatment led many of them to emigrate to the U.S., beginning in the late 19th century. Millions were killed by the Nazi Germans during World War II. Of those who remained under Communist rule, many have emigrated — or tried to emigrate — to the U.S. or Israel.

Today people of all religious beliefs face a stiff challenge as Communist governments seek to discourage religion. Still, religion remains strong in many areas. In Poland, for instance, most people belong to the Roman Catholic Church. Some Soviet citizens attend services of the Russian Orthodox Church, while smaller numbers go to Baptist or Lutheran services. Many people in Soviet Asia are Moslems.

SECTION REVIEW

1. What are minority peoples?

2. Are all Slavic languages similar when spoken? What two different alphabets are they written in?

3. Which of the following peoples are *not* Slavs: Poles, Romanians, Czechs, Russians, Georgians? Where do these non-Slavic peoples live?

4. Which ethnic group in the Soviet Union dominates all the others, and why?

Economic Geography

Maria Thalvieser's family has strong ties to the land. Her father was born in the house where Maria now lives in Hungary. When her father was born, his parents owned not only the house but also the farmland around it. After the Communists came to power, they began to take the farms from the private owners. Today the house still belongs to the Thalviesers, but the land has been pooled with other lands to form one big government-owned farm.

The 20th century has brought many changes to the farms of Eastern Europe and the Soviet Union. After the Communists seized power in Russia, the government launched a drive to bring the peasants' farms under its control. When millions of peasants resisted, the government seized their land and executed many of them. The survivors were forced to work on large government-run farms.

Later when Communist governments came to power in East Europe, they also started to bring privately owned farms under their control. Today only two countries in the region, Poland and Yugoslavia, still have private farms. In fact, most farms in these two countries remain privately owned.

Throughout the region, there are two types of government-controlled farms. A **state farm** is run directly by the government, and the workers receive a regular wage just like factory workers. On a **collective farm**, the workers share in the day-to-day management — and also in the farm's earnings.

Most of these government-controlled farms were created by joining together a number of private farms. Because most private farmers had only small areas of land, few of them used modern machines. Thus the governments planned to mechanize their new, big farms and boost production. Today giant combines clatter across big

Most farms in the region are controlled by the Communist governments. This corn harvest is on a Hungarian farm. Because Hungary is in the region's southwest, it can grow warm-weather crops like corn.

fields in many parts of Eastern Europe and Siberia — just as in Nebraska or Kansas. However, production has not climbed as much as was hoped. People do not seem to work so efficiently when the profits do not come directly to them.

Most families who work on collective and state farms are allowed to grow crops and raise livestock on the land surrounding their homes. They work on these plots in their spare time, producing food for their own use — and to sell in town. Private plots play a key role in helping to feed the people of the region.

In the Modern World

Old ways and new ways often exist side by side in Eastern Europe and the Soviet Union. In Tashkent, Amin lives in a house made of adobe (mud and straw) bricks. The house has a courtyard in back, with high walls that shield it from the street.

Similar houses have lined the streets of Tashkent for centuries — since the city first grew as a stopping place on trade routes from China to Europe. Yet tall, new apartment buildings are found here too, and a modern subway runs beneath the city's streets.

As you learned in Unit 1, nations that belong to the modern world use resources from far-off places. Thus transportation is important for any modern nation. It is especially important for the nations of Eastern Europe and the Soviet Union, since many of their natural resources are hard to reach. Some resources are in remote areas of Siberia; others, in southern deserts; and still others, in rugged mountains. Ways have had to be found to move these resources from one place to another.

Railroads have been a help. As early as 1904, the Trans-Siberian Railway linked European Russia with the Pacific Coast. In recent years, oil pipelines have been built to move oil from one area to another. Much of the oil used in Eastern Europe comes from the Soviet Union by pipeline.

Most of the nations in the region are highly industrialized. Communist leaders have favored **heavy industry** over **consumer goods**. This means that most money has been poured into big factories that produce steel, machinery, and military equipment. Much less money has been spent on producing goods such as shoes, shirts, and cars for the people. In recent years, most nations in the region have paid more attention to consumer goods, but these are still in much shorter supply than in the West.

What's Ahead

In this chapter, you have taken a broad overview of the region as a whole. In the next two chapters, you will take a closer look at two individual nations.

Chapter 21 focuses on the Soviet Union, which has more variety of land and people than most other nations in the world.

Chapter 22 deals with Poland. After the Soviet Union, Poland is the nation with the most people and the most land in this region. While living in the shadow of the Soviet giant, Poland has kept strong cultural ties with Western Europe and the United States.

Finally in Chapter 23, you will look again at the region as a whole to see what new picture has emerged from the details.

SECTION REVIEW

1. When the Communists came to power, what changes did they make in the way farms were run in most countries in the region?

2. Under communism did farms tend to become bigger or smaller? Why?

3. What kind of industries have Communist governments emphasized? Has this emphasis changed at all? Explain.

4. Do most countries in the region belong to the modern world or the traditional world? Explain.

YOUR LOCAL GEOGRAPHY

1. You have read about the great variety of growing conditions within the Soviet Union and Eastern Europe. As you know, temperature and rainfall play a large part in determining what can be grown where. Suppose you were going to plant a vegetable garden outside your home, similar to the private plots of many farm workers in the region. Would you choose cool-weather plants such as potatoes or cabbages? What kinds of vegetables do people grow in your locality? What kinds cannot be grown? Why?

2. The Soviet Union and Eastern Europe are populated by many ethnic groups such as Germans, Magyars, Albanians, Romanians, and Slavs, to name just a few. Do you know what ethnic groups are represented in your own community? To get a general idea, take a survey of your classmates. What countries did their ancestors come from? Can any of your classmates speak a foreign language? Report the results of your survey to the class.

Reading a Polar Map

The Soviet Union and Eastern Europe together form a vast region that extends halfway around the globe. You can see this clearly if you study the map on the opposite page. The North Pole is positioned in the center of this map. Any map that is centered on either the North Pole or the South Pole is called a **polar projection**.

On most maps, the northernmost part of Earth is at the top. On this north polar map, however, the northernmost part of Earth is shown at the center. What would be shown at the center on a *south* polar map?

On this projection, notice also that parallels of latitude are circles and are smallest near the map's center. Meridians of longitude are straight lines that radiate outward from the map's center like the spokes of a wheel. Can you figure out directions and locations on this map?

Write your answers to the following exercises on a separate sheet of paper.

1. On this map, North America is (north; south; east) of the North Pole.
2. The city of Yakutsk, Soviet Union, is (north; south; west) of Anchorage, Alaska.
3. If you traveled from the Soviet Union over the North Pole to North America, in which direction(s) would you travel?
4. The latitude 90°N is at the (center; top; bottom) of this map.
5. The Prime Meridian (0° longitude) is at the (center; left; right) of the map.
6. Which country is located at: (a) 50°N, 20°E? (b) 60°N, 110°W? (c) 60°N, 110°E?

From the polar map, you can easily compare the latitudes of the Soviet Union and Eastern Europe with the latitudes of North America. Answer the following questions on a separate sheet of paper.

7. The Soviet capital of Moscow is located almost halfway between 50° and 60° north latitude. On the map, find the city in the United States closest to the same latitude as Moscow. In what state is this city located?
8. When Russians go south to find a mild vacation climate in their own country, they often go to the Crimean Peninsula on the northern shore of the Black Sea. Between what two parallels of latitude on the map is the Crimean Peninsula located? What city on the west coast of North America is located at about the same latitude?
9. In latitude most of the Soviet Union compares with what country in North America? In latitude most countries of Eastern Europe compare with what area of North America?
10. Which of the following would you expect to find in most of the Soviet Union? (a) long winters. (b) long summers. (c) short growing seasons. (d) high temperatures.

High-Latitude Countries of the North

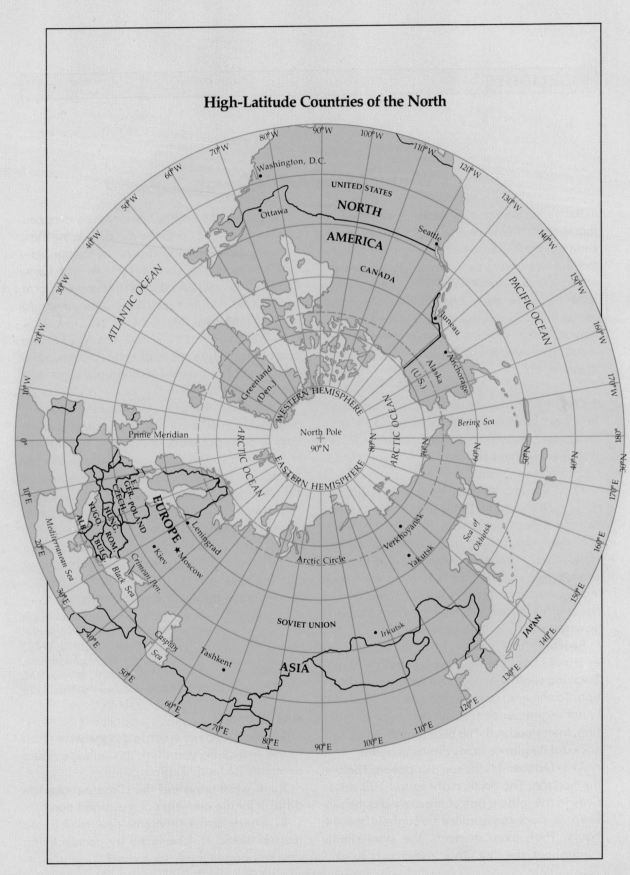

Winter in the Arctic Ocean

Before the Communists seized power, the land that stretched from Russia to the Pacific Ocean was known as the Russian empire. Early in this century, the Russian government decided to explore the empire's Arctic coast. In 1914–1915, an expedition spent the winter there. The following description comes from a journal kept by one of the scientists. As you read it, note the climate and the environment. Think what it would be like to live and work in this region.

Our ship headed for an island that until now had been unknown. We could see dark gray and brown rocky shores and a flat summit. On it strolled a large polar bear, which quickly disappeared from sight.

I had to climb to the summit of the island to determine its height. Three times I tried to climb up the steep rocks, and failed. The frost-shattered rock ledges broke away under my hand. I went around a headland and found a more gentle slope. At the foot, three enormous walrus lay fast asleep. When I shouted, one of the walrus raised itself on its front flippers, yawned widely, and lay down again.

Reaching the summit, I caught sight of an enormous bear. He was looking straight at me. Swaying slightly he took two steps in my direction. To show an animal that one is scared of it is not recommended. I took two steps to meet him, then stood still. The bear turned away. Not once did he glance in my direction again.

After October 31, the sun disappeared below the horizon. The Arctic night lasted 103 days. Only in the middle part of the night did the sky keep its dark appearance throughout the 24 hours. Then even at noon, the stars shone throughout the sky. This lasted two weeks.

On December 26, the temperature dropped to $-45°F$ ($-43°C$), and the mercury in the thermometer froze. A wind was blowing from various points of the compass the whole time. Snow fell several times, and then a snowstorm would rage. Under such conditions, we would not leave the ship.

To welcome in the New Year 1915, we brought out some cases of canned pineapple. Each member of the expedition got a glass of champagne.

Meanwhile the air temperature frequently dropped to $-40°F$ ($-40°C$) and lower. It is not easy to carry out work in the open air at these temperatures. The barometer would quickly freeze up; the kite would break, stopping us from sending up the meteorological [weather] instruments. Then too we could not manage to bore through the thick ice.

On April 23, a living herald of spring was fluttering and chirping aboard our ship: a snow bunting [a kind of bird]. From April 24, the sun ceased to set; the Arctic day had begun.

— From "Charting the Russian Northern Sea Route: The Arctic Ocean Hydrographic Expedition 1910–1915," by L.M. Starokadomskiy, translated by William Barr. McGill-Queen's University Press, 1976

Ask Yourself . . .

1. What kinds of wildlife did the writer find in the Arctic? Do you think this wildlife relied on plants for food? Explain.

2. In what ways did the climate make life difficult for the members of the expedition?

3. Which of the problems described in the journal would still be faced by people living and working in the Arctic today?

CHAPTER REVIEW

A. Words To Remember

From the following list, select the term that best completes each sentence below. Write your answers on a separate sheet of paper.

collective farms   permafrost   taiga
deciduous   state farms   tundra
minority peoples   steppe

1. _____ trees are those whose leaves fall each year.

2. _____ is the name given to permanently frozen ground.

3. _____ belong to different ethnic groups from most of a nation's peoples.

4. The _____ is a wide belt of cool, forested land.

5. Government-owned farms whose workers receive fixed wages are called _____.

B. Check Your Reading

1. Which area of the Soviet Union and Eastern Europe is generally wetter and warmer? Why?

2. Do most factories in the Soviet Union and Eastern Europe produce heavy industry or consumer goods?

3. In what ways do oceans affect the temperatures of nearby land?

4. Is the region rich in mineral resources? Are the resources shared equally among the different nations? Explain.

5. How does the Soviet Union compare with Eastern Europe in area and population?

C. Think It Over

1. The Soviet Union is the nation with the largest area in the world. In what ways is this an advantage to its people? In what ways is it a disadvantage?

2. Why is a north-to-south journey through the region more difficult in Soviet Asia than in Soviet or Eastern Europe?

3. Are cultures more varied in Eastern Europe or in the Soviet Union?

D. Things To Do

With other students, prepare a bulletin-board display based on a large hand-drawn map of the Soviet Union and Eastern Europe. After the map is drawn, locate and mark the following places mentioned in the chapter: Moscow (the capital of the Soviet Union); Warsaw (the capital of Poland); Prague (the capital of Czechoslovakia); Budapest (the capital of Hungary); Tashkent (where Amin lives); Verkhoyansk (the world's coldest spot); Yakutsk (a city built on the permafrost). From the map on page 280, measure the following distances: Warsaw to Moscow; Prague to Moscow; Budapest to Moscow; Moscow to Tashkent; Moscow to Yakutsk; Moscow to Verkhoyansk. On your own map, draw lines joining these cities and write in the distances. What do they suggest about population distribution in the region?

Chapter 21

The Soviet Union
Many Nations in One

It is the year 1242 A.D. Ice covers the lake. German knights charge onto the ice, one after another, their heavy armor clanking. Russian soldiers beat off the attack. Suddenly the lake surface cracks and the ice begins to break beneath the German forces. Screaming knights topple into the water and sink like lead weights.

In this famous battle, Russian soldiers smashed an invasion by Teutonic (German) knights. The fighting took place on what is now Lake Peipus (PIE-puhs) in Estonia.

What month of the year do you think this battle took place — December? January? February? The date was April 5. In much of the Soviet Union, ice lasts far into spring, and is a key feature of the environment. Ice blocks lakes and rivers for months on end and keeps the Soviet seacoast icebound — in some places, all year long.

The Soviet Union does have a few outlets to warmer waters. The most important is the Black Sea, which lies at the southwestern edge of the Soviet land. Ships can pass from the southwestern end of the Black Sea into the Mediterranean. However, the ships must first pass through two narrow straits. The Soviet Union does not control the land flanking these straits. (The land belongs to Turkey.) In wartime an enemy could block the outlet and bottle up Soviet ships inside the Black Sea.

Of all the world's nations, the Soviet Union has the longest seacoast. Yet it is almost a landlocked nation — locked in by ice and the narrow Black Sea straits.

The barriers of land and climate that ring the Soviet Union have had a strong influence on people's lives. As you read in Chapter 20, more people live in the west, where the climate is best for farming.

Now look at the population map on page 294. You will see that settlement becomes thin not only toward the east but also toward the icy north and the deserts and mountains of the south. The most densely populated area forms a rough triangle lying on its side with its tip in the east.

This area of the Soviet Union is known as the "Population Triangle." About 80 percent of the nation's people live there. The area outside this triangle is rich in resources. Because of its harsh environment, however, much of it remains undeveloped.

Physical Geography

In the distant past, much of the Population Triangle was covered by glaciers. In the north, the glaciers were the thickest and acted like massive bulldozers. They scraped off the richest soil, leaving behind only a thin topsoil.

The thin soil, together with the flatness of the land, causes poor **drainage**. In other

Centuries-old church buildings are surrounded with snow in a small town north of Moscow.
In most of the Soviet Union, temperatures dip below freezing at least half the year.

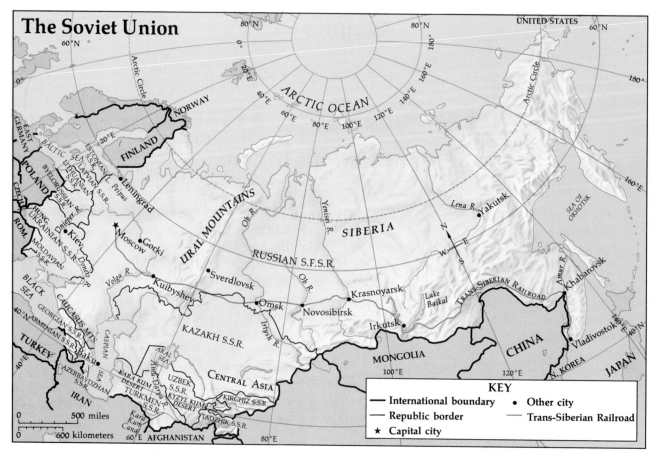

The Soviet Union

ARCTIC OCEAN
SIBERIA
RUSSIAN S.F.S.R.
URAL MOUNTAINS
CENTRAL ASIA
KAZAKH S.S.R.
MONGOLIA
CHINA
JAPAN
N. KOREA
IRAN
AFGHANISTAN
TURKEY
FINLAND
NORWAY
POLAND
UNITED STATES

KEY
— International boundary
— Republic border
★ Capital city
• Other city
— Trans-Siberian Railroad

words, rain and melted snow cannot sink easily through the soil or run off into rivers. As a result, many northern areas are marshy and poor for farming.

Other areas, especially in the south of the Population Triangle, have much better soils. In the past, when the glaciers melted, the soil they scraped off was left behind. Much of this soil was fine enough to be spread around by the wind. Such wind-spread soil is known as **loess** (less), which is excellent for growing crops.

The Ukraine, in the south of the Population Triangle, has a rich black soil that is one of the world's best. This soil is known as **chernozem** (CHEHR-nuhz-yawm), a Russian word meaning "black earth." A similar soil can be found in the U.S. Great Plains, from North Dakota to Kansas. This soil forms in grasslands where rainfall is not

The Soviet Union is divided into the huge Russian Soviet Federated Socialist Republic and 14 Soviet Socialist Republics (S.S.R.). The area between the Caspian Sea and China is known as Soviet Central Asia; the rest of Soviet Asia is called Siberia. Most of the people live in the Population Triangle shown below.

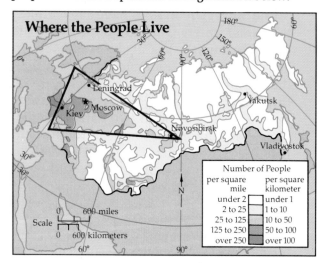

Where the People Live

Leningrad
Moscow
Kiev
Novosibirsk
Yakutsk
Vladivostok

Number of People	
per square mile	per square kilometer
under 2	under 1
2 to 25	1 to 10
25 to 125	10 to 50
125 to 250	50 to 100
over 250	over 100

Scale
600 miles
600 kilometers

quite enough for forests to grow. Over many centuries, grass keeps growing and dying. The dead grass adds more and more minerals and organic matter to the soil, and the rainfall is not heavy enough to wash away these nutrients. Winds add loess, making the soil even richer. In parts of the Ukraine, the rich black earth goes as deep as five feet (1½ meters). Some of this soil extends eastward as far as the tip of the Population Triangle.

Climate controls what crops can be grown in the different parts of the Population Triangle. In the north, the growing season is only three months long, and summers are too cool for wheat or corn. This area is something like Wisconsin or New England. A key crop is grass, which becomes hay when dried, and is fed to dairy cows and goats.

Root crops are another specialty. The potato is king in Byelorussia, and people there take pride in the hundreds of dishes that can be made from potatoes. On festive days, they dance a lively step called *Bulba* (BOOL-buh; "The Potato"). The Soviet Union grows more than one out of every three potatoes grown in the world.

Farther south, summers are longer and hotter, and soils are richer. Fields are bigger, as in Iowa or Illinois. This area can easily grow wheat and corn. Moreover, **winter wheat** can be planted in the fall, survive the winter, and mature the next summer. It yields more grain than **spring wheat**, which is planted in spring and has less time to grow before the summer harvest.

The Ukraine, which covers much of this fertile area, produces more grain than other parts of the country. It is known as the "breadbasket" of the Soviet Union.

The "Other" Soviet Union

Outside the Population Triangle, the environment runs to extremes. Much of it is bitterly cold. The warm parts are mainly dry.

Icy conditions are found not only to the north of the Population Triangle but also to the east. Beginning at the Mongolian border, nearly the whole of the eastern half of the Soviet Union has long, cold winters. The main exceptions are found in the areas south of the Population Triangle.

East of the Caspian Sea, in Soviet Central Asia, is one of the world's largest deserts — or rather, two deserts. Not much grows in the shifting dunes of the Kara Kum (KAR-uh KOOM; black sands) or the rocky ledges of the Kyzyl Kum (kuh-ZEEL KOOM; red sands).

High mountains form the southern rim of Soviet Central Asia. During the winter, snow falls on the mountains. In the spring and summer, the snow melts and feeds mountain streams, which join into rivers that flow into the desert. Two of these rivers reach an inland sea — the Aral (AR-uhl) Sea — but the rest dwindle to nothing in the desert sands. Because water is so important to life, most people in Soviet Central Asia live close to the mountains — where the rivers still flow.

One small area south of the Population Triangle does have an environment that is neither too cold nor too dry. This area lies along the Caucasus Mountains between the Black and Caspian seas. In Soviet Georgia, on the Black Sea coast, rain is plentiful and the climate is almost tropical. Such crops as tea, oranges, and lemons can grow here.

Natural Resources

Each summer groups of Soviet citizens travel to Siberia to look for gold. Like prospectors in the U.S. Old West, they head into the hills to search for the precious metal. However, these prospectors are tightly organized by the Soviet government.

Gold is also mined in Soviet Central Asia and other parts of the country. The Soviet Union is second only to South Africa in world output of gold.

Yet gold makes up only a small part of the mineral wealth of the Soviet Union. This

The Soviet Union is governed from Moscow, the capital city. Here, across Red Square, you see the wall of the Kremlin, headquarters of the Soviet government.

wealth is found not only inside the Population Triangle but also, like gold, in the vast, thinly settled areas in the north, east, and southeast.

Inside the triangle, two areas are especially rich in minerals and have thus become centers of Soviet industry. One area, in the southeastern Ukraine, is called the Donets-Dnieper (doh-NEHTS-NEE-pur) region, after two rivers that run through it. Here are found coal, natural gas, iron, manganese, and many other minerals.

The second area stretches along the Ural Mountains. Together with coal and iron, the mountains contain such metals as copper, lead, and zinc. One of the country's biggest oil fields lies under the western side of these mountains.

Since the 1960's, another mineral-rich area has been opened near the eastern tip of the Population Triangle. Large coal and iron deposits have been found in western Siberia. Even bigger deposits of oil and natural gas have been found in the Ob River Valley, north of the Population Triangle in the west Siberian taiga. The Soviet Union is the world's leading producer of oil, ahead of Saudi Arabia and the United States.

Although much of the Soviet Union is locked in by ice for many months each year, the country is rich in water resources. The Volga and many of its other rivers are used to generate electric power.

The Soviet Union also contains the deepest freshwater lake in the world — Lake Baikal (bie-KAHL), in south central Siberia. It is more than one mile (1½ *kilometers*) deep and holds almost enough water to fill all five of the Great Lakes of North America.

SECTION REVIEW

1. What are the main differences between the Population Triangle and the area outside it?

2. Why is the Ukraine good for growing grain crops?

3. Name one of the Soviet Union's leading mineral resources and describe where it can be found.

4. Why is the Soviet Union considered a landlocked nation when it has the longest seacoast of all the world's nations?

Human Geography

In a green mountain valley, six shepherds sit cross-legged on the ground. They are Soviet citizens, but they neither look nor talk like most people in other parts of the country. They are Kirghiz (keer-GEEZ) — a group that has lived near the Chinese border for centuries. They speak their own Kirghiz language.

More than 2,000 miles (*3,200 kilometers*) northeast of the Kirghiz is another group of herders. The animals they tend are reindeer. In winter these herders live in a small town in the taiga, north of the Arctic Circle. In summer they move farther north into the tundra, where the reindeer can feed on the scrubby plants that shoot up in the brief growing season. These herders belong to a group that has lived in this area for thousands of years. They too speak their own language.

About 1,600 miles (*2,500 kilometers*) west of the Kirghiz, students sit in a classroom writing an exercise. Most people in the Soviet Union would not be able to read a word of it. These students live in Soviet Georgia, and classes are taught in the Georgian language. Russian is learned as a second language. Georgians have lived in the area for many hundreds of years.

The Georgians live closer to Jerusalem, the capital of Israel, than to Moscow. The Kirghiz live closer to New Delhi, the capital of India, than to Moscow. The reindeer herders live closer to Anchorage, Alaska, than to Moscow. They are just three of many groups that live in the vast area of the Soviet Union.

The Rise of Russia

How did so many different peoples and homelands come under the rule of one central government? A Slavic group known as the Rus (roos) built a powerful state, Russia, about 1,000 years ago. By 1500 the city of Moscow had become the center of the Russian state. The city grew around a fortress near the headwaters of four rivers. The Russian word for fortress is *kreml* (KREH-muhl), and the fortress in Moscow came to be called the Kremlin (KREHM-lin). Today's Soviet government has its headquarters in the buildings of the Kremlin.

Moscow's ruler was known as the *czar* (zar) or, if a woman, *czarina* (zah-REE-nuh) of all Russia. (The words came from *caesar*, a title given to emperors of ancient Rome.) Under these rulers, the Russian borders were pushed slowly outward.

The first areas to come under Russian rule were in the north and northeast. For a long time, trappers and traders had visited northern Siberia in search of sables, silver foxes, and other fur-bearing animals of the taiga. Now traders and settlers moved into central and southern Siberia.

In some ways, Siberia was to Russia what the Far West was to America. Siberia was a wild land settled by scattered peoples — native Siberians in the northern forests and Mongols in the southern steppes. By 1800 Siberia was part of Russia.

In the 19th century, the southernmost parts of today's Soviet Union were added to the Russian empire. One part was the small area lying south of the Caucasus Mountains known as the Trans-Caucasus (meaning "beyond the Caucasus"). The Trans-Caucasus includes some peoples with ancient Christian traditions, such as the Georgians, and others who are Moslems. The largest area added to Russia in the 19th century is now known as Soviet Central Asia. Its peoples, including the Kirghiz, all have Moslem traditions.

The many different peoples who live in the far corners of the Soviet Union make up only a minority of the population. The great majority of the Soviet people are Slavs, and most of them live in Europe. Here are some glimpses into their daily lives.

City Life

Nina Lyubimova (NEE-nuh lew-BEE-moh-vuh) is 15 years old and lives in Moscow. Her father writes for the morning newspaper *Pravda* (PRAHV-duh; truth). People all over the Soviet Union read *Pravda*, which is published by the Communist party. There are no privately owned newspapers in the Soviet Union.

Nina's mother works as a bus driver. Most Soviet women work — in fact, they hold more than half of all jobs in the country.

The Lyubimov family rents a three-room apartment with a small kitchen. Nina shares a room with her sister Elena. Nina doesn't mind sharing because she knows her family is lucky. Housing is in short supply in Soviet cities — especially Moscow. One fourth of Moscow's residents must share their kitchen and bathroom with other families.

Nina's parents do not have a car, but they are saving for one. For a long time, very few cars were made in the Soviet Union for private use. Then the Soviet government decided to allow more consumer goods. In the 1970's, the government got Fiat, an Italian auto firm, to set up a plant on the Volga River, near sources of iron, steel, and power. The plant turns out 650,000 small cars a year. Now, though most Soviet citizens still do not own cars, there are sometimes traffic jams in Moscow, and on calm days the air may be clouded with exhaust fumes.

Nina thinks she is lucky to live in Moscow. The stores here have a better choice of goods than stores in smaller cities. There are lamps from East Germany, toothpaste from Bulgaria, and shirts from Czechoslovakia.

Some days Nina has meetings to attend after school. Last year she joined Komsomol (KOHM-soh-mohl), a Communist group for young people aged 14 to 28. More than half of all Soviet children join this group.

On weekends Nina sometimes joins her

More than half of all jobs in the Soviet Union are held by women. Here, in Novosibirsk (noh-voh-see-BEERSK), bricklayers work on apartment houses.

Komsomol friends on a volunteer project. If she works hard enough, she may be invited to join the Communist party, which is the only political party permitted. One Soviet citizen in 15 belongs to the party, and its members have an inside track to advancement in Soviet life.

Country Life

Viktor Murashov (VEEK-tor moo-RAH-shohf) drives tractors and trucks on a collective farm in the Ukraine, near the Black Sea. The Rossiya (roh-SEE-yah; Russia) Collective Farm specializes in corn, wheat, and cattle. The farm is enormous — 55,000 acres (*22,000 hectares*) — with 1,800 families who live on it and another 1,000 farmhands who live off the farm.

The families live in three villages that are scattered about the farm. Each family has a small house with a private plot of land for growing vegetables. Viktor Murashov also keeps a flock of chickens. Many other families also keep chickens or a cow or goat, and some have beehives to provide honey.

Viktor hopes that Rossiya will have a good harvest, since a profit for the collective means a profit for all. Once a year, if there is a profit, part of it is set aside to be invested in new equipment or buildings. The rest is divided among the farm families.

Families on the Rossiya Collective earn less than most city workers. Rural earnings are roughly 60 percent of city earnings. Farm workers may eat well, thanks to their private plots, but the stores in small towns and villages offer only a limited choice of goods.

The farm does have a medical clinic and several doctors. It also has three elementary schools and a secondary school. Movies are shown twice a week in village halls.

Viktor has seen movies about drilling for oil in the Siberian taiga and growing cotton on the edge of the Central Asian desert. But he finds it hard to imagine living anywhere different from the fertile Ukraine.

SECTION REVIEW

1. Name two ways in which the Kirghiz and Georgians differ from most people in the Soviet Union.

2. Why were Russians attracted to Siberia?

3. Which of these is *not* a problem in Moscow: air pollution, housing shortage, lack of jobs for women? Explain.

4. Why may workers on a collective farm be able to eat well when their stores do not offer a wide variety of goods?

Economic Geography

"Crisp cabbages today! . . . Ripe pears! . . . Freshly killed rabbits!" Men and women from different parts of the Soviet Union fill the air with shouts as they display their farm goods.

The place is a farmers' market in Moscow, one of 8,000 such markets in the Soviet Union. The goods sold here are grown in private plots that range up to an acre (*nearly ¹/₂ hectare*) in size. Farmers like Viktor Murashov work these plots in their spare time. Their official working time is spent on collective or state farms.

At a farmers' market, the choice of goods is much greater than in regular Soviet stores. Quality is higher — and so are prices. Some shoppers haggle, since there are no official prices at the farmers' market. In contrast, prices at stores are set by the government.

These farmers' markets are a colorful exception to the way the Soviet Union runs its economy. Overall the Soviet government decides what kinds of work will be done — and where. It has a system of **centralized planning** for farming as well as industry.

Of course, making a plan is one thing, while carrying it out is something else. For example, the environment and the location of resources set limits on the kinds of work that can be done in different parts of the country. Like other nations in the modern world, the Soviet Union often tries to push back those limits. However, in the extremes of cold and dryness found in much of its area, overcoming the environment can be difficult and costly.

Inside the Population Triangle

It's easy to see the influence of the environment in the western half of the country. Why is the Soviet government located in Moscow rather than in Siberia or on the Black Sea? As you read earlier, Moscow became a center of trade, industry, and government because of its location near several rivers. Moscow also stands near the middle of Soviet Europe — the area most favorable for settlement and therefore most densely populated.

Over the years, the city's links with the rest of the region have multiplied. Today, thanks to canals, ships can pass directly to Moscow from the Baltic, Black, and Caspian seas. Moscow is also a hub of railroad and air traffic, with routes running out in all directions like the spokes of a wheel.

The Population Triangle has nearly all of the good farmland in the Soviet Union. The northern part can grow potatoes and other root crops, while the southern part can grow wheat and other grains. Soviet planners still rely on these areas to provide the **staple** (basic) foods eaten by the Soviet people.

The Soviet Union's biggest industrial centers are in the Population Triangle. These centers developed where key mineral resources were found. Both the Donets-Dneiper area and the Ural Mountains (see page 296) have coal and iron, the basis of steel making and other heavy industries.

When the Communists first came to power, they pushed to develop heavy industry. To do so, they cut down on consumer goods. As a result, steel production soared, and today the Soviet Union is the world's leading maker of steel. While production of consumer goods still lags far

The Soviet Union relies on the fertile plains of the Ukraine, in the southwest, for most home-grown food grains. Above, workers on a collective farm winnow wheat (let the wind blow away the husks).

behind the West, the factories of the Donets-Dneiper and the Urals are busy turning out steel, machinery, and military equipment.

Outside the Triangle

A few years ago, an oil field was discovered in a swamp in Siberia. To get to the field, construction crews had to build paved roads. They did this in winter, when the muddy ground was frozen. To keep the ground from thawing in summertime, they first had to cover it with a layer of dry earth.

Since each drilling tower needed a solid base, the workers built little islands of packed earth. To cut down on construction, they drilled up to 20 shafts from each tower, slanting outward in all directions.

Once the oil field was tapped, the oil had to be moved to market. The Soviets built a long pipeline to a rail line, where the oil is loaded into tank cars. Other Siberian pipelines snake their way across swamps, rivers, and the Ural Mountains to the industrial centers of Soviet Europe.

Outside the Population Triangle, the Soviet Union is rich in resources. In addition to oil and other minerals underground, there is timber from the forests and hydroelectric power from the rivers. But the harsh environment makes it difficult to develop those resources. Transportation is the key to development. The need for roads and railroads is always greater than the supply.

The first modern route across Soviet Asia — the Trans-Siberian Railroad — is still the busiest. This railroad was completed early in the 20th century. It is so long that a trip from end to end by train takes almost 10 days.

Cities grew at points where the Trans-Siberian crossed major rivers. Look again at the map on page 294. What Siberian cities are located on both a railroad and a river?

Highways are not common in the Siberian wilds. Unpaved roads may turn to mud in summer, while paved ones may buckle from the frost. With vast distances and few highways, much of the Soviet Union relies on air transportation. The government-owned airline, Aeroflot (ehr-oh-FLOHT), is the busiest airline in the world.

To ease the demand on the transportation system, places outside the Population Triangle are expected to provide as much of their own food as they can. This causes problems — some of which remain to be solved.

In Soviet Central Asia, which is warm but dry, irrigation is the answer. A vast network of canals has been built. The biggest of these canals carries river water some 400 miles (650 kilometers) across the desert. Some canal water is used to grow rice, a staple food in Central Asia.

In the north, extreme cold and the brief growing season make food growing difficult. In Yakutsk, scientists have developed special varieties (types) of wheat, rye, and barley. These varieties grow quickly — they can be harvested 60 days after sowing.

The Soviet government would like to develop its vast frontier lands even faster. However, it still has to face economic problems in the Population Triangle — lagging food production and the demand for more consumer goods. At present the lands of the north and east serve mainly as suppliers of raw materials to more developed areas. These areas include not only the Population Triangle but also the nations of Eastern Europe. In the next chapter, you will read about the biggest of these East European nations, Poland.

SECTION REVIEW

1. Are farmers' markets typical of the Soviet economy? Give reasons for your answer.

2. Give two reasons why Moscow became an important center of trade and government.

3. What are staple foods? Which part of the Soviet Union provides most of them? Why?

4. Why is transportation a problem outside the Population Triangle? What transportation routes are most used there? Why?

YOUR LOCAL GEOGRAPHY

1. Imagine that you have been given the chance to become a pen pal of Nina Lyubimova or Viktor Murashov. Write a letter, introducing yourself and your community. In describing your community, use the three main headings in the chapter: Physical Geography, Human Geography, and Economic Geography.

2. Few places on Earth can rival Siberia when it comes to harsh living conditions. But no matter where you live, chances are that the environment also causes some problems — too little water, poor soil, landforms that make transportation difficult, etc. What would you consider the main environmental drawbacks of your area? Can you name some of the ways in which they might be overcome? Has any action been taken to deal with the problems? What action?

Reading a Loss/Gain Line Graph

Suppose a Soviet family lives near the Black Sea in a mild climate. Then they get jobs in the eastern city of Irkutsk (eer-KOOTSK). Would they have to adjust to a different climate? The line graph below can help give answers.

The graph shows the average monthly temperatures for the city of Irkutsk. You will find some features like those of other line graphs. A vertical axis on the sides shows the temperatures in degrees. A horizontal axis on the bottom shows 12 months. An indicator line shows the rise and fall of temperatures.

Unlike many other graphs, however, this one has a "floating axis" across the middle that represents zero. Where the indicator line falls below the middle axis, it represents an amount *less* than zero. Where it goes above the floating axis, it represents an amount *more* than zero. This type of graph is sometimes called a **loss/gain graph**.

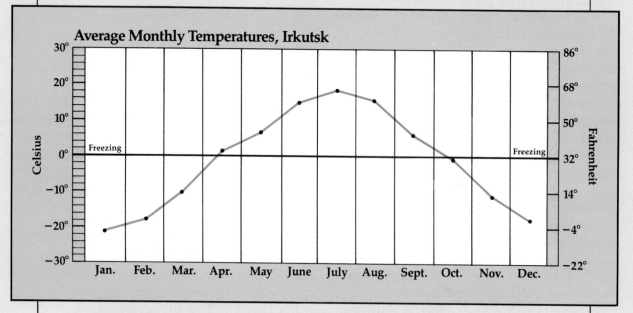

Notice that temperatures are given in the Celsius scale at the left of the graph. Matching Fahrenheit temperatures are shown on the right. Study the graph and answer the following questions on a separate sheet of paper.

1. The floating axis on the graph represents "freezing" (the temperature at which water freezes to ice). This is represented by what temperature in: (a) Celsius degrees? (b) Fahrenheit degrees?

2. Study the left vertical axis. What numbers are printed for Celsius temperatures: (a) above freezing? (b) below freezing?

3. What is the average January temperature in Irkutsk?

4. Average temperatures in Irkutsk are below freezing for how many months?

CHAPTER REVIEW

A. Words To Remember

Three of the following terms are defined by the numbered phrases below. On another sheet of paper, write each term next to the number of its definition. Then write the two terms that are *not* defined and give a definition for each.

chernozem   spring wheat   winter wheat
loess   staple

1. Basic, as the kind of food that people rely on for much of their diet.
2. Rich black soil found mainly in the Ukraine.
3. Wheat that is planted in the fall and harvested the following summer.

B. Check Your Reading

1. Why does the majority of the Soviet population live in the western area known as the Population Triangle?
2. Land in the northern Population Triangle has poor drainage. What does this mean? How does it affect farming in the area?
3. Why did the Donets-Dnieper and Ural Mountain areas become centers of heavy industry?
4. Outside the Population Triangle, the environment tends to run to extremes. What are these extremes?
5. What was the first modern transportation route across Soviet Asia? Why is it still important?

C. Think It Over

1. "In some ways, Siberia was to Russia what the Far West was to America." In what ways did the two areas *differ*?
2. What argument might a resident of Moscow give for preferring city life in the Soviet Union? What argument might a resident of the Rossiya Collective give for preferring farm life?
3. How do the farmers' markets differ from the way the rest of the Soviet economy is run?

D. Things To Do

Using an atlas, encyclopedia, or other source, find a list of the world's longest rivers. From your source, make your own list of rivers longer than 1,000 miles (*1,600 kilometers*) that flow partly or entirely through the Soviet Union. Find these rivers on a map of the Soviet Union. Write a brief report on the importance of the Soviet rivers.

Chapter 22

Poland

A Nation of Dramatic Changes

From the controls of her crane, Ewa Lubowska can look out on the bay at Gdansk (guh-DAHNSK). She can see the ships waiting to unload at this Polish port. Ships are Ewa's business — the crane she operates is helping to build them. Poland depends on ships to bring in goods from the rest of the world — and to carry Polish goods to other nations.

Ewa makes good money by Polish standards. Poland is trying to build up heavy industry. So its Communist government sets a wage scale that favors workers in shipping and coal mining. Ewa makes more money than a doctor or a teacher.

Her husband, Jerzy Lubowski, also makes good money. He is a crewman on a Polish merchant ship that travels between Poland and Central America. Jerzy gets part of his wages in U.S. dollars rather than in Polish currency, the *zloty* (ZWOH-tee). Dollars can be used to buy imported goods — like the Japanese color TV in the Lubowskis' living room.

In 1971 Poland began an all-out push to build up industry. For a time, the Polish people saw their standard of living rise. But they demanded other improvements in their lives. This led to conflict with Poland's Communist rulers — and with its giant neighbor, the Soviet Union.

In this chapter, you will look at the lives of the Polish people — and at the way the environment has helped shape those lives.

Physical Geography

Poland's only natural borders are a sea on the north and mountains on the south. The nation sits in the middle of a plain that stretches to the horizon on both the east and the west. Hostile armies can easily march in — and they often have. Polish armies can also march out — and *they* often have.

As a result, Poland's borders have shifted with the tides of battle. One thousand years ago, Poland was roughly where it is now. Then it expanded to the east. For a time in the Middle Ages, Poland (linked with Lithuania) was the largest nation in Europe; but then it shrank. At the end of the 18th century, Poland was carved up by three neighbors — Russia, Austria, and what today is Germany — and vanished from the map.

For more than a century, the Polish people kept alive the dream of an independent Poland. World War I (1914–1918) weakened Germany, Russia, and Austria. When the war ended, Poland was back on the map.

Then at the beginning of World War II, Nazi Germany and the Soviet Union each seized half of Poland. Later in the war, Germany invaded the Soviet Union. When the

---

**Though Poland is Communist-ruled, the great majority of its people are Roman Catholics.
At left, Polish workers wait for a priest to begin an outdoor service in Gdansk.**

Poland

BALTIC SEA

20°E

24°E

16°E

LITHUANIA
(SOVIET UNION)

54°N

54°N

Gdansk •

EAST GERMANY

Poznan •

Vistula R.

★ Warsaw

N

Odra (Oder) R.

Lodz •

S I L E S I A

Wroclaw •

SUDETEN MTS.

Katowice •

50°N

16°E

Krakow • Nowa Huta

50°N

Oswiecim •

CARPATHIAN MTS.

KEY

International
boundary

★ Capital city

• Other city

CZECHOSLOVAKIA

0   75   150 miles

0   75   150   225 kilometers

20°E

Mountains and the Baltic Sea form natural borders in the south and north of Poland. However, a plain runs all the way across the country from east to west. On these two sides, Poland's borders have often shifted.

tide of battle turned, the Soviets rolled the Germans back and occupied all of Poland. The Soviets overthrew the democratic Polish government and replaced it with a Polish Communist government.

New Polish borders were drawn. The Soviet Union gave back very little of the land it had seized, while Germany had to give back all of its half and more. These changes brought Poland to its present shape and size.

Today Poland has about as much land as New Mexico. It has 36 million people — more than 25 times as many as New Mexico. In both land and people, Poland is the largest nation of Eastern Europe outside the Soviet Union.

The Polish Plain

Poland is the only country in Europe that is mainly on a plain. (The term *Poland* comes

from the name of a Slavic tribe that meant "plains-dwellers.") Three fourths of Poland is made up of lowlands, mostly in the center and north. These lowlands form part of the vast European plain, which stretches far into the Soviet Union (see page 281).

Thousands of years ago, glaciers scraped across the Polish plain. They retreated in several steps, leaving strips of differing types of land.

Farthest north are the swamps and dunes of the Baltic coast. The coast has a few inlets that are protected from the open sea and therefore make good harbors. There are many sunny beaches. In summer the beaches are packed with swimmers and sunbathers.

South of the coast is a belt of hilly land, consisting of **moraines** (hills made up of rocks dropped by the glaciers). The soil here is too thin to be good for farming. However, the hills are heavily wooded, and the area is dotted with lakes. This is prime vacation land for Polish campers, fishers, and hunters.

The central Polish plain forms a third belt. Glaciers pulled back from this zone at an early time, and the soil in some places is richer. This region is the Polish heartland. It includes several major cities, including the capital of Warsaw.

To the west is a strip of lowlands known as Silesia (sie-LEE-zhuh). The Silesian plain has a rich black earth built up by loess, as in the Ukraine (see page 294). This area is Poland's wheat belt, containing some of the country's best farmland.

Most of Poland's farming is done in the plains and other flat places. Poland has a slightly milder and moister climate than much of Soviet Europe, since Poland is closer to the Atlantic Ocean. However, Poland lies farther north than any part of Maine, so the growing season is short. The main crops are those that will grow in the cool summer — rye, oats, barley, potatoes, and sugar beets. As you have seen, Poland also grows wheat — mainly in the south, where summers are longer. In a few places, summer temperatures are hot enough to grow corn.

Only in the south does the Polish land make a complete break with the plain. Two scenic mountain ranges — the Sudeten (soo-DEH-tuhn) Mountains in the west and the Carpathians (kar-PAY-thee-uhnz) in the east — divide Poland from Czechoslovakia. These mountains are made up of a series of ridges with green valleys in between.

In these valleys, flocks of sheep graze in pastures, and streams rush down the wooded mountainsides. As the streams reach the lowlands, they join into rivers. Many small rivers wind northward across the Polish plain to the Baltic Sea. The two biggest are the Vistula (VIS-tew-luh), which flows through Warsaw, and the Odra (OH-druh), which is also known as the Oder. However, these rivers are small compared to major European rivers such as the Danube.

Mineral Resources

Poland's key mineral resource is coal. The richest coal deposits are in the southwest, in Silesia. These deposits lie in seams that vary up to as much as 30 feet (*10 meters*) in thickness. This is the second largest coal field in Europe. Only in the Ruhr Valley of West Germany is there known to be more coal.

Compared to coal, Poland's other mineral resources are poor. There are some deposits of iron, lead, zinc, and copper; but most metals have to be imported.

Poland's salt mines are worth noting. One of the most unusual mines can be found near Krakow (KRAH-koov), in southern Poland. Deep in the ground, miners have carved a chapel into the walls of salt. This chapel contains delicate salt sculptures of Roman Catholic saints. Much of the work

Most farms in Poland are small and privately owned — unlike the state farms and collectives in other parts of the region. Polish farmers usually make do with horses rather than buy tractors and other machinery.

was done two to three centuries ago, but the mine is still in production, turning out 700 tons of salt each year.

Today salt mines are worked from the surface. Water is forced into the ground, causing the salt to dissolve. Then the briny water is pumped out. The water is evaporated, and salt remains.

SECTION REVIEW

1. What feature of Poland's environment has had an influence on its shifting borders? How has it influenced them?

2. How does Poland compare in size and population with the other nations of Eastern Europe?

3. In what part of Poland are there mountains? What neighboring nation shares these mountains?

4. What is Poland's major mineral resource? How do its deposits compare with those in other parts of Europe?

Human Geography

For three years, Roman Catholics in a small town northeast of Lodz (wooj) spent their evenings building a new church. They had many difficulties to overcome. For one thing, they could not openly buy lumber and bricks. Government rules place limits on the number of churches that can be built each year. The reason, say officials, is that materials are in short supply. Poland has a housing shortage, and officials claim that the materials are needed for home building.

Sometimes the police came and tore down part of the new building, but the church members kept on with their work. At last the authorities gave in, and the church was completed.

Although the Communist government opposes religion, most Polish people remain Roman Catholics. The Church has played a central role in their lives for many centuries. During the years when Poland vanished from the map, the Polish priests of the Catholic Church reminded Poles of their unity.

Today the Church still plays a key role. In some ways, it competes with the government for the loyalty of Polish people. No other Communist nation has such a strong organization to balance the power of the state.

Few Minorities

Unlike most nations of Eastern Europe, Poland today has few minority groups. Before World War II, however, only two thirds of Poland's people were considered Polish. There were large groups of Ukrainians, Byelorussians, and Germans; and about 10 percent of the people were Jewish.

The war all but wiped out Poland's Jewish population. Some died fighting the Nazis, while millions were killed in Nazi death camps. One of the worst death camps was at Oswiecim (ohsh-VEE-cheem), west of Krakow. It was known by its German name Auschwitz (OWSH-veets).

After the war, as you read earlier, Poland's boundaries were changed. As a result, some minority groups in Poland became residents of another country without leaving their homes. Others migrated across the new borders. Minority groups make up barely one percent of Poland's population today.

You would meet a unique minority group if you visited the Polonia (poh-LOHN-yuh) Club in Warsaw. The club's members are all Polish Americans.

Stan Lopatko (woh-PAHT-koh; not his real name) is one of those Americans. He was born in a suburb of Cleveland, Ohio. His parents had immigrated from Poland and spoke Polish at home.

When Stan retired, he felt a desire to return to his family's roots. So did his wife Dora, whose parents had also come from Poland. Therefore, they moved to Poland — as have some 4,000 other retired Polish Americans — but remained U.S. citizens.

In Poland their Social Security pension stretches quite far. As in other Communist countries, wages are very low by Western standards. Although most consumer goods are expensive, items such as rent, utilities, and medical treatment cost very little. Thus Stan and Dora pay just 50 dollars a month for rent on their two-room apartment. When Stan spent a week in a medical clinic for treatment, his bill came to $1.56 a day.

Stan's chief complaint is about the food, especially the shortage of meat. Often he or Dora must wait two hours in line at the butcher shop. Even then the meat may be sold out when their turn comes.

Poland and the U.S. have had many ties over the centuries. A Polish patriot, Thaddeus Kosciusko (kohsh-CHOOSH-koh), helped Americans win the American Revolution. In the late 19th and early 20th centuries, about one million Poles emigrated to the U.S. Today more than six million Americans trace their roots to Poland.

Farm Life

Until well after World War II, Poland was mainly a nation of farmers. Now Poland has built up industry, but its three million farms still play a key role in the nation.

Lech Dornowski (lek dor-NOHF-skee) and his wife Anna own one of these farms, which covers some 17 acres (*seven hectares*) west of Warsaw. This farm is quite different from the Soviet collective farm you read about on page 299.

The Dornowskis have one cow, one horse, two pigs, and a flock of chickens. They use the horse to pull a wagon, when they ride to a nearby village. They also use the horse to pull the plow. They could hire a tractor and driver from the state, but Lech and Anna prefer the old ways.

Warsaw and other Polish cities were destroyed or severely damaged during World War II and have been rebuilt since. Here, girls in communion dress walk home to a freshly constructed apartment building.

Most of what the Dornowskis grow is for themselves and their animals. They grow a little rye, a few oats, a little hay. Since they eat potatoes at every meal and also feed potatoes to their animals, they have a large potato patch. They also grow beets and cabbage, which are key ingredients in a soup the Poles call *barszcz* (barsh), known in the U.S. as borscht.

Trees grow on part of the farm. Since the soil there is poor and sandy, Lech has not tried to clear it. Instead he lets his pigs root for food under the trees. Many Polish farmers keep pigs because pigs eat so many things — even table scraps. Polish pigs are noted for fine, lean meat. Polish hams are exported, and are found in many U.S. stores.

Anna milks the cow every morning at dawn, and every afternoon when the village church bell rings. She also collects the eggs from the chickens. Since the Dornowskis cannot use all of the milk and eggs, Anna sells the surplus to villagers. The money she receives forms the bulk of their income.

Lech is 74, and Anna is 73. In a few years, they will have to retire. They have no children, so they will probably turn the farm over to the government. This will get them a small pension, and they will be allowed to stay in their home.

City Life

Many Polish people, like people elsewhere, are leaving the country to move to cities. The growth of Polish industry has opened many new jobs. The government tries to spread these jobs around, setting up factories in areas that had little industry.

Nowa Huta (NOH-vuh HOO-tuh) is a new factory town. It was built in the 1950's on the outskirts of Krakow, and is now a major steel-making center. More than 100,000 people live there, in row after row of tall apartment blocks.

Hanna Dobkowska (dohb-KOHF-skuh) lives in one of those buildings. Her parents both work in a steel mill. She has just finished high school and is studying for a national college entrance exam. Hanna hopes to attend the University of Krakow. Founded in 1364, it is the oldest university in Poland — and one of the oldest in the world. It is one of 12 universities in Krakow.

Hanna spends every moment she can in Krakow, 30 minutes away by streetcar. She thinks it's a marvelous city. The bombings of World War II largely spared its ancient buildings. Hanna's favorite place is a youth club in the cellar of an old palace.

Sometimes Hanna sips tea in a grand café dating from the 19th century. Sometimes she visits an art gallery to see a show of modern art. She also likes to browse in the many book stores in this city of learning. Some of the books are American, and she often looks through them to try out her English.

SECTION REVIEW

1. What role does the Roman Catholic Church play in Poland today?

2. Why does Poland have few minority groups today?

3. How does the Dornowskis' farm differ from a collective farm?

4. What are the major differences between Krakow and Nowa Huta?

Economic Geography

In the mountain valleys of southern Poland, farmers and sheepherders dress in colorful costumes on Sundays. The men sew fancy embroidery down the fronts and sides of their wool trousers and on their jackets. The women wear brightly colored skirts, white blouses, and a sort of vest embroidered with fancy beads.

In the past, this sort of "peasant dress" was widely worn in Poland. Today, however, most Polish farmers go in for the same type of factory-made clothing that people wear throughout the modern world.

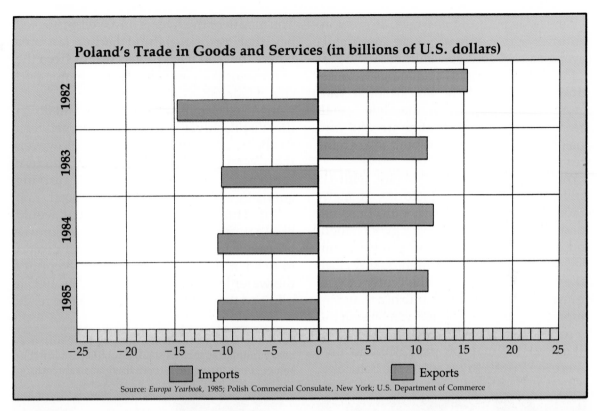

Poland's Trade in Goods and Services (in billions of U.S. dollars)

Source: *Europa Yearbook*, 1985; Polish Commercial Consulate, New York; U.S. Department of Commerce

□ Imports   □ Exports

All the same, there have been fewer changes in Poland's farms than in those of the Soviet Union and the other Communist nations of Eastern Europe. Like the Dornowskis, most Polish farmers still own their own farms. In the early years of Communist rule, the government began a drive to bring all farms under its control, but ran into stiff opposition from small farmers. Today there are relatively few state farms.

Many of Poland's farms are like the Dornowskis' in another way. They are small, and make little use of tractors and other farm machinery.

Booming Industry

While few machines clatter across Poland's fields, they are roaring away in the nation's cities. Poland is among the world's top 10 industrial nations.

As you have read, coal is Poland's main natural resource. Thus many of Poland's industries first developed in the coal-mining

Poland spends money on goods imported from other countries and earns money from Polish goods sold abroad. This graph shows both amounts of money for four recent years. It is a loss/gain bar graph, similar to the line graph on page 302. Has Poland spent more than it has earned? Why do you think trade decreased between 1982 and 1983?

areas of Silesia. Over two million people live in a cluster of industrial cities along a 30-mile (*50-kilometer*) Silesian valley.

The valley was once beautiful, but now it is clouded with smoke and covered with slag heaps. (**Slag** is the waste left when metal is extracted from ore.) Because iron as well as coal was found in this area, steel making became important. Today the area turns out products such as locomotives, heavy machinery, and farm equipment.

There are other traditional centers of industry — but also new ones. Hanna Dobkowska's hometown, Nowa Huta (see page 310), is just one of the new ones.

Warsaw is both old *and* new. It grew up centuries ago at a spot where the Vistula River was easy to cross. During World War II, four fifths of Warsaw's buildings were destroyed. Afterward the Poles rebuilt Warsaw's "old town" in the original style — with low buildings, cobblestone streets, and quaint cafés. A tunnel was built to keep the area free of heavy traffic.

Warsaw is Poland's biggest city, with 1.5 million people. It has factories of many kinds. Some factories turn out consumer goods, such as clothes, books, TV sets, food, and automobiles; others turn out cement, electrical equipment, and high-quality steel. Like Moscow, Warsaw is the center of government planning and the hub of a transportation network. Its products can be sent out by road or rail in all directions.

Look at the map on page 306 and note where the Vistula River enters the Baltic Sea.

The mouth of the river is on a well-protected bay that forms a **natural harbor** (an area where the land itself protects ships from the open sea). As you can see, the major port city of Gdansk is on that bay.

In the past, rafts would float up and down the Vistula River, carrying goods between the Baltic ports and inland cities. However, the river is too shallow for most modern cargo boats. As you read in Unit 1, carrying goods by ship is cheaper than by road, rail, or air. Thus the Polish government would like to turn the Vistula back into a major waterway. It has made plans for a vast project to deepen the river channel and to control the water level by building locks. By 1995 it is hoped that coal barges will be able to reach

Poland is a major industrial nation. In addition to heavy industries such as steel, it produces many consumer goods for export — as in this glassware factory.

Silesia. Some of the coal these barges carry will go to cities such as Warsaw, but most will be exported.

Challenge and Change

The Polish people have often shown that they were not afraid to protest against the Communist government. In an attempt to avoid such protests, the government has sometimes raised wages or lowered prices. However, these moves put a squeeze on the nation's economy.

The government was most concerned about the prices of meat and other foods. It was buying these foods from farmers at one price and selling them to consumers at a lower price. Thus it was losing money with every meal of 35 million people.

At first the government simply tried to raise food prices, but riots broke out. It then tried a more subtle approach. For example, the best cuts of meat were withdrawn from butcher shops and put on sale in fancy stores — at much higher prices.

Meanwhile more far-reaching changes were sweeping through Poland. Polish workers and farmers set up independent labor unions. Unlike the "official" unions that exist in the Soviet Union and most of the other East European countries, these Polish unions were controlled by the workers, not by the government. Showing their power with brief nationwide strikes, the unions forced the Communist government to recognize them. They also pressed for more freedom in everyday life.

In 1981 the Communist government cracked down on the movement for more freedom. The Poles had known that they were running great risks. The Soviet Union had seen the events in Poland as a challenge to its own power, and pressed the Polish government to take stronger action. There was always the threat that the Soviet Union itself would move in. Twice before, the Soviet Union had used force to crush popular opposition to Communist governments in Eastern Europe — in Hungary in 1956 and in Czechoslovakia in 1968.

The drama of Poland overshadowed most other events in the region. At the same time, the Soviet Union and Eastern Europe have undergone changes of their own. The following chapter will look at these changes.

SECTION REVIEW

---

1. Why did many of Poland's industries develop in Silesia?

2. Why would the Polish government want to turn the Vistula River into a major waterway for cargo ships?

3. What is the main difference between "official" labor unions and the independent unions formed by Polish workers?

4. In what way did food prices cause problems for the Polish government?

YOUR LOCAL GEOGRAPHY

---

1. In Chapter 20, you read that rivers serve as important routes for trade in parts of Eastern Europe and the Soviet Union. However, Poland does not at present make much use of its rivers. What use is made of rivers in or near your community? Are there any rivers that are navigable? If not, can you say why (e.g., too little rainfall, or the land is too mountainous)? If there are navigable rivers, do they carry much traffic? Do they link your community with a seaport or industrial center? Would you say that rivers play an important or unimportant role in your local transportation?

2. Because Poland is on a plain, its borders have changed many times in the past, and different peoples have moved back and forth across them. Has the setting of your community made it a "highway" for different peoples in the past? Did American Indian groups move across it? Did early European visitors — traders, fur trappers, explorers, missionaries, etc. — travel through your locality? If so, can you say why?

313

Reading a Tree Chart

The English word *three* is very similar to the Polish *trzy* and the Russian *tri*. Most languages of Eastern and Western Europe belong to the same language family. They developed from an ancient language called Indo-European.

The Indo-European language family is illustrated on the **tree chart** below. This type of chart shows people or things branching off from a common source, like limbs from a tree trunk. Tree charts can also show how one event, such as the discovery of a new resource, is connected to later events.

You read a tree chart from the trunk toward the branches. In this chart, the trunk represents Proto-Indo-European, the common ancestor of the Indo-European language family. As you read up the chart, notice how this ancestor divided into two branches. One branch is called Centum. What is the other branch? The two main branches divide into more branches. A division of a division is known as a **subdivision**. Answer the following on a separate sheet.

1. The Centum languages are split into which four subdivisions? The other main branch of the family is split into which four subdivisions?

2. Find English on the tree chart. English belongs to: (a) which major language branch? (b) which subdivision? (c) which smaller branch?

3. Find Polish on the tree chart. Polish belongs to: (a) which major language branch? (b) which subdivision? (c) which smaller branch of that subdivision? What other language is the closest relative to Polish?

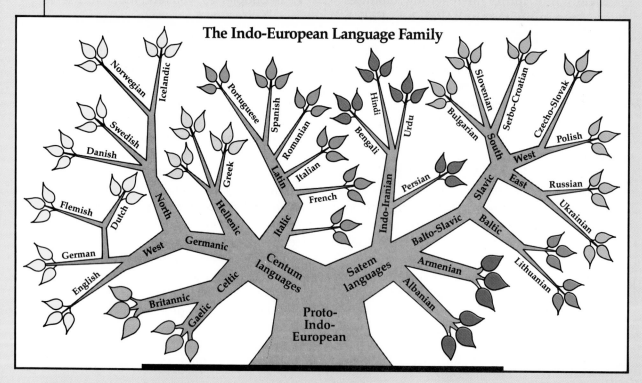

The Indo-European Language Family

CHAPTER REVIEW

A. Words To Remember

From the following list, select the term that best completes each sentence below. Write your answers on a separate sheet of paper.

Kosciusko   moraine   slag
Krakow   natural harbor   subdivision
loess   plain   Warsaw

1. _____, the capital of Poland, is also the hub of its transportation network.
2. A _____ is a hill made of rocks dropped by glaciers.
3. The Vistula and Odra rivers flow northward across the Polish _____.
4. The oldest university in Poland is located in _____.
5. Poland's major port city, Gdansk, is located on a _____.

B. Check Your Reading

1. What natural forces shaped most of Poland's land surface: earthquakes, floods, or glaciers?
2. Which of the following crops is grown least in Poland: rye, potatoes, wheat, or corn? Why?
3. Changes in Poland led to the fear of Soviet invasion. Why might the Polish people have expected such an invasion?
4. Would it be most accurate to call Poland an industrial nation, a nation of farmers, or a developing nation? Explain.
5. Why did Poland's Communist government set a wage scale that favors workers in shipping and coal mining?

C. Think It Over

1. Why have Poland's borders changed less in the north and south than in the east and west?
2. Describe three features of Warsaw that make it an important city.
3. Changes in Poland in the 1970's and early 1980's led to the fear of Soviet invasion. Why might the Polish people have expected such an invasion?

D. Things To Do

Construct a three-dimensional map of Poland and the area just around it, using the physical description given in this chapter for reference. (Cardboard, colored paper, styrofoam, pebbles, sand, and glue should be sufficient to make a good geographical model.) Add labels for the Baltic coast, the Polish plain, Silesia, the Sudeten Mountains, and the Carpathian Mountains.

Chapter 23

A Shared World, A World Apart
The Region in Perspective

On still days, a thick haze floats over Krakow, Poland. People cough and rub their smarting eyes. Each year the factories around Krakow spew 150,000 tons of waste into the air. Soft coal burned to heat homes and offices adds more smog, as do auto fumes. Along with the benefits of the modern world, Poland now has to deal with its problems — including pollution.

Lake Baikal in Soviet Siberia is famous for its pure blue waters. The waters are still blue, but they're no longer pure. Nearby factories dump their wastes into the lake. The chemicals threaten to make the lake's fish unfit for eating. Not even remote Siberia has escaped pollution.

These are two examples among many to show that Eastern Europe and the Soviet Union share some of the problems faced by the rest of the modern world. Yet the region is not *just* like the rest of the world. Its political and economic system sets it apart, and so, in many ways, does its geography.

A Tightly Knit Group

In the course of this unit, you have met young people from different parts of the region. Suppose that Maria of Hungary, Nina of Moscow, Amin of Tashkent, and Hanna of Poland all happened to meet. What language would they speak?

The answer is almost certainly Russian. Though English and other languages are taught in high schools throughout the Soviet Union and Eastern Europe, Russian is the most widespread second language. This is just one sign of the Soviet Union's tremendous influence in the region.

Since the end of World War II, the Soviet Union has kept most of the Communist nations of Eastern Europe under tight control. Only Yugoslavia stayed outside the Soviet bloc (group of nations), and only Albania has broken from it.

Suspicion and hostility divided the region from the Western world. This conflict, which stopped short of fighting, was known as the Cold War. There were few contacts in travel or trade between East and West. The Soviet bloc relied as much as possible on its own resources. It **isolated** itself (cut itself off) from the outside world.

As you read earlier, the Soviet system is highly centralized — that is, most decisions are made by the central government in Moscow. The Soviet Union set up a similar kind of centralized system for the nations of Eastern Europe. This system is run by an organization called the Council for Mutual Economic Assistance — known as Comecon. (Comecon now includes three Communist countries from outside Europe — Cuba,

"Warning! You are now leaving West Berlin," says this sign at the wall of Communist East Berlin. To what extent do the Soviet Union and Eastern Europe live in their own world?

317

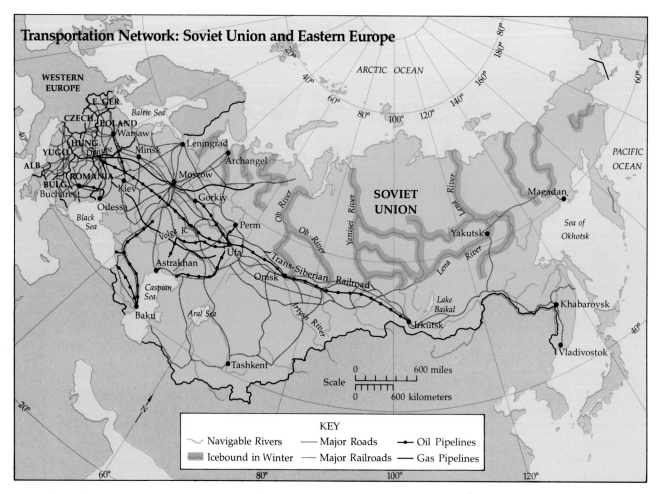

Transportation Network: Soviet Union and Eastern Europe

KEY

~ Navigable Rivers — Major Roads •— Oil Pipelines
▒ Icebound in Winter — Major Railroads — Gas Pipelines

Vietnam, and Mongolia.) The Soviet Union controls Comecon. As you will see, however, some of the other nations have taken a more independent line in recent years.

Comecon works to promote trade among Communist nations, and also serves as a planning agency. For example, member nations plan their future energy use together and then pool some of their resources. Comecon has set up a grid that links the electric power systems of many nations in Eastern Europe. When Maria switches on a light, the electricity she uses may have been generated in Poland.

Can one part of the world be self-sufficient and survive on its own? Though the region is vast and has a tremendous variety of resources, there are still some needs that can-

The region has a dense transportation network in the west, where most people live. The routes there also tie Eastern Europe closely to the Soviet Union. But moving resources from Soviet Asia is still a problem.

not easily be met. For example, basic foods such as wheat and other grains are often in short supply. The Soviet Union has large expanses of good farmland, similar to the Great Plains of the U.S. However, that Soviet farmland has a slightly drier, colder climate, which means that a bad year — extra dry or extra cold — will hurt crop production in the Soviet Union more than in the U.S.

In addition, no part of the region lies within the tropics, and only small areas have a subtropical climate. Thus hot-weather crops such as cacao, bananas, and sugarcane are either scarce or not grown at all. However, this problem has been partly solved now that Comecon includes Cuba and Vietnam, which are tropical countries.

As you saw in Chapter 21, the region has many resources that it cannot readily use. Sugar from Cuba can be transported to Moscow more easily than minerals from northern Siberia. Although the Soviet Union is pushing its transportation links outside the Population Triangle, it still faces the problems of immense land distances and frozen waterways.

SECTION REVIEW

1. Which of the following is frequently imported by the Soviet Union: minerals, grains, potatoes? Why?

2. What does it mean to say that the Soviet system is highly centralized?

3. What is Comecon?

4. The region cannot grow crops that need a certain kind of climate. What climate is that? How can Comecon make up for some of this lack?

Looking to the West

Some of the bread that Nina eats in Moscow is made from American wheat. For many years, the Soviet Union has been buying large amounts of wheat, corn, and other grains from the U.S.

The banana you eat for lunch may have come to the U.S. in a ship that Ewa Lubowska of Poland helped to build. The shipyards at Gdansk have made refrigerated vessels for U.S. shippers.

These are just two of the ways in which links between the region and the rest of the world have grown in recent years. The region has not been self-sufficient. Instead it has become interdependent with other parts of the world. (With **interdependence**, countries rely on trade with one another to help supply their needs.) East European nations, in particular, have stepped up their trade with the West — especially with their West European neighbors.

This increase in trade is due in part to a need for goods that are in better supply in the West. For example, the Communist nations lag in high technology (complex machines and systems such as computers and modern oil-drilling equipment).

At the same time, some East European nations see another advantage in trade with the West: It makes them less dependent on the Soviet Union. Yugoslavia has traded with the West and stayed independent of the Soviet Union all along. Poland began a drive for Western trade in the 1970's. Romania has kept a tight lid on dissent among its people, but has quarreled openly with the Soviet Union over foreign trade.

Signs of Change

When young people in the region go to buy clothes, they find a wider selection of styles and colors than their parents had as young people. In recent years, most of the countries in the region have produced larger amounts and varieties of consumer goods. In the past, Communist countries put heavy industry and bare necessities first, and made few plans for producing consumer goods. Shortages of these goods led to increasing opposition from people in the region, and governments had to change their policies.

Some other changes have had to be made to the Communist economic system. Centralized planning of all details of the economy has led to many mistakes. For example, one factory may be directed to produce too many shoes of one size. Another factory may receive more raw materials than it needs for its production quota. (A **quota** is a planned share or amount.) Yet factory managers have nothing to gain by questioning the plans or making improvements.

Under the Communist system, the planning agency also sets prices. In the U.S. and other free-enterprise countries, the price of a product is based on the cost of making it. Suppose two companies produce shoes of the same quality, but one makes them at less cost. This company can set lower prices and thus sell more of its shoes. If the other company does not lower its costs and prices, it may go out of business. This way of setting prices helps to make business efficient.

In a Communist economy, however, prices may not be based on cost at all. Some consumer goods may be priced high because the government does not want them produced in large amounts. Other goods may be priced less than they cost to make. Even if a factory finds a way to cut costs, it cannot lower its prices. Thus there is no pressure for factories to be more efficient.

In recent years, several East European nations have tried to improve their economies by bringing in some ideas that are familiar in the West. In Hungary, for example, factory managers no longer have to follow orders from the top in detail. Instead they are encouraged to find new ways to cut costs and increase profits. There are **incentives** (rewards) for becoming more efficient. Those who succeed are given higher pay and special benefits, and the workers receive more money.

Even the Soviet Union is taking a second look at its rigid centralized system. Without going so far as Hungary, the Soviet Union has given some leeway to local managers.

Increasing contacts with the rest of the world have helped to spread new ideas within the region. So has the drive to develop new resources and new industries. For example, opening up the resources of Soviet Asia needs the most up-to-date equipment and methods — even if they come from the West.

As you saw at the beginning of this chapter, the region shares the problems of the modern world, including air and water pollution. Winds blow polluted air across borders, and polluted water drifts from one country's coast to another's. These problems cannot be isolated within one region.

As you read in Chapter 22, the Soviet Union has used force or the threat of force to keep East European nations under its control. At the same time, most of these nations are less dependent on the Soviet Union than they were 20 years ago. Even the Soviet Union trades more with the West than it did in the past. These and other changes have made the region increasingly a part of our world. For this reason, Maria, Nina, and the other young people you have met may lead lives quite different from their parents'.

SECTION REVIEW

1. What is high technology? What role does it play in the region's trade?

2. How are prices set in Communist countries?

3. What are incentives? Why have some countries in the region started to use them in industry?

4. In what ways is the region still a "world apart"?

YOUR LOCAL GEOGRAPHY

1. Is any kind of pollution (air, water, noise, etc.) a problem in your community? If so, what kind is most serious? Is the pollution created in your community, or does it spread from other areas? Is any action being taken to reduce pollution? Are there any ways in which people your age can help?

2. Suppose you were given the opportunity to live for a year with either Maria (Hungary), Nina (Moscow), Amin (Tashkent), or Hanna (Poland) in their native lands. Where would you choose to go? Why? Then imagine that you have been there for one month and are now writing home. Try to tell your family how your new environment compares with the community you left behind.

Comparing a Graph and a Table

There is much trade between the Soviet bloc and Western nations—including the United States. To learn more about this trade, you might look at the amount of money spent over a number of years by Soviet bloc countries on goods from the United States.

Suppose you wanted to display this information graphically. One way would be on a complex line graph such as the one below. The same information can also be shown in a table, such as the one on page 322. Facts about a subject can often be shown in more than one way. The way information is presented depends on its *purpose* — on how the information will be used.

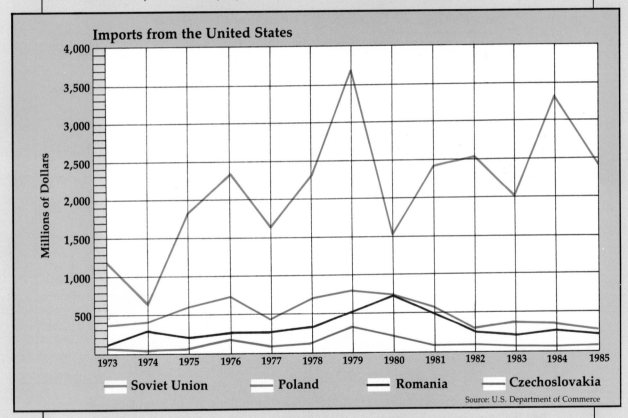

Imports from the United States

Millions of Dollars

— Soviet Union   — Poland   — Romania   — Czechoslovakia

Source: U.S. Department of Commerce

On a separate sheet of paper, tell whether each of the following statements describes the line graph, the table, or both.

1. The information presented is about imports from the United States.
2. Information is given for each year between 1973 and 1985.
3. Countries are represented by colored lines.
4. Information is shown for the Soviet Union, Poland, Romania, and Czechoslovakia.

(Turn page.)

Does the table show the same facts as the graph? Below are several factual statements. Compare each statement to both the graph and the table. On your answer sheet, tell whether each fact is found in the graph, the table, or both.

5. During the years shown, the Soviet Union was the largest importer of U.S. goods among the four Soviet bloc nations.

6. In 1978 Poland spent close to 700 million dollars on imports from the U.S.

7. Between 1976 and 1977, Czechoslovakia's imports from the U.S. declined from 149 million dollars to 75 million dollars.

Imports from the United States (in millions of dollars)

	1973	1974	1975	1976	1977	1978	1979	1980	1981	1982	1983	1984	1985
Soviet Union	1,194	608	1,835	2,310	1,628	2,252	3,607	1,513	2,431	2,587	2,003	3,284	2,423
Poland	350	396	583	623	439	680	793	714	682	295	324	318	238
Romania	117	278	191	250	260	319	501	722	504	224	186	249	208
Czechoslovakia	72	49	53	149	75	106	281	185	83	84	59	58	63
Totals	1,733	1,331	2,662	3,332	2,402	3,357	5,182	3,134	3,700	3,190	2,572	3,909	2,932

Source: U.S. Department of Commerce

If you had to choose *one* way to present this information, which form would it be? For some purposes, the table might be best; for others, the graph might be more useful. For each of the following purposes, tell whether you would choose the graph or the table, and explain why.

8. You want your audience to understand general trends quickly.

9. Your reader needs to know specific amounts to prepare a trade budget.

10. You must be able to find totals easily for all four countries together.

CHAPTER REVIEW

A. Words To Remember

Three of the following terms are defined by the numbered phrases below. On another sheet of paper, write each term next to the number of its definition. Then write the two words that are *not* defined and give a definition for each.

bloc   high technology   self-sufficient
centralized system   interdependence

1. Able to supply one's own needs.

2. Complex machines and systems, such as computers and oil-drilling equipment.

3. A method of governing in which most decisions are made by a central body.

B. Check Your Reading

1. What is the most widespread second language in Eastern Europe? Why?

2. Which country has the closest economic ties with the Soviet Union: Albania, Bulgaria, or Yugoslavia? Explain.

3. Where has trade increased most sharply in recent years: between the Soviet Union and Eastern Europe, between the Soviet Union and the U.S., or between Eastern Europe and Western Europe?

4. Why have some countries in the region made changes in their centralized planning system?

5. What is the difference between the way prices are set in the U.S. and in the Soviet Union?

C. Think It Over

1. How do you think the physical environment of a nation or group of nations affects its ability to be self-sufficient?

2. How does trade with the West make East European nations less dependent on the Soviet Union?

3. In what major ways are the Soviet Union and Eastern Europe linked to the rest of the world?

D. Things To Do

Prepare a list of questions that might be sent to a travel agency to find out whether there are many transportation links between the region and the world. For example, are there many direct air flights between the U.S. and Moscow? Warsaw? How do these compare with the number of flights to Stockholm and Rome? Is it possible to travel by railroad from Western Europe to Moscow? Can you buy tickets for the Trans-Siberian Railroad?

UNIT REVIEW

A. Check Your Reading

1. On your answer sheet, write the letter of each description. After each letter, write the number of the area that best matches that description.

(a) Much of the ground here remains frozen all year long.

(b) Densely settled area of the Soviet Union.

(c) Only East European nation located mainly on a plain.

(d) Location of deserts and mountains.

(e) Remained independent of Soviet domination after World War II.

(1) Yugoslavia
(2) Population Triangle
(3) Siberia
(4) Czechoslovakia
(5) Poland
(6) Soviet Central Asia
(7) Eastern Europe

2. Fill in the blanks in the following paragraph by writing the missing term or name on your answer sheet.

The landscape of Eastern Europe and the Soviet Union is dominated by a vast __(a)__. However, a north-south mountain range known as the __(b)__ forms the border between Europe and Asia. There are enormous expanses of permanently frozen land called __(c)__ in the north and east of the region. Climates are wetter and warmer in the __(d)__. The warmest area of Soviet Europe is around the __(e)__ Sea.

B. Think It Over

1. The following statements may be true of any one or more of the following: Poland (P); Romania (R); Soviet Union (S); Yugoslavia (Y). Write the initial letter of the nation(s) to which you think each statement applies.

(a) Its borders have changed many times during the past 1,000 years.

(b) A great number of minority groups live here.

(c) There are many mineral resources yet to be exploited.

(d) The Soviet government dominates the political life of this nation.

(e) There are large areas of undeveloped or uninhabited land here.

2. Choose any one of the five statements in exercise 1 above, and write its letter on your answer sheet. Then explain how that statement applies to the nation or nations you chose.

Further Reading

Eastern Europe, by Peter Parker. Silver Burdett, 1979. Describes the geography, people, and culture of eastern Europe.

Journey Across Russia: The Soviet Union Today, by Bart McDowell. The National Geographic Society, 1977. A tour focusing on culture and geography.

Russia: The People and the Power, by Robert G. Kaiser. Washington Square Press, 1984. A reporter describes the daily lives and opinions of Russians.

6

The Middle East and North Africa

Chapter 24

Short on Water, Long on Oil

If there is one thing more important than all else to Ali ben-Ali, it is water. Sometimes Ali even dreams of water, which is not surprising, for Ali lives in one of the world's driest regions. He is a camel herder in the Great Arabian Desert, which covers most of Saudi Arabia (SOW-dee uh-RAY-bee-uh).

Ali is one of a group of people known as Bedouins (BEHD-winz). The Bedouins are mostly nomads, moving from place to place. Their way of life is determined largely by water — or the lack of it.

Ali is part of a clan of some 70 Bedouins. (A **clan** is a group of families who have a common ancestor and who live and work together.) The clan has no permanent home, but it does have a central base to which it returns each summer. The base is near an **oasis** (an isolated source of water that allows some crops to grow). This oasis has a deep well.

In winter rains come to the desert. Less than three inches (*eight centimeters*) fall each year. This is too little for growing crops, but it is enough — in places — to cause grasses to sprout. Ali's clan depends on these grasses to feed its animals. The clan moves frequently in winter, seeking out good grazing land.

Ali's clan keeps several dozen camels, which are well adapted to desert life. In summer camels can go for days without drinking. In winter they get all the water they need from the greenery they eat. Camels serve as transportation for Ali's clan and provide food in the form of milk and cheese. The clan also keeps goats for their milk and cheese, and sheep for their wool and meat.

Ali's hope is to buy a pickup truck some day. To earn the money, he will have to leave his nomadic life for a time and find a job. Many Bedouins have done so. Some have returned, using pickup trucks for herding sheep and goats. Other Bedouins have settled in one place and taken up a new way of life.

Finding a job is not hard in today's Saudi Arabia. Oil was discovered here in the 1930's. Since then the oil wealth has been used to build up industry. Today there are more jobs than there are Saudi Arabians. The government has had to bring in foreign workers from as far away as Pakistan and the United States.

Ali's country may be short on water, but it is long on oil. Saudi Arabia's known oil reserves are among the largest in the world. Many nearby countries are also rich in oil, natural gas, or both. Oil has brought rapid changes to the region in which Ali lives.

This unit is about that region — the Middle East and North Africa. The region in-

With irrigation, crops can grow in the desert. In Libya, parts of the Sahara are irrigated by water from deep artificial wells. The high cost of these wells is paid for by the nation's oil.

cludes more than two dozen countries. Some of those countries have oil, but none has much water. The two substances, water and oil, play a key role in the region's life, as you will see in the pages that follow.

Overview of the Region

A look at the map on the opposite page will show you that the Middle East and North Africa cover a broad expanse of tropical and subtropical land. The region touches the Atlantic Ocean in the west. In the center, it borders on the Mediterranean Sea. In the east, it meets the Red Sea, the Persian Gulf, and the Indian Ocean. These bodies of water have helped to make the region a crossroads of peoples for thousands of years. Today more than 300 million people live in the Middle East and North Africa.

Geography and history have divided the region into three main areas, each with its own special features. The three parts are:
■ *The Middle East.* This area lies between two great river systems. On the west, in Egypt, is the Nile. On the east, mostly in Iraq, are the Tigris (TIE-gris) and Euphrates (yoo-FRAY-teez) rivers. The area extends from Syria in the north to include all of the Arabian Peninsula in the south. The Middle East overlaps two continents. Most of it is in Asia — but Egypt is mainly in Africa.
■ *North Africa.* Four of the countries in this area — Morocco, Algeria, Tunisia, and Libya — run along the Mediterranean coast. South of these countries is a second belt of countries that runs from Mauritania in the west to Somalia in the east.
■ *The Northern Mountains.* While there are mountains in other parts of the region, Turkey and Iran are made up almost entirely of tall mountains and highland plateaus. These two countries extend across the north of the Middle Eastern countries.

If you look at the Checklist of Nations on pages 584–585, you will see that countries in the region vary widely in area. Bahrain (bah-RAYN) is smaller than Memphis, Tennessee, while Sudan and Algeria are the ninth and tenth largest countries in the world. Populations vary too — from fewer than half a million people on up. Yet no nation, even the largest in area, has as many as 50 million people. One reason why populations have not grown bigger can be found in the region's physical environment.

Physical Geography

Ali ben-Ali was three years old before he first saw a raindrop. This is unusual, but not unheard of. Parts of the region may go for several years without a drop of rain.

The deserts of the Middle East and North Africa are the most extensive in the world. Combined they span 6,000 miles (*10,000 kilometers*). The Sahara stretches all the way across North Africa. The Great Arabian Desert, where Ali ben-Ali lives, is east of the Red Sea. Desert areas are also found in Syria, Iraq, Iran, Jordan, and Israel.

More than half the land in this region is too dry for humans. In places the deserts are made of barren rock. Elsewhere there are shifting sands. The sands come from rocks that have been eroded century after century by fierce desert winds.

In most places, the deserts do not end abruptly. They shift gradually into less dry areas known as steppes. These receive enough rainfall for short grass to grow.

South of the Sahara, the steppe area is known as the *Sahel* (SAH-hehl), an Arabic word meaning "border." Grazing animals have eaten away much of the grass in this area, which is slowly turning into desert.

Deserts and Farmlands

As you have seen, the desert areas may go for months or even years without rain. What causes this dryness?

One reason is the general movement of Earth's air. You read in Unit 1 (page 43) that air tends to rise at the Equator, move toward

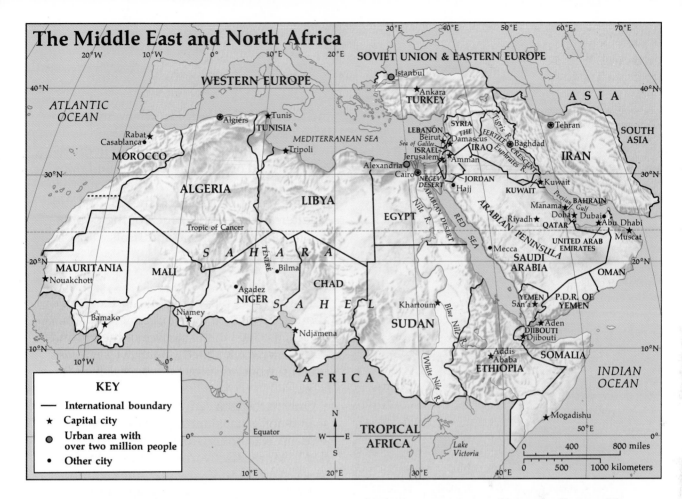

The Middle East and North Africa

KEY
— International boundary
★ Capital city
◉ Urban area with over two million people
• Other city

the poles, and fall to the surface around 30° north and south. This falling air gives up nearly all of its moisture when it first rises, and it is then too high up to gather any more on the way. As a result, much of the land area around 30° north and south is dry. Notice where the 30°N line of latitude runs through the region in the map above.

Still, some land areas around 30°N do receive moderate rainfall. Why does so little rain fall in the Middle East and North Africa? The answer is that these areas are also kept dry by the region's mountains. As you can see on page 42, air masses north of the Equator usually move from west to east. These air masses pick up moisture as they pass over large bodies of water. As you can see from the map above, there are mountains near the western coasts in North Africa and in the Middle East. These mountains "steal" moisture from the eastbound air masses. Rain falls in the mountains — so mainly dry air is left to pass over the deserts.

Dryness is one feature found in nearly all parts of the region. Not only does a vast area of desert stretch across North Africa and the Arabian Peninsula, but much of the other land receives little rain. One cultural feature is also common to the region: the Islamic religion. Except in Israel (mainly Jewish) and Ethiopia (largely Christian), most people are Moslems.

At certain times of the year, however, moist air masses sometimes do reach the deserts. Since the desert is hotter than the air masses, the air rises and is cooled. The result is a cloudburst that runs off in temporary streams called **wadis** (WAH-diz).

When the water in these wadis reaches soft sand, it sinks into the ground. Thus even under the driest deserts, there may be layers of **groundwater** (water underground). It is this water that feeds wells such as the one on which Ali ben-Ali's clan depends.

The Middle East and North Africa have a limited amount of rich farmland. One belt of such land is shaped like a new moon, a crescent. This land stretches from Egypt north-

A few parts of the region do receive adequate rainfall. In winter, for example, winds blow from the sea over coastal areas of Morocco, producing rich vegetation.

the U.S. Middle West. Elsewhere irrigation is necessary.

Most of the other good farmland in the region is found on the coastal plains and rainy upland areas. In North Africa, a narrow farming belt stretches along the Atlantic and Mediterranean coasts. Much of the population of Morocco, Algeria, and Tunisia lives in this belt.

Oil and Other Resources

The early peoples of the Middle East knew about petroleum. For example, in some places, it oozed to the surface of the ground in a sticky form called "pitch," which was used on wooden ships to make them watertight.

Only in the 20th century, however, did the search for petroleum begin in earnest. In 1908 oil was found in Persia (present-day Iran). More discoveries followed in Iraq, Saudi Arabia, and Kuwait. Other countries of the region that have large amounts of oil are Libya, Algeria, and various small countries along the Persian Gulf.

Aside from oil, the region is not very rich in mineral resources. Deposits of phosphates (nutrients used in fertilizer) are found in some places, especially Morocco. Iron and copper are found in several places, and salt deposits are widely distributed. But few other resources are known to be present in useful amounts.

east along the Mediterranean coast, then southeast through the valleys of the Euphrates and Tigris rivers to the Persian Gulf. The area is sometimes given the name of the *Fertile Crescent*.

The Fertile Crescent has a suitable climate for agriculture. As in nearly all of the region, temperatures are quite warm and the growing season is long. Moreover, the Fertile Crescent has water. Along the eastern coast of the Mediterranean, there is enough rainfall for farming without irrigation. For example, the area around Beirut (bay-ROOT), the capital of Lebanon, gets about 33 inches (*85 centimeters*) of rain a year. That's about as much as the fertile grain-growing lands of

---

SECTION REVIEW

1. What is a steppe?

2. How do mountains in the Middle East and Northern Africa cause part of the region's dryness?

3. What features of the Fertile Crescent make it good for agriculture?

4. What is the most valuable mineral resource that has been found in the region? Is this resource found in just one area, or is it fairly widely scattered through the region?

Human Geography

Nomads like Ali ben-Ali have lived in the deserts of the Middle East and North Africa for many thousands of years. The region was also the home of some of the world's first farmers. Around 8000 B.C., some nomads settled down and planted seeds. In other words, they invented agriculture.

The great rivers of the region made this invention possible. The Tigris, Euphrates, and Nile rivers flooded every year. Their muddy waters covered the surrounding land with a rich soil in which crops grew readily.

Among the earliest farming peoples were the Sumerians (soo-MEER-ee-uhnz). They settled in the dusty plain where the Tigris and Euphrates join and then flow into the Persian Gulf. Today this area is part of Iraq.

As the Sumerians settled down from their nomadic life, they were able to build a civilization (an organized society with advances in technology and art). Their villages grew into towns that grew into cities with as many as 50,000 people.

The Sumerian civilization produced the first known system of writing. Using reeds from the river, Sumerians wrote symbols on slabs of damp clay. Then they dried the slabs in the sun to make them hard.

A little later, the ancient Egyptians built another civilization along the Nile. Some of their monuments still remain — including the pyramids they built as tombs for their kings. The ancient Egyptians also created an accurate calendar and their own way of writing. The Egyptians made a type of paper out of a reed called *papyrus* (puh-PIE-ruhs), which grew along the riverbanks.

Three Religions

The region also gave birth to three of the world's great religions — Judaism, Christianity, and Islam (is-LAHM).

The Jews were a group of people who settled in an area known as Palestine, on the eastern shore of the Mediterranean, about 3,500 years ago. Their nation was based on the belief in only one God. This belief led to a strong code of morality and justice. Although various conquerors of the region forced many Jews to leave, the Jewish people carried their religion with them. Those who live in Israel today consider themselves direct descendants of the ancient Jews.

The Christian religion also grew up in Palestine. Jesus Christ was a Jew whose followers believed he was the Savior promised to the Jewish people. After his death, Christianity spread to Europe and flourished most strongly there. However, it still has some followers in the Middle East and North Africa. Christians make up half the population of Lebanon, and have long been the strongest religious group in Ethiopia. Egypt also has a large number of Christians. The Christian church in Egypt and Ethiopia, which is known as the Coptic Church, grew up separately from the church in Europe and has its own traditions.

More than 600 years after the birth of Jesus, the prophet Mohammed was born in Mecca (MEH-kuh), a city in what is today Saudi Arabia. He considered himself a leader in the tradition of Moses (the Jewish leader) and Jesus. Mohammed founded a third religion based on the belief in one God, but different from both the others. Islam is today the major religion in the region. Its followers are called Moslems (MOS-luhms) or Muslims (MOOZ-luhms).

Islam divided into different branches too. Today there is one main branch to which the great majority of Moslems belong. In Iran, however, most Moslems belong to a different branch.

Islam began among the Arabs, who then spread their religion as they conquered most of the land that now makes up the Middle East and North Africa. The region came under another influence in the 19th century, when European nations began to expand

Every devout Moslem tries to make a pilgrimage to the holy city of Mecca, Saudi Arabia. This aerial view shows the central shrine and surrounding mosques.

their empires. Britain and France took the biggest shares in the region. For example, Algeria and Morocco were French colonies, while Egypt and Sudan came under British control. The colonies began to gain their independence after World War II.

Many Influences

Today different peoples and cultures overlap throughout the region. In the Middle East and in North Africa, the majority of the people are Arabs — an ethnic group descended from nomads who lived on the Arabian Peninsula. In North Africa, there are also many people called Berbers (BUR-burs). They are descended from the people who lived in North Africa before the Arab conquest.

In the Middle East, besides the Arabs, there is a wide variety of peoples, such as Turks, Armenians, and Israelis. Hebrew, the language of Israel, belongs to the same family as Arabic, the language of the Arabs. Both Hebrew and Arabic are written and read from right to left.

In the Northern Mountains, the Turks belong to a different ethnic group from the

Arabs and Israelis. The people of Iran belong to yet another group.

In the North African countries of the Sahel, the people reflect a variety of influences. Their skins tend to be dark like the peoples of tropical Africa. Some speak African languages and share the traditional African religions. Others speak French like the colonists who once ruled the area. However, many more speak Arabic. And Islam is the religion of most people in the Sahel.

Throughout the region, in fact, the most widespread influence today is that of Islam. The following sections take a closer look at this influence.

Pilgrimage to Mecca

Ali ben-Ali is not yet 20. But as a devout Moslem, he has already planned to make an important journey before he dies. He will make a pilgrimage to the holy city of Mecca, the birthplace of Mohammed. For Moslems this pilgrimage is a religious duty. Each believer is expected to make the *hajj* (hahj; pilgrimage) at least once. The *hajj* takes place once each year, and often more than a million pilgrims take part.

Ali has heard his grandfather, Hussein (HOO-sayn), tell over and over of his own *hajj*. It was a week's journey across the desert. For the journey, Hussein wore his usual garment — a hooded white robe made of several layers of cloth. The white robe reflected the sunlight and absorbed perspiration, making it a cool garment.

Only Moslems were allowed to enter Mecca, which was jammed with pilgrims from many parts of the world. Some had come by ship and bus, some by air, some by bus alone, and some by camel. Hussein spent many days in prayer at Mecca and visited the most sacred of Moslem mosques (mosks; places of worship).

Hussein is happy to have made the holy pilgrimage. He is now known as Hajji Hussein. Hajji is the title used by Moslems everywhere who have made the *hajj*.

Women in the Moslem World

Ali's grandmother, Nura (NOO-rah), went with her husband on the *hajj*, but she could not have gone alone. Men play a dominant role in Moslem society, and Moslem tradition sets strict rules for women.

Desert life may have helped form this pattern. Perhaps because the harsh environment took a high toll in human life, Moslems permitted **polygamy** (the marriage of a man to more than one wife). The limit in Moslem lands was (and sometimes still is) four wives. Today most Moslem men cannot afford more than one wife.

Because Nura lives in the desert, with few strangers about, she does not cover her face with a veil. But in towns and cities, many Moslem women do wear veils. Traditional Moslems believe it is immodest for a woman to show her face to a man who is not part of her family.

Some Moslem women have rejected the old ways. They wear Western-style clothes without a veil. They take jobs outside their homes and appear in public with men to whom they are not related. This shocks many old-style Moslems.

In Saudi Arabia, the government enforces many traditional restrictions on women. For instance, women are required to wear a cloak when appearing in public. Newspapers are forbidden to print pictures of women. And women are barred from driving automobiles. But in recent years, the government has opened higher education to women.

Other countries that once followed strict Moslem traditions have gone much further in easing restrictions. Turkey, which is partly in Europe and has long been subject

Old traditions continue amid modern ways in much of the region. Moslem women veil their faces in public as they pass new construction and late-model cars.

to European influences, is one example. In the 1920's, Turkey banned the wearing of the veil and outlawed polygamy.

Israel: A Special Case

Hanna Salinger would not think of wearing a veil — nor would her family expect her to wear one. Hanna lives in Israel, where both women and men lead lives that differ in many ways from those of people in Moslem countries. When she finishes high school, she will be drafted into the Israeli army.

Hanna is a *sabra* (SAH-bruh), a native-born Israeli. Her parents, like most older Israelis, immigrated from other lands. Hanna's father came from Germany as a child in the late 1930's. Her mother came from Poland about the same time.

They settled on a farm near Jerusalem, which was then part of Palestine. Palestine was under British rule, and a majority of its population was Moslem.

Since the late 19th century, large numbers of Jews had been emigrating from Europe to Palestine. They longed to create a Jewish state. They hoped to reunite Jews who had scattered to many countries. Jews who shared this goal were called *Zionists* (ZIE-uh-nists). Zion was a name for the original Jewish homeland in biblical times.

During World War II, the Nazis killed millions of European Jews. To many Jews, it now seemed more important than ever to have their own homeland. The United Nations approved a plan to split Palestine into two states — one for Jews, one for Moslems. Moslems rejected the plan, but the state of Israel was formed in 1948.

This touched off the first of several wars between Israel and its Arab neighbors. Many Moslems fled from Israel and took refuge in Arab countries. To this day, large groups of Palestinian refugees live in "temporary" camps outside Israel's borders. Hostile feelings continue to divide Israel from most of its Arab neighbors.

There are many cultural contrasts between Israel and its Arab neighbors. Compare people and clothing at this Tel Aviv cafe with those in the photo on page 333.

The Jews of Israel reflect the diversity of the lands from which they came. The dominant group is from Europe. Its members share the culture of other Europeans and eat foods such as *borscht*, the beet soup common in Poland and Russia. Other Israelis have come from parts of North Africa and the Middle East. These Israelis share much of the culture of Moslem lands. They are more likely to eat *moussaka* (moo-suh-KAH), a Middle Eastern dish blending ground meat with nuts, eggplant, and spices.

SECTION REVIEW

1. What kind of land or water form was the setting of the region's earliest civilizations? How did this land or water form encourage the development of civilization?

2. Which three great religions began in the Middle East and Northern Africa? Which of these religions is now the major religion of the region?

3. How does Moslem tradition influence women?

4. For what purpose was Israel created?

Economic Geography

As you might expect, the natural resources of oil and water play major roles in the economy of the region. So too do the people. The human resources of a nation determine how well the natural resources will be used — and how well a nation that lacks those natural resources can survive.

Some countries in the region have more water than others, but none has a lot of water. With oil and human resources, however, there are wide differences among nations. Some nations are overflowing with oil, some have small or moderate amounts, and some have none. Some nations have enough skilled workers to develop their resources, while most have far too few.

You can see what these differences mean by looking at three extreme examples — Saudi Arabia, Mali, and Israel. All three nations are short of water. Saudi Arabia is rich in oil, Israel has very little, and Mali has none. Israel is rich in skilled workers, Saudi Arabia has few, and Mali has even fewer. To sum up their differences in another way, Mali belongs to the traditional world, Israel belongs to the modern world, and Saudi Arabia has a foot in both worlds.

You can see from this map that the region's mineral resources cluster around the Persian Gulf and along parts of the Mediterranean coast. Elsewhere, nations have little or no mineral wealth. If you look more closely, you can see that there are few minerals other than oil. (Chromite is the ore of chromium, a metal used for plating other metals and hardening steel.)

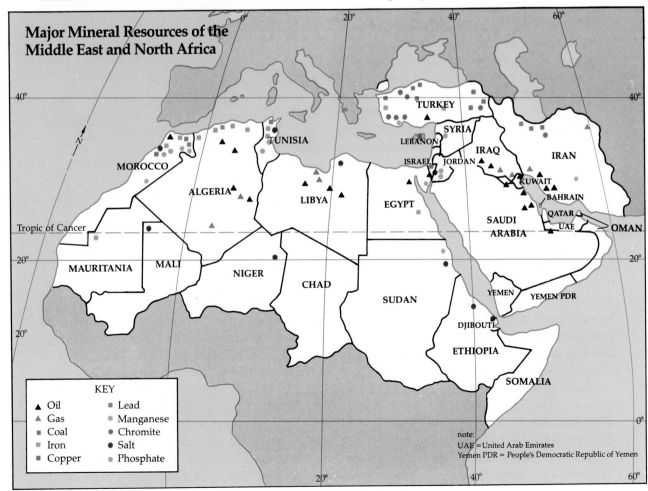

Major Mineral Resources of the Middle East and North Africa

KEY
▲ Oil
▲ Gas
■ Coal
■ Iron
■ Copper
■ Lead
● Manganese
● Chromite
● Salt
● Phosphate

note:
UAE = United Arab Emirates
Yemen PDR = People's Democratic Republic of Yemen

Saudi Arabia

Although Saudi Arabia is the richest nation in the region, it is also in some ways the most traditional. Oil has given Saudi Arabia tremendous wealth. But the Saudis are still trying to find how to make the best use of it. In the cities, mud houses are being bulldozed to make way for high-rise buildings. Superhighways and huge airports are being built to join the new cities.

Because oil brings in so much wealth, Saudi Arabia tends to solve problems by spending money. For example, since water is scarce, most Saudis drink bottled mineral water. In a recent year, gas cost 31 cents a gallon while the same amount of mineral water cost five dollars.

Because only about one percent of Saudi Arabia's land is suited for farming, food is a major problem. Up to 85 percent of what is eaten is brought in from other countries.

Long-term solutions to the shortages of water and food require both technology and skilled workers. As you have read, Saudi Arabia has more skilled jobs than there are Saudi Arabians to fill them. Yet half of the population — the women — are barred from such jobs. Some women do go to work in the cities, but they are limited to teaching, welfare, and nursing. Strict Moslem tradition affects Saudi Arabia's development as a modern nation.

Mali

The poverty of Mali contrasts sharply with the wealth of Saudi Arabia. Mali was the center of a great Islamic empire several centuries ago, and the majority of its people are still Moslems. But today Mali is one of the poorest countries in the world.

Mali has no oil and few other natural resources. The French who ruled Mali until 1960 left little in the way of industry or an educational system.

Much of Mali stretches across the Sahel. Over recent years, the herders who live in

Some countries in the region lack both oil and water. Much of Mali, in the Sahel, is dry and barren, as in this mountainside village. Lacking oil, Mali cannot afford big irrigation projects like those on page 326.

the Sahel have increased their flocks of goats, sheep, and camels. As a result, much of the scanty grass has been eaten away. Worse still, there have been years of drought that brought starvation and disease to thousands of the herders. The government of Mali is now trying to prevent further overgrazing. It is planting more trees to hold back the advancing desert.

To escape from poverty, Mali needs more industry. It needs the technology of the modern world to help improve farming and water control. Yet Mali cannot export enough to buy that technology. Moreover, 95 percent of its people cannot read or write, and only 10 percent of the children go to school. Mali has no easy way out of its lack of natural and human resources.

Israel

A skilled work force is perhaps the greatest resource of Israel. Like some other countries in the region, Israel has only small amounts of oil, and is short on water. But Israel is far more industrialized than its neighbors. Its many immigrants brought with them skills learned in other parts of the world. All but eight percent of Israelis can read or write.

Skill in technology has boosted farming in Israel's arid land. Its irrigation projects are among the most successful in the Middle East. For example, a 150-mile (*240-kilometer*) pipeline carries water from the Sea of Galilee (GAL-uh-lee) to farms in the Negev (NEH-gehv) Desert. Israeli farms produce enough to export citrus fruit and grapes to other parts of the world.

In many ways, Israel's biggest economic problem is its tense relations with its Arab neighbors. The high cost of maintaining armed forces is a continual drain on the nation's economy.

What's Ahead

In the chapters that follow, you will look at two other countries in the region — Egypt (Chapter 25) and Turkey (Chapter 26). These countries are not as rich as Saudi Arabia, as poor as Mali, or as developed as Israel. But to a greater or lesser extent, they share the problems and hopes of those three nations.

The concluding chapter of this unit looks at what may lie ahead for the region as a whole. Will oil and water be put to their best uses? The answer is important for people in the region and all over the world.

SECTION REVIEW

1. Is it true to say that all of the Middle East and North Africa is rich in oil? Explain.

2. Why does Saudi Arabia import most of its food?

3. How did Mali's environment change for the worse in recent years?

4. What kind of resource has enabled Israel to develop? How does its development compare with that of neighboring countries?

YOUR LOCAL GEOGRAPHY

1. Oil is the major natural resource in many nations of the Middle East and North Africa. What is the major natural resource in your community? It may be a mineral resource, a crop, hydroelectric power, etc. Is it the only major resource? Is it a renewable resource (such as a crop), or will it be used up one day? Does its market price vary from year to year, or does it remain steady (or increase steadily)? Are there enough skilled workers in the locality to develop the resource? Does its development pose any threat to the environment? You can find answers to most of these questions from your local government, Chamber of Commerce, or a labor union. Then write a brief report on the economic outlook of your community.

2. Animals play an important role in the life and work of people from the Middle East and North Africa. Camels are used for transportation, and like goats, they provide milk and cheese. Sheep are valuable for their wool, and both sheep and goats are used for meat at times. Do animals play an important role in your community? If you live in a rural area, perhaps your own family keeps livestock — such as cattle, horses, sheep, goats, pigs, or chickens. If so, in what ways are your animals used? If you live in an urban or suburban area, you may find that many people keep animals for protection or companionship. Can you think of any other services these animals provide?

Comparing Wind and Rainfall Patterns

For centuries peoples of the Middle East and North Africa have described paradise as a fertile garden. In dry lands swept by hot winds, it is natural to imagine paradise as green and well watered. But why is so much of this region so dry? Part of the answer has to do with the patterns of winds that sweep over the region.

The two maps in this lesson show both the wind and rainfall patterns for the region in January and July. By comparing the two maps, you can see how rainfall and wind patterns are related.

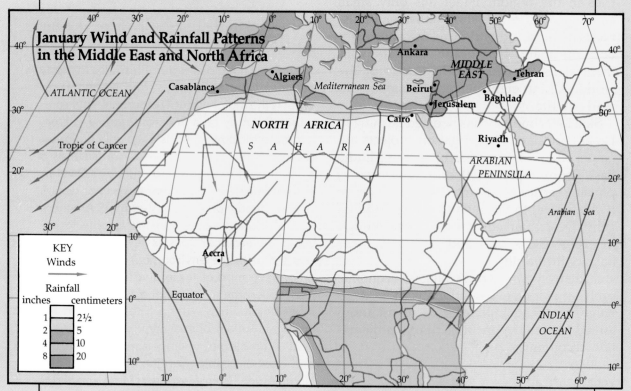

First check the map keys to find out what kinds of information the maps show. What do colors represent on the maps? What do the arrows stand for? With the help of the rainfall key, compare the two maps to see where and when most of the rain falls. Answer these questions on a separate sheet.

1. In January how much rain falls in most of the region between 5°N and 30°N latitude? In July areas with more than one inch (2½ centimeters) of rain are south of what latitude?

2. In which month, January or July, does it rain: (a) in areas near the Mediterranean Sea? (b) in areas near 10°N, 10°W?

You can see that not all of the region is dry all of the time. Do wind patterns account for the differences? Compare the wind patterns on both maps to answer the following questions.

3. In January winds generally blow over the region from northeast to southwest. How is the pattern different in July?
4. In which month, January or July, do winds blow northward toward parts of North Africa? In which month do the same parts have more rainfall?
5. In which month, January or July, do winds travel northward over the Arabian Sea toward land? In which month do winds travel southward over the Arabian Sea away from land? In which month is there more rain in parts of the Arabian Peninsula?

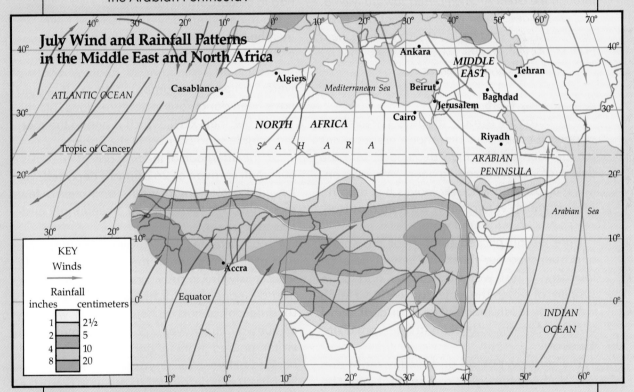

July Wind and Rainfall Patterns in the Middle East and North Africa

Winds blowing in from the sea can pick up moisture and deposit it as rain on the land. Dry winds that stay over land areas do not bring rain. Compare the wind and rainfall patterns on each map, and answer the following questions.

6. On the July map, compare the wind and rainfall patterns around the cities of Casablanca and Accra. How might wind patterns affect differences in rainfall near these two cities?
7. Compare the wind and rainfall patterns over the Sahara and near the cities of Riyadh and Tehran on both maps. How might wind patterns account for the dryness of these areas in both January and July?

Desert Crossings

These two descriptions of desert travel were written 650 years apart. The first is by Ibn Batuta (EE-buhn buh-TOO-tuh), a Moroccan who lived in the early 1300's. It describes part of his pilgrimage through the Great Arabian Desert toward Mecca. The second description is by a present-day European, who crossed a barren stretch of the Sahara in Niger. As you read the two descriptions, look for both similarities and differences.

Ibn Batuta, 1326

The caravan [group of desert travelers] stopped for four days and made preparations for the journey. We then descended into the desert, of which the saying goes: "Whoever enters it is lost, and whoever leaves it is reborn." After a march of two days, we halted at Hajj, which has underground water.

Here we all pitched camp near the spring and quenched our thirst. We stayed here for four days to rest, water the camels, and lay in supplies of water. The water carriers have large waterbags (each made of several goat skins) and the ordinary waterskins (made from a single goat).

The caravan pushed on speedily night and day. Halfway through is a valley that might be in hell. One year pilgrims were caught in a *simoom* — a hot, dry wind. As their water supplies dried up, the price of a drink of water rose to 1,000 dinars — but both seller and buyer died.

The pilgrims reached a large and pleasant village and stayed here for four days. They bought supplies and washed their clothes. They also left any extra possessions they might have, taking on with them only absolute necessities.

— from *The Travels of Ibn Batuta*

F. Paolinelli, 1980

In Agadez (ah-guh-DEHZ) a month was spent in preparation for the journey. Three camels, pack-saddles, and other accessories had to be bargained for.

On the eleventh day after leaving town, we met the track to Bilma. We followed it as far as a well, where we refilled our waterskins.

A caravan was camped nearby, and we were invited to accompany them. It would have been dangerous to venture alone into the Ténéré (teh-neh-REH), which is one of the most formidable sand barriers in the Sahara.

On the afternoon of the fifth day at this oasis, our small caravan of 26 animals departed. We traveled across small wadis and over a few rocky hills. We then entered a seemingly endless sand-sheet.

I soon accepted the caravan's routine. We rose at first light, when the men immediately faced eastward to pray. Breakfast was cooked over a meager fire, for firewood was a luxury. Loading took almost an hour. By 9 A.M. we were on our way. Lunch was a porridge of millet and water poured from a goatskin into a bowl handed between us. It was taken on the move because it was unfair to the camels to break the rhythm of the march.

— from "Saharan Caravan," by F. Paolinelli. *The Geographical Magazine,* February 1980.

Ask Yourself . . .

1. For both caravans: What was the main means of transportation? What was the main means of carrying water?

2. Why do people who cross the desert travel in caravans (groups)?

3. Are there any major differences between the two journeys? Explain.

A. Words To Remember

From the following list, select the term that best completes each of the sentences below. On a separate sheet of paper, write your answer next to the number of each sentence.

clan	oasis	steppe
Fertile Crescent	papyrus	wadis
nomad	petroleum	

1. The _____ is a belt of Middle Eastern land in which the soil is rich, the growing season is long, and there is plenty of water.

2. A group of families who have a common ancestor and who live and work together is called a _____ .

3. A _____ is land that gets enough rainfall for short grass to grow.

4. _____ is one of the most valuable resources found in the Middle East and North Africa.

5. Egyptians made the first paper out of a reed called _____ .

B. Check Your Reading

1. What are the two major deserts of the Middle East and North Africa? Where is each located?

2. What is the Sahel? How has it changed recently?

3. Name three of the inventions or technological advances made by the ancient peoples of this region.

4. Among what group of people did Islam originate? How was it spread throughout the region?

5. In what way does Saudi Arabia belong to both the modern and the traditional worlds?

C. Think It Over

1. Can natural resources alone make a nation prosperous? If so, how? If not, what else is necessary?

2. How did the rise of agriculture lead to the development of civilization?

3. How might a peaceful settlement of Arab-Israeli tensions improve life in the region?

D. Things To Do

Imagine that you are either Ali ben-Ali or Hanna Salinger. What does the future hold in store for you? What would you like to accomplish in your lifetime? Write a brief essay expressing your hopes and plans. Explain how such things as culture, historical events, patriotism, and religion have influenced your outlook.

Chapter 25

Egypt
The Gift of the Nile

More than 60 years have passed since Halim Amin (hah-LEEM ah-MEEN) helped to find the fabulous treasure buried with King Tutankhamen (too-tuhn-KAH-muhn). But the memory is still alive. As one of an army of boys hired to dig in the ancient ruins, Halim was paid in bread and beans for all the rocks and gravel he could scrape into his basket and carry away.

Halim remembers the shouts of joy when an Englishman pushed a lighted candle through a small hole to reveal a room full of golden treasure. He is glad to have helped with the work that made the discovery possible.

Since 1922 the treasure buried with King Tutankhamen more than 3,000 years ago has been displayed in museums around the world. It has also been studied by archaeologists (people who seek to learn how humans lived long ago).

Halim Amin lives in a village in southern Egypt near the Valley of the Kings. Not only King Tut (as Tutankhamen has come to be called) but also many other Egyptian pharaohs (FAY-rohs; kings) were buried in the Valley of the Kings. Ancient Egyptians regarded their kings as gods and made sure they would be comfortable in their tombs.

Halim has taken his grandchildren to visit other ruins nearby. On the walls are pictures of Egyptian farmers from long ago. These *fellahin* (feh-luh-HEEN), as Egyptian farmers are called, look very much like Halim and his sons and grandsons. They use the same kinds of tools and wear the same kinds of clothes.

In many ways, Halim Amin and his family seem to lead the same kinds of lives as the fellahin of 3,000 years ago. However, if Egypt is going to feed its fast-growing population, the traditional ways of farming will not do the job. As in much of North Africa and the Middle East, modern ways of farming must take the place of traditional ones. Industry must be developed also to make the most of Egypt's limited resources.

As you have already read, Egypt has found one resource — oil. Perhaps oil will help build a better future for Egypt.

Physical Geography

The Valley of the Kings is a wasteland of rock and sand. Years may go by without rain. Beyond the valley lies the vast Sahara. Yet just a short distance away, Halim Amin's village sits amid green fields of clover, sugarcane, and wheat. These crops can grow because the village lies close to Egypt's major water source — the Nile River. In the words of an ancient Greek writer, Egypt exists only as "the gift of the river."

In Egypt most cities and farms are crowded along the banks of the Nile River, which keeps the land green. As you can see at left, the desert begins only a short distance away.

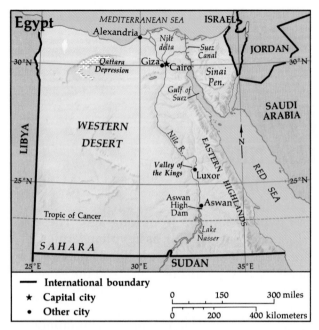

Egypt

MEDITERRANEAN SEA — ISRAEL — Alexandria — Nile delta — Suez Canal — JORDAN — 30°N — Qattara Depression — Giza — Cairo — Sinai Pen. — Gulf of Suez — SAUDI ARABIA — WESTERN DESERT — LIBYA — Nile R. — EASTERN HIGHLANDS — RED SEA — 25°N — Valley of the Kings — Luxor — Aswan High Dam — Aswan — Tropic of Cancer — Lake Nasser — SAHARA — SUDAN — 25°E — 30°E — 35°E

— International boundary
★ Capital city
• Other city

0   150   300 miles
0   200   400 kilometers

Nearly all of Egypt's land is desert. The only fertile areas lie along the banks of the Nile and in its delta. The delta begins north of Cairo, spreading along the Mediterranean from Alexandria to the Suez Canal.

The Nile flows from south to north through the entire length of Egypt. The river itself is as much as half a mile (*800 meters*) wide. It has cut a gash in the desert that is up to 14 miles (*22 kilometers*) wide. This gash is the Nile Valley, which contains almost all of Egypt's farmland.

Egypt is about as big as Texas and New Mexico combined. The Nile Valley takes up less than four percent of Egypt's surface. Yet most Egyptians live in this valley. It is as if all Texans lived crowded together in a narrow strip of land along the Gulf of Mexico.

The Importance of the Nile
The Nile is the longest river in the world — more than 4,150 miles (*6,650 kilometers*) from its source to the Mediterranean. Only one fifth of this length lies in Egypt.

The Nile actually begins as two separate rivers — the White Nile and the Blue Nile (see the map on page 329). The White Nile flows north from Lake Victoria in East Af-

rica. The Blue Nile starts in Ethiopia. The two rivers join in the Sudan. The Nile then tumbles over a series of low waterfalls and rapids, called **cataracts** (KAT-uh-rakts), as it flows north into Egypt. The northernmost cataract is near a place called Aswan (ahs-WAHN). Egypt's farmlands begin here.

The Nile brings Egypt water from the East African highlands, where rainfall is fairly heavy. A drop of rain that falls on these highlands may take as much as three years to reach the Mediterranean. However, most raindrops never finish the journey. Much of the Nile's water evaporates along the way, sinks into the ground, or is used to irrigate crops. Unlike Brazil's Amazon River, much of the Nile runs through thousands of miles of desert, unprotected by vegetation. Thus the Nile carries less water than the Amazon — but every gallon is welcome in Egypt.

Egypt depends almost entirely on the Nile for its water. No part of the country gets more than eight inches (*20 centimeters*) of rain a year, the amount that falls near the Mediterranean coast. (The wettest part of Egypt gets less rain than one of the driest U.S. cities, Albuquerque, New Mexico.) As you can see from the rainfall maps on pages 338 and 339, there is far less rain toward the south. Most of Egypt gets less than two inches (*five centimeters*) of rain a year.

Beginning the third or fourth week of June each year, the Nile begins to swell. It will run high until September. The swelling is due to heavy rains that have fallen on the mountains of Ethiopia. The rains also erode the mineral-rich mountain soil and wash it into the Nile.

For centuries the river spread across the valley during the flood season, covering low-lying fields and even some homes. Then the river dropped slowly. When the fields dried, they would have a new layer of silt and mud — the perfect fertilizer. This kind of rich soil, left behind by running water, is called **alluvium** (uh-LOOV-yuhm).

Since 1971, however, the Nile has been tamed by a huge dam at Aswan. Yearly floods no longer bring their gift of alluvium. Farmers have had to find other ways to fertilize their fields. But there are big advantages to the dam. It has created a large lake that serves as a "storehouse" of river water. This means that water is available for irrigation all year round, whether the rains in East Africa are light or heavy. In addition, the dam provides hydroelectric power.

From Aswan north to Cairo (KIE-roh), Egypt's capital, farms crowd the narrow strip of the river valley. At Cairo the river fans out. Its waters take several different channels to reach the Mediterranean. This region is called the delta. It has some of the richest alluvium of all and has long been a prime Egyptian farm area.

The temperature and soils of the Nile Valley are excellent for growing many crops. Summers are hot, and winters warm. In fact, most of the valley never gets a hard frost — only an occasional light one. Therefore, sugarcane, which requires a full year to mature, will do well here. It is grown from Aswan to the delta.

Egypt's most important crop is cotton, which also is grown in the Nile Valley. Egypt is one of the world's leading producers of high-quality cotton.

Desert Regions

Outside the Nile Valley, Egypt's vast deserts can be divided into three distinct parts. They make up most of the land area of Egypt, but contain little of its population. From west to east, they are:

■ *The Western Desert.* This desert stretches from the Nile Valley westward into Libya.

Since the Aswan High Dam began operation in 1971, farm production in Egypt has doubled. This picture shows Lake Nasser, the huge lake created by the dam.

The Western Desert is part of the Sahara. It includes a few scattered oases and a large marshy area called the Qattara (kuh-TAH-ruh) Depression. This depression lies as much as 440 feet (*135 meters*) below sea level.

■ *The Eastern Highlands.* This is an area of rocky ridges that separates the Nile Valley from the Red Sea to its east. There is almost no vegetation in this area. The only human settlements are along the Red Sea.

■ *The Sinai (SIE-nie) Peninsula.* This peninsula is the only part of Egypt that lies in Asia. It is joined to the rest of the country by a narrow neck of land near the Mediterranean. The Suez Canal cuts through this neck of land from north to south, connecting the Mediterranean to the Gulf of Suez and the Red Sea. The population of the Sinai Peninsula is clustered near the canal.

The deserts are where most of Egypt's mineral resources are found. The most valuable discovery is oil, which has been found in the Sinai, in the Western Desert near the Mediterranean, and offshore in the Red Sea. Other minerals include phosphate (in the Sinai) and iron (near Aswan).

SECTION REVIEW

1. In what way does the Nile change downstream (north) of Cairo?

2. What resources are found in Egypt's deserts? Which resource is the most valuable?

3. What is alluvium, and how is it formed? What is its importance?

4. Sugarcane is grown in the Nile Valley. What does this tell you about the growing season in this valley?

Human Geography

The ancient Egyptians set a record for long life among civilizations. Their society, which was established about 3100 B.C., lasted 2,800 years. During that time, one man or woman was the ruler of all Egypt. These leaders, the pharaohs, regarded themselves as gods. They believed that life after death was very much like that before death. That is why they furnished their tombs so elaborately.

The pharaohs left huge monuments, such as the pyramids where they were buried. There are 35 still standing along the west bank of the Nile. The largest, near Cairo, is as tall as a 40-story building.

The rule of the pharaohs ended in the seventh century B.C., when invaders swooped down on Egypt from the Fertile Crescent. Later conquerors included the Persians, the Greeks, the Romans, the Arabs, the Turks, the French, and the British. Egypt did not rule its own country again until the 1950's, when the British left.

Of all the conquerors, the Arabs, who swept into Egypt in 640 A.D., had the greatest influence. They brought what is now the official language of Egypt — Arabic — and the dominant religion — Islam. Today only one in 15 Egyptians is Christian.

Farms and Cities

The population of Egypt is almost evenly split between farm and city. Fifty-one percent of Egyptians (including a small Bedouin population) live in rural areas. Forty-nine percent live in cities. Since so much of Egypt's land is barren, both its farm areas and its cities are overcrowded. Egypt is one of the most densely settled lands in the world. Egypt's population of nearly 44 million is the largest of any Arab country.

Like the great majority of fellahin, Halim Amin lives in the Nile Valley. His village is located on the west bank of the river, across from Luxor (LUHK-sor). The village is a collection of mud-brick houses, mostly of one or two stories.

After Halim married his wife Khudaira (koo-DIE-ruh), he built their house himself. He used the Nile's alluvial mud, with branches and leaves from nearby palm trees for the roof. At one corner, he added an

open terrace above the first floor. The terrace is reached by an outside stairway. Khudaira serves supper on the terrace to catch the evening breeze. The family squats on the floor as it eats.

A typical evening meal consists of broad beans, onions, and disks of a coarse, flat wheat bread. The family eats meat (chicken, lamb, or goat) at least once a week — usually on Thursday night, before the Moslem Sabbath on Friday. Hot, sweet tea is served with the meal.

The big events in the village are weddings, funerals, holidays, and feasts. When Halim's father died, Halim hired professional mourners to chant prayers far into the night. This custom dates back to the days of the pharaohs.

Halim's son still relies on ancient methods of irrigation to spread water released from the Aswan High Dam among his crops. But

Just over half of all Egyptians live on farms. Here, in the delta, a boy acts as a weight on his father's harrow. Plowed soil is being broken up for planting.

the day may come when hydroelectric energy from the Aswan High Dam or oil from Egypt's newly found resources will do this work. And most likely, Halim's grandsons will read by electric lights some day.

Changing Ways

The Egyptian government is trying to make sure that young people today learn how to read and write. About 90 percent of adult women and 65 percent of adult men are illiterate. But elementary and high school education is free in Egypt now. Students between the ages of six and 15 are legally required to go to school. Many cannot, however, because there are still too few schools, and many students need to work.

347

Although fewer women than men in Egypt can read and write, there has been a steady movement in Egypt to encourage girls to attend school. In ancient Egypt, women were freer than in other countries in the region. When the country became Moslem, Egyptian women were required to wear the veil. Today four percent of Egyptian women work outside the home. This figure will probably rise as more factories are opened.

Life in the Cities

As in most of the region, women can have more freedom in a city than in a village. Life in the city is more varied than rural life. It is most varied in Cairo, by far the biggest of Egypt's cities. With its suburbs and sister city Giza (GEE-zuh) across the Nile, Cairo has about 13 million people.

Egypt's cities are home to a growing middle class of merchants, technicians, and teachers. These people usually have adopted Western ways and live very much as middle-class people do in Europe. They have electric appliances in their homes, along with television, and are likely to read one of Egypt's 40 newspapers.

The large majority of city dwellers, like Egypt's small farmers, are poor. They live in crowded tenements in old parts of the cities. Some even live as squatters in mosques and cemeteries.

Although the Egyptian government has not been able to build enough housing, it does try to make sure that food is available. The price of basic foods is quite cheap in the cities. Each family receives a ration card that lists how much sugar, salt, cooking oil, and other items it may buy at low prices. The government makes up the difference between this price and the cost of the food. Such government payments to keep prices low are known as **subsidies**.

Like other major cities, Cairo has monuments, modern buildings, bustling traffic, and millions of people. Egypt's capital is one of the world's largest cities.

SECTION REVIEW

1. Do Egypt's middle classes live in urban or rural areas? Is their way of life traditional or modern?

2. How has the Egyptian government kept food prices low?

3. Which group of invaders had the greatest influence on Egypt as it is today? Give an example of this influence.

4. How does Egypt's population compare with that of other Arab nations? Why is Egypt called "overcrowded"?

Economic Geography

Like all countries in the region, Egypt is struggling to move from the traditional toward the modern world. Unlike those other countries, Egypt is able to help itself by using resources of both water and oil.

The Aswan High Dam generates electricity from the waters of the Nile. It supplies about half of the electric power used in Egypt. It also provides farmers with constant water supplies. At one time, Nile farmers dug basins to hold the water when the river's flooding was over. Today the flood waters are stored by the dam.

However, farmers still have to lift the river water into irrigation ditches and then lift it again into the fields. For this work, some ancient devices are often used.

One of these devices is the *shaduf* (shah-DOOF; "well sweep"). A thick wooden beam pivots on a shaft connecting two posts set in the ground. On one end of the beam is a weight. On the other, a rope holds a bucket. By pulling down on the rope, the farmer makes the bucket dip into the canal. The weighted beam then lifts the bucket to the top of the dike. The worker empties the bucket over the dike into the irrigation ditch. Then he pulls the rope down so the bucket dips again into the canal.

Today many farmers or groups of farmers have bought diesel pumps to do this never-

Some farmers now have pumps to bring Nile water to their crops, but many use traditional methods. Here, a farmer lifts water from an irrigation ditch onto his field by working a *shaduf* (see the text on this page).

ending work. But the creak of the *shaduf* is still heard in many villages. Egypt's population is so large that there are plenty of people to do farm work by hand.

The average Nile Valley farm covers only about four acres (1½ *hectares*). However, many farmers make do with as little as one tenth of an acre (1/25 *hectare*). Like their ancestors, the fellahin work the fields in their bare feet. Most use wood-and-iron plows pulled by cattle. A simple hoe is used to keep down weeds. Grain is still harvested with a sickle.

The steady water supply from the new dam has been used to open new land for farming. So far, more than one million acres (*400,000 hectares*) of desert have been reclaimed.

Even so, Egypt still must import about 40 percent of its food — largely from the United States. To raise the money, Egypt sells cotton, its main export crop. Some cotton is spun into thread and woven into cloth, but most is sold raw.

Hopes for Development

Egypt could earn more by exporting manufactured goods rather than raw cotton. However, factories and mines together employ only 13 percent of Egyptians. The textile industry is the largest manufacturer. Egypt's other industries include petroleum and sugar refineries, paper mills, and fertilizer plants. An automobile factory is planned. The Egyptian government sees more industry as one of the nation's biggest needs.

Oil may hold the key to Egypt's future. Since its Red Sea oil field went into production in the 1970's, Egypt has had more fuel than it can use. Some is exported, and the profit is used to buy more food.

The Qattara Depression may hold even more oil. Unfortunately the ground is too marshy to be drilled. One plan calls for flooding the area with salt water from the Mediterranean so that oil-drilling platforms could float on the resulting lake. However, some scientists fear that the salt water would filter underground to the Nile delta.

Other plans call for creating new cities in the desert. Industries could be built on land unsuitable for farming.

Halim Amin does not work on his land anymore. This work is done by his sons while he sits in a café talking with his friends. The idea that his grandsons might live in a desert city and work in a factory is to him unthinkable.

Yet such changes could be more valuable to Egypt than King Tut's splendid treasures. Today's Egyptians can find glory in their nation's past. Will they be able to find it in the future as well?

SECTION REVIEW

1. For what purpose is the shaduf used?

2. What is Egypt's most vital import? What recent discovery has helped pay for it?

3. What are the two main ways in which the Aswan High Dam has helped Egypt?

4. What is Egypt's main export crop? What is its major industry? How could Egypt enlarge this industry?

YOUR LOCAL GEOGRAPHY

1. You have read that Egypt depends almost entirely on the Nile for its water. The rainiest part of Egypt receives only about eight inches (*20 centimeters*) of rain a year. Compare this with the rainfall in or near your community by consulting an almanac (see Geography Skills 34 on the opposite page). In the index, look under "Rainfall" or "Precipitation" for the subtopic "U.S. Cities." Find the city nearest to your community. Check its average annual rainfall. How does this compare with the rainiest part of Egypt? Do you think your community has to rely on a river for much of its water supply?

2. You have read about various kinds of technology found in modern-day Egypt, from the Aswan High Dam to the shaduf. These are clues that life in Egypt is modern as well as traditional, and agricultural as well as industrial. What clues are given by the kinds of technology used in your community? Mentally take a "walk" through five or six blocks in your neighborhood (your route home from school, for instance). As you go along, list the kinds of technology you find — anything from lawn mowers to telephone wires. Then compare your list with those of your classmates. What do the lists reveal about life in *your* community?

Using an Almanac

How large is Egypt's Aswan High Dam? How does its size compare with the world's other large dams? To answer these questions, you need some facts. A good place to look for them is in an **almanac.** Modern almanacs have facts about almost everything — from major-league batting averages to major world dams.

To find facts in an almanac, begin with the index. Some almanacs have the index at the front, others at the back. The index lists general topics in alphabetical order, with the page numbers where you will find information. Some general topics have subtopics listed under them with separate page numbers.

Skim the sample almanac index below. Answer the following questions on a separate sheet of paper.

1. On what pages of the almanac would you find: (a) general information about Egypt? (b) facts about the Egyptian government?
2. Is the Aswan High Dam listed as a general topic or as a subtopic? On what almanac pages would you find facts about it?

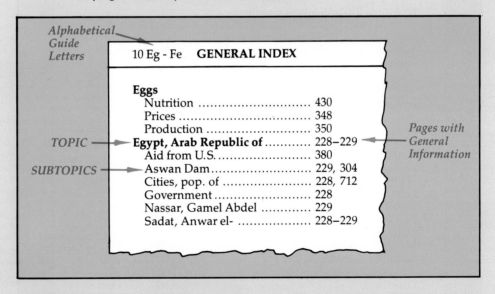

Alphabetical Guide Letters

10 Eg - Fe   **GENERAL INDEX**

Eggs
Nutrition 430
Prices 348
Production 350

TOPIC — **Egypt, Arab Republic of** .... 228–229 ← *Pages with General Information*
Aid from U.S. 380

SUBTOPICS — Aswan Dam.......................... 229, 304
Cities, pop. of 228, 712
Government 228
Nassar, Gamel Abdel 229
Sadat, Anwar el- 228–229

This almanac has facts about the Aswan High Dam on two different pages. Sections of both pages are shown in this lesson on page 352. One gives general information about Egypt. The other shows a table of facts about major world dams. Compare the sample almanac pages to do the following exercise.

3. Are these facts found with the general information, in the table, or both?
   (a) Estimated population of Egypt. (b) Peoples and religions in Egypt.
   (c) Size of the Aswan High Dam. (d) When the dam was completed.
   (e) How the dam is used. (f) The dam's official name.

(Turn page.)

Use information from either almanac section to answer these questions.

4. How much land was made available for irrigation when the Aswan High Dam was built?

5. In the table, the sizes of major world dams are given in what measurement? The world's five largest dams are located in what two countries?

6. How large is the Aswan High Dam? How many dams listed in the table are larger? What country in the Middle East has a dam larger than Egypt's?

228 **NATIONS—Egypt**

EGYPT
(Arab Republic of Egypt)

Population: 43,000,000 (est.). **People:** Egyptians, Bedouins, Nubians. **Language:** Arabic. **Religions:** Islam, Christianity. **Area:** 363,250 sq. mi. (944,450 sq. km). **Cities:** (1976 cen.) Cairo, 5,084,463;

reclaimed through artesian wells.

The Aswan High Dam, started in 1960, was completed in 1970 (cost: approx. 1 billion dollars). It made available more than a million acres of land for irrigation and provides a potential of 10 billion kilowatts of electric power a year.

304 **WORLD FACTS—Dams**

MAJOR WORLD DAMS
Source: Bureau of Reclamation, U.S. Interior Dept.

Name of dam	Cubic yards	Completed
New Cornelia Tailings, U.S.	274,026,000	1973
Tarbela, Pakistan	158,268,000	1975
Fort Peck, U.S.	125,612,000	1940
Oahe, U.S.	92,008,000	1963
Mangla, Pakistan	85,872,000	1967
Gardiner, Canada	85,743,000	1968
Afsluitdijk, Netherlands	82,927,000	1932
Oroville, U.S.	78,008,000	1968
San Luis, U.S.	77,666,000	1967
Garrison, U.S.	66,506,000	1956
Cochiti, U.S.	64,631,000	1975
Tabka, Syria	60,168,000	1975
Kiev, U.S.S.R.	57,552,000	1964
W.A.C. Bennett, Canada	57,203,000	1967
High Aswan Sadd-El-Aili, Egypt	57,203,000	1970

CHAPTER REVIEW

A. Words To Remember

Three of the following terms are defined by the numbered phrases below. On another sheet of paper, write each word next to the number of its definition. Then write the two terms that are *not* defined, and give a definition for each.

alluvium   cataracts   subsidies
almanac   fellahin

1. A rich soil left behind by running water.
2. A series of waterfalls.
3. Payments that support low prices.

B. Check Your Reading

1. Besides water, what did the annual flood of the Nile River bring to the lands it spread across?
2. What two major benefits does the Aswan High Dam provide?
3. Are most farms in the Nile Valley small or large? Are they farmed by modern or traditional means?
4. Describe two ways in which Egypt is encouraging the education of its people.
5. At present does Egypt need more oil or more industry? Explain.

C. Think It Over

1. The Nile is the world's longest river. Why does it carry less water than some other rivers, such as the Amazon?
2. Is Egypt likely to become less dependent on the Nile if it develops more industry and opens more of its desert areas? Why, or why not?
3. Egypt is a crowded nation. In your opinion, will the size of its population be mainly an asset or a problem in the years to come? Why?

D. Things To Do

Imagine that you are an archaeologist who has just discovered one of Egypt's pyramids. Write a brief report explaining what you have found. Be sure to mention what the pyramid is built of, its size, its age, and its purpose. You might also report on the role of geography in its location and in the transportation of its stones. Your "dig" will take place in the library, instead of the desert. Two good sources are *Ancient Egypt* (Time-Life Great Ages of Man Series) and *Ancient Egypt: Discovering Its Splendors* (National Geographic Society).

Chapter 26

Turkey

Where Europe and Asia Meet

Naila Demirkapan (NIE-luh duh-meer-kuh-PAHN) is a daily commuter from one continent to another. Her home is in Asia; her office, in Europe. Every workday she rides a bus back and forth across a bridge from Asia to Europe. Her home and office are both in the same country — in fact, in the same city.

Naila lives in the Turkish city of Istanbul (ee-stahn-BOOL), formerly the Roman city of Constantinople (kon-stan-tuh-NOH-puhl), formerly the Greek city of Byzantium (buh-ZAN-tee-uhm). As its many names suggest, Istanbul has had many rulers. It has a fine location on the Straits that link the Black Sea with the Mediterranean and divide Europe from Asia.

Today as so often in the past, Turkey is a border nation between rival power blocs. From her bus, Naila can sometimes spot gray warships plowing through the waters below. These may be Soviet ships, steaming out of the Black Sea toward the open waters of the Mediterranean. Or they may be U.S. ships, heading northeast to scout out Black Sea waters.

By international treaty, the Straits through Turkey are open to all nations in time of peace. However, another treaty binds Turkey and the U.S. as allies in case of war. Turkey plays a key role in NATO — the North Atlantic Treaty Organization formed by the U.S. and many West European nations. On Turkish soil, U.S. bases are close to the Soviet Union.

Because Istanbul is a meeting place of different worlds, some people think of it as a mysterious city. Many spy novels and movies have been set there. Yet everyday life in Istanbul is much the same as in other cities.

Naila and her husband, Bulent (boo-LEHNT), live in a modern home with central heating, books, and a TV. They read a newspaper to keep abreast of world events. But they are not typical of all Turks. Most people in Turkey are villagers. Many live in mud houses with no plumbing. As many as half of all villagers are illiterate.

Since the 1920's, Turkey has been ruled by governments intent on "modernizing" it — that is, making it more like Western Europe. Kemal Ataturk (kuh-MAHL AH-tuh-turk) was the first leader who aimed to push Turkey out of its traditional ways. Ataturk died in 1939, but his successors have carried on where he left off.

In this chapter, you will learn more about the struggle between traditional and modern Turkey. In doing so, you will also learn about the many ways in which Turkey's geography and culture have worked together to shape life there today.

Turkey straddles both Europe and Asia—and so does its largest city, Istanbul. Here, in Europe, a Moslem mosque towers over ferry boats that will cross the straits to Asia.

Physical Geography

Naila would describe herself as a European, but most of Turkey lies in Asia. The country forms a rough rectangle, slightly larger than Texas. However, the population of Turkey is three times that of Texas.

The two parts of Turkey are called Thrace (in Europe) and Anatolia (an-uh-TOHL-yuh; in Asia). The Black Sea, which lies between Europe and Asia, forms most of Turkey's northern border. To travel from the Black Sea to the Mediterranean, a ship must pass through three waterways, together called the Straits. These waterways separate Europe and Asia. As you saw earlier, both shores of the Straits are part of Turkey.

Imagine you are on a ship heading south out of the Straits. You pass the ruins of the ancient city of Troy on your left. Actually these are the ruins of *nine* cities, for Troy was destroyed time and again over the centuries. Turkey's geography has made it vulnerable both to war and to earthquakes, and many cities rest on the ruins of earlier ones.

Your journey now takes you south through the Aegean (uh-JEE-uhn) Sea. You pass many islands, some within sight of the Turkish shore. All but two of these islands belong to Greece — a fact that many Turks deeply resent.

Farther along the Aegean coast is the mouth of a broad river. This river enters the sea only after winding lazily across a coastal plain. In ancient times, this river was called the Meander (mee-AN-dur), giving us the English word *meander*, meaning "to wander aimlessly." Today the river bears the name Menderes (mehn-duh-REHS). Like all of Turkey's rivers, the Menderes is not navigable (cannot be used by ships).

Finally you round the southwestern corner of Turkey and turning eastward, pass from the Aegean Sea into the Mediterranean. The Mediterranean forms about half of Turkey's southern border.

Ever since leaving the Straits, you have seen mountains rising a short way inland. Now, in places along Turkey's southern shore, those mountains extend right to the sea. It is time to look at the inland areas of the country.

Highlands and Lowlands

Turkey is a land of tall mountains, high plateaus, and coastal plains. Central Anatolia is one of the few relatively flat places in the country. It is on a high plateau ringed by mountains.

Turkey's capital city, Ankara (AHN-kuh-ruh), is set in a bowl of hills on the Anatolian plateau. The hills tend to trap smoke and fumes, so on some winter days, Ankara is covered by smog.

Anatolia, the Asian part of Turkey, is mainly on a plateau. Some areas receive just enough rain for farming.

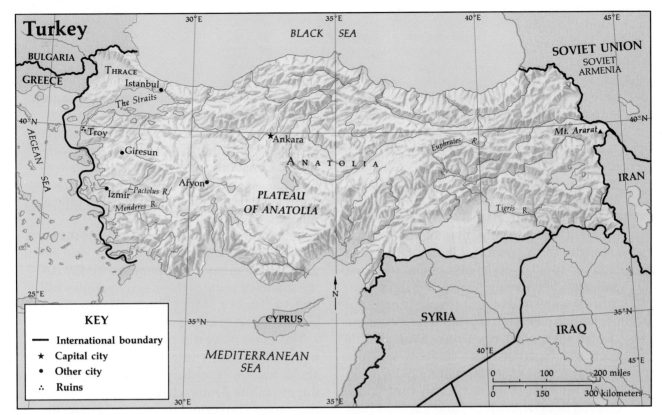

Turkey

Turkey's mountains rise higher and higher toward the east. The tallest of all is Mount Ararat (AR-uh-rat), which pushes to a height of almost 17,000 feet (*5,200 meters*) — higher than any mountain in Europe. Its peaks are blanketed by snow all year round.

Many rivers rise (have their source) in Turkey's mountains. The two longest and best-known of these rivers are the Tigris and Euphrates, which you read about in Chapter 24. Southeastern Turkey forms part of the basin of these two rivers. The basin's rich soil and plentiful rainfall (20 to 40 inches or *50 to 100 centimeters*) are used for growing grains and fruit. However, most of the basin lies in Iraq (see the map on page 329).

Although agriculture is Turkey's main occupation, only one half of the soil is farmed. The remaining land is either too steep or too dry. Outside the southeastern basin, Turkey has two areas of farmland. One area is on the plains along each of Turkey's three coasts. The other is on the Anatolian plateau, which lies inland.

The Anatolian plateau is the driest of Turkey's farm areas. Much of it gets less than 10 inches (*25 centimeters*) of rain a year, and

For centuries Turkey was at the crossroads of trade between Europe and Asia, and different peoples made it their homeland. As a result, Turkish place-names became familiar. The word *cherry* comes from Giresun; Izmir produced Smyrna figs; and Ankara gave its name to the angora cat and angora wool.

few crops besides wheat and barley will grow well without irrigation.

The Mediterranean and Aegean coasts get more rainfall but not always at the right time. All three coastal areas can grow such warm-weather crops as citrus fruit.

Mineral Resources

To the ancients, Anatolia was legendary for its wealth in gold. Like the California rivers of Gold Rush times, the Pactolus (pak-TOH-luhs) River in western Anatolia yielded gold to those who panned for it. But that was long ago. When the Turkish government sent people recently to prospect for gold, they found no useful deposits left.

The Black Sea coast was another source of minerals in ancient times. Here they can still be mined. The mountains near the coast have deposits of copper, lead, and manganese, among other minerals. Above all there is coal, which is used to fuel much of Turkey's industry.

The one mineral that Turkey would like to find in large quantities is oil. So far, however, oil has been found in only limited amounts, mainly in the southeast.

As you saw in Chapter 24, Turkey's environment differs from that of most other countries in the region. Like Iran, Turkey's land consists largely of mountains and high plateaus. Yet Turkey shares the same concern for water and oil as is found throughout the region. Though Turkey receives more rainfall than other countries of the Middle East and North Africa, there is still not enough for all of its farmlands. And as in countries such as Morocco and Tunisia, industrial development in Turkey is held back in part by a limited supply of oil.

SECTION REVIEW

1. Why is only half of the land in Turkey used for farming?

2. What are the names of the two parts of Turkey? Where is each part located?

3. Would you call Turkey rich in mineral resources? Why, or why not?

4. In what two ways does Turkey lie between different worlds?

Human Geography

Pairs of wrestlers, their muscles rippling under a slick coating of olive oil, struggle to keep their grip as a drum beats time and fans cheer from the sidelines. It is early May in Turkish Thrace, time for the annual festival of "greased wrestling." The festival was started centuries ago by a Moslem ruler. It is held every spring around St. George's Day, named for a Christian saint.

There is nothing odd to Turks about this blend of Moslem and Christian traditions. Both religions have contributed richly to Turkish culture. While most Turks today are

Moslems, elements of the Christian past are easy to find. Istanbul has been under Moslem rule for five centuries, but it was a center of Eastern Orthodox Christianity for 15 centuries. Constantinople (as it was then called) was the capital of the eastern Roman (Byzantine) Empire. Christianity was the official religion.

Before the Romans united the land that is now Turkey, it was ruled by Greeks, and before that by many other peoples. Today descendants of some of these early peoples still live in Turkey. Most (but not all) have blended in with later arrivals.

The most powerful of later arrivals were the Turks, who came from Central Asia about the ninth century A.D. Under the influence of the Arabs whom they conquered, the Turks converted to Islam. Slowly the Turks took control of Anatolia from the Byzantine Christians. In 1453 they took Constantinople.

The Turks spread beyond Anatolia to a large area of southern Europe and North Africa. This area became known as the Ottoman (OT-uh-muhn) Empire, from the name of its ruling family. In the 19th century, the Ottoman Empire began to crumble. In the 1920's, the empire finally collapsed, and modern Turkey rose from its ruins.

Change and Resistance

When Naila looks at a Turkish newspaper, she reads the words from left to right. The words are formed with Latin letters — like the words on this page. When Naila signs a check, she uses both a first and a last name. When she goes out with Bulent, both wear Western-style clothes. She wears no veil; he wears no fez (traditional hat).

If they like these aspects of life, they can thank Ataturk, the first president of the Turkish republic. Ataturk led the Turkish forces that defeated a Greek army and prevented Turkey from being carved into separate nations after World War I.

Many cultural traditions blend in Turkey today. Here, Turks in Western clothes watch an Anatolian dance. Some village dances date back more than 1,000 years.

Ataturk saw many of Turkey's ancient customs as barriers to progress. These customs included writing in the Arabic script (from right to left), the use of one name, and the veil for women. He also cut the close ties between Islam and the country's government. Ataturk replaced the old Islamic law with a new legal system based on Switzerland's. Among other things, this did away with polygamy (a man marrying more than one woman).

Many Turks vigorously resisted these changes. Even today there is tension in Turkey between those who want a Westernized country and those who want to return to traditional Islamic law.

Other tensions among the people of Turkey today go further back in history. These tensions are one result of Turkey's position on a crossroads of civilizations, between Asia and Europe. The tensions involve two ethnic minorities — the Kurds and the Armenians.

The Kurds have occupied their present homeland for 40 centuries or more. This homeland is called Kurdistan, but it does not appear on most maps. If it did, it would cover about 30 percent of Turkey, in the southeast.

Kurdistan spills over the borders of five countries — Turkey, Iraq, Iran, Syria, and the Soviet Union. The Kurds, who are mostly Moslems, are the largest minority group in Turkey. They number about six million in a total Turkish population of 46 million. At times the Kurds have tried to rebel against Turkish rule.

Armenians, like the Kurds, have a long history. Before and after the time of Christ, they had their own kingdom, centered near Mount Ararat. Later their land was split between the Russian and Turkish empires. Millions of Armenians lived throughout Turkey. Unlike the Turks and most Kurds, the Armenians are Christians.

In the late 19th century, the Armenians tried to throw off Turkish rule. Hundreds of thousands of Armenians were killed in massacres that followed. Many emigrated to the West, especially to the United States.

In World War I, many Armenians sided with the Allies and welcomed the invading Russian troops. This support aroused the anger of Turks and Kurds, and more Armenians were massacred. Today there are only about 70,000 Armenians living around Istanbul. Most of the world's Armenians live just across the border from Turkey in the Soviet Union. Communities of Armenians also live in the U.S. and other countries. Because of the past massacres, relations between Armenians and Turks are still bitter.

Rural Life

Fifty-seven percent of Turks today live in rural areas. A few roam the countryside as herders of goats and sheep, but the great majority live in farming villages. The life of Elif Eski (eh-LEEF EHS-kee) is fairly typical. Her family owns a five-acre (*two-hectare*) farm near the city of Afyon (ahf-YOHN) in west central Anatolia.

They live in a house made of unbaked clay bricks. It has no running water, and the roof is made of mud. Each time rain washes off some of the mud, Sadi (sah-DEE), Elif's father, climbs up on the roof and slaps on more mud. The roof is now quite heavy. In a major earthquake, people are often crushed by the collapse of such houses.

Twice each day, Elif milks the family buffalo. Kamer (kuh-MEHR), her mother, turns the milk into yogurt by placing a spoonful of yogurt in a batch of milk and letting it slowly sour. When the yogurt is thinned with water, it makes a refreshing drink. Sometimes Kamer takes cream, adds nuts, sugar, and a few other ingredients, and makes a candy known around the world as "Turkish delight."

Elif's village has an elementary school, which is a part of the government's effort to educate more Turks. About 55 percent of adult Turks can read and write. Now that Elif is 13, she spends only the summers in the village. She is one of a handful of villagers who attend higher grades at a school in Afyon. There she lives with relatives, returning home for holidays.

City Life

Elif's older brother, Akib (ah-KEEB), lives in Istanbul. He owns a special kind of taxi that operates much like a bus. The taxi follows a regular route and carries four or five passengers at a time. It is called a *dolmus* (DOHL-moosh), meaning "full" — because it will not start out until it has a full load.

Istanbul is a city of five million — the largest city in Turkey. Akib likes the bustle of the city. At noon he buys lunch from a street vendor. Such vendors sell meat cooked on skewers over a charcoal fire. The most popular is *sis kebab* (SHEESH keh-bahb), which is made of small chunks of lamb and pieces of onion and pepper. There are also kebabs made of fish caught nearby.

Akib likes to stop at a small café for a glass of water and a cup of coffee. Turkish coffee is a thick brew served in a small cup. The bottom of the cup is covered with a layer of fine coffee grounds, and if drinkers aren't careful, they may get a mouthful of grounds.

For several years, Akib and other Turks had to go without coffee. Turkey banned coffee imports when oil prices rose sharply in the 1970's. The money saved was used to pay for oil. But the ban on coffee was lifted in the early 1980's.

SECTION REVIEW

1. Name two of the changes made by Ataturk after World War I. What was his purpose in making these changes?

2. Why has there been continuing opposition to Ataturk's changes among many Turks?

3. Why are there tensions between the Turks and some of the ethnic minorities in Turkey?

4. If Elif Eski wants a job when she finishes her schooling, do you think she could find one in her home village? Explain your answer.

Economic Geography

On their farm, Sadi's family grows oriental poppies — one of the few crops that can grow well in the area's thin, acid soil. The poppies' leaves make feed for cattle, their seeds provide oil for cooking, and their dried stalks can be used as fuel.

Oriental poppies also produce a gum called opium, which can be used legally, to make medicines such as codeine (an ingredient of some cough syrups). But it can also be used illegally, to make heroin and other dangerous narcotics.

In the early 1970's, more and more opium was showing up as heroin on U.S. streets. The U.S. government persuaded Turkey to restrict poppy growing. Now poppies must be grown under the eye of an inspector. The Turkish government buys the seed pods to make legal medicines.

Turkish agriculture has made great strides. At one time, crop failures caused periodic famines. Now this is a thing of the past. More land has been brought under cultivation, and crop yields are slowly rising.

Still, Turkey must overcome some serious problems. One problem is a shortage of fertilizer. Animal manure is often burned to heat village homes, so farmers must buy expensive artificial fertilizer instead. Far too little fertilizer is available. A second problem is the dry climate. Farmers on the Anatolian plateau usually let one third of their land lie **fallow** (without a crop) each year. This rest period allows the soil to store a little moisture and become more fertile again.

A Struggle for Industry

Since the time of Ataturk, Turkish leaders have tried to build up industry away from the coast, in areas where the land is not suitable for farming. But this movement has not been very successful. Today there is far more industry in the western part of Turkey than in central and eastern Anatolia. Although

Most industry in Turkey is found in the large cities, but some is linked to agriculture. Here, in rural Anatolia, rose petals are boiled in water. The steam is then cooled to obtain oils used in making perfume.

the roads in central and eastern Turkey are not very good, air and rail transportation has been improved. This still might be an area of future growth.

About half of Turkey's industries are concentrated in Istanbul. The city's location on the Straits makes it Turkey's busiest port, so that it is convenient for industries to import raw materials and export manufactured goods. In this populous city, workers are readily available.

Like Naila, many Turkish workers are women. Turkey has the highest percentage of working women in the region. More than one third of its women are in the labor force (compared to slightly more than one half in the U.S.).

A challenge to further industrial growth has been the development of sources of power. As you saw earlier, Turkey has many rivers that run down from the mountains and plateaus. Some of these rivers have been dammed to provide hydroelectric power, which has given a boost to industry. But hydroelectric power alone is not enough for the nation's needs.

Private oil firms have done extensive drilling, but have found only limited supplies of oil. Lack of oil has slowed the growth of businesses. In recent years, some factories have had to close for long periods for want of fuel. Heating oil has also been in short supply. In winter Naila's family wears sweaters in the home, since room temperatures are often as low as 50°F (*10°C*).

Toward the Modern World

In the village where he grew up, Akib Eski would have found it impossible to earn enough money to buy his *dolmus*. In order to earn the money, he lived in West Germany for three years. Many Turks today are raising their living standards by working in West European countries.

Workers who return to Turkey rarely go back to their home villages. Instead, they usually settle in a city as Akib did. But they keep in touch with people back in the village. By spreading knowledge of the outside world, they may be helping to bring further changes to Turkey.

The struggle between traditional and modern ways is not new in Turkey. Nor is the struggle to develop economically in the face of limited resources. However, many other nations of the Middle East and North Africa have only recently begun to move toward the modern world. In the next chapter, you will see what the struggle to develop may mean for all of the nations in the region.

SECTION REVIEW

1. In what way did the U.S. affect Turkish farming in the 1970's?

2. How does the proportion of women in Turkey's labor force compare with the proportion in other countries in the region? How does it compare with the proportion in the U.S.?

3. Many former Turkish villagers are spreading knowledge of the modern world among Turkey's villages. How did they acquire this knowledge?

4. Why is so much of Turkish industry near Istanbul?

YOUR LOCAL GEOGRAPHY

1. Find a geographical dictionary or gazetteer in your school or public library. Following Geography Skills 35 (opposite page), look up the nearest large city. First check all the abbreviations and symbols used in the entry, and write out your own key. Then compare the kind of information listed with that given for Istanbul in Geography Skills 35. Has the name of the city ever changed? Is it on a waterway? If so, what waterway? What population is given, and for what date? What are the chief industries? When was the city founded?

2. Turkey's location on the border between contrasting worlds has to a large extent shaped its history. Is there any kind of border that affects your community? It may be political — if your community lies on or close to an international, state, or county boundary. It may be a physical border — between different landforms or climate zones. Or it may be some other kind of border — between time zones, or between urban and rural areas. If there is such a border, describe briefly how the border affects your community. If you do not find such a border, describe what all of your community has in common, politically and geographically. On balance do you think the border — or lack of it — is an advantage or disadvantage? Why?

Using a Geographical Dictionary

Is Istanbul a lake, a city, or a mountain? Where is it, and how big is it? Did it ever have a different name? You can find the answers to such questions in a **geographical dictionary**, also called a **gazetteer**.

Unlike an almanac (see Geography Skills 34), a gazetteer has no index. It does not need one, since names of places are listed in alphabetical order. The places include countries, cities, seas, rivers, mountains, and many other geographic features. The entry for each place usually has a guide for pronouncing its name, plus various geographic facts. To save space, gazetteer entries also use *abbreviations* (shortened forms of words). Keys to pronunciation and abbreviations — similar to those shown on the following page — usually appear in the front of a gazetteer. You can find them by checking the gazetteer's contents.

A page of entries from a geographical dictionary is shown below. Refer to this sample page and to the sample keys on page 364 to answer the following questions on a separate sheet of paper.

1. To aid you in thumbing through the gazetteer, *guide names* are printed at the top of each column on a page. Which guide name on the sample page indicates the *first* entry on the page? Which guide name indicates the *last* entry?

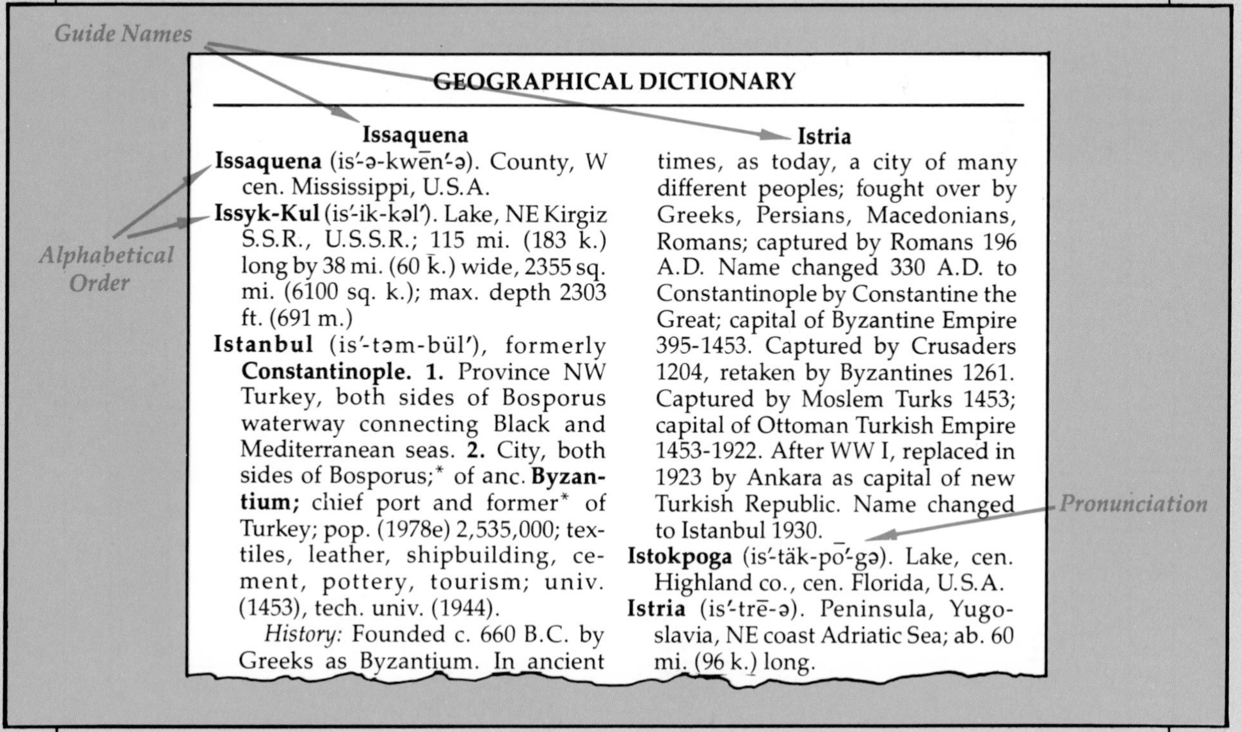

Guide Names

GEOGRAPHICAL DICTIONARY

Issaquena   **Istria**

Alphabetical Order

Issaquena (is'-ə-kwēn'-ə). County, W cen. Mississippi, U.S.A.

Issyk-Kul (is'-ik-kəl'). Lake, NE Kirgiz S.S.R., U.S.S.R.; 115 mi. (183 k.) long by 38 mi. (60 k.) wide, 2355 sq. mi. (6100 sq. k.); max. depth 2303 ft. (691 m.)

Istanbul (is'-təm-bül'), formerly **Constantinople. 1.** Province NW Turkey, both sides of Bosporus waterway connecting Black and Mediterranean seas. **2.** City, both sides of Bosporus;* of anc. **Byzantium;** chief port and former* of Turkey; pop. (1978e) 2,535,000; textiles, leather, shipbuilding, cement, pottery, tourism; univ. (1453), tech. univ. (1944).

   History: Founded c. 660 B.C. by Greeks as Byzantium. In ancient times, as today, a city of many different peoples; fought over by Greeks, Persians, Macedonians, Romans; captured by Romans 196 A.D. Name changed 330 A.D. to Constantinople by Constantine the Great; capital of Byzantine Empire 395-1453. Captured by Crusaders 1204, retaken by Byzantines 1261. Captured by Moslem Turks 1453; capital of Ottoman Turkish Empire 1453-1922. After WW I, replaced in 1923 by Ankara as capital of new Turkish Republic. Name changed to Istanbul 1930.

Pronunciation

Istokpoga (is'-täk-po'-gə). Lake, cen. Highland co., cen. Florida, U.S.A.

Istria (is'-trē-ə). Peninsula, Yugoslavia, NE coast Adriatic Sea; ab. 60 mi. (96 k.) long.

(Turn page.)

2. Check the Key to Abbreviations and Symbols. What does each of the following mean, as used in the geographical dictionary?   (a)*;   (b) anc.;   (c) c;   (d) e;   (e) A.D.;   (f) B.C.;   (g) m.

Notice that after the place-name of each entry on the gazetteer page, letters and symbols show how the name is pronounced. Pronunciation guides vary. The one shown here differs from the guide used in this text. To use the gazetteer guide, study the symbols in the Key to Pronunciation. Then figure out the pronunciation of *Istanbul* in the sample entry. Now answer these questions.

3. When you pronounce *Istanbul*, do you put the main stress on the first, second, or third syllable? Is the third syllable of *Istanbul* pronounced like bool, bull, or ball?

Now read the rest of the entry for Istanbul and answer these questions.

4. Istanbul is located on both sides of what important waterway?
5. Under what name was Istanbul founded? When? By what people? When was this name changed? By whom? When did it get its present name?

KEY TO ABBREVIATIONS AND SYMBOLS (part)

ab.—about	NE—Northeast
A.D.—Anno Domini ("In the year of our Lord"); that is, "after Christ."	NW—Northwest
	pop.—population
anc.—ancient	sq. m.—square miles
B.C.—Before Christ	S.S.R.—Soviet Socialist Republic
c.—about (before a date)	tech.—technical
cen.—central	univ.—university
co.—county	U.S.S.R.—Union of Soviet Socialist Republics
e—estimated (after a number)	
k.—kilometers	W—West
m.—meters	WW I—World War One
mi.—miles	*—capital city

KEY TO PRONUNCIATION

ə—*uh* as in **a**bout or S**u**nday	ō—*oh* as in fl**o**w or t**o**e
ä—*ah* as in f**a**ther	ü—*oo* as in c**oo**l
ē—*ee* as in qu**ee**n	'—syllable with strongest stress
i—*ih* as in **i**s or s**i**ck	'—syllable with second strongest stress

364

A. Words To Remember

From the following list, select the term that best completes each of the sentences below. On a separate sheet of paper, write your answer next to the number of each sentence.

basin	guide name	plateau
fallow	meander	Straits
gazetteer	navigable	

1. In Istanbul, a bridge crosses the _____ that divide Europe from Asia.

2. Some Turkish farmers let part of their land lie _____ each year in order to increase its fertility.

3. A _____ is a type of geographical dictionary.

4. An area of land drained by rivers is called a _____.

5. A Turkish river gave us the word _____, meaning "to wander aimlessly."

B. Check Your Reading

1. What are Thrace and Anatolia?

2. What three kinds of landforms in Turkey are suitable for farming? Which of these three is least suitable? Why?

3. Why do a larger proportion of women have jobs in Turkey than in other countries in the Middle East and North Africa?

4. Name two major causes of tension among the Turkish people today.

5. What change is the Turkish government trying to make in the location of the nation's industry?

C. Think It Over

1. In what way does Turkey have a strategic location? Do you think this is an advantage or a disadvantage?

2. Would you say that Ataturk succeeded in modernizing Turkey? Why, or why not?

3. Why is Turkey eager to find oil within its borders?

D. Things To Do

Perhaps you would like to sample some "Turkish delight." To make it, you need a heavy sauce pan; a candy thermometer; an egg beater; and a small, deep bowl. Different ingredients are used in different parts of Turkey. One easy recipe requires: two cups sugar, ½ cup light corn syrup, ½ cup water, two egg whites, one teaspoon vanilla, one cup shelled walnuts.

First combine the sugar, corn syrup, and water over a stove until it boils. The candy thermometer should reach 265°F (130°C).

Pour this mixture over the already beaten egg whites, beating as your pour. When the mixture is very thick and creamy, add the vanilla and the nuts. Now pour the mixture into the small bowl. Allow it to cool before you slice and eat.

Chapter 27

Not by Oil Alone
The Region in Perspective

Muhammed Takla (moo-HAH-mehd tuh-KLAH) is an Egyptian, but he lives in Saudi Arabia. He works for an oil company, living in a dormitory complex with 10,000 workers from all over the world. Muhammed is one of more than one million people from other countries who hold jobs in Saudi Arabia.

Muhammed has lived in Saudi Arabia for two years. He has only a grade-school education, but he is eager to get ahead. Sometimes he works as a waiter in his spare time. He makes far more money than a man of his skills could make in Egypt. He has bought a tape deck, a portable radio, and a color TV. Every month he sends part of his pay back to his family in Egypt. Hundreds of thousands of Egyptians who work abroad do the same thing. Egypt gets as much foreign money in this way as by selling its oil.

For the present, such arrangements solve problems both for Saudi Arabia and for Egypt. Egypt cannot support its fast-growing population at home. Saudi Arabia has too few people to develop its vast wealth. For the future, however, each country would like to find ways to become more self-sufficient.

In this chapter, you will look again at the countries of the entire region. As you have seen, all must be concerned with making better use of their limited supplies of water.

Some have vast supplies of oil — but oil alone will not guarantee a secure future. In many countries, the size of the population will play a key role in future success. So will the abilities and education of the people. The building of modern industry is crucial to the region. Which countries are best suited to the task?

A simple way to answer this question is to divide the countries into four groups, based on their physical and human resources. In the first group of countries, there is a good supply of raw materials for industry and a population to develop them. In the fourth group, there are neither the raw materials for industry nor a skilled population. The other two groups fall in between.

Where the Resources Are

■ *Group I*. There are four countries in the region with an important raw material and the population to develop it — Algeria, Iraq, Iran, and Morocco. The first three countries export oil and natural gas. Morocco exports phosphates.

All of these countries have taken advantage of their natural wealth. But they have not placed their hopes for the future solely on selling raw materials. Sooner or later, they know their key resources will dwindle. Each is trying to **diversify** (to build up a

Like other nations in the region, Kuwait (koo-WIET) has become wealthy on its oil. But it has hardly any industry or agriculture. What happens if the oil runs out or drops in price?

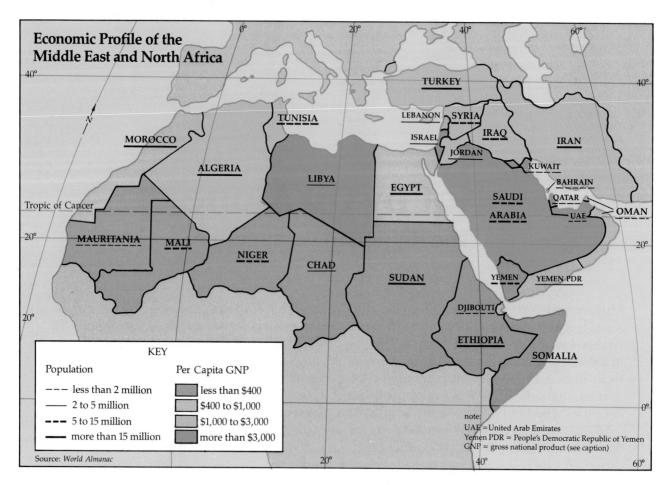

Economic Profile of the Middle East and North Africa

KEY

Population

- - - less than 2 million
— 2 to 5 million
- - 5 to 15 million
— more than 15 million

Per Capita GNP

less than $400
$400 to $1,000
$1,000 to $3,000
more than $3,000

Source: *World Almanac*

note:
UAE = United Arab Emirates
Yemen PDR = People's Democratic Republic of Yemen
GNP = gross national product (see caption)

range of different industries). In Morocco, for example, some factories process the nation's leading mineral resource — phosphates. But other factories turn out glass, bricks, textiles, flour, and chocolate.

■ *Group II*. Eight countries in the region are rich in natural resources (oil and gas) but have small populations — Bahrain, Djibouti, Kuwait, Libya, Oman, Qatar, Saudi Arabia, and the United Arab Emirates. In one sense, these countries are not really economic units at all. They are like the boom towns that sprang up in California in the Gold Rush days. When the oil and gas are gone, will these countries fade as so many of those boom towns did?

Leaders of the countries hope not. They are trying to diversify. For instance, many nations near the Persian Gulf are building petrochemical plants to turn natural gas into plastics, synthetic fibers, and other products.

But costs are high, mainly due to a harsh

Gross National Product (GNP) is the total value of all goods and services produced in a nation in a year. Since large nations are likely to produce more than small nations, their economies can be compared better with the *per capita* GNP. This is the GNP divided by the number of people in the nation. You can see that tiny oil-rich nations have large per capita GNP's.

environment. Food for workers must often be imported, at high cost. Water is in short supply too. Recently Dubai (duh-BIE), in the United Arab Emirates, built a plant to process aluminum ore. It also had to build a plant to remove salt from sea water so the water could be used in making aluminum. The plant puts out 25 million gallons (*95 million liters*) of fresh water a day.

■ *Group III*. There are six countries in the region that have a limited supply of natural resources and an abundance of people — Egypt, Israel, Sudan, Syria, Tunisia, and Turkey. These countries are developing industry, but not so fast as they would like.

The six nations have made progress in

building roads, railroads, and ports. They have workers who are skilled in certain tasks, like food processing and textiles. Most people in these societies get a grade-school education, and some go on to high school and even college. Large populations provide a market for goods, but they also pose the problem of too many mouths to feed. As you have seen, Egypt must import food so its people can have enough to eat.

Israel's population is large only in comparison to its area and resources. With its skilled population, Israel makes the most of its limited environment. Other countries in this group are making less use of their resources.

For instance, Turkey could grow more grain if more of its farms were modernized. One day Turkey and Syria might become the breadbaskets of the Middle East. The Sudan might take a similar role for North Africa.

■ *Group IV.* The countries in this group are the poorest in the region. Most lie in the southern belt of North African countries, along the Sahara. They consist of Chad, Ethiopia, Mali (which you read about in Chapter 24), Mauritania, Niger, Somalia, and two countries on the Arabian Peninsula — Yemen and Yemen P.D.R. (People's Democratic Republic).

Most of the people in these countries are illiterate. Most live in rural areas and lack skills for modern industry. Resources are limited, and there has been little industrial development. Throughout this area, people must depend on themselves and their neighbors for food and clothes. Many people live on the borderline of starvation, and an extra-dry year may mean disaster.

Two other countries, Jordan and Lebanon, are included in this group for different reasons. While these countries also lack oil and other resources, they do have much higher literacy rates and more industrial development. However, both have been weakened by serious conflicts. Jordan has suffered from conflict with neighboring Israel. Lebanon has been racked by a civil war between the two major groups that make up its population — Moslems and Christians.

Obtaining enough water is a challenge in most of the region. Egypt has the Nile, and a few countries (such as Israel) can use smaller rivers for irrigation. A well feeds these irrigation channels in Saudi Arabia.

SECTION REVIEW

1. Why do some Egyptians look for work outside Egypt? Why does Saudi Arabia look for foreign workers?

2. Name two of the countries in the region that have the best prospects for economic development. Why are their prospects good?

3. Kuwait and Libya are two countries that are very rich in oil. What major problem do they face in developing?

4. What two problems make it difficult for the poorest countries in the region to overcome their poverty?

The Challenge of Change

The countries of the region need more than resources and skilled populations in order to develop. They also need a certain amount of **stability** (calm). Political upheavals, wars, and other conflicts drain their physical and human resources.

Yet there are bound to be some conflicts when a nation moves from the traditional world into the modern world. As modern industry develops, many people leave their traditional villages to find jobs in cities. These people tend to drop their old customs.

Along with economic development, there may be social changes that clash with strict religious beliefs. For example, many Moslems are shocked that women in some of the cities in the region have jobs and wear Western clothes. As you read in Chapter 26, some Turks still resist modern customs that were brought in more than a half century ago.

In some countries, traditional Moslem groups seek to wipe out Western influences altogether. Resistance to change has sometimes been violent. In Iran, for example, such a movement helped bring down the shah (king) in 1979. The shah was accused of being a dictator. At the same time, he had brought new industry to Iran, given new freedom to women, and allied himself with the West. Iran's new leaders turned the country toward strict Moslem rule and away from the West.

Conflicts between nations and political unrest within nations have slowed development in some parts of the region. This protest is in Iran, where a 1979 revolution was followed by bitter political clashes.

Iran's revolution did not simply slow industrial development. It also disrupted oil shipments to other nations, reducing the money that Iran was able to earn from its oil. Similar trouble from within could pose problems for other countries in the region.

In addition, there have been many conflicts in the region. These include war between Iran and Iraq, civil war in Lebanon, and Israel's conflict with its Arab neighbors. Much of the resources, skills, and efforts of these nations have gone into fighting and defense instead of development.

The Region and the World

There are two main reasons why the rest of the world is deeply concerned with events in the Middle East and North Africa. The first reason is the fear that conflicts in the region — especially between Israel and its Arab neighbors — might spread and touch off a world war. Thus, for example, the U.S. has given aid both to Israel and to Arab nations, and has also tried to arrange a peaceful settlement between them.

The second reason is oil. The U.S., along with many other countries of the modern world, relies on the region for a large part of its oil supply.

There are other ways in which the region may affect the rest of the world. In order to build more industry, most nations in the region will have to borrow money from outside. Those who have enough money will want to bring in trained engineers and new technology.

Right now the economic problems of the region center on water, oil, and people. How can more water be made available for farming and making electricity? What is the best use of the vast riches that oil has brought to its producers? How can the ever-growing populations of many of these nations be fed? How can they best be educated? The answers will require time and a lot of persistence.

SECTION REVIEW

1. What major conflicts have arisen within some countries in the region as a result of economic development?

2. Give an example of a conflict between countries in the region.

3. What are the two main reasons why the rest of the world is concerned with events in the Middle East and North Africa?

4. What is stability? Is it important for a country to be stable if it wishes to develop? Explain your answer.

YOUR LOCAL GEOGRAPHY

1. One reason for U.S. concern with events in the Middle East is oil. The cars and buses you ride, the furnaces that heat many buildings, the phonograph records you play, and many other things in your community depend on oil. Much of that oil is imported, and part of that imported oil comes from the Middle East. But just how much comes from the Middle East? Use reference sources (such as the *Statistical Abstract of the U.S.*) to find the answer. For the three most recent years listed, note (a) the total amount of oil used in the U.S.; (b) the total amount of oil imported into the U.S.; (c) the total amount imported from Middle Eastern countries. On the basis of this information, how important do you think Middle Eastern oil is to the U.S.?

2. You have read how some nations of the region are better equipped than others for future development. The nations are divided into four groups: those with important natural resources and the population to develop them; those rich in natural resources but small in population; those with limited natural resources and large populations; and those with limited natural resources and small populations. Suppose you were asked to place your community in one of these groups. Where would your community fit best? Give reasons for your choice.

Making Predictions

People who study topics such as economic geography sometimes try to make **predictions**, which are statements about what will happen in the future. To predict future trends, they look for past trends that are likely to continue.

The band graph below shows trends in total world oil production, year by year, from 1975 to 1985. It also shows what part of the world total each year came from the Middle East and North Africa. Look first at the **short-term trends** of only a year or less. Answer the following on a separate sheet.

1. By how many barrels did oil production increase or decrease during 1981: (a) worldwide? (b) in the Middle East and North Africa?
2. In what year did Middle East and North Africa oil production decrease by about a million barrels? What was the world trend for the same year? Did both trends continue?

As you have seen, short-term trends can shift rapidly. To make reliable predictions, you need to take a longer view. A **long-term trend** is the overall direction of change that takes place for five or ten years or more. Study the band graph for long-term trends to make some long-term projections.

3. In 1977 oil production from the Middle East and North Africa was between one third and one half the world total. In the 1980's did the share of oil from the region increase, decrease, or stay the same? If the same trend continues, what would be the regions share in the 1990's?
4. Sometimes events can change the direction of a trend. For example, a drop in oil prices or the discovery of large oil deposits in Southeast Asia might affect oil production in the Middle East and North Africa. Can you think how other events might affect future oil production in the region?

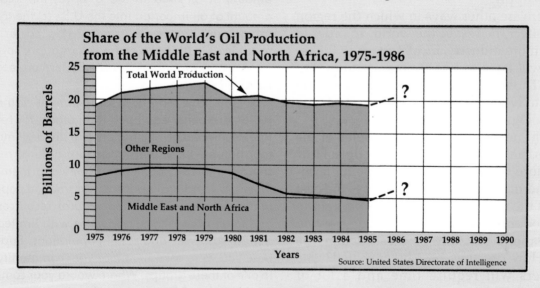

Share of the World's Oil Production from the Middle East and North Africa, 1975-1986

Source: United States Directorate of Intelligence

CHAPTER REVIEW

A. Words To Remember

In your own words, define each of the following terms.

diversify   predictions   short-term trends

long-term trends   self-sufficient

B. Check Your Reading

1. Small countries such as Oman and Qatar are rich in oil. What major problem do they face?

2. Egypt and Sudan have large populations and limited natural resources. What is the biggest problem these countries face in the future?

3. Name two nations in the Sahel. In general, are nations in the Sahel better off or worse off economically than nations in other parts of the region?

4. Countries in the region need stability in order to develop. What does this mean?

5. Give two reasons why countries outside the region are concerned with conflicts inside the region.

C. Think It Over

1. What reasons could you give Muhammed Takla to stay in Saudi Arabia? What reasons could you give him to return to Egypt?

2. If you were the head of an oil-rich nation, such as Saudi Arabia, on the way to development, how would you spend the money coming to your nation? What portions would go for food imports, education, irrigation and agriculture, new industries, defense, research, or something else? Why?

3. In your opinion, should richer nations of the world help poor nations like Chad, Ethiopia, and Mali? If so, how? If not, why not?

D. Things To Do

Select one of the nations in the Middle East or North Africa that you would like to visit some day. Ask the librarian at your local or school library to help you find a book or magazine article on this place. Read the material, looking for information on geography and cultures. Then write a brief description of an imaginary journey you took through this land. Describe the land and water forms you see there, the kinds of people you meet, their occupations and pastimes, etc. Be prepared to read your "travel diary" to the class.

A. Check Your Reading

1. On a separate sheet of paper, write the letter of each description. After each letter, write the number of the nation that best matches that description.

(a) Created in 1948, this nation has an educated population and advanced technology.

(b) This Sahel nation has very limited resources and little industrial development.

(c) This nation has the largest population of any Arab country.

(d) This nation lies in Asia and Europe.

(e) The richest nation in the region.

(1) Egypt
(2) Libya
(3) Mali
(4) Saudi Arabia
(5) Algeria
(6) Israel
(7) Iran
(8) Turkey

2. Fill in the blanks in the following paragraph by writing the missing term on your answer sheet.

The deserts of the Middle East and North Africa are the most extensive in the world. Part of a steppe area known as the __(a)__ is also slowly turning into desert. The location of __(b)__ near the western coasts accounts in part for the dryness of much of the area. There are two great river systems in the Middle East and North Africa. In the west is the __(c)__, and in the east are the Tigris and Euphrates. The __(d)__ (area drained) of these eastern rivers is part of a belt of farmland. This belt, called the __(e)__, stretches from Egypt to the Persian Gulf.

B. Think It Over

1. The following statements may be true of any one or more of Turkey (T), Egypt (E), Israel (I), Saudi Arabia (S), and Mali (M). Write the initial letter of the nation(s) to which you think each statement applies.

(a) It is in North Africa.   (b) It is rich in mineral resources.

(c) Compared to its area and resources, its population is large.

(d) There is a large skilled work force.

(e) The majority of its people are Moslems.

2. Choose any one of the five statements in exercise 1 above, and write its letter on your answer sheet. Then write a paragraph giving reasons to support your choice of nations.

Further Reading

The Blue Nile, by Alan Moorehead. Random House, 1983. Covers the recent history and exploration of the river.

The Land and People of Turkey, by William Spencer. Harper & Row, 1972. Survey of Turkish geography and culture. Revised Ed.

The Middle East Today, by Don Peretz. Prager, 1983. Discusses Middle Eastern history, followed by chapters on individual nations.

7

Tropical and Southern Africa

Chapter 28

Many Climates,
Many Ways of Living

Anicet (ah-NEE-seh) goes to high school in a village near the mouth of the Zaire (zah-EER) River. The area where he lives receives much less rainfall than most of Central Africa. Yet this "dry" coastland has as much rain as the green fields of Wisconsin.

Hendrik (HEHN-drik) grows apples and grapes in a land of rolling hills and mountains. He lives in a two-story house with running water and electricity. The area where Hendrik lives receives scarcely 20 inches (*50 centimeters*) of rain a year — about the same as western Kansas. The rain falls mainly in winter, so irrigation is necessary.

Omatayo (oh-muh-TIE-oh) sees more rain in the summer months than Hendrik sees all year. On rainy days, the drops rattle against the iron roof of her mud house. Omatayo cooks her family's meals outdoors, beneath a shelter of thatch. Many meals include roots (such as yams) and flat cakes made of flour from cassava or corn. Omatayo's husband grows these crops on scattered plots of land.

Sijey (see-JAY) and his family grow a few crops too, but their main interest is their animals. Sijey keeps a herd of cattle, goats, and sheep. The animals often feed on a wide plain that stretches toward a snow-capped mountain. Rainfall is sparse, and thick forests cannot grow here.

Anicet, Hendrik, Omatayo, and Sijey live far apart, in four different climates — yet all live in Africa, south of the Sahel (the southern border of the Sahara). Their skin colors range from Sijey's black to Hendrik's tanned white — yet all are Africans. Their ways of living range from traditional to modern — yet all show the influence of Africa's environment.

African life is so varied that it would be hard to pick a "typical" life-style. These four people can serve as a starting point for learning about Tropical and Southern Africa.

Overview of the Region

Tropical and Southern Africa covers a vast area, stretching from more than 1,000 miles (*1,600 kilometers*) north of the Equator to almost 2,400 miles (*3,800 kilometers*) south of the Equator. Africa has open grasslands and dense rain forests, snowy mountain peaks and flat plains, wooded hills and forbidding southern deserts.

As a whole, the African continent is about three times the size of the United States. This unit deals only with Tropical and Southern Africa — that part of the continent lying south of the Sahel. This region is about twice the size of the U.S. As you can see from the Checklist of Nations on pages 585–586, the nations in the region range in area

Africa's environments include mild highlands and snowy mountains—on the Equator. At left, Tanzania's Mt. Kilimanjaro (kil-uh-muhn-JAR-oh) looms over giraffes in a game reserve.

Tropical and Southern Africa

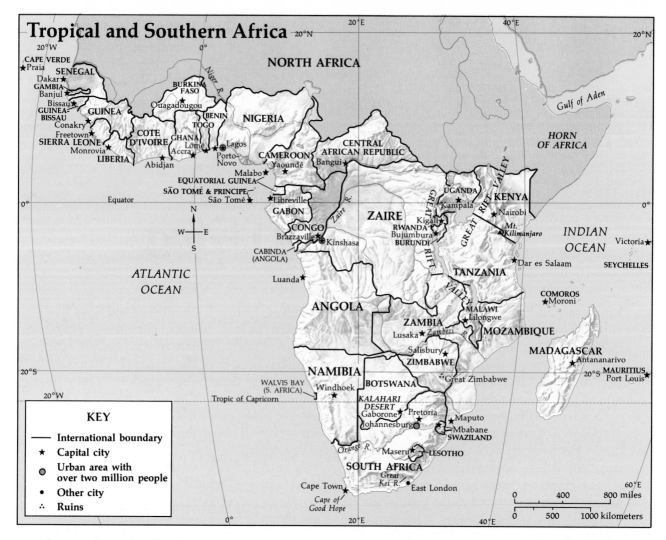

20°N · NORTH AFRICA · 20°N

CAPE VERDE · ★Praia · SENEGAL · Dakar★ · GAMBIA · Banjul★ · Bissau★ · GUINEA-BISSAU · Conakry★ · GUINEA · Freetown★ · SIERRA LEONE · Monrovia★ · LIBERIA · Abidjan★ · COTE D'IVOIRE · GHANA · Accra★ · Lomé★ · TOGO · BENIN · Porto-Novo★ · Lagos · BURKINA FASO · Ouagadougou★ · Niger R. · NIGERIA · CENTRAL AFRICAN REPUBLIC · Bangui★ · CAMEROON · Yaoundé★ · Malabo★ · EQUATORIAL GUINEA · SÃO TOMÉ & PRINCIPE · São Tomé★ · Libreville★ · GABON · CONGO · Brazzaville★ · CABINDA (ANGOLA) · Kinshasa · ZAIRE · Zaire R. · UGANDA · Kampala★ · RWANDA · Kigali★ · Bujumbura★ · BURUNDI · KENYA · Nairobi★ · Mt. Kilimanjaro · GREAT RIFT VALLEY · Gulf of Aden · HORN OF AFRICA · INDIAN OCEAN · Victoria★ · SEYCHELLES · Dar es Salaam · TANZANIA · COMOROS · Moroni★ · MALAWI · Lilongwe★ · MOZAMBIQUE · MADAGASCAR · Antananarivo★ · MAURITIUS · Port Louis★ · ZAMBIA · Lusaka★ · Zambezi R. · Salisbury★ · ZIMBABWE · Great Zimbabwe · NAMIBIA · Windhoek★ · WALVIS BAY (S. AFRICA) · BOTSWANA · Gaborone★ · KALAHARI DESERT · Pretoria★ · Johannesburg · Maputo★ · Mbabane★ · SWAZILAND · Orange R. · Maseru★ · LESOTHO · SOUTH AFRICA · Cape Town · Cape of Good Hope · Great Kei R. · East London · ATLANTIC OCEAN · ANGOLA · Luanda★ · Equator

Equator · 0°
Tropic of Capricorn · 20°S

KEY
— International boundary
★ Capital city
◎ Urban area with over two million people
• Other city
∴ Ruins

0   400   800 miles
0   500   1000 kilometers

from a tiny island such as Mauritius (maw-RISH-uhs) to Zaire, which is one fourth the size of the U.S.

Although Africa lies just south of Europe, Africa's vast interior was largely unknown to Europeans until late in the 19th century. Of course, Africans knew the area in which they lived. In some parts of Africa, there was trade between peoples in different areas. At various times, peoples in different areas joined together to form large empires. But most Africans stayed near home, living in groups whose members were often related.

There are good reasons why the region was largely isolated. Even today forests, deserts, and mountains hamper travel in many areas. Also, diseases like malaria and sleeping sickness kept humans out of some areas. Malaria is spread by mosquitoes that thrive in damp lowlands and flatlands.

The region includes all of Africa south of the Sahel, the southern edge of the Sahara. Three subregions — West, East, and Southern Africa — radiate from a fourth, Central Africa (which covers the Zaire basin).

Sleeping sickness, which affects both humans and livestock, is spread by the tsetse (SET-see) fly. Despite such problems, Tropical and Southern Africa is home to more than 300 million people.

Physical Geography

Africa's environment seemed forbidding to outsiders. Henry Morton Stanley, an American, explored a vast expanse of Central Africa in the 1870's. Here is what he wrote in his diary as he pushed his way on foot through the rain forest: "The trees kept shedding their dew on us like rain in great round drops. Every leaf seemed weeping.

The atmosphere was stifling. The steam from the hot earth could be seen settling in a gray cloud above our heads."

Reports such as this made many Americans and Europeans think of Africa as one vast jungle. Yet only 10 percent of the continent is covered by rain forest (see the map on page 47). In fact, Africa has a wide variety of landforms and climates.

The Shape of the Land

Much of Tropical and Southern Africa lies at high elevations. The land rises slowly from west to east, one plateau following another. Only in West Africa and near the coasts are there broad lowlands.

■ *East Africa*. Much of East Africa consists of high plateaus, broken occasionally by sharp drop-offs or tall mountains. The tallest mountain is called Kilimanjaro, and it rises more than 19,000 feet *(5,800 meters)* above sea level. Kilimanjaro was once a volcano, and serves as a reminder of the powerful forces that shaped the East African landscape.

Another reminder is the Great Rift Valley — a huge Y-shaped scar down Africa's east side. In many places, steep cliffs rise sharply from the valley's floor. The cliffs have hindered east-west travel, but the waters that fill parts of the valley floor have made north-south travel relatively easy. Look at the map on page 378. You will see a string of lakes starting just northwest of Mozambique and stretching northward along the Great Rift Valley.

■ *West Africa*. In contrast to East Africa, West Africa is mainly a low plateau that drops off sharply to coastal lowlands. The Niger (NIE-jur) River starts in one of the few highland sections in the area. It is West Africa's major waterway.

■ *Central Africa*. The main feature of Central Africa is the vast basin of the Zaire River. The basin, which is flat and surrounded by low hills, is on a plateau, roughly 1,000 feet

(300 meters) above sea level. To the west, the plateau descends in a series of steps to the Atlantic Ocean.

■ *Southern Africa*. Most of Southern Africa is on a plateau, somewhat higher than the Zaire Basin but lower than East Africa. Mountains rise close to the coast at the southern tip of the continent. Southern Africa has two major rivers, the Orange and the Zambezi (zam-BEE-zee).

Climate

The Equator passes through the center of Africa, so the climate is warm all year. But remember, altitude also affects climate. High places tend to be cooler, and some parts of Africa are very high.

Kilimanjaro, the highest point in Africa, is closer to the Equator than the place where Stanley wrote of the steamy heat. Yet Kilimanjaro's twin peaks are always blanketed with snow. Many parts of the East African highlands have mild temperatures all year round.

Rainfall varies greatly from one place to another. In general the rainiest places are those nearest the Equator, where the Sun is directly overhead. As you read in Unit 1 (page 40), the heat from the Sun makes the air rise. Any moisture in the air is then cooled and condenses as rain.

In June the Sun is directly overhead north of the Equator, at the Tropic of Cancer. The zone of heavy rain then shifts toward the north, and the southern tropics are dry. By December the zone of heavy rain has shifted toward the south, and the northern tropics are dry. Thus many parts of Tropical and Southern Africa have what is known as a **tropical wet and dry climate** — wet in the summer and dry in the winter. (Don't forget that when it is summer north of the Equator, it is winter south of the Equator.)

However, places on or close to the Equator get the rains both coming and going. The year-round warmth and rainfall create a rain forest zone that stretches through Central Africa and along the West African coast.

Other parts of Tropical and Southern Africa get little rain. There are deserts and semideserts both along the southwest coast of Africa and in the northeast near the Horn of Africa (the sharp point below the Gulf of Aden). The southern tip of Africa has a Mediterranean climate, with mild temperatures and dry summers.

Between the rain forest and the deserts lie broad **savannas** (tropical grasslands with scattered trees). These savannas get less rainfall than the rain forests, but more than the deserts. Many kinds of animals live in these grasslands. In East Africa, wild animals include zebras, giraffes, elephants, lions, leopards, and cheetahs. Most wild animals today live in large game reserves.

Now that you have looked at both land and climate, you can see that the region can be divided into three main subregions:

East Africa	Grasslands. High elevation. Dry season.
West and Central Africa	Tropical rain forest. Low elevation. No dry season.
Southern Africa	Grasslands and deserts. Medium elevation. Dry season.

Resources

As in other tropical lands, soils in much of Africa are not rich. In many places, the soils are made poorer by heavy tropical rains that wash out nutrients or cause erosion. In other places, especially the grasslands, the soils can be used for growing many food crops. Here the warmth and moisture of the region are a help to farming.

There are more riches beneath Africa's soils than above. A young boy is credited with finding the first diamond in what is now the Republic of South Africa. The "pretty pebble" that he showed to a passerby in 1866 set off a diamond rush.

With diamonds, gold, iron, copper, coal, uranium, chromium, and platinum, the mineral wealth of Southern Africa is immense. Some of these minerals, such as copper and iron, have been used for centuries by Africans.

Large areas of Africa have never been fully explored for mineral resources. Major deposits may still be found — just as oil was discovered off the shores of Nigeria in 1956.

At present the Republic of South Africa seems to be the country with the greatest resources. Not only does South Africa have many minerals, but they are found close together. For instance, iron deposits are near coal deposits, where they can be used together to make steel. There are coal and iron in West Africa as well — but the coal (in Nigeria) is 1,300 miles (*2,100 kilometers*) from the iron (in Liberia).

Energy resources are scarce. Major oil fields have been found only along Africa's western coast — in Nigeria, Gabon (guh-BOHN), Cameroon (kam-uh-ROON), and Angola (an-GOH-luh). There is little coal outside of South Africa and Nigeria.

SECTION REVIEW

1. How do the resources of Southern Africa compare with those of East Africa and West and Central Africa?

2. In East Africa, travel between north and south has been easier than between east and west. Explain why.

3. Although Africa is close to Europe, Europeans did not explore its interior until the 19th century. Give one reason they did not explore it earlier.

4. Both East Africa and West Africa are on the Equator. Describe the climates of each. Why do these climates differ?

Human Geography

Anicet in Central Africa, Omatayo in West Africa, and Sijey in East Africa are all black Africans. Hendrik in Southern Africa is a white African of European origin. There are also other groups of people in the region, including Arabs (in East Africa) and Asians (in East and Southern Africa). But the majority of people in the region are blacks.

Around 1,500 years ago, groups of black Africans began building cities and empires in West Africa. Later blacks moved east and south to other parts of the continent, where new empires grew.

The names of some of these empires are on the map today. The nation of Ghana (GAH-nuh) took its name from the earliest of the West African empires. Ghana flourished thanks to a mineral resource — gold. This gold was traded with peoples in North Africa, and some of it even reached Europe.

The nation of Zimbabwe (zim-BAHB-way) took its name from an empire that developed in Southern Africa 800 years ago. Ruins of stone buildings known as Great Zimbabwe still remain from that empire.

As you have read, the people of the Ghana empire traded with North Africa. There were other early contacts between Tropical and Southern Africa and the outside world. Before the time of Christ, Arabs and other Asians traded with peoples on the East African coast. Then came European explorers and traders, starting with the Portuguese in the late 15th century.

For a long time, a cruel slave trade domi-

The most populous nation in the region is Nigeria, in West Africa. Its capital, Lagos (below), is the region's largest city. Different ethnic groups in Nigeria speak different languages; the official language is English.

nated relations between Africa and the rest of the world. Europeans and Arabs took countless numbers of slaves, buying many of them from African traders. The slaves were carried away in boats to be sold in distant parts of the world.

Large numbers of Europeans settled in Southern Africa. In 1652 the Dutch started the colony of Cape Town at a natural harbor on the Cape of Good Hope. Cape Town served as a supply post for Dutch ships bound for Asia. As time passed, Dutch and other white settlers moved inland.

By the late 19th century, slavery was ending. But Europeans found new interests in Africa: rubber, ivory, gold, diamonds. Starting in the 1880's, the European governments of Britain, France, Germany, Portugal, Spain, Belgium, and Italy divided Africa into a patchwork of separate colonies.

Now more Europeans came to live in Africa — mostly in the East African highlands. This area has a mild climate and its soil is richer than in most of Africa. As a result, European-style farming is possible.

The Europeans brought many Asians to Southern Africa to work on sugar plantations. In East Africa, Asians were put to work on building railroads. In time many Asians became small shopkeepers or business owners.

Colonial rule began to break up in the 1950's. Almost all of Tropical and Southern Africa is now independent. Many whites left Africa as the colonies became independent nations.

Today the largest group of whites is in the Republic of South Africa, an independent country of 33 million people where power is in the hands of about five million whites. The white minority keeps its power by a system of **apartheid** (uh-PAR-tayt; separation of the races). Only whites can vote for the national government; nonwhites are limited by law in their choice of jobs and areas where they can live.

Differences of Culture

When Sijey's father died, his body was left on top of the ground for nature to dispose of. Sijey's people, the Masai (mah-SIE), believe that life comes from above the ground on which they and their animals walk. To the Masai, placing a dead person under the ground would be a terrible insult. Sijey's father was laid out with a pair of sandals in his hand, ready to begin his journey to the next world.

Omatayo's people, the Yoruba (YOH-roo-buh), would be horrified at the thought of leaving a human body unburied. The Yoruba have long been farmers. For them the soil gives life in the form of crops. Therefore they place the dead in the soil. When Omatayo's father died, he was wrapped in fancy clothes and buried with great ceremony.

Ideas about how to deal with death are part of each people's culture. Many separate cultures are found among African peoples. Often, as in the above example, a group's culture is closely linked to the way that group makes a living.

The kinds of food that Africans eat vary with the environment. In a dry area, such as northern Nigeria, the staple food is sorghum or millet. This grain is cooked with water and eaten with vegetables such as okra, tomatoes, and hot peppers. In a wet area, such as southern Nigeria, the staple food is rice or cassava. This is eaten with vegetables such as yams and plaintains (which are like bananas). But the ingredients and the spices used vary widely from place to place.

Meat is not an everyday part of most Africans' diets. Even people who keep cattle, such as the Masai, rely on milk rather than meat. People who live near the sea, a lake, or a river make fish a part of their diets.

Languages differ widely throughout the region. The peoples of Tropical and Southern Africa speak more than 1,000 languages.

The largest group of whites in the region is found in the Republic of South Africa, where they rule a black majority. This "whites only" beach is in Cape Town.

People often learn several languages to make day-to-day dealings easier.

In a few cases, a single African language has become a **lingua franca** (LIN-gwuh FRAN-kuh) — that is, a common language spoken over a broad geographic area. Such a language is Swahili (swah-HEE-lee), which is widely spoken in East Africa. Swahili is an African language that includes many Arabic words. It has been shaped by long-time contacts between East Africans and Arab traders.

Some African nations, lacking an African *lingua franca*, have chosen a European tongue as their official language. Usually it is the language of a former colonial ruler. For instance, Portuguese is the official language of Mozambique.

Africans practice a variety of religions. Like American Indians, African peoples had their own concepts of spiritual forces long before outside religions were introduced. Followers of most traditional African religions believe that the world was created by God and is under the control of various spirits. If people offer these spirits prayers,

ceremonies, and good behavior, life will go well for them in this world and after death.

Today about half of all people in Tropical and Southern Africa follow these ancient beliefs. But Islam and Christianity have won many followers. Many of the people north of the forest zone and along the East African coast are Moslems (followers of Islam). Many of the people in countries farther south are Christians.

Where the People Live

If you visited the Central African rain forest or the Kalahari Desert, you might travel quite a long way before seeing signs of human life. In fact, most of Africa has a fairly low population density — but there are important exceptions. Three areas in particular are centers of population. Look at the population map on page 546 and see if you can spot them.

One population center is in Nigeria, in West Africa. Nigeria has the most people (80 million) of all African countries. It was an ancient center of civilization.

A second population center extends along and inland from the eastern coast of Southern Africa. This area includes many of South Africa's mines, cities, and industries.

A third center of population is in the East African highlands, on either side of the Great Rift Valley. This area has a favorable climate, being cooler and wetter than many other parts of Africa.

SECTION REVIEW

1. You have read that the region has three densely populated areas. Describe two of the three.

2. Why did many Asians settle in Africa?

3. What role does Swahili play in East Africa?

4. In the late 19th century, European nations began to divide Africa into colonies. What caused this new interest in the region?

Economic Geography

In the United States, most families have a refrigerator, running water, and other comforts. In Tropical and Southern Africa, this is not the case. Most Africans make do without modern conveniences — as did many Americans as recently as the 1930's.

Many Africans do not feel a need for such things as refrigerators. Earlier you met Anicet, who lives in Zaire. Anicet's father is a fisherman who goes out in his boat every day. He keeps just as many fish as his family needs and sells the rest at a market or tourist hotel. The hotel uses a refrigerator, but Anicet's family gets its food fresh every day.

Omatayo would like very much to have a refrigerator in her Nigerian village. A refrigerator would cut down on the time she spends harvesting perishable food. She and her family would also like to buy other things, such as a transistor radio.

In most African countries, such manufactured goods have to be imported and are very expensive. There are relatively few industries, and most factories produce basic items such as clothing and construction materials.

Most African leaders are eager to build up industry. They see fast economic growth as the key task facing Africa. A few African countries are well on their way to development, but most are only beginning.

Traditional and Modern

In the villages, most Africans still follow traditional ways. They grow food crops for their immediate needs. They are subsistence farmers, with little surplus to sell or trade.

But the growing of crops sold for cash is gaining in importance. Some crops, such as rubber and palm oil, are grown on large plantations. Other crops, such as tea and cacao (for chocolate), are grown mainly by

Cacao, which grows well in a hot, humid climate, is a major cash crop along the West African coast. Here workers spread cacao beans out to dry.

small farmers. In some areas, cash crops have been grown for many decades.

In southwestern Nigeria, Omatayo and her husband have two small plots of land. On one plot, they grow yams, cassava, corn, and pumpkins for their own use. On the other, they grow cacao as a cash crop.

The cacao they grow may end up as cocoa or chocolate ice cream in your neighborhood grocery store. Cacao starts out as a bean growing in a football-shaped pod on the trunk of a tree (see page 437). At harvest time, the beans are "cured" (fermented) for a week before being sent to market. Like many Nigerian women, Omatayo is a small-scale dealer in cacao. She buys pods from neighbors and ferments the beans along with her family's beans.

Omatayo handles a great deal of money. She plays an active part in the exchange of goods and services that makes up the cash economy of Nigeria. Most African nations are trying to draw more and more people into the cash economy. Leaders see this as a key step in building a modern nation.

A Struggle To Develop

Many African countries got a start toward economic development when they were under colonial rule. European nations wanted Africa's raw materials (crops and minerals). Europeans went to Africa to start plantations and mines. However, the aim of the colonial nations was to satisfy their own needs. They were not interested in developing balanced economies in their colonies.

Therefore the economies of many African countries are lopsided. Zambia (ZAM-bee-uh) depends almost entirely on exporting copper. Ghana depends heavily on cacao. Burundi (boo-ROON-dee) depends on coffee. Relying on one product can be risky. Prices of raw materials rise and fall on the world market. A sudden drop in prices can throw many Africans out of work and cause a sharp drop in a nation's income.

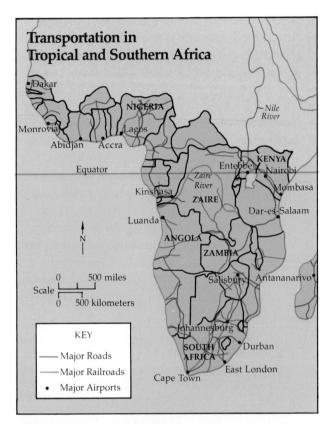

African nations are working to extend railroads and complete long-distance highways, but many routes date from colonial times. Thus there are few links between some neighboring nations that had different colonial rulers, such as Congo (France), Zaire (Belgium), and Angola (Portugal). Also, the only routes for the foreign trade of some black-ruled nations are railroads leading to ports in white-ruled South Africa.

Colonial rulers also left African nations with a lopsided system of ports and railroads. Transportation routes were designed to ship products overseas. It is often harder to travel or ship goods within Africa than to a country outside.

African nations are trying to improve their economies in various ways. Most nations have programs to help boost farm production. Some have set up "model farms" to show farmers how they could successfully grow rice. Others have organized cooperatives to help farmers sell their coffee. Many nations are resettling families on new farms.

African nations have tried to boost industry as well. In many cases, they can build on mining operations begun in colonial days. One goal is to set up more processing plants, so that a raw material like bauxite can be turned into aluminum before being shipped overseas. In this way, African countries can earn more from their resources.

Other industries are being developed, but not as fast as African leaders would like. One difficulty is a shortage of skilled workers and technicians. Anicet of Zaire will be going to college to study chemistry when he graduates from high school — but he is an exception. In most of the region, as in developing nations around the world, only a small minority of young people go to

Much African industry is based on minerals. This fertilizer plant in Senegal uses local phosphate deposits.

school through sixth grade. The majority still learn the skills of the traditional world rather than the modern world.

Africa and the World

Africa has a mineral resource that is essential to modern aircraft — cobalt. By mixing cobalt with other metals, aircraft makers can build engines that withstand the extreme heat of modern jets. Some 97 percent of the cobalt used in the U.S. comes from Zaire and Zambia.

Cobalt is one of several key minerals that are in short supply in the world — and that Tropical and Southern Africa can provide. Other minerals important to the U.S. and other modern nations include chromium, platinum, and manganese. The Republic of South Africa is a key supplier of these.

Nigerian oil is important too. The United States imports more oil from Nigeria than from any other nation except Saudi Arabia.

Africa's vital resources are one reason why tensions on the continent can send shock waves around the world. Perhaps the biggest tensions are those involving race. Racial tensions have complicated Africa's relations with the outside world, as well as relations among many African nations. For example, leaders of most outside nations (including the U.S.) have condemned South Africa's system of apartheid. Many African nations have vowed to help bring about "majority rule" in the Republic of South Africa. Since the majority of South Africans are black, majority rule means black rule.

Meanwhile, South Africa has close economic ties with many black-ruled African nations. For instance, the neighboring nation of Zimbabwe conducts 90 percent of its trade through South African ports. This pattern of trade was set when Zimbabwe was a colony. It depends in part on transportation routes built in colonial times. Thus the major seaports in Southern Africa are in the Republic of South Africa, and the major railroads lead to those ports (see the map on page 385).

It is difficult for Zimbabwe and other former colonies to change their patterns of trade without damaging their economies. And to many African leaders, the most urgent task is to *improve* their nation's economy.

Improving African economies means changing Africa's role in world trade. As in colonial days, most of Africa's exports are raw materials. In contrast, most of Africa's imports are factory-made goods. Africa's leaders believe this pattern of trade works against their interests.

The problem with selling raw materials is that the seller has little control over the price. If the world supply of a raw material is less than the demand, the price will drop. On the other hand, the price of manufactured goods is set to cover the cost not only of raw materials but also of labor, machinery, and other expenses.

It is not easy to start manufacturing goods. Much time, effort, and money are needed. However, as you read earlier, African nations are striving to build up their own industries. These industries are planned to produce many of the goods now imported from abroad.

Looking Ahead

The rest of this unit will focus on two countries in Tropical and Southern Africa. Chapter 29 deals with Kenya in East Africa. Like many developing countries, Kenya is short on natural resources, but its economy could do well on agriculture and tourism.

Chapter 30 deals with Zaire in Central Africa. Rich in minerals, Zaire is the largest nation of Tropical and Southern Africa in terms of area. It is also one of the poorest.

Chapter 31 will look more closely at the ways in which Kenya, Zaire, and other countries in the region are trying to break out of their past and present poverty.

SECTION REVIEW

1. In what way are the region's resources important to the economy of the United States?

2. What is a cash crop? Why are most African nations trying to increase the use of cash among their people?

3. One result of the colonial period has been the lopsided growth of many African economies. What does this mean? Describe one problem this lopsided growth has caused.

4. Describe one way in which African nations are expanding industry.

YOUR LOCAL GEOGRAPHY

1. Geography Skills 37 (pages 388–389) compares photographs and a map of Africa. Now try this for your own community. From a local newspaper, postcards, or other source, choose two photos showing broad views of an area in or near your community. Then look in your school library or public library for large-scale maps of the area showing one or more of the following: relief, population density, industry, land use. Suppose someone who lived in a different area was thinking of moving to your community. Would the maps or the photos be more helpful in giving them an idea of the area? Why? How would both be helpful?

2. Use an almanac to find the area and population of your state. Then consult the Checklist of Nations on pages 585–586. Look down the area column to find the two nations whose area is closest to that of your state on the larger side (if any), and the two nations that are closest to that of your state on the smaller side. List these nations and your state, with areas, from largest to smallest. Then look down the population column to find the two larger nations (if any) and the two smaller nations closest to the population of your state. List these nations and your state, with populations, from largest to smallest. Overall, how does your state compare in size with the nations of the region?

Relating Photographs to a Map

The photographs on the opposite page show two of the three main types of vegetation areas you would see in Tropical and Southern Africa. From photographs such as these, you can find out what a vegetation area looks like, but not where you would find it. A special-purpose map, like the one below, can tell you where each vegetation type grows, but not what it looks like. To find out what vegetation areas look like at the places shown on the map, you can **synthesize** (combine) information from both photographs and the map.

Notice that the map key lists three major classes of vegetation: forest, grasslands, and desert. The photographs match two of these vegetation classes.

1. Find the letter A or B in the lower left corner of each photograph. On a separate sheet of paper, write the letter of each photograph with a matching category from the map key.

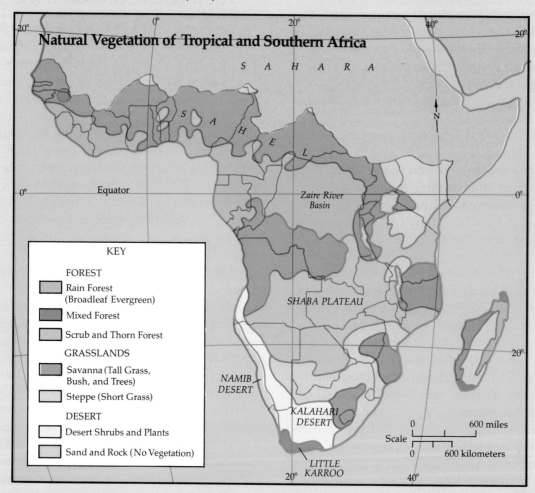

Natural Vegetation of Tropical and Southern Africa

S A H A R A

S A H E L

Equator

Zaire River Basin

SHABA PLATEAU

NAMIB DESERT

KALAHARI DESERT

LITTLE KARROO

KEY

FOREST
- Rain Forest (Broadleaf Evergreen)
- Mixed Forest
- Scrub and Thorn Forest

GRASSLANDS
- Savanna (Tall Grass, Bush, and Trees)
- Steppe (Short Grass)

DESERT
- Desert Shrubs and Plants
- Sand and Rock (No Vegetation)

Scale
0   600 miles
0   600 kilometers

You can see from the map key that each class of vegetation is subdivided into more than one type of desert, grassland, or forest. Under desert, for example, one type has shrub vegetation; another has no vegetation at all. Grasslands range from areas of short grass to wooded savanna with tall grass. While some forests have widely spaced trees, treetops in a tropical rain forest come together to block out the sun.

2. Compare each photograph with the types of vegetation described for its class in the map key. Then on your answer sheet, tell which type of desert, grassland, or forest you think is shown in each photograph. Where would you see these types of vegetation if you were traveling in Tropical and Southern Africa? Can you find them on the map?

3. Six areas shown on the map are listed below. Find each area, and use the map key to identify the type of vegetation in that area.

Kalahari Desert   Little Karroo   Shaba Plateau
Namib Desert   Sahel   Zaire River Basin

4. Each of the photographs matches *one* of these six areas. Tell which one on your answer sheet.

A Bus Ride in South Africa

Noni Jabavu (NOH-nee JAH-bah-voo) belongs to a group of black South Africans known as Xhosa (KOH-suh). She left South Africa to live in England, but returned to visit her family. Here she describes a bus trip from a city into the countryside to visit her uncle. All the passengers are blacks. As you read, note how Noni Jabavu describes not only how the countryside looks but also what it means to the passengers.

We set off through the seaside town along paved roads bordered by European bungalow-houses washed white or cream. Morning glory cascaded over porches, and roofs were painted or tiled red or green. The dwellings looked inviting, cared-for. Soft lawns lay in front of them, and black gardeners held hoses, watering before the sun should rise in the sky. East London is a clean, orderly, bustling port and residential town with a long frontage on the Indian Ocean. Almost all of its pleasant sandy beach-line is reserved for "Europeans Only." It was not long before the bus was heading out of the town and inland.

We climbed up, then down and round the spreading hills of now rather bare countryside, sometimes dotted with huge granite boulders and thorn trees. Sometimes the land stretched out smooth and featureless, with isolated houses in the distance protected by windbreaks of tall trees. There were occasional windmills whose shining metal wheels were immobile.

The chatter on the bus was different from some I had heard on country buses in England, my other homeland. Here people discussed history, politics, land reform, property, sociology. Conversation hinged on these topics because people were preoccupied by the history of "the migrations" of tribes and the encounters that these had brought. The bus was now crossing country that had been migrated into and settled by Chief Ndhlambe (uhnd-LAHM-beh). It was his descendants whose huts we were passing and whom we saw walking and bicycling.

I wondered how differently the countryside must appear if you did not know the language and could not understand what was being said. You would note the scenery to be sure, with its roadside advertisements for "South African Beers," or "Cigarettes." You would note the hamlets at which the bus stopped at gas pumps labeled in English, "Pegasus Oil," or in Afrikaans, *"Ry mit die Rooi Perd"* [Ride with the Red Horse]. You would see the white-owned stores standing in dust and rubble. . . .

The bus came to the Great Kei. The river lies in the folds of a breathtaking valley, between hills covered with bush and brown boulders. The road twists back on itself down steep inclines [slopes] until level with the bridge.

Excitement mounted, for we were approaching a landmark indeed and the old-time migrations rose again in our minds. As we reached the bridge and drove slowly along it, we looked from side to side at the waters spreading on to the dunes on either side. A hush fell on us all. We were now in Xhosaland.

— From *The Ochre People: Scenes from a South African Life,* by Noni Jabavu. New York: St. Martin's Press, 1963.

Ask Yourself . . .

1. What natural feature forms one border of Xhosaland? Why were the passengers excited when they reached that border?

2. Does the land that the bus passes through have a dry or moist climate? How can you tell?

3. Why might the land look different if you could not understand what the passengers were saying? How do you think the "pleasant sandy beach-line" of East London looked to the passengers? Why?

CHAPTER REVIEW

A. Words To Remember

Three of the following terms are defined by the numbered phrases below. On another sheet of paper, write each term next to the number of its definition. Then write the two terms that are *not* defined and give a definition for each.

apartheid   majority rule   synthesize
lingua franca   savanna

1. A common language that is spoken over a broad geographic area.

2. A system that would allow blacks to take part in the government of the Republic of South Africa.

3. To combine.

B. Check Your Reading

1. What kind of landform covers most of the region? Where are the major lowlands?

2. What two features of the climate in Central Africa produced the rain forest zone?

3. Describe one way that Sijey's life-style differs from that of his fellow African Omatayo.

4. What is the basic difference between a subsistence farmer and a farmer who grows cash crops?

5. Name one of the region's mineral resources that is vital to the modern world. Why is this resource important?

C. Think It Over

1. How does a *lingua franca* help an African state build a modern economy?

2. During the colonial period, railroads and ports were built by the colonial powers. What have been the advantages of these transportation systems? What have been the disadvantages?

3. How do relations between the Republic of South Africa and other African nations affect the region's development?

D. Things To Do

Black-ruled African nations object to the Republic of South Africa's policy of apartheid and would like to cut off all trade relations with South Africa. Nevertheless, nine of these nations do trade with South Africa or rely on its ports and railroads for transporting their goods. These nine nations are Angola, Botswana, Lesotho, Malawi, Mozambique, Swaziland, Zaire, Zambia, and Zimbabwe. Using the map on page 378, locate the Republic of South Africa and the nations listed above. Can you suggest a geographic reason why many of the nine nations trade with South Africa?

Chapter 29

Kenya
Making Do Without Minerals

In the East African highlands, the white blossoms of the coffee trees sparkle as if snow has frosted them. Fortunately there is no snow here. Cold weather would quickly destroy the coffee trees. The clear light of the highlands only creates the illusion of snow.

Muumbi Kigana (moo-OOM-bee kee-GAH-nuh) loves the sparkling hillside of the Kenya highlands. With her husband, Wisdom, she owns a small coffee farm a short way north of Kenya's capital city of Nairobi (nie-ROH-bee). Their cement house perches on a hilltop. Similar coffee farms and homes dot the hills all around.

Coffee trees thrive in the volcanic soil of the highlands. The area receives about 40 inches *(100 centimeters)* of rain a year — the same as Pennsylvania. The climate is mild — not too hot in summer, not too cold in winter. Even in the coolest month of July, the temperatures for day and night average about 60°F *(16°C)*. January temperatures in Tampa, Florida, are about the same.

Muumbi looks after the farm while Wisdom teaches at a nearby school. After the petals fall from the coffee trees, green berries (beans) appear. Slowly the berries ripen and turn to red. The harvest lasts a long time, because the berries do not all ripen at once. Therefore, picking must be done by people instead of machines.

It takes a lot of work to put a steaming cup of coffee on someone's breakfast table. A coffee tree about the size of a tall man yields perhaps three pounds *(1½ kilograms)* of beans in a good year. Thus it takes about eight trees to supply two cups of strong coffee daily for a year.

Wisdom helps take the beans to a nearby coffee factory, which is owned jointly by the small farmers of the area. Factory machines remove the pulp from around the beans. The beans are then washed, dried, and trucked to Nairobi. There they are either roasted for local use or put on trains to be taken to the coast for export. Months later they may wind up on the shelf of a grocer in the U.S. or Europe.

Coffee is important to the Kiganas — and to Kenya as a whole. Coffee brings more money into Kenya than any other export. Since Kenya has no known major mineral resources, it must find other ways to develop its economy.

Physical Geography

Kenya sits astride the Equator in East Africa. It is roughly the size of Arizona and Nevada combined.

Most of the nation is hot and dry, but parts of southwestern Kenya form part of the East African highlands, which Kenya shares

---

Moderate warmth and rainfall make the highlands of East Africa ideal for growing such crops as coffee and tea. At left, workers pick tea leaves on an extensive plantation in Kenya.

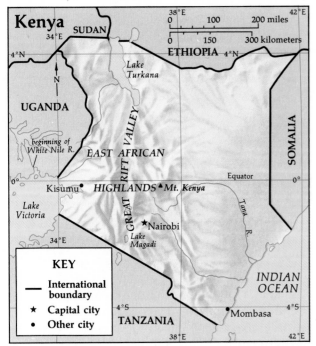

Although Kenya lies on the Equator, its highlands have a mild climate. Nairobi is at an elevation of 6,000 feet (*1,800 meters*). Most Kenyans live in the south, which receives more rain than the rest of the country.

with neighboring countries (see the map above). Here, because of the high altitude, temperatures are mild all year round, and rainfall is high. The highlands have richer soils than most surrounding places, and are an important farming area. The Kiganas' coffee farm is here.

As you saw in the last chapter, the highlands lie on either side of the Great Rift Valley. The valley's width generally ranges from 30 to 40 miles (*50 to 65 kilometers*). In many places, the bottom of the valley lies more than 5,000 feet (*1,500 meters*) above sea level. In such places, the valley itself is part of the highlands.

Several volcanic mountains tower over the East African highlands. Within Kenya the highest peak is Mount Kenya, which is more than 17,000 feet (*5,000 meters*) above sea level. There are glaciers on Mount Kenya all year round — although the mountain is close to the Equator.

The highlands are one of four distinct areas within Kenya. To look at these areas, imagine you are taking a trip on the railroad that crosses Kenya from east to west.

A Varied Land

The trip begins in Mombasa (mohm-BAH-suh), Kenya's main port on the Indian Ocean. A low-lying coastal plain is the first of Kenya's four areas. Here it is hot and humid, as one expects the tropics to be. Bushy plants and coconut trees are common. Cultivated crops include cashew nuts and sisal (a fibrous plant that is used for making rope).

The weather in the coastal plain is strongly affected by winds that change direction at regular times of the year. For part of the year, the winds bring rainy periods that may last up to six months in some places. For another part of the year, the winds blow from a different direction and do not bring rain. The shifts are caused by the changing positions of Earth and the Sun. The Sun is directly overhead in the tropics north of the Equator at one time of year, and south of the Equator at another time (see pages 36–37). Winds that make such regular changes are known as **monsoons**, which are common in coastal regions of eastern Africa and southern Asia.

Within an hour, your train has begun to climb a series of plateaus. Monsoon rains do not reach the plateaus, and little grows there but brush and grass. Three fourths of Kenya lies on these arid plateaus, which make up the second major area of the country.

In some northern stretches, the plateaus are covered by desert. Elsewhere there are savanna grasses that provide food for domestic (tame) and wild animals. From the train windows, you may see elephants and giraffes, lions and cheetahs. In most places, these plateaus are too dry for much farming.

After several hours of travel, the land rises more sharply. You are entering the highlands, and the air becomes cooler. The highlands make up the third major area and cover about one fifth of Kenya. In valleys and on fairly level land, crops like corn, peas, cabbages, and potatoes can be grown.

Coffee grows on hilltops around Nairobi; tea, on the higher hills to the northwest. Another popular crop is pyrethrum (pie-REETH-ruhm), a daisylike flower whose petals are dried and ground up to make a natural insecticide.

Next the train descends into the Great Rift Valley and climbs out again. After passing over the high western wall of the valley, the train descends into Kenya's fourth area — the Lake Victoria Basin. This area is a low plateau that surrounds Lake Victoria — the second most extensive freshwater lake in the world. Lake Superior in North America is the first.

Lake Victoria lies mainly within Uganda and Tanzania, but Kenya's share of the basin is one of the nation's best farming areas. Moisture from the lake evaporates, forms clouds, and eventually falls as rain on nearby land. Therefore, rainfall is high — between 40 and 70 inches (*100 and 180 centimeters*) per year. Corn and other food crops grow well here, as do cotton and sugarcane.

Your trip ends at Kisumu (kee-SOO-moo), a port city on Lake Victoria. There you see fishing boats — mostly small sailboats — docking with their day's catch. The lake contains 200 kinds of fish and provides much of the protein in the local people's diet.

Where Are the Minerals?

On your railroad trip, you would see freight trains loaded with copper ore and other minerals, headed for Mombasa to be loaded on ships. These trains come from Uganda, Kenya's neighbor on the west. Kenya has few mineral deposits of its own. The only mineral found in Kenya in any quantity is soda, which is used in making glass and chemicals. Soda is found around Lake Magadi (muh-GAH-dee) in southern Kenya.

The lake is fed by hot springs that bubble up around it. These springs are heated by molten rock deep underground — the same

Africans who live near oceans or lakes rely on fishing to provide much of their protein. Kenya has coasts on the Indian Ocean and two large lakes — Lake Victoria in the west and Lake Turkana (above) in the northwest.

kind of rock that erupts from volcanoes. Kenya is dotted with volcanoes — for instance, Mount Kenya. Over the ages, the lava from Kenya's volcanoes has crumbled to form rich soils along the Great Rift Valley.

Kenya has searched for oil, gas, and coal without success. The only fuel found widely in Kenya is wood. Many kinds of trees grow on the slopes of the highlands.

Of Kenya's rivers, only the Tana (TAH-nuh) flows strongly enough to provide hydroelectric power. But the strength of the Tana's flow varies greatly from season to season. The river offers one other resource — water for irrigating the dry plateaus.

SECTION REVIEW

1. What are Kenya's main fuel resources?

2. How do monsoons affect the weather of the coastal plain in Kenya?

3. Why are the East African highlands a good area for growing coffee?

4. What type of environment lies between the coastal plain and the fertile highlands? Why aren't there many farms in this area?

Human Geography

An election campaign is under way in Kenya and the candidates are out in force. A man in a three-piece suit moves through the streets of Nairobi, kissing babies and shaking hands.

In Kenya there is only one legal political party — the Kenya African National Union (KANU). One-party states are common in Africa. But in Kenya this does not prevent lively primary elections. As many as a dozen candidates may seek an office at one time. Very often voters replace half of Kenya's legislature in an election.

Kenya has a one-party government because its leaders want to play down ethnic divisions. The leaders want people to think of themselves as Kenyans, not as members of an ethnic group. If different political parties were allowed, each group might form its own party. It would then be more difficult to bring the peoples of Kenya together.

There are about three dozen ethnic groups in Kenya. Each speaks its own language, although many people also speak Swahili or English. Swahili is Kenya's official language. The Kikuyus (kee-KOO-yooz) are the largest group, numbering about 3.5 million people. Muumbi and Wisdom Kigana are Kikuyus. Most Kikuyus are farmers who live north of Nairobi. Other groups, like the Masai, are mainly herders.

About three percent of Kenya's people do not belong to African groups. These people include Asians, Arabs, and whites of European origin. Most live in cities. The Asians

---

The largest ethnic groups in Kenya live in the highlands, but many other Kenyans are found in different areas. These Kenyans are commuting across a creek near the port of Mombasa, in the coastal lowlands.

and Arabs tend to be clerks, shopkeepers, and businesspeople. Many of the whites hold high positions in business and agriculture.

Kenya's people follow a variety of religions. A majority of blacks and whites are Christians, with Protestants outnumbering Roman Catholics two to one. Islam has many followers among blacks, Arabs, and Asians. The Hindu religion is practiced by a majority of Asians. And about 40 percent of the blacks follow traditional religions of Africa.

From Colony to Nation

Kenya became an independent nation in 1963. Before that it was a British colony. Long before the British arrived, Arabs and Asians traded with people in the area. In the 16th century, the east coast of Africa came under Arab rule. But for centuries, outsiders had little influence over the interior.

This changed in the late 19th century. As you saw in the last chapter, European nations divided Africa among themselves. Kenya and neighboring Uganda came under British rule. Borders drawn by Europeans often cut through the middle of a group's lands. For example, the Somali (so-MAH-lee) people of northeastern Kenya were cut off from other Somali people to the east, who came under the rule of Italy. This caused disputes that continue to this day.

The British built a railroad through Kenya partly as a means of "opening up" Uganda to trade. But the railroad "opened up" Kenya too, by allowing easy access to the highlands. As you read earlier, the climate in the highlands is mild all year round. Europeans who visited the highlands found the climate delightful. White settlers began to arrive. Even larger numbers of Asians came, many from British India.

More whites settled in Kenya than anywhere else in East, West, or Central Africa. Colonial officials set aside the best parts of the highlands for whites only. Asians settled mainly in the cities. Native Kenyans found their best land occupied by foreigners.

Kenyans began a drive for independence in the 1950's. For a time, a terrorist group spread violence that hurt other Kenyans far more than whites. In the end, Kenya became independent peacefully.

Since independence many whites have remained as owners and managers of large tea and coffee plantations. But many whites who owned small farms have sold their land and moved away. Meanwhile Asians became so important in trade and business that many black Kenyans resented their influence. Kenya's government expelled many Asians in the 1970's.

Country Life

More than four out of every five Kenyans live in rural areas. You have already seen something of a farming family. The life of a herding family is quite different.

In the last chapter, you met Sijey, a man of the Masai group. Sijey lives in the dry grasslands of southern Kenya. His home is in a village of mud huts surrounded by a thorn fence. The village is moved every six months or so to a place with greener pastures.

Sijey is a man of some wealth. In his group, one sign of his wealth is the many cattle that he owns. Another sign is his three wives. In Africa most groups approve of a man having more than one wife. In reality, only about 10 percent of African men can afford it. Sijey had to give the family of each of his wives a number of animals. This payment is called "bride wealth," and is meant to compensate each woman's family for the loss of her work. One reason for **polygyny** (puh-LIJ-uh-nee; a man having more than one wife at a time) is the need for many children to help with the work.

Traditionally the Masai lived solely by herding. They did not hunt. They did not grow crops. If they needed corn or a giraffe

Nairobi, the capital city of Kenya, is one of the most modern in the region. Here, before the rush hour, a trainee police officer is learning how to direct traffic.

tail to switch away flies, they traded an animal from their herds. Today, however, Sijey's wives tend small patches of vegetables, while he does the cattle herding.

The Kenyan government is trying to encourage women to seek more education and good jobs. But many rural families cannot afford to have women go to school, since the women are needed for work at home. This same poverty sends many young men to the city to look for work. Without wealth they cannot afford to marry.

Still, several changes are reaching the rural areas of Kenya. Some of Sijey's children now go to school, although school is not compulsory in Kenya. For these children, school is a challenge. Lessons are given in Swahili, not in the Masai language.

Even bigger changes may be on the way for the Masai. The government is urging groups of herders to settle down and build ranches. These ranches would enable the herders to build bigger herds of cattle, which they would then sell for money. But although the Masai would like to have bigger herds, few would want to sell them. To the Masai, cattle are far more valuable than money.

City Life

For the one Kenyan in five who lives in a city, life moves at a faster pace. Mwai (muh-WIE) and Mwaka Waiyaki (muh-WAH-kuh wie-YAH-kee) and their four children live in Nairobi. Their home is a small apartment in a drab four-story building.

Nairobi also has modern hotels and tall office buildings. Fancy restaurants serve crabs caught in the nearby Indian Ocean. Discos and nightclubs stay open late.

This side of Nairobi plays little part in the Waiyakis' lives. Mwai takes a crowded bus to and from his job in a railroad repair yard. He and Mwaka have very little money. All the same, they manage to enjoy life. They have a radio on which they can hear the latest pop tunes, many of them by singers based in Zaire. They eat well, since Nairobi markets carry all kinds of vegetables and meats from nearby farms. Their favorite dishes are *ugali* (oo-GAH-lee; a stew of corn and beans) and goat stew.

For entertainment there are movies and sports. A popular spectator sport is cricket, which is played in many former British colonies. Nairobi also has a museum featuring stuffed animals from the Kenyan countryside. The children love the elephants, rhinoceroses, storks, zebras, and birds with brightly colored feathers.

The Waiyakis were born in neighboring villages north of Nairobi. Mwai and Mwaki still keep in touch with people in their home villages. Sometimes relatives from the country come to stay with the Waiyakis. Sometimes the Waiyakis take a bus and visit their home villages. They talk of moving back to the country for good sometime in the future.

Nairobi is the home base of many business firms that operate throughout East Africa. The city has a population of more than 800,000. Kenya's next largest city is Mombasa, with 400,000 people.

These cities are growing rapidly, as is Kenya as a whole. The nation has one of the highest population growth rates in the world. This is not surprising, since Kenyan women give birth to an average of eight children in a lifetime.

SECTION REVIEW

1. Why did many Europeans choose to settle in the East African highlands?

2. Mwai and Mwaka live in Nairobi. Are they typical of most Kenyans? Explain.

3. The traditional life-style of Sijey and other Masai herders has been changing in recent years. Describe one change.

4. Why did Kenya's leaders decide it would be best for Kenya to have only one political party?

Economic Geography

Kenya is one of the few African nations that has encouraged free enterprise. The nation's leaders hope that private businesses will help raise income per person from its present low level of 270 dollars a year.

Until 1977 Kenya and its neighbor Tanzania were linked with Uganda in an East African economic union. The three nations ran joint services such as an airline and a postal system. They hoped that by sharing the costs of such services they would all benefit. But tensions grew among the three nations, and the union collapsed.

Farm Production

The Kiganas' coffee farm covers just five acres (*two hectares*), which is tiny compared to most U.S. farms. But by Kenyan standards, it's a fair-sized piece of ground — especially since it's all in one place.

In many parts of Africa, farms tend to be scattered far and wide in many small parcels. Such farms are suited only to subsistence crops. Until quite recently, most farms in Kenya were like that too. Now, however, government programs have helped bring small parcels together into larger plots of land. The result has been higher output and a switch to cash crops.

Some farms in Kenya are very large. Extensive tea and coffee plantations in the highlands date from colonial days, and some are still owned by whites. Kenya's government has allowed the plantations to continue because the system is very productive. But the government urges landowners to build modern homes for their workers.

Making farms bigger is just one of the ways in which Kenya's government has tried to boost output. It has also encouraged farmers to plant special types of corn that yield more grain. It has built irrigation dams, so that crops can be grown in drier areas. New settlements along the Tana River are growing cotton and rice with water from a series of dams.

Building Up Industry

The United States has spent 200 years developing its industries. In Kenya, as in many colonies, there were almost no industries until after World War II (1939–1945). Since then hundreds of large and small factories have been started in Kenya.

Most factories process Kenya's main resources — farm products. Thus many factories are located in farming areas rather than cities. For instance, large coffee plantations often have their own pulping factory.

Most consumer goods are made in the Nairobi area, where most of Kenya's people live. The factories there make everything from clothing to cement and from furniture to beer. Mwaka Waiyaki wears colorful print dresses made locally from Kenyan cotton.

Mombasa and a few other cities also have major industries. An oil refinery near Mombasa refines crude oil imported by ship from Arab states on the nearby Persian Gulf (see the map on page 329). Much of the oil is used in Kenya to generate electricity.

This power is greatly needed, since hydroelectric dams on the Tana River cannot meet Kenya's needs. Kenya also imports electric power from dams on the White Nile River in neighboring Uganda.

Kenya has made great strides in developing transportation. The cross-country railroad now has branches to many areas. Kenya is also at the eastern end of a cross-Africa highway, which is under construction from Lagos (LAY-gos), the capital of Nigeria, to Mombasa.

The Importance of Tourism

Since colonial days, tourists have visited Kenya to hunt elephants, lions, and other big game. Kenya added a new word to the English language — *safari* (suh-FAR-ee; hunting trip), from the Swahili word for

Elephants and other big game help to support Kenya's economy. Much of the nation's foreign earnings come from tourists who flock to the many game reserves.

"journey." Today some 400,000 tourists come to Kenya yearly. After coffee, tourism is Kenya's biggest industry. Safari-goers are now armed with cameras, not guns. Kenya banned big-game hunting in 1977, fearing that its animals were being wiped out.

Kenya has set up many national parks and game reserves where visitors can watch Kenyan wildlife in its natural setting. Rather than trudge cross-country in long caravans, most tourists now ride in vans or jeeps.

Some Kenyans say tourism is bad because it creates only "servant" type jobs such as those for waiters, hotel maids, and drivers. Others say the trade greatly benefits Kenya. By one estimate, tourism provides jobs for some 40,000 Kenyans.

Kenya and the World

Kenya's exports go into the flow of world trade, and the U.S. does much business with Kenya. The U.S. has another reason for interest in Kenya — the country's location not far from the major oil-producing areas around the Persian Gulf. Since the late 1970's, by agreement, Kenya has provided air and port facilities for U.S. forces. Ships of the U.S. Navy make frequent calls at Mombasa.

Ships bringing imports to Kenya are even more frequent visitors. Above all, Kenya depends on outside sources of oil. As much as 90 percent of Kenya's energy must be imported. In recent years, this has been a great burden on Kenya's economy. In order to pay its oil bills, Kenya must earn as much as possible through exports or tourism. Many African nations face this problem.

But Kenya has some advantages. Unlike some other African nations, Kenya has not suffered severe political upheavals. The nation's economy grew steadily for many years after independence. In recent years, growth has slowed but not stopped. Kenya has taken some important steps on the road from the traditional to the modern world.

SECTION REVIEW

1. Why are many factories in Kenya located near farming communities?

2. Why has the Kenyan government encouraged tourism?

3. Describe two ways the Kenyan government has tried to boost farm output.

4. Kenya depends on outside sources of energy for its industries. Describe two ways that Kenya gets this energy.

YOUR LOCAL GEOGRAPHY

1. Geography Skills 38 tells how to find books using a library catalog. Use this skill to locate books dealing with various aspects of your local geography. Make a list of 12 topics that you would look under in the subject catalog in order to get this kind of information. ("Climate" would be one example; "vegetation," another.) Note features you might want to find in books about each subject. (For instance, you might want a book on climate to have a climate map of your region. Perhaps in another book, you would like to see photographs of different kinds of vegetation.) Then take your list with you to the library. Using the subject catalog, try to locate books that would fit your needs. Remember to check in the catalog to see if the books have illustrations, photos, etc. Check out the book that interests you most, and prepare a report on your chosen topic.

2. The chapter describes a railroad journey that would take you through all of the main environments found in Kenya. Suppose a visitor wanted to get a quick idea of the variety to be found in your community (county or city). Plan the shortest possible journey by car or bus that would show (a) any major variations in landforms and water forms, and (b) major types of land use (farming, industry, residential, recreational). Make a rough sketch map of the route, labeling one example of each major physical feature and land use.

Using a Library Catalog

A library **catalog** is a type of index. It lists all of the nonfiction books in the library alphabetically, with "call-numbers" telling where you will find each book. (You will learn more about call numbers on page 453.)

If your library has a card catalog like the one pictured below, each book will be listed in three ways — on separate cards — by author, by title, and by subject. The cards are filed alphabetically in drawers. To find any card, first check the **guide letters** on the front of each drawer.

Suppose you want to read *Born Free*, by Joy Adamson, the true story of living with lions in East Africa. You can look for the book by title in the drawer marked "B." You would also find a card for the book under the author's name in the "A" drawer. As you thumb through a drawer, refer to the guide cards. Each one tells you the first several letters of the cards filed behind it.

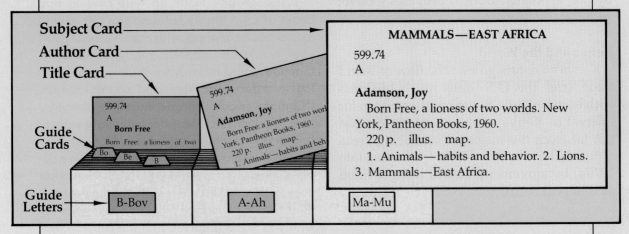

Subject Card
Author Card
Title Card

Guide Cards

Guide Letters

599.74
A
Born Free
Born Free: a lioness of two

Bo   Be   B

B-Bov

599.74
A
Adamson, Joy
Born Free: a lioness of two worlds
York, Pantheon Books, 1960.
220 p.   illus.   map.
1. Animals—habits and beh

A-Ah

MAMMALS—EAST AFRICA

599.74
A

Adamson, Joy
　　Born Free, a lioness of two worlds. New York, Pantheon Books, 1960.
　　　220 p.   illus.   map.
　　　1. Animals—habits and behavior. 2. Lions. 3. Mammals—East Africa.

Ma-Mu

Refer to the sample drawers and cards, and follow the directions below.

1. Which card drawer would you open to find a card for each subject, title, or author on this list? Write the matching guide letters on your answer sheet.   (a) Mammals—East Africa.   (b) *Born Free*.   (c) Adamson, Joy. (d) Kenya.   (e) Animals.

　　　　　A–Add   Al–Ape   B–Bov   Ke–Kite   Ma–Mu

2. To find a title card for the book *Born Free*, would you look for listings after the guide card letters "B," "Be," "Bo," or "Bu"?

3. At the bottom of each card is information called "tracings," which include all subject headings where the book is listed with other books on the same subject. Under what three subject headings could you find *Born Free* listed?

Check the catalog of your own school or local library. Is *Born Free* in its collection? Then use the catalog to find: (a) two other books by the same author; (b) two books by different authors on the subject of East African mammals.

CHAPTER REVIEW

A. Words To Remember

From the following list, select the term that best completes each of the sentences below. On a separate sheet of paper, write your answer next to the number of each sentence.

bride wealth   Mombasa   pyrethrum
glacier   monsoons   safari
Kikuyus   Nairobi   sisal

1. _____ is the capital city of Kenya.
2. The _____ are the largest ethnic group in Kenya.
3. In Swahili _____ is the word for "journey," but in English it means "hunting trip."
4. _____ is a popular crop that makes a natural insecticide.
5. _____ are winds that change direction at regular times of the year, bringing rainy and dry seasons.

B. Check Your Reading

1. Why did the Kenyan government ban big-game hunting in 1977?
2. Kenya is a trading partner of the United States. What other important tie is there between the United States and Kenya?
3. Is the coffee that Kenya exports grown on the country's small farms or on larger farms? Explain your answer.
4. Both Sijey and Muumbi live in rural areas. Describe one way in which their life-styles are the same. Describe one way in which they are different.
5 "Kenya has taken some important steps on the road from the traditional to the modern world." What does this mean?

C. Think It Over

1. Tourism is Kenya's second largest industry, after coffee. Yet some Kenyans object to it. Why? If you were a Kenyan, how would you view tourism? Explain your answer.
2. Why did Kenya's leaders want to play down ethnic divisions?
3. What advantages does Kenya have for economic development? What disadvantages does it have? Do you think the advantages outweigh the disadvantages? Explain.

D. Things To Do

You have read about the various measures the Kenyan government has taken to change the traditional herding life-style of the Masai. Imagine you are a member of the Masai. Decide whether you would prefer to continue the herding life or settle on a ranch. Then write a short dialogue between yourself and a friend who decides on the opposite course. Include two arguments on each side.

Chapter 30

Zaire
Rich Land, Poor People

The Zaire River flows in a huge curve across Central Africa. Its tributaries pour in from right and left along a course of more than 2,900 miles *(4,700 kilometers)*. By the time it reaches Kinshasa (keen-SHAH-suh), Zaire's capital city, the river has grown to an immense size. Then squeezed between the cliffs of the Crystal Mountains, the Zaire begins its drop to the Atlantic Ocean 300 miles *(475 kilometers)* away. Its waters roll and crash over dozens of cataracts (waterfalls and rapids).

These waters are being used to produce electricity. Engineers say the Zaire River below Kinshasa could generate more than enough electricity to meet the needs of an industrial nation like Britain.

The Zaire River is the biggest of the many rivers in the nation of Zaire. All told, the nation has about one sixth of the world's total capacity for hydroelectric power. But so far, little of this water power has been harnessed.

Zaire is a nation rich in many resources. It has minerals in abundance. It has land suitable for many kinds of tropical crops. It has lions, elephants, hippos, and other wild animals. It has people who can be as resourceful as any in the world.

Yet Zaire is one of the world's poorest nations. In a year, the average Zairean earns less than an American would earn in a week on a job that pays the minimum wage. Per capita income in Zaire is about half that of Kenya. This chapter will show why Zaire is both rich and poor.

Physical Geography

Zaire is the largest nation in area in Tropical and Southern Africa. It is roughly the size of the United States east of the Mississippi River.

From a satellite far above Central Africa, you could pick out almost the entire outline of Zaire, because the nation's boundaries run along large physical features — such as the Zaire River. In places the river itself is the boundary. Elsewhere the boundary follows the edges of the Zaire Basin.

As you saw in Chapter 28, the Zaire Basin is shaped rather like a saucer — a broad, flat area with a rim around its edges. This rim is highest in the east, where there are highlands and mountains at the edge of the Great Rift Valley. One peak, Mt. Margherita (mar-guh-REE-tuh) rises above 16,000 feet *(5,000 meters)*. In the south, the rim forms high plateaus and divides the basins of the Zaire and Zambezi rivers. Most of Zaire's minerals are located around the rim.

A "copper belt" roughly the size of Pennsylvania stretches across the border be-

The Zaire River serves as a highway for the Zaire nation. On barges like the one at left, you find people traveling to nearby villages and goods being exported to the world.

Zaire has the largest area in the region, and the third largest on the African continent (after Sudan and Algeria). The nation extends over much of the basin of the Zaire River. This basin is mainly on a plateau 1,000 to 2,000 feet (*300 to 600 meters*) high. In the narrow coastal strip, the river falls rapidly to sea level.

tween Zaire and Zambia. This area contains a large portion of the world's copper and cobalt. Zaire's part of the copper belt lies in the province of Shaba (shah-BAH). Zinc, lead, silver, and uranium (a key raw material for nuclear reactions) are also found in this area.

Coal, tin, and tungsten are found north of the copper belt; gold, far to the northeast. Elsewhere in Zaire are diamonds and oil. The nation is indeed rich in mineral resources.

Climate

The Equator plays a key role in Zaire's climate. One third of the nation lies north of the Equator; the rest, south. As you read in Chapter 28, Equatorial areas of Central Africa receive regular and heavy rains. These rains, together with continuous warmth, create the conditions for a tropical rain forest. A broad band of rain forest stretches across north central Zaire.

North and south of this rain forest are areas that receive less rain and have dry periods each year. These areas are largely savanna lands, covered with grasses and scattered trees. In the highlands of eastern Zaire, the climate resembles that of the Kenya highlands (see Chapter 29). Temperatures are mild, and rainfall is plentiful.

The Zaire River packs tremendous power because it catches runoff waters from a vast area on both sides of the Equator. You read earlier that a rainfall belt moves back and forth across the Equator each year (see page 379). When the southern savanna is having a dry season, the northern savanna is having a wet one. Later the seasons are reversed.

It is always the wet season for the Zaire River. The river's flow varies little from one month to the next, which is perfect for turning out hydroelectric power. A number of large rivers flow into the Zaire. By the time the Zaire reaches the Atlantic Ocean, it carries more water than any other river in the world except the Amazon.

A Variety of Vegetation

Differences in climate bring differences in vegetation. In Zaire's tropical areas, soils tend to be poor, because nutrients are washed away by the rains. Better soils are found near rivers, where floods have deposited alluvium (fine river mud). The soil in the eastern highlands is even better, because it comes from volcanic lava.

Zaire shares the rain forest zone with the Congo, Gabon, Nigeria, and a few other countries of Central and West Africa. In this dense, steamy area, many types of trees grow side by side, often so thickly that no sunlight can reach the ground. Balsa, rubber, ebony, and mahogany trees send their leafy branches upward, groping for light.

406

In many places, the rain forest has been cut down. Large plantations of trees such as rubber, cacao, banana, and oil palm may grow in place of the original vegetation. In smaller clearings, people grow crops like cassava and yams.

In northern Zaire and in the southwest, the land is low, with dry grasslands that may be planted with crops like cotton. In Shaba, in the southeast, land is higher, and crops include coffee, citrus trees, and beans. In the eastern highlands, as in Kenya, coffee is the leading crop.

Most Zaireans live close to a river. Nearly all of the nation's cities and many of its villages are found alongside the Zaire River or its tributaries. This fisherman mends his net by a river in the savanna.

SECTION REVIEW

1. What two types of natural vegetation are most common in Zaire? Where are they located?

2. Why does the Zaire River carry so much water?

3. What is the copper belt? Name two other resources besides copper that can be found in this belt.

4. Describe one area of Zaire that is good for farming. What feature makes this area good for farming?

Human Geography

The rapid-fire beat of drums sent a shudder of fear through 19th-century travelers along the Zaire River. The drums, imitating speech, might be sounding a call to war. Yet often they were simply announcing the travelers' arrival.

"Talking" drums are still heard along the river. Today they are likely to be sending news of a riverboat's progress. When the boat makes its next stop, crowds of people will be waiting with food and other goods for sale.

The beat of a drum may also furnish the rhythm for a musical tune. Zairean jazz is famous throughout Africa. The drumbeat may also form the backdrop for a service of Christian worship. Even in schoolrooms, students sometimes pass messages by drumming fingers on desks.

People of Zaire

Zaire has 30 million people — about as many as New York and Pennsylvania combined. Because Zaire is much bigger in area than those two states, it is not a crowded nation. Most people live around the edges of the rain forest — in the river valleys, the eastern highlands, or the mining belt of Shaba. The major cities are also found in these areas.

Ninety-nine percent of Zaire's inhabitants are blacks. The population also includes several thousand whites. Zaire was once a colony of Belgium, and most of the whites are Belgians.

A majority of Zaireans practice traditional African religions. About one percent are Moslems. The remaining 40 percent are Christians, mainly Roman Catholics.

It is not unusual for Zaireans to speak five or six languages. At home they might speak a local language like Kiwoyo (kee-WOH-yoh). At school they use French, the official language. On the road, it is good to be able to speak Lingala (leen-GAH-luh). This is the language spoken by most soldiers — and travelers in Zaire must often stop at army roadblocks. Most Zaireans also speak a *lingua franca* that might be Kikongo (kee-KOHN-goh) in the west or Swahili in the east.

The first Europeans reached what is now Zaire in 1482. With the help of sign language, Portuguese sailors learned that the people called themselves the Bakongo (bah-KOHN-goh). The Bakongo ruled a large empire along the coast and inland. Europeans called this empire the Kingdom of Kongo.

Kongo soon became *Congo,* and this name was given both to the land and to its big river. Eventually France and Belgium got control of the area and carved two colonies out of it. One colony was known as the French Congo; the other, the Belgian Congo.

In the 1960's, the two colonies gained their independence. If you look at the map on page 378, you will see a nation called the Congo, which is the old French Congo. The nation of Zaire is what used to be called the Belgian Congo. Today the Bakongo people live mainly in the Congo, although some of them live in western Zaire.

A Stormy Past

For two decades, starting in the 1880's, the Belgian Congo was the personal property of King Leopold II of Belgium. In order to make the colony pay off, he allowed his agents to use great cruelty on its people. When news of this cruelty reached Europe, Leopold had to turn his colony over to the Belgian government.

Belgian rule lasted until 1960. The Belgians improved public health and introduced a system of elementary education. But they gave black Africans no say in the colony's government. When Belgium suddenly pulled out in 1960, the Belgian Congo became an independent nation with no trained leaders and only two dozen college graduates.

Several years of chaos followed. Finally in 1965, an army leader named Joseph Mobutu (moh-BOO-too) seized power and restored order.

Mobutu sought to build pride in the African past. He changed many place-names to African names. Thus the nation became Zaire, from the name that the Bakongo gave to the river. Among other changes, the name of the capital city changed from Leopoldville to Kinshasa.

Like several modern African leaders, Mobutu created a one-party state. He took a personal share in ownership of key Zairean businesses and became very wealthy.

The great majority of Zaireans are not wealthy. A visit to a rural family will give you an idea of what life is like for many people in Zaire today.

Country Life

Mbombo (uhm-BOHM-boh) sleeps on a mat on the floor of her family's hut. At daybreak small children beat a drum in the village center. The drum serves as the village "alarm clock."

Three out of four people in Zaire live in rural areas. Mbombo's village lies in a river valley. The scrubby bush around the village is broken here and there by farm plots.

To attend school, Mbombo must walk

seven miles (*10 kilometers*) to another village, carrying her books on her head. She leaves home without eating breakfast.

Many rural families eat just once a day. Mbombo's family sits on the floor and passes bowls of food around. Most meals contain no meat. Cassava (also known as manioc) is usually part of the meal. There may also be beans for protein.

Malnutrition (lack of a proper diet) is widespread in Zaire. In some areas, one child in two dies before the age of five. The average person lives less than 50 years.

Mbombo and her sisters help their mother tend the family crops. But they have time to play too. Sometimes the younger children pretend to be wild animals — while watching out for real lions and hyenas.

There is no paved road, no telephone, no radio nor television in Mbombo's village. At school, however, she has heard about life in Kinshasa and other big cities. She knows that the city streets are busy with cars, trucks, and buses. She knows that many city people own bicycles and play or watch soc-

cer. Some city people even watch television.

Many Zairean farm families have moved to the cities in search of a better life. But for unskilled workers, city life can be hard too. Wages are low, and food is expensive. Still, by hard bargaining at the food market, city people may eat better. Often they can add fish to their cassava or rice for more protein.

SECTION REVIEW

1. Why do many Zaireans speak more than one language?

2. Does Mbombo's family have a good diet? Explain.

3. How many people live in Zaire? About how many Zaireans live in rural areas?

4. What nation ruled Zaire before its independence? Was Zaire prepared for independence? Explain.

Economic Geography

People in what is now the Shaba area of Zaire were mining copper before Europeans came. The Shaba people placed copper-bearing rock in giant anthills. Then they built fires, using the anthills as ovens in which to smelt the copper. Much of the copper was shaped into bracelets and rings.

David Livingstone, a British explorer of the 19th century, observed that the Shaba people preferred farming to mining. "Those who cultivate the soil," he wrote, "get more wealth than those who mine the copper."

Today much has changed. Copper is needed around the world for things such as electrical wires and cooking pots. Giant open-pit mines scar the landscape of Shaba. Zaire has become one of the world's leading producers of copper, which brings in four fifths of the nation's foreign income.

For most workers, though, the mines are not a source of wealth. Unions are barred by law. Wages are often less than 50 dollars a month.

Could a person "get more wealth" today by farming the land? Probably — with the right tools. But most Zairean farmers have only the simplest tools and raise only subsistence crops. As a result, farm incomes today average far less than the wages of Zairean mine workers.

Farming

Cassava is the main subsistence crop in Zaire, because it will grow in the poorest of soils and can hold its own against weeds.

The modern sections of big cities tend to look much the same in all parts of the world. This boulevard with modern apartment buildings is in Kinshasa, the capital of Zaire. The Zaire River flows past in the distance.

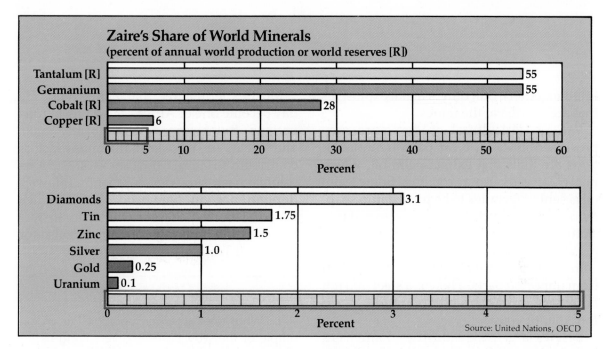

Zaire's Share of World Minerals
(percent of annual world production or world reserves [R])

Mineral	Percent
Tantalum [R]	55
Germanium	55
Cobalt [R]	28
Copper [R]	6

Percent (scale 0–60)

Mineral	Percent
Diamonds	3.1
Tin	1.75
Zinc	1.5
Silver	1.0
Gold	0.25
Uranium	0.1

Percent (scale 0–5)

Source: United Nations, OECD

To show a wide range of percentages, this graph is divided into two parts with different scales. Reserves are amounts remaining in the ground. Tantalum and germanium are used in electronics. Most of Zaire's diamonds are for industrial use (cutting and grinding).

Corn requires better soils to do well. Farmers tend to grow corn mainly in newly cleared fields where the soil is not yet worn out.

With chemical fertilizer, a farmer can grow seven times as much corn as before. But fertilizer is expensive, and most Zairean farmers cannot afford it.

Zaire could produce enough food to supply all of its people and then export some. But for now, Zaire must import food. Vegetables are flown in regularly from South Africa to supply the mining towns of Shaba. So far the government of Zaire has done little to improve agriculture.

Zaire's leading cash crops are coffee, cotton, palm oil, and rubber. Coffee and cotton are grown mainly on small farms. Palm oil and rubber are grown mainly on large plantations.

Mining
Zaire has used its mineral wealth to start up industry. It has several hundred mines, ranging from vast open pits to underground mines. Some mines — especially surface mines — are highly mechanized. Mining provides jobs for some 70,000 people.

Shaba is the leading mining area, since most of the minerals are found there. Because of Shaba's inland location, however, there is the problem of transporting minerals to ports so they can be sold abroad.

Zaire's highway system is poor. The nation has fewer trucks in use than Kenya, which is only one fourth as big in area. However, Zaire does have a vast system of inland waterways. A total of 7,200 miles (11,500 kilometers) of river are navigable. Thus much of the nation's trade goes by riverboat.

To get around the cataracts downriver from Kinshasa, some 300 miles (480 kilometers) of railroad track have been built. Cargo is brought downriver to Kinshasa, transferred to railroad cars for the trip to the coast, and then put on ships.

Industry
Zaire is just beginning to diversify its industry — that is, create a variety of different industries. In Shaba the only industry used to be mining. The ore was shipped straight out from the mines. Today some plants in Shaba process minerals for export. Others produce prefabricated buildings (walls, roofs, and other parts that need only to be

fitted together). Still others make explosives that are needed in the mines.

Various agricultural products are now processed in Zaire. Near palm tree plantations are mills that extract palm oil, which many Zaireans use in cooking. Some palm oil is sent to factories to be turned into soap, margarine, or candles. Cotton is turned into cloth at textile mills in various cities. Clothing and shoes are produced in factories.

In Kinshasa automobiles are assembled from imported parts. There is a steel mill not far from Kinshasa and an oil refinery near the Atlantic coast.

Besides large-scale industries, Zaire has a range of "cottage industries" (small-scale manufacturing done by individuals). For example, a dressmaker may set up a sewing machine near a marketplace and make dresses to order. Or a carver may turn a piece of wood into a graceful statue.

Zaire and the World

Zaire does a brisk trade with the United States and nations in Western Europe. Copper is its leading export; but diamonds, cobalt, and palm oil are also important. One third of the world's industrial diamonds come from Zaire. Exports of cobalt are so valuable that it is sometimes shipped by air to avoid delays in ground and sea transport. As for palm oil, look at the labels of processed foods in the grocery store. You will find that palm oil is a common ingredient.

Although Zaire has tried to diversify, its economy still depends heavily on copper. This can lead to serious problems. Not long ago, the price of copper dropped by 60 percent in one year. Zaire had to borrow heavily from other countries — and then had trouble meeting its debt payments.

Another problem has been sharply rising prices for oil and manufactured goods. Zaire must import most of its oil. It also buys modern equipment for industry from nations like the United States.

Most experts agree that Zaire could one day become one of Africa's most prosperous nations. But for now, it can barely make ends meet. You have seen that many factors contribute to Zaire's poverty — such as its colonial past, slow development, and over-dependence on one product. What lies ahead? The next chapter will try to answer this question, not only for Zaire but also for the rest of Tropical and Southern Africa.

SECTION REVIEW

1. What does *diversify* mean? Give an example of the way Zaire is diversifying its industry.

2. Why is Zaire considered a valuable trading partner by the United States and many other countries?

3. Zaire has plenty of farmland but at present has to import food. Give one reason why farm production is not well developed.

4. In spite of a wealth of natural resources, Zaire remains a poor country. Give two reasons why.

YOUR LOCAL GEOGRAPHY

1. Geography Skills 39 on the next page tells you that the length of a longitude degree varies with latitude. In Zaire, around the Equator, a longitude degree is about 70 miles (*110 kilometers*). How does this compare with the length of a longitude degree at the latitude of your community? Find a large-scale map of your locality that has a grid showing degrees of latitude and longitude. (Most road maps have such a grid.) Using a standard ruler and the map's scale, calculate the length of a longitude degree at your latitude.

2. Imagine that Mbombo is coming to live with you and go to your school for six months. Prepare a list of five places to see and five things to do that Mbombo would enjoy. What would be the most unusual thing she would see and do? Why would they be unusual for her?

Estimating Distance from Latitude or Longitude

A boat trip down the Zaire River in central Africa would be an adventure. How far would such a journey take you? There are no signs marking distances between places. But most of your trip would take you near the Equator.

At the Equator, a degree of latitude and a degree of longitude are both close to 70 miles (*110 kilometers*). So in this part of the world, if you know the latitude or longitude of places, you can estimate the approximate distances between them. (Of course, the actual distances you would have to travel on Earth's surface might be much longer.)

Suppose you begin traveling along the Zaire River near Kibombo and continue north to Kindu. Find Kibombo on the map at about 4°S, 26°E. Kindu is about one degree north at 3°S, 26°E. Since one degree of latitude is about 70 miles (*110 kilometers*), how far is Kindu from Kibombo?

You know that lines of latitude are parallel around the globe (see Geography Skills 2 on page 21). Therefore, a degree of latitude equals approximately 70 miles (*110 kilometers*) anywhere on Earth. You can use latitude information to figure north-south distances between places anywhere. First count the number

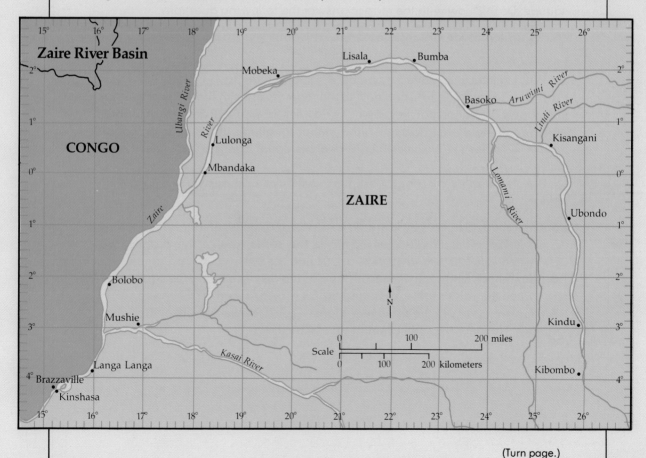

(Turn page.)

of degrees of latitude between points. Then multiply this number by 70 to find the approximate distance in miles.

Use this method and the map on page 413 to answer the following questions.

1. From Kindu you travel north to the river town of Ubondo. There you are about: (a) how many degrees north of Kindu? (b) how many miles north of Kindu?

2. As you continue traveling north the next day, you check the sun at noon and find you are directly on the Equator. Now Kindu is about: (a) how many degrees to the south of you? (b) how many miles?

3. Your next stop is at Kisangani, which is 30 minutes (half a degree) north of the Equator. Kisangani is about how many miles north of the Equator?

You remember that lines of longitude are not parallel, but meet at the North and South poles. At the poles, a degree of longitude is zero miles. Only close to the Equator, where lines of longitude are farthest apart, does a degree of longitude equal about 70 miles (*110 kilometers*). And only close to the Equator can you use the same method you used above to figure east-west distances by longitude. Do this now with the map to answer the following questions.

4. What longitude distance (in degrees) would you travel between Bumba and Lisala? About how many miles is this journey?

What distance is covered by a degree of longitude as you get farther away from the Equator? To find out, turn to the map of Africa in the Atlas of this text on page 578. Notice that the map has grid lines for every 20 degrees of latitude and longitude. If one degree of latitude, or longitude at the Equator, equals about 70 miles (*110 kilometers*), 20 degrees equal about 1,400 miles (*2,220 kilometers*).

Use the map's scale line to estimate distances and answer the following questions.

5. First check distances by latitude. Is the distance between 0° and 20°N close to 1,400 miles (*2,220 kilometers*)? Is the same true of the distance between: (a) 20°N and 40°N? (b) 0° and 20°S?

6. Next check longitude distances. What is the distance along the Equator between meridians for: (a) 0° and 20°E? (b) 20°E and 40°E?

7. How far apart are the same meridians: (a) 20 degrees south of the Equator? (b) near the southernmost tip of Africa?

CHAPTER REVIEW

A. Words to Remember
In your own words, define each of the following terms.

alluvium
cataracts

lingua franca
malnutrition

Zaire Basin

B. Check Your Reading
1. What area of Zaire is most similar to part of Kenya? What two main kinds of natural vegetation does Zaire have?

2. In what parts of Zaire do most of its people live? Why?

3. What is a cottage industry? How does it fit in with Zaire's plans for future economic development?

4. How are most mineral resources transported to the coast? Why are roads used very little? Why are most minerals shifted from one means of transportation to another at Kinshasa?

5. Can a wide variety of food crops be grown in Zaire? Explain your answer.

C. Think It Over
1. The city of Kinshasa used to be called Leopoldville. What does the change tell you about Zaire's history?

2. How do you think the life of most Zaireans today differs from life under Belgian rule?

3. Is Zaire's physical environment more of an advantage or a disadvantage for economic development? Explain your answer.

D. Things To Do
Imagine that you have taken a riverboat trip down the Zaire River from Shaba Province to Kinshasa. Describe some of the interesting sights along the river in a letter to a friend. Tell your friend about the different kinds of terrain, vegetation, and wildlife. Describe some of the people you met and their ways of life.

Chapter 31

Breaking Out of Poverty
The Region in Perspective

Poverty is a fact of life all over Tropical and Southern Africa. Poverty may mean going without food, or depending on food that is not nutritious. Or it may mean crowding into a one-room apartment with five or six other people. In this unit, you have had a glimpse of some of Africa's poverty-stricken people — people like Mbombo, the Zairean farm girl.

Try to put yourself in Mbombo's place. Her daily meal is made up mostly of starchy cassava. An American equivalent might be oatmeal flavored with vegetable oil or with a handful of lima beans, or occasionally a piece of fish. Imagine living on oatmeal, oatmeal, and more oatmeal — day after day. Such a diet fills the belly, but it has few of the nutrients a body needs.

How much work could *you* do on such a diet? Would you be up to farm work like hoeing weeds? Would you feel like concentrating on your school lessons? It requires a strong will for someone like Mbombo to apply herself to tasks such as these. But she does so — day after day.

Mbombo and others like her are caught in a cycle of poverty. A **cycle** is a series of events that is repeated over and over. For example, many of the effects of poverty lead to more poverty. Poor diets leave people with little energy. This means that work is

dragged out or left undone. This means that crop output is low. This means that a farm family cannot grow much extra to sell. This means that the family can get little income. This means that the family cannot buy better tools or fertilizer. This means that crop yields will remain low year after year.

Nations — like individuals — can be caught in a cycle of poverty. Many nations in Tropical and Southern Africa seem to be trapped in this way. Somehow the cycle must be broken. But how?

Finding a Way Out

For an individual African, the effort to break out of the cycle of poverty often means moving to a city. There are jobs in the city. A job means money and a chance for a better life. Cities all over Africa are swelling with country people hoping for jobs.

Many people do succeed in bettering their lives in this way. But there are never enough jobs. How could there be, when people pour into the cities in such massive numbers? Kinshasa, Zaire, has become as big as Los Angeles. The metropolitan area of Lagos, Nigeria, is even bigger, having mushroomed from a population of 350,000 in the 1950's to more than four million today.

Crowded cities are one more problem for struggling nations. In Lagos seven families

Many businesses in the region are run by one person, like this tailor in northern Nigeria. African nations want to encourage larger businesses that will open up more jobs.

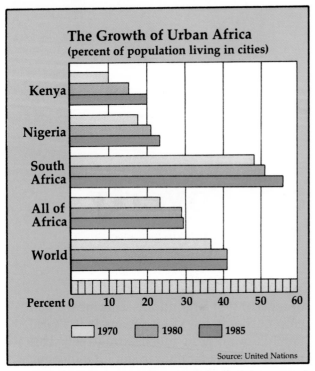

The Growth of Urban Africa
(percent of population living in cities)

Kenya
Nigeria
South Africa
All of Africa
World

Percent 0 10 20 30 40 50 60

☐ 1970 ☐ 1980 ☐ 1985

Source: United Nations

Because total populations are rising, the *number* of African city dwellers is increasing more rapidly than the *percentage.* For example, Kenya's urban population doubled between 1970 and 1985.

out of 10 have only a single room to live in. Such crowded conditions may lead to further problems, such as disease or violence.

What works for an individual, then, may not work for a nation. A national effort to break the cycle of poverty must do two things. First, it must help rural people to grow more food. Second, it must open up jobs for city people and for rural people who are leaving farms and plantations. The next section will show how those two tasks are being approached.

SECTION REVIEW

1. What is meant by a cycle of poverty?

2. How does malnutrition play a part in the poverty cycle for many African families?

3. How do many individual Africans try to break out of the poverty cycle? What major problems do they face?

4. Some African nations have cities with very large populations. Do you think the governments of these nations are pleased to have such cities? Explain your answer.

The First Step: More Food

Almost all nations in Tropical and Southern Africa are trying to boost agricultural production. The more food a nation grows, the better its people can eat.

Many African nations have enlisted the aid of outside advisers from the U.S. and other modern nations. These advisers can share useful knowledge about modern techniques. They can explain the advantages of fertilizer and of new plant varieties. Farmers like Mbombo's mother in Zaire and the Kiganas in Kenya are often eager to try anything that can help them get ahead.

But some farmers don't always follow the advice. They may not be able to afford needed items such as fertilizer. Those who cannot read may be discouraged by printed instructions that come with the new items.

Sometimes too, modern techniques need to be specially adapted to suit African conditions. Some machines that are suitable for large modern farms are out of place on small African-style farms. Increasingly African nations want small, cheap, sturdy machines that can be built locally and repaired by farmers themselves.

Slow progress has been made in many nations. Irrigation projects have provided more water for dry areas. More land has been brought into production. Government help has made it possible for farmers to pool their efforts. They can buy equipment jointly at a low cost to each farmer.

An improvement that seems small may make a big difference to a poor farm family. For instance, a new type of crop seed may boost the family's yield by 20 percent, which may give the family a surplus to sell. With the money this brings, the family can buy things it needs — including better tools to boost crop yields still more. Eventually such improvements may spread through an entire nation's economy, bringing a better life for all.

Opening Up Jobs

At first glance, opening up new jobs seems an even tougher task than growing more food. It takes money to start new businesses. Where will the money come from?

Some money can come from savings within the country. Private individuals may put together a little cash and invest in a small business. In Zaire, as you have seen, some people become tailors by buying a sewing machine and taking space at an open-air market. If the business does well, the tailor may rent a shop and hire helpers. Eventually the tailor may build up enough profits to start a clothing factory that will provide jobs for dozens or even hundreds of people.

On a larger scale, some African businesses may raise money by selling **stock** (a share in ownership). People who buy stock give money now in exchange for a share in any profits later. Such people are known as **investors**, and the money they put into stock is known as an **investment.** Kenya has its own stock exchange (a place where business shares are traded) in Nairobi.

Under colonial rule, however, few Africans had the chance to become experienced in big business. Today there are two main ways in which large businesses are set up in African nations.

Large symbols show major deposits. Note that the richest deposits are found in South Africa and, to the north of these, in Zaire and Zambia. Fairly rich deposits are scattered along the west coast, from Guinea to Gabon. Other countries in the region, however, have found few minerals or almost none at all.

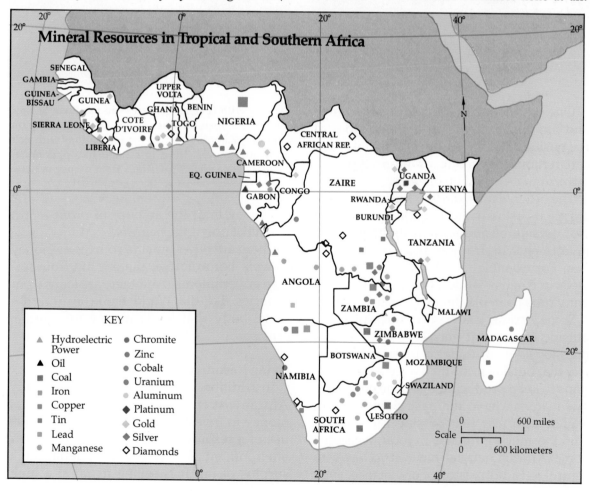

Mineral Resources in Tropical and Southern Africa

KEY

- ▲ Hydroelectric Power
- ▲ Oil
- ■ Coal
- ■ Iron
- ■ Copper
- ■ Tin
- ■ Lead
- ● Manganese
- ● Chromite
- ● Zinc
- ● Cobalt
- ● Uranium
- ● Aluminum
- ◆ Platinum
- ◆ Gold
- ◆ Silver
- ◇ Diamonds

One way is for the government to own and run the businesses. The money to start the businesses comes from taxes. Most African nations have one or more types of government-owned businesses.

The other way of starting businesses is to attract investors from outside nations. Today outside business interests have many investments in Tropical and Southern Africa. For example, a U.S. oil company has wells in Angola, while a British-Dutch company runs palm plantations in Zaire.

Some African leaders are concerned that outside investment may keep their economies lopsided, as under colonial rule. They fear that outside firms may want African nations to keep producing the same crops and raw materials as in the past. Some African leaders also fear that outside firms might gain too much influence in local decisions. Some African nations have strict laws to control outside investors. For instance, such

Nations with special resources can use them to earn money and open up jobs. As you saw in the map on page 419, Zambia is one of the few nations with rich mineral resources. Here, in a Zambian copper mine, a supervisor answers questions from trainee miners.

laws may limit the amount of profit a firm may take out of the country.

There are other outside sources of money as well. For example, there is aid from foreign governments. And there are loans from agencies like the World Bank (an international body that lends money for specific projects).

Using Resources

In addition to all of these sources, many African nations have another way to get money to invest in growth. They can make use of a resource or special feature of their environment. For example, Nigeria has a surplus of oil. It is selling its oil at a high price and investing the proceeds in industrial development.

No other nations in the region are as rich in oil as Nigeria. But many have *something* to build on. As you have seen, Kenya has good potential for farming and tourism. Zaire has many minerals that are vital to modern industry and are in short supply worldwide.

So far Kenya has been able to make its resources pay off better than Zaire. Zaire has more potential wealth, but has also had more political troubles. Similar problems have affected other African countries as they moved from colony to independent nation.

Raising money for investment is not the only way that African nations are trying to develop their economies. Just as some African villagers are developing their farms by pooling their efforts, many African nations are going in for economic cooperation. In 1975, for example, 16 West African nations formed a group called the Economic Community of West African States. Its aim is to increase trade among member nations and do away with all tariffs (taxes on imports). By increasing trade, these nations will be able to increase production. This in turn will open up more jobs and give a boost to the economy as a whole.

At present Tropical and Southern Africa is still trying to move into the modern world. There are modern outposts within the region, but the old ways are still strong. Most people depend largely on their immediate surroundings for what they eat and wear. They use local resources — including their own muscles — for much of their energy. They use the simplest of tools. They do not yet have the skills needed for jobs in the modern world.

Changes are coming, however slowly. You have read how Africans and their leaders are striving to move their nations into the modern world. They see this as a way to break the cycle of poverty in which many are now trapped.

SECTION REVIEW

1. What are the two main tasks that face African nations trying to break out of poverty?

2. Describe two improvements being made in agriculture in the region.

3. Starting new businesses requires money. Describe the two main methods used in African countries to raise money for new businesses.

4. "Today outside business interests have many investments in Tropical and Southern Africa." What does this mean?

YOUR LOCAL GEOGRAPHY

1. Many nations of Tropical and Southern Africa rely on a resource or special feature to get ahead in the modern world. In Kenya, for instance, safari tourism may be helping to break the cycle of poverty. Consider your own community. What is its outstanding resource? Draw an illustrated diagram showing the benefits that this resource has provided for you and your neighbors. (A diagram depicting how safari tourism may help Kenya might start with *wild game* and show it leading to *tourism,* which in turn leads to jobs and money, which lead to more *food, clothing, schools, industry,* etc.) Draw pictures to illustrate the different parts of your diagram.

2. You have read that in many parts of Africa, people are pouring into the cities in massive numbers. Is this a current trend (general pattern of change) in your part of the world as well? Using the *U.S. Statistical Abstract* or some other source, find out how the population of the major city nearest to your community has changed in the past four decades. Use these figures to make a line graph showing population growth or decline over a period of time. (Geography Skills 11 on page 78 gives information on line graphs.) When the graph is completed, answer this question: Based on the trend shown, what is the city's population likely to be in the year 2000?

Evaluating a Graph

The bar graph below (Graph A) compares the numbers of tourists visiting three African countries during a recent year. The table gives facts on the numbers of tourists who visited the same countries during the same year. Is the graph misleading in any way? You can use facts in the table to evaluate the bar graph. First check the right end of each bar in the graph with the scale on the horizontal axis. Do the bars agree with the figures given in the table?

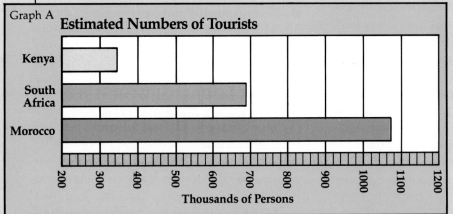

Graph A

Estimated Numbers of Tourists

Thousands of Persons

Estimated Numbers of Tourists	
Country	Tourists
Kenya	347,300
South Africa	682,198
Morocco	1,078,678

Now use a ruler to measure the length of the bar for Morocco. Measure the bar for Kenya. Answer the following questions on a separate sheet.

1. According to the table, how many tourists visited Morocco? How many tourists visited Kenya? Was the number of tourists who visited Morocco (two times; three times; five times) as many as visited Kenya?

2. How many times longer than the bar for Kenya is the bar for Morocco? Does this agree with the proportion between the numbers of tourists?

3. Why do the bars give a false impression of the facts in the table?

For a graph to give an accurate picture, the number scale should begin with zero. On Graph A, the scale begins with what number? Since the scale is not complete, the bars are shorter than they should be. This gives you a misleading picture when you compare them.

When information must be left off a graph to save space, a careful graphmaker draws an **interrupted graph** like Graph B below. Compare Graph B with Graph A and answer the following questions.

4. Does the number scale on Graph B begin with zero? What numbers are omitted? Why is it easy to notice these omissions?

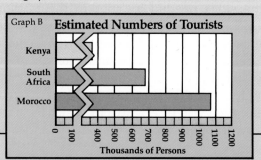

Graph B

Estimated Numbers of Tourists

Thousands of Persons

CHAPTER REVIEW

A. Words To Remember
Three of the following terms are defined by the numbered phrases below. On another sheet of paper, write each term next to the number of its definition. Then write the two terms that are *not* defined and give a definition for each.

colonial rule   investors   World Bank
cycle   irrigation

1. A system under which most African countries were governed by foreign countries.

2. People who pay money to start a business in return for a share in any profits the business may make.

3. A series of events that is repeated over and over.

B. Check Your Reading
1. What reason is given in the text for the rapid growth of such cities as Lagos and Kinshasa?

2. Describe one way in which modern agricultural techniques may not always work in African nations.

3. Zaire has greater potential wealth than Kenya, but Kenya has developed faster. What problem has affected Zaire but not Kenya?

4. The text notes that many nations can use a special feature of the environment to raise money for new businesses. What does this mean? Give an example.

5. What problems do some African leaders see in outside investments?

C. Think It Over
1. Suppose Mbombo were thinking of moving to a city. Give two reasons she might have for going, and two reasons for staying. Which do you think are stronger? What would you do?

2. Most businesses in African nations are run either by the government or by outside investors. Why do you think there are few private investors inside African nations?

3. The text states that a small improvement, such as a new type of seed, may make a big difference to a poor farm family. Could the result of such improvements be called a cycle of prosperity? Why, or why not?

D. Things To Do
Choose three cities in Tropical and Southern Africa, one on the west coast, one on the east or south coast, and one inland. Locate them on a globe. Then locate your area of the United States. Which direction would you travel to get from your community to each city most directly? If you were to travel by land and sea rather than by air, which route should you take? Why? What major land and water barriers would you have to pass over on the way?

UNIT REVIEW

A. Check Your Reading

1. On your answer sheet, write the letter of each description. After each letter, write the number of the area that best matches that description.

(a) It is one of Africa's leading agricultural areas.

(b) This area has year-round warmth and almost continuous rainfall.

(c) With almost no mineral resources, this nation has still built up its economy.

(d) Once the center of an ancient civilization, it is now densely populated.

(e) This nation is very poor but has many mineral resources.

(1) Great Rift Valley
(2) Shaba Province
(3) Zaire
(4) Kenya
(5) tropical rain forest zone
(6) savanna
(7) East African highlands
(8) Nigeria

2. Fill in the blanks in the following paragraph by writing the missing term on your answer sheet.

In the 19th century, European nations were interested in Tropical and Southern Africa for its raw materials: crops and __(a)__ . Today raw materials are still the major exports of most nations in the region. Copper and cobalt are among the leading exports of __(b)__ in Central Africa. For Kenya in __(c)__ Africa, the leading export is coffee. After that Kenya earns most from __(d)__ . Many nations in the region hope to break out of the cycle of __(e)__ by developing industry.

B. Think It Over

1. The following statements may be true of any one or more of Zaire (Z), Kenya (K), and South Africa (S). Write the initial letter of the nation(s) to which you think each statement applies.

(a) It is on the Equator.
(b) Much of it is on a plateau.
(c) The majority of its people are blacks.
(d) Large numbers of Europeans settled in the area and still live there.
(e) It provides key minerals needed by the U.S. and other nations.

2. Choose any one of statements (c), (d), or (e) in exercise 1 above and write its letter on your answer sheet. Then write a short paragraph explaining the statement.

Further Reading

Africa: The People and Politics of an Emerging Continent, by Sanford J. Unger. Simon & Schuster, 1985. Focuses on social conditions in Africa—particularly Liberia, Kenya, Nigeria, and South Africa.

Land and People of Kenya, by Edna M. Kaula. Lippincott, 1968. Survey of Kenya geography and culture. Revised Ed.

8

South and Southeast Asia

Chapter 32

Land of the Monsoons

It is late May in India, and intensely hot. In much of the country, the daily temperatures rise well above 100°F (*38°C*). For nearly three months, the land has been scorched by the Sun. Day after day, there has not been a cloud in the sky. In thousands of farm villages, the fields are cracking and turning to dust. The streams and wells that supply the villagers with water are almost dry.

Now the people begin to count the days until June, because June is the month when "the rains" come to most of India. Everything depends on them. When the rains come, they last through September. Crops will grow, and there will be enough food for everyone. The streams and wells will fill with water again.

But in some years, the rains don't come, or they come too late to do the farmers any good. Then there will be little to eat — perhaps one meal a day, or none at all.

Early in June, the villagers look anxiously at the skies. When the first dark clouds appear, the people begin to shout with joy: "It will rain. There will be plenty of it!"

Suddenly there is a sound of thunder, and a gust of wind sweeps through the villages. Big drops of rain begin to fall. Some people pray that the clouds will not blow away, as they sometimes do. But this time the clouds stay. Soon the rain is pouring down.

The heat breaks, the rains let up, and the farmers rush into the fields to plant their seeds. Soon the land will turn green again, and life will be renewed. There will be a good crop in the fall.

These rains are brought by seasonal winds called *monsoons*. (These are the same monsoons that play a part in East Africa, as described in Chapter 29.) In the summers, the monsoons are full of moisture and produce heavy rainfall. In the winters, the monsoons are dry and cause drought. How does this happen?

As you know, land heats more rapidly than water and also cools more rapidly than water. When the huge landmass of Asia becomes heated in the summer, the air above it also becomes heated, and rises. At this time, the air over the Indian Ocean is cooler. The cooler, moist air from the Indian Ocean moves toward the land to replace the rising warm air. This movement causes the "wet" monsoon that arrives in June.

In the winter, the continent of Asia becomes colder than the waters of the Indian Ocean. When the warmer air above the ocean rises, the cool, dry air from the continent moves toward the sea. This movement causes the "dry" monsoon. The two maps on page 430 show how the wet and dry monsoon winds move across the region.

Much of the region is close to the sea and receives plenty of rain when the wind blows from that direction. Here, a boat dodges showers in Brunei (BROO-nie), on Borneo (BOR-nee-oh) island.

Overview of the Region

The monsoons are one feature that most — but not all — of the nations of South and Southeast Asia share in common. Most — but not all — of the nations are within the tropics. Most — but not all — of the nations are on the sea, with lengthy coastlines.

The map on page 429 will give you an idea of the diversity of the region. From Afghanistan (af-GAN-is-tan) and Pakistan (PAK-istan) in the northwest, the region sweeps across the large peninsulas of India and Southeast Asia to the thousands of islands that make up Indonesia (in-doh-NEE-zhuh) and the Philippines (FIL-i-peenz).

The distance across the mainland from Afghanistan to Vietnam (vee-eht-NAHM) is more than 3,000 miles (*5,000 kilometers*) — about the same as from San Francisco to Halifax, Nova Scotia. It is another 3,000 miles (*5,000 kilometers*) along the chain of islands that make up Indonesia. From north to south, the Philippine Islands cover 1,200 miles (*2,000 kilometers*) — the distance from San Diego to Vancouver, British Columbia.

About 1.50 billion people live in the region — about one fourth of all the people in the world. As you can see from the Checklist of Nations on pages 586–587, nations that are fairly small in area may have large populations. Bangladesh (bahng-luh-DEHSH) is the same size as Iowa, but has 30 times as many people. India is the second most populous country in the world (after China), with three times the population of the U.S.

The people of the region represent an extraordinary variety of ethnic groups and cultures. Much of that variety can be found within nations. In nations such as India and Malaysia (muh-LAY-zhuh), there are many groups that have different ethnic backgrounds, speak different languages, and follow different religions from their neighbors.

The region's cultural diversity is due in part to the environment. Because so much of the region is exposed to the sea, people have traveled from one part to another for many hundreds of years. People from outside the region have also had easy access to many parts of South and Southeast Asia. The strongest cultural influence within the region has come from India. The strongest outside influences have come from China, the Islamic Middle East, and Europe. Many of these overseas visitors traveled by sailing ships that were driven by the monsoons.

Physical Geography

In most of South and Southeast Asia, the shifting monsoons cause a sharp difference in the amount of rainfall from one season to another. However, northwest India, Pakistan, and Afghanistan are hardly touched by the wet monsoons. The summer winds there come from the southwest, but hardly at all from the Indian Ocean. As you can see from the map on page 430, the winds come instead from the Arabian Peninsula, which is hot and dry.

As a result, this area gets less than 10 inches (*25 centimeters*) of rain a year. The people who live there depend on irrigation from rivers to grow food. Where there is too little water for irrigation, nomads raise herds of sheep and goats.

East of India, the Southeast Asian mainland is also strongly affected by monsoons. This area includes Burma, part of Malaysia, and the Indochina peninsula — Thailand (TIE-land), Laos (lows), Kampuchea (kampoo-CHEH-uh), and Vietnam. Heavy rainfall during the summer monsoons and heat all year long makes vegetation grow rapidly there. Much of this area is covered with tropical forests.

The island nations of Sri Lanka (suh-ree LAHN-kuh), Indonesia, and the Philippines also have monsoon seasons. But the difference between the seasons is small. Because of the surrounding seas, these nations get plenty of rain year round —

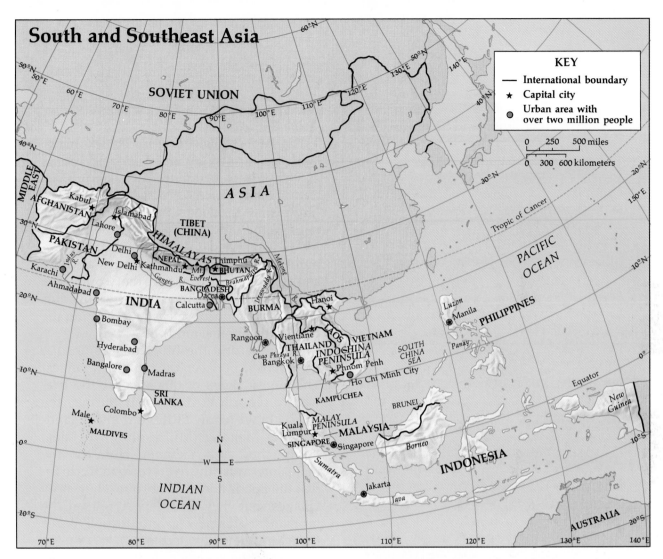

South and Southeast Asia

from 50 to 150 inches (*125 to 375 centimeters*). During the wet monsoon, they simply get more rain than during the dry monsoon.

The largest part of the region lies within the tropics. As you know, in tropical lands, food can be grown all year round if there is enough moisture. In Indonesia and the Philippines, there is enough rainfall to produce two or three crops a year from a single field.

Landforms

South and Southeast Asia are made up mostly of highlands — mountains, plateaus, and hills. An almost continuous wall of mountains forms India's northern boundary, which extends about 1,500 miles (*2,500 kilometers*). The mountains include the Himalayas (him-uh-LAY-uhz), which have the highest peaks in the world. The tallest peak,

Most of the region lies in the tropics and is affected by monsoons — shifting winds that bring rain for part of the year. But the region also has great variety. Different ethnic groups have often used the seas to travel, producing a wide range and mix of cultures.

Mount Everest, soars nearly 5½ miles (29,000 feet or *8,700 meters*) above sea level. Everest lies on the border between Nepal and Tibet. Fifteen other peaks in the mountain wall rise above 25,000 feet (*7,500 meters*).

The southern half of India forms a peninsula that juts into the Indian Ocean. Most of the peninsula is covered by a huge plateau.

The Indochina peninsula has several mountain ranges that extend north to south. These ranges are not so high as the mountains that form India's mountain wall, but they are quite rugged. Some have peaks of more than 10,000 feet (*3,000 meters*). The is-

429

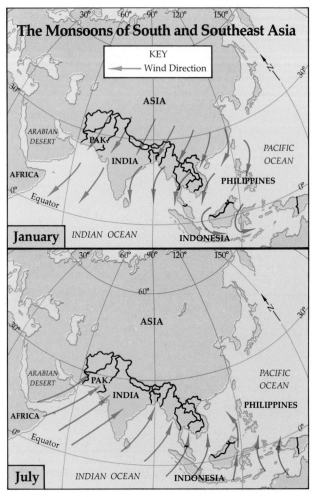

The Monsoons of South and Southeast Asia

KEY
← Wind Direction

ASIA

ARABIAN DESERT

PAK.

INDIA

PACIFIC OCEAN

PHILIPPINES

AFRICA

Equator

January   INDIAN OCEAN   INDONESIA

ASIA

ARABIAN DESERT

PAK.

INDIA

PACIFIC OCEAN

PHILIPPINES

AFRICA

Equator

July   INDIAN OCEAN   INDONESIA

These maps show how the monsoon winds change in summer and winter. Note that in the northwest, the winds come from dry land in both seasons. In the islands of the southeast, the winds come from the sea all year. In the center, especially in India, the monsoon winds bring distinct wet and dry seasons.

land nations of Indonesia and the Philippines are also mountainous and contain many active volcanoes.

Most of the 1.50 billion people in South and Southeast Asia are farmers. Thus most of them are crowded into the lowland areas, where farming is good. In many places, some 1,700 people live on each square mile of land (*about 700 on each square kilometer*). By contrast, the most densely populated state in the U.S., Rhode Island, has fewer than

1,000 people per square mile (*about 400 per square kilometer*).

The largest lowland areas on the mainland are formed by river valleys. India's great northern plain stretches about 1,500 miles (*2,500 kilometers*) across the country. This plain is formed by three large river systems — the Indus (EEN-duhs), the Ganges (GAN-jeez), and the Brahmaputra (brah-muh-POO-truh). The soils of this area are among the most fertile in the world. The Ganges and the Brahmaputra overflow during the wet monsoon season, enriching the plain with silt (fine mud).

In Burma and the Indochina peninsula, the rivers lie between the mountain ranges. The largest rivers of this area are the Irrawaddy (ee-ruh-WAH-dee) in Burma, the Chao Phraya (chow PRIE-yuh) in Thailand, and the Mekong (MEH-kohng) in Laos, Kampuchea, and Vietnam. Here too the monsoons cause the rivers to flood. As in India, the floods may damage or even destroy villages, but they also deposit silt on the farmlands. This new layer of topsoil makes the river valleys very fertile.

Most of the islands of Indonesia are covered with forests. Sumatra (soo-MAH-truh) and Borneo also have large swamps. Thus most islands do not have many people. But one island, Java (JAH-vuh), is crowded.

Almost two thirds of the people of Indonesia live on Java because the mountains there include many active volcanoes. The lava and ashes that pour from these volcanoes have made the soil very black and rich. As a result, thousands of farm villages cover most of the island. Some people on Java actually live on the slopes of volcanoes. But most people farm the wide, fertile plains that lie north of the island's mountains.

Of the 7,000 islands that make up the Philippines, fewer than 500 are larger than one square mile. The great majority of the islands are uninhabited. Most of the people live on two lowland plains, one on the island of Luzon (loo-ZOHN) and the other on the island of Panay (puh-NIE). Volcanoes have enriched the soils of these islands.

Natural Resources

The fertile soils of this vast region are its most valuable resource. But areas with poorer soils provide another important resource, timber — especially in Indonesia and the Philippines. Indonesia's forests supply large amounts of valuable hardwoods, like teak and ebony. In both Indonesia and the Philippines, bamboo and rattan are important forest products. Bamboo is a huge, tough kind of grass that can grow up to three feet (*one meter*) per day. Rattan is a kind of palm tree that climbs like a vine. Both have long canelike stems. Both are used to make houses, furniture, baskets, hats, and many other products.

South and Southeast Asia are also rich in mineral resources. India has huge deposits of high-grade iron ore, perhaps the largest in the world. It also has large deposits of coal, manganese, tungsten, and many other minerals. Both Pakistan and Bangladesh have natural gas.

The region has large deposits of tin and oil. The tin comes mainly from Malaysia, Indonesia, and Thailand. Indonesia is the leading producer of oil in the region, and also has supplies of bauxite (used to make aluminum). The mineral wealth of the Philippines includes gold, silver, and iron ore.

SECTION REVIEW

1. Name three important products of the forests of Indonesia and the Philippines.

2. Pakistan, Afghanistan, and parts of northern India rarely receive the heavy rains brought by the wet monsoons. Why is this so?

3. Why is Java better for farming than other parts of Indonesia?

4. What causes the monsoon winds to blow toward the land in summer?

Human Geography

You are walking along the busy streets of Singapore, an independent island-city at the tip of the Malay peninsula. Singapore is a modern city, with tall buildings of glass and concrete. What surprises you is the great variety of its people. You see Indian men wearing turbans, Malay and Indonesian women wrapped in sarongs (loose skirts), and Chinese women wearing the latest London fashions.

You hear many languages and dialects, including several Chinese dialects, Tamil (an Indian dialect), Malay, Arabic, English, and Hebrew. About three out of four people you see are Chinese, and most of the stores have signs written in Chinese characters. But some storekeepers came originally from southern India and northern Sri Lanka (formerly Ceylon). Their signs are written in Tamil. Nearby you see a number of Malay banks and shops. Their signs may be written either in the Roman alphabet (like our own) or in the Arabic alphabet.

In every country of South and Southeast Asia, you would also see many ethnic and cultural differences among the people. How did this mixture of cultures come about?

As you can see from the map on page 577, most of the region lies close to India and China. The influence of these two giant nations spread and often blended. Most of the region also touches on two oceans, the Indian and the Pacific. People from many lands came across these oceans.

Early Settlers

The first people to migrate within the region in large numbers were the Malays. They spread out from the Asian mainland about 3,000 years ago and settled in the islands that now belong to Malaysia, Indonesia, and the Philippines.

The next major groups to spread through much of the region were the Indians and the

Chinese. Almost 2,000 years ago, Indian sailors learned how to use the seasonal monsoon winds to travel the seas. Starting from ports along India's southeast coast, their ships were swept by monsoon winds to Burma, the Malay peninsula, Thailand, Kampuchea, and Vietnam. Others sailed to the Indonesian islands of Sumatra, Borneo, and Java. Indian traders and settlers began to live in all of these areas. They taught many of the local peoples to irrigate their soils, to build canals, and to develop seaports for trade.

The Indians also brought their religions, Hinduism and Buddhism, to these lands. Hindus believe that all living creatures are reincarnated (reborn after death). If people become more spiritually aware during their lives, they will be rewarded when they are reborn. They may become superior beings, like Hindu priests. But if they are too proud or greedy, they will be punished. They will be reborn as inferior creatures, perhaps as insects. (For more on Hinduism, see the next chapter.)

Buddhists believe that people are reborn many times after death. But, they say, life is full of sorrow and suffering. The only way for people to escape endless rebirth and suffering is to "do right." Then their souls will find everlasting peace and happiness in the universe.

Hinduism became the main religion in India, while Buddhism lost most of its Indian followers. But Buddhism remains the chief religion of the people of Burma and the Indochina peninsula.

The Chinese moved into Southeast Asia in even greater numbers than the Indians, entering the Indochina peninsula by overland routes from southern China. (The name *Indochina* comes from *India* and *China*, the two countries that had the greatest influence in the area.) Chinese traders and settlers also sailed to the coasts of the Malay peninsula and Indonesian islands.

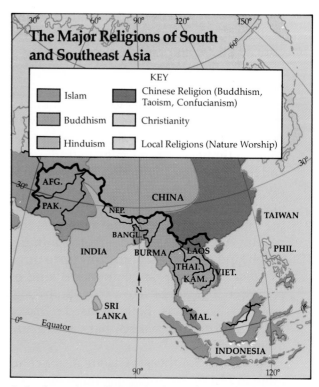

Only the major religions in the region are included. Note the wide influence of Buddhism and Islam. "Chinese religion" is a blend of Buddhism and Chinese beliefs. Taoism is a belief that humans should be in harmony with nature, rather than try to control it.

Chinese immigration continued into modern times. Like the Indians, the Chinese often went into business. Many became shopkeepers, or rose to high positions in banking, industry, and shipping. Today more than 10 million Chinese live in Southeast Asia, many of them in the cities.

The next powerful group that swept into this area were Moslems from the Middle East. Moslems are followers of the religion known as Islam, which was founded by an Arab named Mohammed in the seventh century A.D. Like Jews and Christians, Moslems believe in one God, and they recognize Moses and Jesus as prophets. However, they believe that Mohammed was the chief prophet of God.

Moslems first entered India from Afghanistan, moving through passes in the high mountains. They converted large numbers of Indians to Islam. At the same time, Arab trading ships sailed across the Indian Ocean to the northwest coast of India. The Arabs and other Moslems became important links in trade between India and Europe.

From India, Moslem ships extended their trade to the coasts of Southeast Asia. They were soon followed by Moslem settlers and missionaries. Islam triumphed over Buddhism in Malaysia, and over Hinduism in almost all of Indonesia. Today Islam is still the chief religion in these areas. Indonesia is the largest Moslem nation in the world, with almost 150 million people.

In the 15th century, Moslem traders brought to Europe the spices that grew on the islands of Indonesia. These spices came from various tropical plants and included pepper (the berry of a vine), cloves (the dried flowers of a tree), and cinnamon (the bark of a tree).

To Europeans spices were much more than a flavoring. They could be used to preserve food in an age when refrigeration did not exist. Some spices could also be used as drugs or disinfectants. Even today oil of cloves is sold as a pain-killer for toothache. In the past, spices were so valuable that only the wealthy could afford them.

Colonies and Conflicts

Soon European explorers were lured to the "Spice Islands" (Indonesia). Portuguese explorers led the way, sailing around the tip of Africa to reach India and Southeast Asia. Later the Dutch and the English entered the race for the spices and rare woods of Indonesia. Christopher Columbus hoped to reach

The influence of China is strong in Southeast Asia. This Chinese section is in Kuala Lumpur (KWAH-luh LOOM-poor), the capital city of Malaysia. More than one third of the Malaysian people are Chinese.

the "East Indies" (a former name for Indonesia) by sailing west across the Atlantic. Instead he discovered the Caribbean islands of the New World. He called these islands the "West Indies."

In time European nations took control of both South and Southeast Asia. India, Burma, and the Malay peninsula became colonies of the British empire. The islands of Indonesia became mainly Dutch colonies. Except for Thailand, the Indochina peninsula became a colony of France. Thailand was the only nation of Southeast Asia that remained independent.

The Philippines came under the rule first of Spain; later, of the United States. Spanish missionaries converted many of the Philippine people to the Roman Catholic religion. Today the Philippines are the region's only nation that has a large Christian majority.

European rule influenced the entire region in many ways. Europeans set up large plantations and organized the building of roads and railroads. They introduced new crops and some industries. They also introduced modern medicine and public health measures. Diseases that used to kill millions of people every year were brought under control. As a result, the number of people in this region greatly increased. The Europeans also built many schools. Education spurred many of the people of South and Southeast Asia to press for independence after World War II.

In much of the region, independence did not bring peace. In 1947 British India became independent as two nations, largely Hindu India and Moslem Pakistan. As many Hindu

Western influence is strong in large cities throughout the region. Opposite, only the signs in Thai script place this city in Thailand—it is Bangkok, the capital. Above, in Singapore, high-rises soar over houseboats and tenements. Both a city and a nation, Singapore fills an island smaller in area than El Paso, Texas.

was in the west, and the people in the east felt that they were treated as second-class citizens. In 1971 East Pakistan broke away to form the separate nation of Bangladesh.

Meanwhile open warfare had broken out in Indochina. After the French left in the 1950's, Vietnam was divided. Vietnamese Communists took control of the north. Non-Communists controlled the south. As fighting spread between the two sides, the United States sent troops to support the non-Communists. During the 1960's, warfare spread not only in Vietnam but also in Laos and Kampuchea. As the war dragged on, and many thousands of Americans died, public opinion in the U.S. grew in favor of U.S. withdrawal. In 1975 separate Communist forces took control of Vietnam, Laos, and Kampuchea. Under Communist rule, millions of Kampucheans starved to death or were killed. In 1979 most of Kampuchea came under Vietnamese rule.

Today the Western influence in South and Southeast Asia is seen mostly in the large cities. There you will see Western-style buildings, clothing, and entertainment like movies and TV. But three out of four people in the region still live in small farming villages. They have hardly been touched by Western ways. Most of them follow traditional life-styles that existed long before the Europeans arrived. Many of them have never been more than 25 miles (40 kilometers) from their homes.

SECTION REVIEW

1. What parts of the region were settled by Malays?

2. What two religions spread from India to other parts of the region? Which one no longer has many followers in India?

3. What is the major religion: (a) in the Philippines? (b) in Malaysia? How did these religions first reach the two nations?

4. How was Bangladesh created?

and Moslem refugees tried to move to their new homelands, riots broke out and hundreds of thousands of people were killed. At that time, Pakistan was divided into two parts, one to the west of India and the other to the east. Pakistan's seat of government

435

Economic Geography

For weeks the people of Kampuchea have watched the Mekong River rise higher and higher. The Mekong is fed by streams that start in the snow-covered mountains of China. When the snow melts in the spring, the streams rush down into the Mekong. The monsoon rains add even more water. Finally the river overflows, and the land on either side turns to mud.

Now the rice farmers of Kampuchea have plenty to do. They start by planting seeds in plots near their village homes. Soon bright green shoots of rice begin to appear. After the young plants sprout, however, they must be transplanted into fields that are filled with water. (Rice grows more fully in water than it does in dry fields.)

The rice fields are surrounded by earth dikes from one to two feet (*30 to 60 centimeters*) high. Water from the flooding Mekong flows into ditches dug by the farmers, and enters the fields through gates in the dikes. By July all the fields have filled with water. At this time, the farmers hitch plowlike implements to pairs of water buffalo and stir up the heavy mud at the bottom. Now the fields are ready for the rice. Boys carry the young rice plants from the plots to the fields.

The job of transplanting the rice is usually done by women. Stooping down, women push each shoot of rice into the thick, soft mud of the rice fields. Broad pointed hats protect their heads from the hot sun. The women labor for days to plant all the shoots. It is very hard and back-breaking work.

Gradually the rice grows until it is several inches above the water. When the crop ripens, the fields are drained. The farmers use curved knives to cut the rice stalks. The stalks are then piled into bundles to dry. Afterward the women toss the grains of rice into large bamboo trays. Finally they beat the grains with sticks to separate the kernels

Rice is the staple food crop in the wetter parts of the region, since it grows best in flooded fields. The fields are drained when the plants are ripe. Above, farmers in Burma are harvesting ripe heads of rice.

from their shells. Each family now has much of its food supply for the coming year. The rice is stored in clay jars or baskets of straw. Some families may have extra rice that they can sell for cash.

Farming and Fishing

That farming scene could be found in many parts of South and Southeast Asia. Rice is the staple food throughout much of the region — wherever there is enough moisture. It is grown on farms that are usually less than five acres (*two hectares*) in area. Some are much smaller.

Often fish are kept in the flooded rice fields as an additional "crop." Fish are an important source of protein in the people's diet. Many fish are also caught along the seacoasts and in the rivers.

Southeast Asia is a leading producer of plantation crops for export. Most of the big plantations in the area were set up by Europeans. The chief plantation crops are rubber, tea, coconuts, palm oil, and sugar.

Rice will not grow in the drier parts of the region. There the most important crops are millet, barley, wheat, and corn. South Asia also produces cash crops like cotton and jute (for making burlap sacks).

India leads the world in growing and exporting tea. Most of the tea is grown on large plantations that were set up by the British. Sri Lanka also has large plantations that produce tea, coconuts, and rubber.

Industry and Mining

In India manufacturing has been growing rapidly in recent decades. About five million Indians now work in factories. Their chief products are cotton textiles, jute goods, iron, and steel. India now produces

---

These crops are grown widely in the tropics and subtropics, including South and Southeast Asia. The top row shows staple foods. They are all grains (edible seeds) except cassava, a root (it provides tapioca). Most crops come in different varieties for different needs. A few such varieties are shown for wheat.

Some Major Food Crops Grown in Tropical Regions

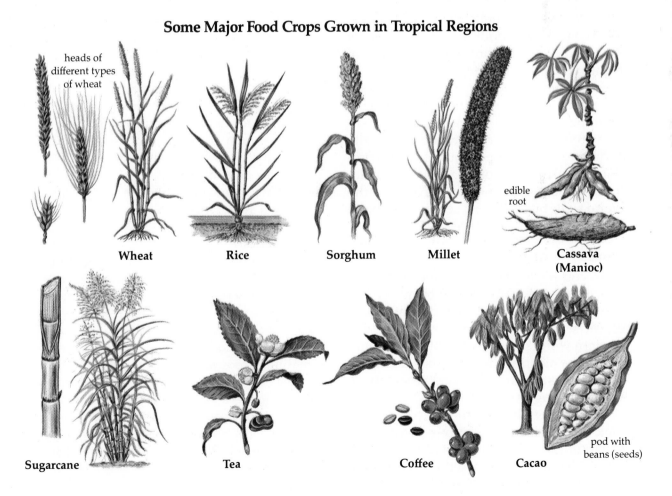

heads of different types of wheat

edible root

Wheat   Rice   Sorghum   Millet   Cassava (Manioc)

Sugarcane   Tea   Coffee   Cacao

pod with beans (seeds)

437

more steel than either Sweden or Australia. More than 20 million other Indians work on handicrafts at home. They produce items such as rugs, silk fabrics, and jewelry.

Pakistan and Bangladesh have also been expanding their industries. Pakistan's greatest success has been its large cotton textile industry. Bangladesh's jute industry accounts for more than half of its exports.

In Southeast Asia, on the other hand, most factories are small and supply only a fraction of the people's needs. Usually these factories process food, or produce cotton cloth, cement, and fertilizers. Southeast Asian nations must import most manufactured products.

As you have seen, both South and Southeast Asia have a wealth of mineral resources. India has more than 3,000 mines, and is one of the world's 10 top producers of coal. India also exports large amounts of manganese ore to the United States.

Some mining industries have also been developed in Malaysia, Indonesia, and the Philippines. Tin and oil are the most important products, but large amounts of bauxite and iron are also mined. Southeast Asia produces about one half of the world's tin. Indonesia supplies about two percent of the world's oil, but is almost certain to expand its production in the future.

What's Ahead

As you have read, South and Southeast Asia has faced many big problems. It is a crowded region. It has suffered many recent conflicts. It is moving only slowly into the modern world. Yet these problems are only part of the picture. The region has also made tremendous progress in recent years, above all in the struggle against hunger.

The chapters that follow give you a closer look at problems and progress in India (Chapter 33) and Indonesia (Chapter 34). Then Chapter 35 surveys the prospects of the region as a whole.

SECTION REVIEW

1. Rice is not grown in some farming areas in the region. Why not?

2. Why does Southeast Asia need to import most of its manufactured goods?

3. When the Mekong River overflows, are farmers who live in its valley pleased or displeased? Why?

4. What part do plantations play in the economy of the region?

YOUR LOCAL GEOGRAPHY

1. In this chapter, you read that many parts of South and Southeast Asia get most of their rainfall between June and September. Check in an almanac to find out how many inches of rain fall each month in your city or the large city nearest to you. Then use this information to make a single-bar area graph similar to the one in Geography Skills 41 (page 440).

The whole bar should represent the total amount of rainfall in a particular year. The scale at the top of the bar can be marked off in inches — from 0 inches to the year's total inches. Divide the bar into 12 segments representing the amount of rainfall for each month. Below the bar, add labels showing which parts of the bar represent summer, winter, spring, and fall. Which season was the wettest? Which was the driest?

2. Although this chapter is titled "Land of the Monsoons," you have read that some parts of the region are hardly affected by the monsoons. Now consider how the geography of your state is usually described. Is it a plains state or a mountain state? Is it a desert state or a maritime (coastal) state? Is it in the corn belt or the wheat belt? Find at least three generalizations (general statements or descriptions) that are made about the physical and economic geography of your state. Then compare these generalizations to your community. Is your community typical of the state, or is it at least partly an exception?

Folk Songs from Southeast Asia

Folk songs are songs created by ordinary people in their daily lives. Some of the songs date back hundreds of years. They may give a glimpse of life in the past — and also in the present.

The song below left comes from the Malay peninsula. The second comes from Vietnam. You will note that each of them was sung for a special purpose. Note also how each tells you something about life in its area.

A persimmon is a kind of fruit. A kite is a bird of prey. The buffaloes mentioned here are water buffaloes. Farmers in the region rely on water buffaloes to pull plows and carts and do other heavy work.

Tin Ore

Peace be with you, O Tin Ore.
At first it was dew that turned into water,
And water that turned into foam,
And foam that turned into rock,
And rock that turned into tin ore.
O Tin Ore, lying in solid rock,
Come out of this solid rock.
If you do not come out,
You will be a rebel in the sight of Allah.
Ho, Tin Ore, you island in the rock,
Float up to the surface of my tank
Or you will be a rebel to Allah!

— From *A World Treasury of Oral Poetry*,
edited by Ruth Finnegan.
Indiana University Press, 1978.

Lullaby

When will it ever be March?
Then frogs will bite snakes and drag them
   to the fields,
Tigers will lie down for pigs to lick their fur,
Ten persimmons will swallow an old man,
A handful of steamed rice will devour a
   10-year-old child,
A chicken and wine jar will gulp down a drunk,
Eels will lie still, swallowing the bamboo traps,
A band of grasshoppers will chase after the
   fish,
Rice seedlings will jump up and eat the cows,
Grasses will crouch and ambush the buffaloes,
Chicks will chase kites,
Sparrows will track down pelicans
And break their feathered necks.

— From *A Thousand Years
of Vietnamese Poetry*
edited by Nguyen Ngoc Bich, Knopf, 1975.

Ask Yourself . . .

1. What kind of persons would sing the "Tin Ore" song? How did they try to extract the ore from the rock?

2. "Lullaby" may seem full of strange happenings. How do you read it so that it becomes normal? Give one example of something that might really happen in spring in Vietnam.

3. From reading "Lullaby," what four foods can you name that are eaten in Vietnam?

Comparing a Special-Purpose Map and a Graph

When you eat a dessert topped with shredded coconut, you are probably biting into a piece of South or Southeast Asia. Most of the world's **copra** (dried coconut) comes from the region. The map below shows the region's main copra-producing areas. The graph shows the percentages of world copra produced by these countries in 1986. First use the map to answer the following questions.

1. The copra-producing countries are located in which climate zone?
2. Copra is produced throughout which two large islands of Indonesia?
3. Which countries in this region produce little or no copra?

The type of graph with the map is called a **single-bar area graph.** Like a pie graph (Geography Skills 13, page 90), the whole bar represents 100% — in this case, 100% of world copra production. The scale at the top of the bar is marked off in percentages from 0% to 100%. Labels below show which parts of the bar represent Asia and other areas. Bar segments show the percentages of world copra produced in certain places. Use the graph to answer the following.

4. Which country produced the largest percentage of the world's copra in 1986? What was the percentage? The Philippines and Indonesia together produced what percent of the world's copra? The countries of Asia altogether produced what percent of the world's copra?

Use the map and graph together to answer the following questions.

5. Is most of the region's copra produced in mainland or island countries?
6. Which countries are included in the "Other" bar segment *within* Asia?

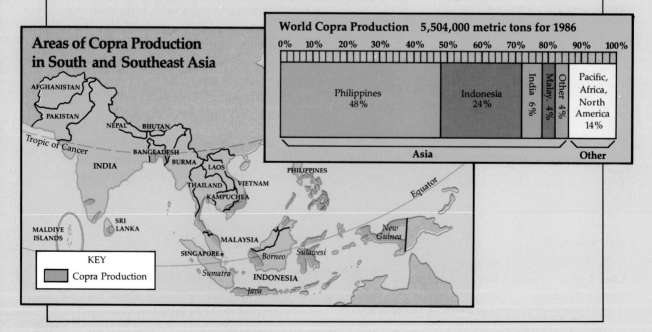

Areas of Copra Production in South and Southeast Asia

KEY
▨ Copra Production

World Copra Production   5,504,000 metric tons for 1986

| Philippines 48% | Indonesia 24% | India 6% | Malay. 4% | Other 4% | Pacific, Africa, North America 14% |

Asia — Other

A. Words To Remember

In your own words, define each of the following terms.

copra	monsoon	silt
Indochina	river system	

B. Check Your Reading

1. Which receives more rain — Pakistan or Indonesia? Why?

2. What kind of plant resource is found in areas with poorer soils? Give one example of this resource and describe one use that is made of it.

3. Name two peoples who have had strong cultural influences on the region. Give an example of that influence.

4. Name one nation in the region that has expanded industry. What is an important industry in that nation?

5. Give three examples of the region's diversity.

C. Think It Over

1. Where is the West's influence most visible in many countries? Give two examples of Western influence in the region.

2. "The region is underdeveloped because it exports mainly farm products and raw materials." Do you agree with this statement? Give reasons for your answer.

3. In what ways do you think events since World War II have affected development in the region?

D. Things To Do

Because of the many cultures in the region, there are many different religious and national festivals. Choose one country in the region. Use reference materials from your library to find out what is the country's most important festival. Write a short essay describing the festival. End your essay by answering this question: Does the festival show the influence of more than one culture? If so, what cultures?

Chapter 33

India
An Awakening Giant

It would be impossible for anyone to visit all of India's small farming villages. Altogether there are more than 500,000 villages, and many are linked only by cart tracks. Even at the rate of three villages a day, it would take 500 years to visit all of them.

These villages contain about three fourths of the nation's 770 million people. The villages produce the food for all of India's huge population.

As in many developing countries, India's villages cling to ways of life that are hundreds, even thousands, of years old. During India's long, dry season, for example, people still depend on the village well — or wells — for water. Women carry jars of water to drink, bathe, and wash clothes. In many areas, village wells are the only source of water for irrigating the fields.

As you read in the last chapter, the winter monsoons bring drought to India. The summer monsoons bring rain. When the rainfall is heavy, food can be grown for the people's needs. But sometimes the summer monsoons bring little rain. In the past when this happened, famine was certain. Without rain there was no water for crops for an entire year. The appeal for food would go out around the world. But the problem was to transport the food to the people in all those scattered villages.

In 1979 the summer monsoons brought drought — the worst in 60 years. This time, however, India made it through on its own. True, the meals were poor by Western standards. But there was no mass starvation, because India's villages were learning to use their land and water resources better. In many villages, electricity rather than oxen supply the power to pump water. Farmers are using modern irrigation methods to grow crops year round.

India is in many ways like a giant awakening from a long sleep. Fully awake it could become the leading food supplier of the region. India has huge potential, but it also has problems to overcome. In this chapter, you will see some of both.

Physical Geography

Suppose that you are standing on top of India's tallest mountain, Nanda Devi (NAHN-duh DEH-vee). It rises more than 25,000 feet (*7,500 meters*) in the Himalaya Mountains on India's northern border. Also suppose that there is a large telescope on top of the mountain. Looking through this telescope, you can see clearly for about 2,000 miles (*3,200 kilometers*) in all directions. You have a good picture of the physical region known as the Indian subcontinent. (A **subcontinent** is a major division of a continent.)

With more than 700 million people, India is the world's second largest nation (after China). At left, pedestrians stream across a main street in Calcutta (kal-KUH-tuh), India's largest city.

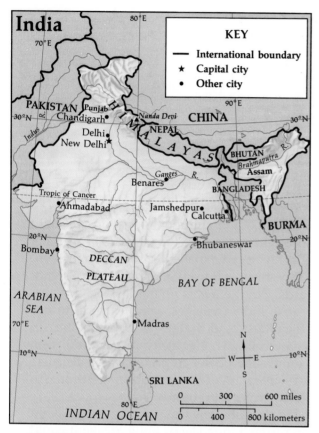

India

KEY
— International boundary
★ Capital city
• Other city

The Himalaya Mountains form India's northern border. South of the Himalayas is the broad fertile plain of the Ganges River. The rest of the country lies mainly on a plateau that is also suitable for farming. Thus nearly all of the land area is densely populated.

This area consists of India, Pakistan, Bangladesh, and the island country of Sri Lanka. India is the largest of these countries by far. It is about two fifths the size of the continental United States.

Peering through your telescope, you notice that very high mountain chains almost wall off Pakistan and India on the north. The Himalayas are the greatest of these mountains, and the highest in the world (see Chapter 32).

Now turn your telescope toward the south. Beyond the mountains, a great plain extends all the way across Pakistan, northern India, and Bangladesh. It is about 1,800 miles (3,000 kilometers) from east to west and

120 to 200 miles (200 to 320 kilometers) from north to south. This lowland plain is formed by three great rivers and their branches. The rivers are the Indus (which lies mostly within Pakistan), the Ganges, and the Brahmaputra.

The Ganges and the Brahmaputra are India's longest and most important rivers. They are fed by streams that run down from the Himalaya Mountains. In the spring, the streams are swollen by melting snow. The Ganges and the Brahmaputra begin to rise.

Then the summer monsoons bring heavy rainfall to India's northern plain — from 40 to 100 inches (100 to 250 centimeters). Now the rivers flood, spreading silt over the land and making it very fertile. There is ample water to fill the rice fields all along the plain. This plain is India's largest and best farming area.

South of this fertile plain rises a vast plateau that covers most of the southern half of India. By turning your telescope from side to side, you will notice that southern India is a peninsula surrounded by three seas. The Arabian Sea is on the west, the Bay of Bengal on the east, and the tip of the peninsula juts into the Indian Ocean. From the most northern part of India to its southern tip, the distance is about 2,000 miles (3,200 kilometers).

Mountain ranges extend along both sides of the V-shaped peninsula and come together at its tip. The mountains slant down toward coastal plains that contain good soils. The combination of fertile soils, tropical climate, and heavy rainfall make these coastal lowlands fine for farming.

Climate

As you know, most of India has wet and dry seasons that are caused by monsoon winds. The wet season lasts four months, from June through September. The heaviest rainfall occurs on the mountainsides of the northeastern state of Assam (ah-SAHM). This area gets an average of 425 inches (1,080

444

centimeters) of rain a year — and is one of the wettest places in the world. The least amount of rain falls in the northwest. Parts of this area get only two inches (*five centimeters*) of rain a year, and are desert.

When the dry monsoon starts, it brings cooler weather to most of India, especially in the north. There the temperatures will range between 60° and 70°F (*15° and 21°C*). This comfortable weather lasts through February. In March a period of stifling heat begins and lasts until the rainy season arrives. During this period, temperatures in most of India can rise as high as 120°F (*about 50°C*).

As you read in the last chapter, India has a great wealth of mineral resources. So far, however, India has not discovered any large deposits of oil. To help make up for this, India is using water power to develop electricity. Many dams have been built in recent years on the country's great rivers. The dams not only generate electricity. They also store water so it can be used for irrigation during the dry season, and they help control floods in the wet season.

SECTION REVIEW

1. What landform extends along the northern border of India? What is outstanding about this landform?

2. When does India have its hottest season? Base your answer on the monsoon periods.

3. Some of the best farming land in the country lies alongside the Ganges and Brahmaputra rivers. Why is this so?

4. What usually happened in the past when the summer monsoons failed to bring rain to India? What happened in 1979?

Northern India shares the world's tallest mountains with China and Nepal. The Himalayas have about 60 peaks higher than the highest in North America. This peaceful Himalayan valley is in India's far northwest.

Human Geography

About four out of every five people in India are Hindus. In the Hindu religion, all rivers are sacred. But the most sacred river of all is the Ganges. Hindus have more than 100 names for this river. Among the names are Staircase to Heaven, Creator of Happiness, and Destroyer of Sin.

The most holy place on the Ganges is the city of Benares (buh-NAR-eez). All religious Hindus wish to visit Benares at least once in their lifetimes. They believe that bathing in the waters of the Ganges washes away their sins and purifies their souls. Many Hindus take some of the water home in jars. They use the water in their daily prayers to Brahma (BRAH-muh), who is the Supreme Power, and to other Hindu gods.

Hinduism shapes the everyday lives of its followers. Hindus are divided into about 3,000 different groups called **castes.** Traditionally the members of each caste follow a special set of rules. They do the same kind of work, wear the same kind of clothes, and eat the same kind of foods. There is a caste for almost every kind of job or occupation. A person who is born into one caste is supposed to remain in that caste for the rest of his or her life.

In the past, people of a higher caste did not mix with, or marry, people of a lower caste. Some Indians were outside the caste system altogether. These outcastes, known as Untouchables, suffered discrimination and could get only the lowest-paid jobs. The Indian government banned such discrimination in 1950.

Today the caste system is not so strong as it used to be. Members of different castes — and former outcastes — mix more freely, and sometimes even marry one another. But many people still follow the old rules, especially in the villages. Few people want the caste system to die out completely. Castes give their members a sense of belong-

ing. And castes help pass on skills in arts and crafts from one generation to another.

Hindus believe that people may be reborn as animals. As a result, Hindus respect all living creatures. Some, like monkeys and snakes, are considered sacred. But the most sacred of all are cattle, especially the cow.

India has more cattle than any other country in the world. The cattle may be used to pull plows or carts, and the cows may be milked. But devout Hindus will not kill them for food.

A Meeting Place of Cultures

Hinduism dates back more than 3,000 years. Its complex details are due in part to the fact that it was developed by two very different peoples. Warlike invaders from the northwest seized control of much of India from its more peaceful inhabitants. Over the centuries, the two groups settled together, and were often linked by marriage. But even today, there are ethnic differences between many Indians in the south (descended from the original inhabitants) and in the north (descended from the invaders).

Another wave of invaders began to arrive from the northwest in the 12th century A.D. These invaders were Moslems. By the 16th century, much of India was under the rule of a Moslem emperor. During most of its history, India had been divided into many separate states and kingdoms. The area was too vast, and transportation was too slow, for India to be ruled as one country.

Before long the Moslem empire began to collapse. By that time, European nations were eyeing the riches of India. Britain took the lead. In the mid-19th century, with the help of new inventions such as the railroad and the telegraph, Britain was able to bring all of the subcontinent under its rule.

As you read in Chapter 32, when British India became independent in 1947, it was divided into India and Pakistan. India's leaders set up a democratic form of government

based partly on the British parliamentary system and partly on the U.S. Constitution.

Although many Moslems moved to Pakistan when India became independent (see page 435), many others remained. Today more than 60 million Indians are Moslems, making Islam the second largest religion in the nation. Moslems live by different rules from the Hindus. Moslems do not believe in the caste system. All people, they say, are equal in the eyes of Allah, the Supreme Being. They do not believe in rebirth. They eat meat, but not pork. Most Moslems live in the north of India. In some cities and villages, they live apart from the Hindus.

Other important religions in India include three that developed out of Hinduism in much the same way that Christianity has its roots in Judaism. The followers of the three religions are known as Buddhists, Jains (jiens), and Sikhs (seeks). There are also about 14 million Christians in India today.

India has 14 official languages and more than 150 local dialects. People from one part of the country often do not understand the speech of people from another part. English was the official language of India when the

India's great variety of cultures can be seen in its big cities. On this street in Bombay, Moslem names (on the building at right) mingle with Hindu names. Signs are in different Indian languages and English.

country was under British rule. Today English is India's *lingua franca* (unofficial shared language). Government leaders speak English because it is the one language they can all speak and understand.

Cultural differences in India also exist between people who live in villages and those who live in large cities. As you have seen, most of India's people live in small farm villages. But one fourth of them live in big cities. Calcutta, which is India's largest city, has almost 10 million people. In the following sections, you will be able to compare village life with life in a big city.

Village Life

The Gokhali (goh-KAH-lee) family lives in a farm village in the northeastern part of the peninsula, near the city of Bhubaneswar (boo-buh-NEHSH-war). The village was settled only a few years ago by 370 people who came from a much drier part of India.

447

For two years before the move, they had had little or no rain. Faced with hunger, they left their old village and moved to an area where the rainfall was more dependable.

Pritam (PREE-tuhm) and Surjit (SOOR-jeet) Gokhali live in a hut made of mud, with a straw roof. They have six children. The oldest is Krishna (KREESH-nuh), age 11. As in most Indian homes, the family includes more than parents and children. Pritam's parents and one of his brothers live in the same house. A family that includes the parents' parents and brothers or sisters is known as an **extended family**.

The family gets up each day at dawn. Surjit prepares rice for breakfast, and Krishna helps feed the two youngest children, who are babies. Later Hindu prayers are chanted at a small shrine in the hut.

After morning prayers, nearly all the villagers older than 12 go to work in the rice fields. Pritam's parents are too old to work, so they stay home. Krishna also stays in the village to take care of his younger brothers and sisters. When it is time for lunch, he prepares rice flavored with curry powder. This powder is made of ground spices.

The meals are much better when Pritam and Surjit are not working in the fields. Then Surjit prepares chicken and eggs, or rice pancakes stuffed with vegetables.

The village does not have electricity, so there is no radio nor television. But in the evening, the Gokhali family gathers around a kerosene lamp in their hut. They sing traditional Hindu songs, and listen to ancient Hindu legends told by Pritam's father.

Although India's population is vast and scattered, nearly half of its young people receive schooling. Most schools make do with simple equipment, as in this village classroom in the plain of northern India.

Industry and construction are expanding in India, but they often use labor (which is plentiful) more than machinery. Human muscles are raising these homes.

The village does not have a school yet. But in four years, there will be enough money for a one- or two-room schoolhouse. The villagers will build the school with their own hands. (This has been done in thousands of villages throughout India.) Krishna will be among the hardest workers. For when he becomes 15, his parents will start making arrangements for him to marry. Krishna wants to make sure that his children will get the education that he missed.

City Life

Bombay (bom-BAY), which is India's second largest city, has about seven million people. It is a busy seaport on the west coast of the peninsula. In one of Bombay's many apartment houses, you will find 14-year-old Bharti Acharya (BAR-tee uh-CHAR-yuh) and her parents.

Bharti is in the ninth grade of a new private school. Like most schools in crowded Bombay, it has double sessions. The subjects that Bharti studies include four languages. One of these is Sanskrit, the ancient language of India. Writings in Sanskrit include Hindu scriptures dating back more than 2,500 years, as well as classical plays and poetry. The other three languages that Bharti studies will all be useful in her everyday life. They are: Marathi (muh-RAH-tee), the chief local language; Hindi (HIN-dee), a language spoken by about half the people of India; and English, which is spoken by educated people in all parts of the country.

In her free time, Bharti likes to read mystery stories or listen to music on the radio. Her favorite music comes from movies. Bombay is the center of India's large movie industry, and Bharti's father is a musician who plays the *sitar* (see-TAR) for movie sound tracks. The sitar is an Indian instrument that is something like a guitar.

Before dinner, which is served at 9 P.M., Bharti washes and prays. A typical evening meal is rice and chapatis (thin flat disks of wheat bread), with vegetables or lamb. Like other Hindus, Bharti's family avoids beef.

Some day Bharti's parents will choose a husband for her. This has been a tradition in India for centuries. Like most Indian girls, she does not object. Her parents will find a better husband for her than she would herself, she says, because they are older and wiser, and know more about life.

But some traditions are changing in India. For example, Bharti plans to be more than a housewife. She wants to become a doctor.

SECTION REVIEW

1. Why wouldn't you see beef served at a meal in a Hindu home?

2. Describe two major cultural differences found among people in India.

3. Describe two ways that Bharti's life in the city is different from Krishna's life in the country. Describe one way that their lives are similar.

4. During most of India's history, the country was divided into many separate states. Why was it difficult to unite these states?

Economic Geography

A typical village near Chandigarh (CHAHN-dee-gur) in northern India has changed a great deal in the past 20 years. Once all the houses were made of mud. There was only one well for drinking water. In nearby fields, bullocks (steers) pulled wooden plows over fields that were often parched. Wheat grew in thin, ragged rows.

Today more houses are built of brick than mud, and the number of wells has grown to 50. There are now as many tractors as bullocks in the village. Electricity has also been brought in. The homes are still lit by kerosene lamps, but electric pumps are used to irrigate the fields. As far as the eye can see, the fields are covered with healthy green stalks of wheat.

What accounts for these great changes? In this Indian village, as in many others, a "green revolution" is taking place. Farmers are using new "miracle" seeds developed by scientists in Japan, the Philippines, the U.S., and Mexico. These seeds grow faster and stronger than the old ones. The farmers are also using large amounts of chemical fertilizers. Many more wells and irrigation ditches have been dug since it became possible to pump water by electricity.

As a result, wheat harvests have increased tremendously in the village near Chandigarh. The village now produces three times as much wheat to sell in city markets as it did in the past. The money the villagers get is used to build better homes and new schools. A few farmers have small Indian-made cars, and others own motor scooters. Most importantly, in the late 1970's, India became able to feed its people without importing large amounts of food.

The green revolution is spreading in India, but most farm villages still use traditional methods of growing food. The work is done by hand, or with the help of bullocks. Farms are usually very small — less than five acres (*two hectares*). Most farmers cannot afford to buy seeds or fertilizer. They have trouble just growing enough food for their families. There is nothing left over to sell.

While the green revolution holds off starvation, about 300 million people still eat only two meals a day. Sometimes a day's wages buy only one meal.

The government of India is trying to change these conditions. Farm experts are sent into the villages to teach modern farming methods. The government broadcasts programs that explain the new methods. In villages that have electricity, people gather around a radio or TV set and listen. Many now face the future with new hope.

The Indian government has an even bigger project for increasing food production. This project is to build a long canal that will extend from the Ganges River to the southern half of the country. Most of the water of the Ganges is now unused — it simply flows into the sea. The canal would make it possible to irrigate large areas that do not have enough water. It will take about 25 years to build the canal and will cost billions of dollars. But it may be the answer to India's problem of drought during the dry monsoon season.

Despite drought and farming methods that are often inefficient, India's total production of crops is huge. India leads all other countries in growing tea, sugarcane, pepper, and peanuts. India also ranks very high in growing jute, cotton, and tobacco. These crops are raised chiefly for export or for industry. India's chief food crops are rice, wheat, and vegetables. In all the world, only China grows more rice than India.

Industry

When India became independent in 1947, modern industry was centered in three cities — Calcutta, Bombay, and Madras (muh-DRAHS). Since then industry has greatly increased in these cities, and has spread to other areas. In Calcutta jute processing is the leading industry, as it was under British rule. But the city now also produces a great variety of machines, tools, and transportation equipment.

About 150 miles (*250 kilometers*) inland from Calcutta are deposits of coal and iron ore, the basic ingredients for making steel. As a result, giant steel mills have been built nearby. The city of Jamshedpur (JAHM-shehd-pur) has become the "Pittsburgh of India."

Bombay and Ahmadabad (AH-muh-duh-bahd), two cities on the west side of the peninsula, specialize in cotton textile and chemical industries. India's cotton industry was first developed by the British. Today only the United States and the Soviet Union produce more cotton cloth than India. Madras is the leading center of textiles and other light industries in southern India.

Transportation is better in India than in most developing nations. India's railroad system is the largest in Asia and the fourth largest in the world. Much of the system was built under British rule, and is now being modernized. India's great rivers are also an important means of transportation. About 5,000 miles (*8,000 kilometers*) of these rivers are navigable. Some are linked by canals.

Road transportation is not so good, however. Paved highways do link the cities. But as you have seen, most Indians are scattered among half a million villages, and the quickest way of traveling from one village to the next may be by ox cart.

India is trying hard to develop modern industry and scientific farming methods. With its varied environment, India has the potential to produce a broad range of crops and industries. As in the U.S., different areas could specialize in different products.

India imports food, oil, machinery, and manufactured goods. Its chief exports are mineral ores and agricultural products. The harbor below is at Calcutta.

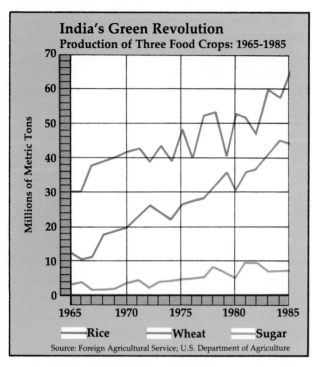

India's Green Revolution
Production of Three Food Crops: 1965-1985

Millions of Metric Tons

Rice   Wheat   Sugar

Source: Foreign Agricultural Service; U.S. Department of Agriculture

Since the 1960's, new irrigation projects and better farming methods have helped to boost production. New fast-growing varieties of wheat and rice enable farmers to bring in at least one extra harvest a year.

However, developing this potential is not an easy task. A dictatorship may *force* people to accept changes that the government wants. But India is a democracy that must *persuade* people to accept its decisions. For example, large families are a Hindu tradition, and many village parents have eight or more children. To make the food supply go farther, the Indian government wants parents to have fewer children. Yet so far, government efforts to promote small families have not been very successful.

The task of modernizing India is made even harder by the many differences among its people. The people are often separated by differences of religion, caste, and language.

The surprising thing is not that India is moving slowly into the modern world. It is that India has made such progress in the past two decades.

SECTION REVIEW

1. Describe two ways that technology has improved the lives of many Indian villagers in recent years.

2. Is it true to say that India is a major world producer of farm crops? Explain.

3. Why did steel making develop at the city of Jamshedpur?

4. What are the strengths and weaknesses of India's transportation system?

YOUR LOCAL GEOGRAPHY

1. The Gokhali family lives in traditional housing — a mud hut with a straw roof. But in the village near Chandigarh, houses are now built of brick rather than mud. Is the kind of housing found in your community much the same now as it has been for many years? Or have changing life-styles, population shifts, and technology affected the homes in your area? At your library or local historical society, do some research into the "look" of buildings in your community in years gone by. Find photographs of a few local streets 25 to 50 years ago. Compare the photos with those same streets today. In what ways have they changed? How have they remained the same? Can you tell why they have changed or remained the same?

2. In Geography Skills 42, you learn how the Dewey Decimal System works. Use this knowledge to locate books dealing with the geography of your area of the United States. First select several aspects of geography that are of special interest to you. You may want to find out more about your area's landforms, industry, culture, religions, etc. After selecting two or three topics, refer to the list of 10 major subject groups to see in which section your topic belongs. This will tell you what call numbers you would look under to locate books on the subjects you have picked. When you go to the library, find the shelves where those numbers are located.

Finding a Library Book

A holy man stands by the sacred Ganges River. Fans crowd around a national movie star. These are some images of India. To learn more about this country, you need only look on your library shelves. But where do you look?

In Geography Skills 38 (page 402), you learned that every nonfiction library book is identified by a call number. A **call number** is a book's shelf address — it tells you where, on the library shelves, a book is located. Books on the same subject are usually shelved together under the **Dewey Decimal System**, which is illustrated in the diagram below.

How the Dewey Decimal System Works

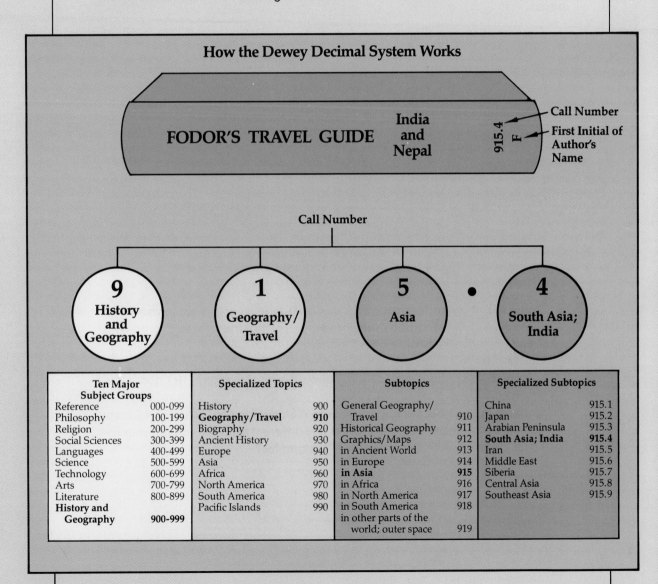

Ten Major Subject Groups		Specialized Topics		Subtopics		Specialized Subtopics	
Reference	000-099	History	900	General Geography/ Travel	910	China	915.1
Philosophy	100-199	**Geography/Travel**	**910**	Historical Geography	911	Japan	915.2
Religion	200-299	Biography	920	Graphics/Maps	912	Arabian Peninsula	915.3
Social Sciences	300-399	Ancient History	930	in Ancient World	913	**South Asia; India**	**915.4**
Languages	400-499	Europe	940	in Europe	914	Iran	915.5
Science	500-599	Asia	950	**in Asia**	**915**	Middle East	915.6
Technology	600-699	Africa	960	in Africa	916	Siberia	915.7
Arts	700-799	North America	970	in North America	917	Central Asia	915.8
Literature	800-899	South America	980	in South America	918	Southeast Asia	915.9
History and Geography	**900-999**	Pacific Islands	990	in other parts of the world; outer space	919		

(Turn page.)

A travel guide to India is shown in the diagram on page 453, with a call number printed on its spine. Notice that each *digit* (numeral) of the call number is also shown over part of a chart that explains what the digit represents. For example, the first digit of the call number refers to its "major subject group."

In the Dewey Decimal System, all knowledge is divided into 10 *major subject groups*, and each group is assigned a series of call numbers. The 10 major subject groups, with their number categories, are in the left-hand column of the diagram on page 453. Look down the list at the different subjects and their number categories. Then answer the following questions on a separate sheet of paper.

1. Cooking is an "applied science," or *technology*. Would you find a book of Indian recipes on the shelves marked 600's, 700's, or 800's?
2. Under what number category would you look to find an atlas with a map of India and other reference books?
3. Under what subject group and number category would you find books about: (a) Indian movies and sitar music? (b) India's 14 official languages and 150 dialects? (c) the beliefs and practices of Hinduism?
4. For general books about India, 9 is the first digit of the call number. Does this indicate "arts," "social science," or "history and geography"?

Each of the 10 major subject groups is subdivided into 10 *specialized topics*; and each of these, into 10 *subtopics*. For each subdivision, there is another set of call numbers. When subtopics are subdivided even further, a decimal point is placed after the third digit, and more digits are added to the right of the decimal point. Study the pattern illustrated in the diagram on page 453, and answer the following questions.

5. In the call number for *Fodor's Travel Guide to India and Nepal*, what digit indicates: (a) "Geography and Travel"? (b) that the book has to do with Asia? (c) that the book is about India?
6. Under what call number would you find books with geographic and travel subtopics: (a) about Africa? (b) about North America? (c) about the moon?
7. Richard Lannoy's book *The Speaking Tree: A Study of Indian Culture and Society* has the call number 915.4–L. How is this different from the call number for *Fodor's Travel Guide*? What does the "L" stand for?

A. Words To Remember

Three of the following terms are defined by the numbered phrases below. On a separate sheet of paper, write each term next to the number of its definition. Then write the two words that are *not* defined, and give a definition for each.

caste   green revolution   subtopic
extended family   subcontinent

1. An increase in farm production because of new "miracle" seeds.
2. A subdivision of a subject.
3. One of the groups into which Hindus have traditionally been divided by job or occupation.

B. Check Your Reading

1. What is the population of India? About what fraction of those people live in farming villages?
2. India has many large cities. What are the two largest? What is the population of these two cities?
3. How does India hope to eliminate the problems of drought in the future?
4. Describe one way that a Moslem's life differs from a Hindu's.
5. The Ganges River is important to Indians for two different reasons. What are they?

C. Think It Over

1. What are the main advantages and disadvantages of India's environment?
2. What are the main difficulties faced by India as it tries to develop further?
3. The chapter says, "India is in many ways like a giant awakening from a long sleep." What does this mean?

D. Things To Do

Imagine that you are either Krishna Gokhali or Bharti Acharya. Write a letter to a friend outlining your hopes for the future. Make sure to mention any possible changes in the country as a whole that could affect those hopes.

Chapter 34

Indonesia
Rich in Soil and Oil

For months the great volcano on the island of Krakatoa (krah-kuh-TOH-uh) had been pouring ashes into the air. People on nearby Java and Sumatra could feel the earth tremble from time to time. But no one was worried. The volcano had not erupted in 200 years, and a few **tremors** (brief shaking movements of the earth) were harmless. People in this part of the world, now known as Indonesia, were used to volcanic activity. Those who lived on Krakatoa simply moved to a nearby island, expecting to return home soon.

But on August 26, 1883, the volcano began to erupt. Explosions shook houses 100 miles (*160 kilometers*) away. Rocks were tossed high into the air, and the sky turned black with ashes. The only light on the island came from glowing lava.

Soon after 10 o'clock in the morning, a tremendous explosion took place. The sound was heard as far away as Burma, a distance of 1,500 miles (*2,500 kilometers*). The explosion wrecked the volcano and sent two thirds of the island plunging into the sea. However, the worst was yet to come. A **tidal wave** (a huge ocean wave created by the explosion) traveled away from the island at a speed of 350 miles (*550 kilometers*) an hour. The tidal wave crashed upon the coast of Java and destroyed hundreds of villages within minutes. Parts of Sumatra were buried under 80 feet (*25 meters*) of water. In some villages, not one person was left alive.

The dust from the explosion was carried around the world at least 12 times in the next few months. The dust created strange blood-red sunsets in many parts of the planet. Years passed before the last of the dust fell to Earth.

Physical Geography

Indonesia, which is made up of more than 13,500 islands, is the most volcanic area in the entire world. The main center of volcanic activity is the island of Java, which still has about 50 volcanoes that erupt from time to time. One of them, Mount Fire, erupts at least once every five years. Active volcanoes are also found on Sumatra, Sulawesi (soo-luh-WEH-see), and other islands.

Volcanoes can be very destructive, yet people continue to live near them. One reason is that lava from the volcanoes has often made the nearby soils very fertile. Java's most important natural resource is its volcanic soil. Combined with a hot climate and ample rainfall, this soil enables farmers to grow three crops of rice a year.

Java has by far the richest farmlands in Indonesia. This island makes up only seven percent of the total land area of the country.

Warmth, plenty of rain, and volcanic soil enable crops to be grown in most parts of Indonesia. These terraced rice fields have been cut out of a forest hillside.

457

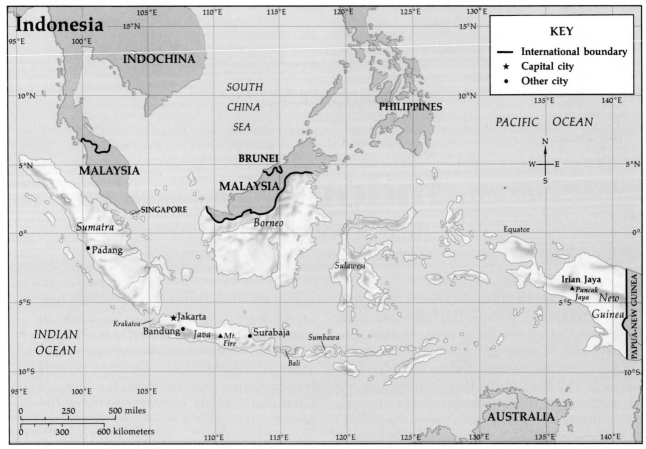

Indonesia

INDOCHINA

SOUTH CHINA SEA

PHILIPPINES

PACIFIC OCEAN

MALAYSIA

SINGAPORE

BRUNEI

MALAYSIA

Sumatra

• Padang

Borneo

Sulawesi

Equator

Irian Jaya
▲ Puncak Jaya

New Guinea

PAPUA-NEW GUINEA

INDIAN OCEAN

Krakatoa ★Jakarta
Bandung• Java ▲Mt. Fire • Surabaja

Sumbawa

Bali

AUSTRALIA

0 250 500 miles

0 300 600 kilometers

Yet more than half the people of Indonesia live on it. Java is one of the most densely populated places in the world. Most other islands of Indonesia are thinly populated or uninhabited.

The islands of Indonesia lie along the Equator and extend farther than the distance between Maine and California.

Indonesia is an archipelago (a large group of islands). The islands in the west are surrounded by shallow seas, often only 200 feet (60 meters) deep. Geographers believe that these islands were part of the mainland of Asia until a few thousand years ago. However, the islands to the east are the tops of very steep mountains that rise from the ocean floor. In some cases, the distance between the ocean floor and an island mountaintop may be 30,000 feet (9,000 meters).

Varied Islands

The Indonesian islands vary greatly in size. Five of them make up more than 90 percent of the country's total land area. Many islands cover less than a dozen city blocks.

The nation of Indonesia is spread over a chain of islands that extend from the Southeast Asian mainland to Australia. More than half of all Indonesians live on Java, where volcanoes have made the soil fertile.

Four of the largest islands are in the west. Starting with the largest, in order of size, they are Borneo, Sumatra, Sulawesi, and Java. Borneo is the third largest island in the world, after Greenland and New Guinea (GIN-ee). Not all of Borneo belongs to Indonesia, however. About one quarter of the island is part of Malaysia (see the map above). Sumatra is the fifth largest island in the world. If it were placed over the Mississippi Valley, Sumatra would cover both Minneapolis and New Orleans.

At the eastern end of the archipelago lies the island of New Guinea. Half of it, known as Irian Jaya (EE-ree-ahn JIE-yuh), belongs to Indonesia. Much of Irian Jaya is covered with rain forests, and is still unexplored.

The Indonesian islands generally have mountainous or hilly backbones. Some mountain peaks on Sumatra and Java are more than 12,000 feet (*3,500 meters*) above sea level. The highest mountain in Indonesia, Puncak Jaya (POON-chahk JIE-yuh), lies on Irian Jaya. Puncak Jaya is more than 16,000 feet (*5,000 meters*) high, and capped with snow. There are also large lowland areas on the bigger islands, especially near the coasts. The coastal lowlands are often swampy. Northern Java is a large plain made fertile by lava that long ago spurted from volcanic peaks to the south.

As you read earlier, Indonesia lies within the monsoon zone of Asia. From December to March, monsoon winds blow from the Asian mainland toward Indonesia. As the winds cross the South China Sea, they pick up considerable moisture and bring heavy rainfall to the islands. From June to October, monsoon winds from the desert areas of Australia blow toward Indonesia. The effects of these dry winds are not very great, however. Only eastern Java and some smaller islands near Australia have true dry seasons, ranging from three to five months.

Because of the surrounding seas and the warm climate, most of Indonesia gets a lot of rainfall all year round. Sumatra and Borneo, which lie the farthest from Australia, get only slightly less rainfall during the dry season than they do in the wet season. They get a total of 120 to 145 inches (*300 to 370 centimeters*) a year. On the Indian subcontinent and in Indochina, the wet and dry seasons contrast sharply. In most of Indonesia, as well as the Philippines, rain is plentiful and evenly distributed.

Natural Resources

Heavy rainfall and warm temperatures throughout the year have produced a great abundance of vegetation in Indonesia. All of the large islands except Java are thickly covered with rain forests. (Most of Java's forests have been cut down to make room for farmland.) In these forests, you will find many kinds of wildlife. There are tigers, rhinos, pythons up to 30 feet (*nine meters*) long, crocodiles, monkeys, and birds with brilliant feathers. The variety of plant life is almost endless. Giant ferns grow as tall as trees. Climbing vines grow hundreds of feet long and many inches thick.

Although the islands of Indonesia lie on or close to the Equator, temperatures are moderated by the surrounding seas. The highest temperature ever recorded in Jakarta (juh-KAR-tuh), the nation's capital city located on Java, was 96°F (*35°C*). The average yearly temperature in the lowlands is only about 80°F (*27°C*). In the highlands, of course, temperatures are lower. Indonesia's location at the Equator makes the temperature in any one place almost the same all year round.

Indonesia's nearness to the Equator also means that the hours of daylight change very little around the year. On the shortest day, the Sun sets less than one hour earlier than on the longest day. (In Chicago the difference is almost six hours.)

The islands of Indonesia are rich in a number of mineral resources. The most valuable are oil and tin, but there is also some natural gas, coal, manganese, bauxite, copper, and nickel.

Indonesia's forests are another important resource. As you read on page 431, they contain valuable hardwoods like teak and ebony, as well as two canelike plants, bamboo and rattan.

SECTION REVIEW

1. What two large islands does Indonesia share with other nations?

2. Indonesia lies very close to the Equator. Why don't temperatures there become extremely hot?

3. Why doesn't Indonesia have a long dry season?

4. Why is there less forest on Java than on other islands?

Human Geography

Outside a village theater in Indonesia, a large crowd waits impatiently for the doors to open. The show that they are so eager to see will start at 9 A.M. and last until 6 P.M. It is a puppet show, a form of entertainment that has been popular in Indonesia since ancient times. It has developed into an art that fascinates Indonesians and visitors alike.

Unlike a puppet show in the Western world, the figures are seen only as shadows on a screen. For the screen, a white cloth is stretched on a bamboo frame. Behind the screen is a lamp. A small orchestra sits behind the screen to provide music. The orchestra uses gongs, drums, and instruments like xylophones. There may also be a stringed instrument and a kind of flute.

The puppets themselves are flat cutouts made of buffalo hide. The puppeteer sits at the bottom of the screen and holds up the puppets so their shadows appear to the audience. As he moves the puppets around, he tells a story. This story may be drawn from history or religion.

These shadow plays have a long history. Their stories were introduced to the islands by settlers from India who arrived in large numbers more than 1,500 years ago. These Indians influenced the people of the islands in many ways. The Indian religions of Hinduism and Buddhism spread rapidly. Magnificent Hindu and Buddhist temples were built that still stand today.

As you read on page 433, Moslems later settled in the islands, and today nine out of ten Indonesians are Moslems. But the Indian influence remains strong. Hinduism is still the chief religion on the island of Bali (BAH-lee). Statues of Hindu gods are often visited by local Moslems. Hindu legends live on in the puppet shows.

After the Moslems, Europeans began to arrive — Portuguese, English, and Dutch. The Dutch ruled the islands almost continuously from the 1700's until 1949. Mainly as a result of past European influence, there are more than five million Christians in Indonesia.

The seas made the islands open to immigrants and colonists from many nations. In one way or another, these peoples have helped to shape modern Indonesia.

"Unity in Diversity"

Indonesia today has about 150 million people — the fifth-largest population in the world. The motto of the nation, which won its independence in 1949, is "Unity in Diversity." (This motto is very similar to the U.S. motto *E Pluribus Unum*, which means "One Out of Many.")

This diversity is partly the result of foreign influences. But it is also due to the great distances that separate the islands. For many centuries, people in one part of the archipelago had little contact with people in another part. Sometimes they were not even aware of the many islands that lay beyond their own. As a result, different customs and languages developed among the islands. Today there are more than 200 cultural groups in Indonesia, each with its own language.

When Indonesia became an independent nation, its leaders wanted a common language to help unite the people. Which one should they choose? The Javanese language was spoken by more Indonesians than any other. But many people on other islands thought that Java already had too much influence over the rest of the country. After much debate, the government chose a form of the Malay language. It had long been used by traders from different parts of the country to talk to each other.

The new official language, called Bahasa (bah-HAH-suh) Indonesia, was taught in the schools, and spread very quickly over the whole country. Today most Indonesians still speak their local languages, but they can also talk to each other in Bahasa Indonesia.

The national language has many words that come from Western tongues, including English. For example, a motorcycle is called a *speda motor.*

Like most developing countries, Indonesia is a land of sharp contrasts. In the cities, for example, you will see high-rise buildings of glass and steel, luxury hotels, and modern shopping centers. In large areas of the country covered with rain forests, people usually live in houses built on stilts to protect them from floods caused by heavy rains.

Village Life

Large wooden houses on stilts are common in the western part of Sumatra. The sides of these houses are richly decorated and carved. The roofs, which are steep and thatched with palm leaves, have peaks that resemble buffalo horns. Such houses are called *Minangkabau* (mee-NAHNG-kuh-baw). *Minangkabau* means "triumphant water buffalo." The word comes from an ancient Indonesian legend about a baby buffalo that defeated a much larger buffalo in a fight.

In one of these colorful Sumatran houses lives a 16-year-old farm girl named Fatimah (FAH-tee-muh). She and her parents live in a village not far from Padang (PAH-duhng), where they grow rice and corn. During the busy harvest season, all the neighbors help each other. This is a tradition in all parts of the country — and in farm villages in many other countries too. Helping others at harvest time is so common in Indonesia that it has a special name, *gotong-rojong* (GOH-tong ROH-jong).

At sunrise Fatimah begins walking to school, which is several miles from her village. Classes start at 7 A.M. Fatimah is in the second year of a girls' high school for home

Most of Indonesia's 150 million people are Moslems. On Sumbawa (soom-BAH-wuh) island, children read the Koran (koh-RAHN), the holy book of Islam.

Jakarta, Indonesia's capital, is one of the world's largest and most crowded cities. In some areas, people live as best as they can in shacks. The downtown area, however, is full of modern buildings and cars. The buildings in this picture include a department store.

economics. She studies Bahasa Indonesia, history, English, and geography. She is also taught homemaking and traditional Indonesian cooking.

School ends at 1 P.M., when Fatimah goes home for lunch. There is always a lot of work for her in the afternoon. Because her mother helps her father in the fields, Fatimah prepares dinner. She also cleans the house, does her homework, and if she has time, bakes a dessert. The main dish is rice, plus fish or meat, vegetables, and peppers.

Like most of the people of Indonesia, Fatimah is a Moslem. She goes to religious classes each week, and learns the prayers and the history of her faith. Her great ambition is to make a *hajj* — a pilgrimage to Mecca, the holy city of Islam, in Saudi Arabia. Many thousands of Indonesians make this pilgrimage every year.

City Life

The capital city of Indonesia, Jakarta, is found on Java. Jakarta is Indonesia's largest city, with a population of about six million. If you visited one of Jakarta's quieter resi-

dential areas, you might meet Ali (AH-lee) and Siti Hurkens (SEE-tee HOOR-kehns) and their 14-year-old daughter Donnagaby (doh-nuh-GAH-bee).

Ali Hurkens works in a government office where records of births, marriages, and deaths are kept. Most Indonesians have only one name, generally a given name. But some of Ali's ancestors were Dutch, so his family uses both given names and their Dutch surname, Hurkens.

The Hurkens family lives in a Dutch-style house with big windows and doors to allow the air to circulate freely. The roof is made of red tiles that help cool the house by reflecting the rays of the Sun.

Donnagaby, her three younger sisters, and her younger brother all get up about 5 A.M. Donnagaby helps her mother with the younger children, and afterward has a cup of tea. Then she goes to school for classes that start at 7 A.M. and last until 12:10 P.M.

After school Donnagaby goes home and sets the table for lunch. Her favorite meal is a mixture of lobster, tomatoes, and red peppers with rice. Later she plays cards with her younger sisters and brother, or tells them stories. Everyone takes a nap from 2:30 until 4 P.M., because the afternoon is the hottest part of the day.

Swimming used to be Donnagaby's favorite pastime, but now she has a new one. She likes to visit Jakarta's modern department stores, looking at all the goods on display and riding the escalators. Many Indonesians who visit Jakarta have never ridden an escalator before. So guards stand at the top and bottom to keep anyone from falling.

SECTION REVIEW

1. Indonesia is made up of islands. How has this influenced its cultural development?

2. Why does Fatimah wish to make a journey to Saudi Arabia?

3. What is Bahasa Indonesia? Why was it created?

4. Describe one way that Donnagaby's life in the city differs from that of Fatimah's. In what way are their lives similar?

Economic Geography

In Indonesia, where every bit of food is precious, farmers have various ways of driving off wild birds that may eat entire crops. Sometimes the farmers place little windmills in the rice fields. The windmills make noise as they turn, scaring the birds away.

But the most common way to keep birds away is an example of *gotong-rojong* (cooperation; see page 461). A platform on stilts is built in the middle of a group of rice fields. A boy or girl standing on this platform can watch over the entire area. Overhead strings extend from the platform to every field in the group. At the end of each string is a bell

or a rattle. If birds enter any of the fields, the young guard pulls the proper string and scares them away.

Farming

About four out of five Indonesians live in villages and are farmers. In areas with volcanic soils, irrigated rice farms are most common. These farms can be found on Java, parts of Sumatra and Sulawesi, and Bali.

As you have seen, the combination of volcanic soils, heavy rainfall, and warm climate allows the farmers in these areas to grow as many as three crops a year. Multiple cropping makes it possible for Java, which is smaller than Florida, to support more than 80 million people — more than the combined populations of Florida, California, New York, Texas, and Illinois. The average farm covers less than 2½ acres (*one hectare*), yet it has to feed a family of five or more.

In recent years, the government of Indonesia has encouraged modern methods of farming that increase the amount of food that can be grown. Many small farmers now use new "miracle" seeds, fertilizers, insecticides, and improved irrigation systems.

These methods doubled the total production of rice on Java between 1969 and 1978. As a result, most people on Java are better off today than they were before the green revolution. Many are buying TV sets, transistor radios, bicycles, and motorcycles.

However, there is a limit to the number of people that Java can support. The government of Indonesia has urged Javanese farmers to move to less-crowded islands. So far this program has had little success. The people of Java have strong ties to their villages and their families. Few of them want to live anywhere else.

In the Rain Forests

In large areas of Indonesia where there are no volcanoes, the soils are generally poor. This is especially true in the rain forests. As in the rain forest of the Amazon Basin, the heavy rains wash away the soil nutrients.

Farmers in these areas use a farming system known as **slash and burn.** First they cut down the trees in a section of forest at the end of the rainy season. Then they burn the underbrush. The ashes help make the soil fertile. The farmers then dig holes with pointed sticks, and plant their seeds.

Slash-and-burn farmers plant a wide variety of crops, which are harvested at different times of the year. This assures a steady supply of food, and also provides a balanced diet. Among the crops they grow are rice, corn, sweet potatoes, cassava, and green vegetables. They also grow many kinds of fruit, including bananas and pineapples.

When the soil wears out, the farmers move to another part of the forest. The process of slash and burn begins again.

Slash-and-burn farming requires much more land than irrigated rice farming to support the same number of people. But in the tropical rain forests, there are few people and land is plentiful. Borneo, which is heavily covered with rain forests, has only about 25 people per square mile (*10 people per square kilometer*). Java has more than 1,500 per square mile (*600 per square kilometer*).

Indonesia has many large plantations that grow cash crops for export. Plantation farming was introduced to the islands by the Dutch in the 18th century. Java once provided much of the world's coffee supply. Later the Dutch introduced other plantation crops, including rubber, sugar, and cacao. Indonesia produces more natural rubber than any other country except Malaysia.

While some plantations are on Java, most are on the larger and less populated island of Sumatra. Rubber trees grow well in the hot, wet lowlands of the island.

Money from Minerals

As you have seen, Indonesia has many mineral resources. It is one of the world's leading oil-producing countries. Oil accounts for almost two thirds of the value of Indonesia's exports. Most of the oil fields are on the island of Sumatra, and some are on Borneo. In recent years, large deposits of oil and natural gas have been discovered in the shallow seas off Sumatra, Borneo, and Java.

Tin is Indonesia's second most valuable export. Together with the neighboring countries of Malaysia and Thailand, Indonesia is among the world's leading tin producers. Most of the tin mines are located on three tiny islands east of Sumatra.

Before World War II, plantation products accounted for about 80 percent of Indonesia's total exports. But in recent years, there

A traditional craft industry in Indonesia is the making of a fabric called *batik* (buh-TEEK). Patterns are drawn on the fabric, which is then plunged into dyes.

has been a boom in oil, metal ores, and timber. The value of these exports is now much greater than that of farm products.

Transportation and Industry

Travel in Indonesia usually is difficult. The islands are separated by wide stretches of sea, and heavily covered with rain forests, mountains, and swampy coastal areas. Outside Java there are few railroads. Modern highways connect the large cities, especially in Java. But elsewhere most roads are unpaved. Many areas have no roads at all.

Although highways are being improved and an airline serves the islands, boats are still the chief means of traveling long distances. Boats carry passengers and freight between the islands and along the coasts.

Lack of good transportation is one reason why modern industry does not yet play an important role in Indonesia's economy.

In the villages, small, traditional craft industries produce such items as carved statues and fine metal pitchers and bowls. The best known of all Indonesian handicrafts is the making of a colorful cloth called *batik*. The cloth is dyed with designs and used to make clothes for women.

Most modern industry is concentrated in the large cities of Java. Much of this industry involves processing farm products for export. Other plants are based on Indonesia's mineral resources. These plants manufacture petroleum products, or tires and other rubber goods. Several plants assemble automobiles and trucks, using imported parts. But Indonesia has no steel mills or factories that produce heavy machinery. So far industrial development has had little effect on the lives of most Indonesians. The whole country uses less electricity than Connecticut.

Unlike many other developing nations, Indonesia is moving into the modern world in its villages as well as in its cities. Villagers who once farmed with age-old methods have quickly seen the benefits of new seeds and modern farming techniques. Trees and grass will have to be replanted on Java to halt erosion and increase the fertility of the soil. But Java and other islands can increase their food supply greatly as the use of modern farming technology continues to spread. A brighter future for Indonesia may come from its rice fields rather than the high-rises of its cities.

SECTION REVIEW

1. In areas with poor soils, how do Indonesian farmers improve the soil?

2. Give two reasons why farms in areas such as Java can grow three crops a year.

3. Why would the Indonesian government like farmers to move from Java to other islands?

4. What were Indonesia's leading exports before World War II? What is its leading export today?

YOUR LOCAL GEOGRAPHY

1. Using Geography Skills 43 (page 466) as a guide, draw your own flowchart. Choose a process that plays an important part in your community. For example, if you live in an agricultural area, you might show the steps by which a local crop is planted, cultivated, and harvested. If you live in an industrial area, you might show the steps involved in manufacturing a local product. Or you could illustrate the process by which water is brought from lakes and streams to reservoirs and, finally, to homes and offices. Make use of flowchart symbols like the ones in Geography Skills 43 to call attention to the different kinds of steps.

2. Suppose that some students from Indonesia are visiting your community. What would seem most different or unfamiliar to them about your climate, seasons, hours of daylight, vegetation, and means of transportation? Write part of a letter home in which one of the students describes such differences.

Reading a Flowchart

On the other side of the world, an Indonesian farmer is *tapping* (cutting) the bark of a rubber tree for a milky white liquid, called *latex*. This liquid may end up as crepe soles for shoes, tire treads, rubber cement, or another of the many rubber products used in the U.S. and other parts of the modern world.

The *process*, or series of steps, that turns latex into crepe soles is shown in the **flowchart** below. In a flowchart, activities or steps are described in separate boxes and arranged in the order that they happen. In some flowcharts, such as this one, boxes also come in different shapes to stand for different things.

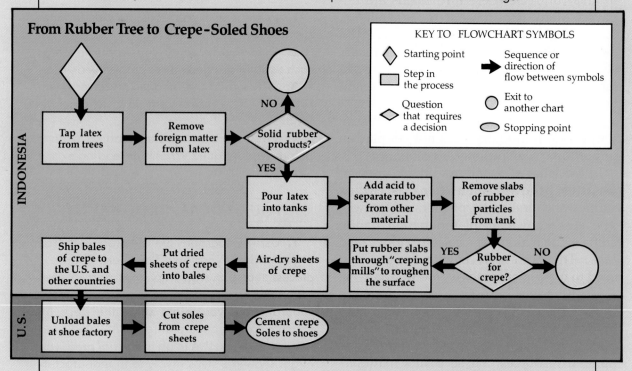

From Rubber Tree to Crepe-Soled Shoes

Study the key. As you read the flowchart, refer to the arrows and follow the steps in order. Answer the following questions on a separate sheet of paper.

1. What is meant by a box shaped: (a) like this ☐ ? (b) like this ◇ ?
2. In the process shown by this flowchart, what happens: (a) at the Starting Point? (b) at the Stopping Point?
3. What happens in: (a) the second step of the process? (b) the third step?
4. Can you tell from this chart what happens to latex which will *not* be made into solid rubber products? Why?
5. What is the *last* step that takes place in Indonesia? What is the *first* step that might take place in the United States?

CHAPTER REVIEW

A. Words To Remember

In your own words, define each of the following terms.

archipelago   flowchart   tremor
Bahasa Indonesia   slash and burn

B. Check Your Reading

1. Why is there little difference between the shortest and longest days of the year in Indonesia?

2. Why do Indonesians in some rural areas build their homes on stilts?

3. What is the major religion in Indonesia? What is the major religion on the island of Bali?

4. What is the most common means of transportation in most of Indonesia? Why?

5. Why does slash-and-burn farming take up more land than irrigated rice farming?

C. Think It Over

1. What advantages and disadvantages does Indonesia have as a result of being an archipelago?

2. How does the distribution of Indonesia's population affect the nation's development?

3. Indonesia is a nation of great diversity. Do you think that diversity is more of a help or a hindrance to Indonesia's economic development? Explain.

D. Things To Do

You have read that puppet shows have an important role in Indonesia's cultural history. Choose some area of Indonesian life that has interested you. Then write a short scene for a puppet show about that area of life. The scene should have two or three characters. Briefly describe the movements and gestures they might make. Write a paragraph of narration to introduce the scene. Then write a short dialogue for your characters.

Chapter 35

A Growing Success?
The Region in Perspective

Mohinder Grewal (moh-HEEN-dur GROO-uhl) works in a V-neck sweater, shirt, and tie. He does not look like a farmer. He does not look like a revolutionary. But he is both.

Mohinder owns a 12-acre (*five-hectare*) farm on India's northern Punjab plain. On the same land, his father grew one crop a year. But Mohinder is part of the green revolution, and grows four or five crops a year. There are rows of vegetables — carrots, onions, and potatoes. And there are fields of high-yield rice and wheat. To keep an eye on his crops, he often works into the night.

The rewards have been great. In addition to his modern, irrigated farm, Mohinder owns an orchard, a poultry farm, a seed farm, and a nursery (for plants). His home would fit into any American suburb.

Mohinder bought all this additional property in 15 years. When he borrowed the first money, the outlook — for him and for India — was not bright. Many experts predicted that India's food supply would never keep up with its growing population. But India — and farmers like Mohinder — are proving the experts wrong.

Deciding How To Develop

As you have seen, India and the rest of South and Southeast Asia have much in common. Except for Pakistan and Afghani-stan, all the countries are affected to some extent by monsoon winds that bring rain. The environment is generally good for growing food. The trouble is that the land must support so many people. Moreover, all countries in the region have high rates of population growth. As a result, they have had great difficulty growing enough food for their increasing numbers of people.

In addition, almost all of the nations in the region were once ruled by Western powers. The Asian countries had to support the economies of the colonial powers, not develop their own. Thus most of the nations had little modern industry while they were colonies.

When these nations became independent after World War II, they wanted to develop industry. Importing manufactured goods is costly. But building new industry is costly too. As nations put money and effort into industry, farm production fell behind the rising need for food.

India was the first country in the region to change its **priorities** (the order of things to be done). India decided to help the nation's farmers first. The reasons were:

■ Farmers make up a large majority of the people. Industry cannot succeed unless farmers have money to buy manufactured products. The only way for farmers to get

---

Nations in the region cannot develop all of their economy at once — they have to make choices.
In Karachi (kuh-RAH-chee), Pakistan, a concrete tower rises behind a camel-drawn cart.

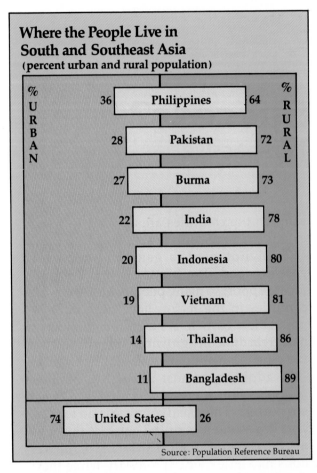

Where the People Live in South and Southeast Asia
(percent urban and rural population)

% URBAN	Country	% RURAL
36	Philippines	64
28	Pakistan	72
27	Burma	73
22	India	78
20	Indonesia	80
19	Vietnam	81
14	Thailand	86
11	Bangladesh	89
74	United States	26

Source: Population Reference Bureau

This graph shows the eight most populous nations in the region. Only one other nation has a higher urban percentage. This is Singapore, which has the same area as Chicago and is all urban. The U.S. has more than twice the urban percentage of the Philippines, which is the region's most urbanized large nation.

more money is to grow extra food. The farmers can sell their extra food in city markets for cash.

■ Extra food is needed to feed workers in industrial cities. Without food, industry and cities cannot grow.

Other countries in the region—especially Indonesia and Pakistan—decided to follow the same course as India. In the preceding chapters, you read about some of the ways in which farming has changed in the region. The next section will focus on those changes and discuss how successful they have been.

SECTION REVIEW

1. Why have the nations in the region had difficulty growing enough food for their people?

2. Why did most countries in the region develop very little industry while they were colonies?

3. Why is Mohinder Grewal described as a revolutionary?

4. What made the government of India decide to develop farming before industry?

Getting More from the Land

The farther west you go in the region, the drier the climate becomes. In areas such as western India and Pakistan, farm production could not increase without a better water supply.

In order to help farmers, India and its neighbors began large programs to provide more water for irrigation during the long dry seasons. Dams were built to store water and to provide electric power. As electricity came to thousands of villages, it became possible to tap large supplies of water that lay underground. Deep wells were dug that brought water to the surface by means of electric pumps.

Irrigation was not the only answer to the need for more food. During the 1960's, scientists in Mexico and the U.S. developed new varieties of wheat, while scientists in Taiwan, Japan, and the Philippines developed new varieties of rice.

These new varieties were tough and better able to survive poor weather conditions. They also produced a larger amount of grain on each plant. Most important, however, they grew more quickly than ordinary wheat and rice, making it possible for farmers to grow at least one extra crop on their fields during the growing season. Thus farmers who previously grew only one crop of wheat each year could now grow two or perhaps three.

This was the beginning of the green revolution. The new "miracle" seeds made it possible for the supply of food to grow at a faster rate than the number of people.

The new seeds required a great deal of fertilizer to get the best results. Chemical fertilizers are expensive, and many farmers couldn't afford them. So the nations of South and Southeast Asia began building chemical fertilizer factories to increase output. The governments of these countries often supplied farmers with fertilizers either free of charge or at low prices.

To grow more food, it was also necessary to educate the farmers. A large majority of them could not read or write. How could they read the instructions for planting the new seeds and using chemical fertilizers? How could they write for help? Thousands of farm villagers helped themselves by building new schools with their own hands. At the same time, teams of farming experts were sent into the villages to teach the new ways. Radio and television broadcasts helped spread the green revolution to villages that had electricity. Even villages without electricity could get information by transistor radios.

Problems and Prospects

Thanks to the green revolution, an Indian village may now produce more than twice as much grain as in the past. There is more

---

The green revolution stemmed from new varieties of grain crops developed by scientists in different countries. Below, in the Philippines, new varieties of rice are being tested for toughness and fast growth.

than enough to feed everyone in the village. This surplus could be sold in a market. From there the surplus might be added to the supplies needed by an expanding city.

But suppose there is no road from the village to the nearest market town. The villagers may try to sell their surplus to other villages nearby. But farmers in these other villages may have their own surplus. Next year the villagers may decide to grow less — just enough for their own needs.

Lack of transportation is a problem throughout the region. Most roads and railroads link the cities, but are scarce in the rural areas where crops need to be moved to market.

Modernization has brought some problems of its own. The need for oil is probably the greatest. Tractors are more efficient than oxen, but they need fuel, not food. The sharp rise in world oil prices during the 1970's was a blow to many nations in the region.

There are wide cultural differences not only between nations in the region but between groups within each nation. Conflicts between groups and nations have often slowed or halted economic improvements. For example, minority groups in India and the Philippines have led armed rebellions against government rule. A brutal Communist takeover in Kampuchea was followed by an invasion from Communist Vietnam.

Yet on the whole, the green revolution has had greater and more widespread success than many experts thought possible. If it continues to spread, the outlook for the region is hopeful. The discovery of oil in the region would make the outlook even brighter. The search for oil goes on in India, Bangladesh, and Sri Lanka.

Today one out of four people in the world live in South and Southeast Asia. If the green revolution continues, success stories like that of Mohinder Grewal will become more and more common.

SECTION REVIEW

1. Why is transportation important in increasing food production?

2. How does modernization of farms increase the demand for oil?

3. Describe two advantages of the new varieties of wheat and rice.

4. Why are "miracle" seeds more costly to use than ordinary seeds? How have governments in the region helped farmers with this extra cost?

YOUR LOCAL GEOGRAPHY

1. The development of new varieties of seeds touched off a green revolution in parts of South and Southeast Asia. What effects have technological advances had on people in your community? Take a survey of a few older family and community members. You might ask such questions as: What invention or discovery has most affected your life, and how? How has transportation changed since you were a child? Has modern technology made your job (as a laborer, parent, homemaker, etc.) easier in the past decade? If so, how? What modern invention would you have the hardest time doing without?

2. You have read that India chose to give priority to developing farming rather than industry. In economic matters, such choices cannot be avoided. Economic resources (including money) are always limited, while needs and wants are unlimited. Thus individuals, businesses, and governments have to choose which of their needs and wants they will meet. Make a list of four needs or wants that apply to your community as a whole. For example, your community might need or want a new road, an improved water supply, a new school building, etc. Then suppose that your community has enough resources to meet just two of those needs or wants. Which two would you choose, and why?

Using Evidence To Make Generalizations

India produces more rice than any other crop. Rice is also a major crop of Burma, Thailand, Bangladesh, and Vietnam. From these facts, you could make this general statement, which is also called a **generalization**: "Rice is a major crop in countries of the South and Southeast Asian mainland."

Can you use geographic facts to make some more generalizations? Below is a physical map of mainland South and Southeast Asia. On page 474 is a map of economic activities in the same region. Compare the two maps and follow the directions below.

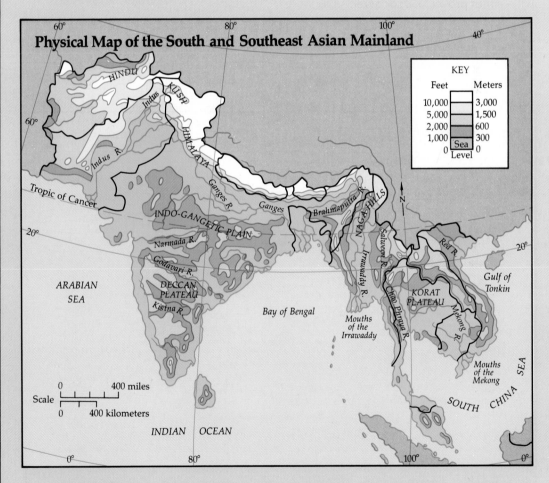

Physical Map of the South and Southeast Asian Mainland

1. On the physical map, find each place in the following list. Then on a separate sheet of paper, tell what kind of economic activity, shown on the other map, is found in the same place.
   (a) Mekong River Delta
   (b) Chao Phraya River Valley
   (c) Irrawaddy River Valley
   (d) Ganges-Brahmaputra River Delta

(Turn page.)

473

2. Which of the following generalizations is true, based on the facts you found in exercise 1?
(a) Rice is a dominant crop on the central plateaus of the region.
(b) Rice is a dominant crop in river valleys and deltas of the region.
(c) Plantation crops are grown in the region's river valleys and deltas.

Any generalization should be based on *evidence* (facts), before you can believe it to be true. Can you believe the statements below? Use evidence from the maps to decide whether each generalization is true. Explain whether each of the statements is true or false, and give at least two facts — from either map — that proves your answer.

3. There are no manufacturing or service industries in the region.
4. Wheat and other grains are grown in the region's higher plateau areas.
5. Rice is the main crop in all of the river valleys and deltas of this region.
6. No economic activity occurs in highland or mountain areas of the region.

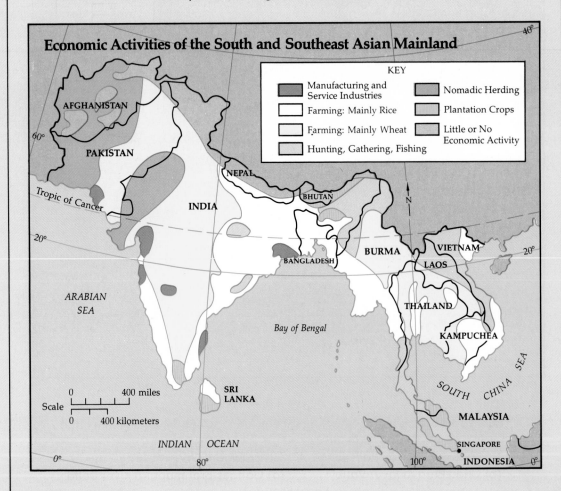

Economic Activities of the South and Southeast Asian Mainland

KEY

Manufacturing and Service Industries
Farming: Mainly Rice
Farming: Mainly Wheat
Hunting, Gathering, Fishing
Nomadic Herding
Plantation Crops
Little or No Economic Activity

AFGHANISTAN
PAKISTAN
Tropic of Cancer
INDIA
NEPAL
BHUTAN
BANGLADESH
BURMA
VIETNAM
LAOS
THAILAND
KAMPUCHEA
ARABIAN SEA
Bay of Bengal
SRI LANKA
SOUTH CHINA SEA
MALAYSIA
SINGAPORE
INDONESIA
INDIAN OCEAN

Scale
0   400 miles
0   400 kilometers

CHAPTER REVIEW

A. Words To Remember

From the following list, select the term that best completes each of the sentences below. On a separate sheet of paper, write your answer next to the number of each sentence.

conflicts   irrigation   surplus
fertilizer   "miracle" seeds   transportation
green revolution   priorities

1. India changed its _____ by deciding to develop farming before industry.
2. The new _____ made it possible for farmers to grow more crops per year.
3. A great deal of _____ is needed to get the best results from the new seeds.
4. Farmers need to grow a(n) _____ of food in order to sell it at the market for money.
5. Dams were built to provide _____ for farms in dry areas.

B. Check Your Reading

1. What country in Southeast Asia was not a colony?
2. Why is irrigation important in India and Pakistan?
3. Why did the green revolution make it important for farmers to be able to read and write?
4. The new varieties of wheat and rice grow quickly. What benefit does this provide?
5. Why is it difficult for nations in the region to develop industry and farming at the same time?

C. Think It Over

1. Modernization has brought both problems and benefits to the region. Which do you think are greater? Explain your answer.
2. How do you think nations in the region should improve transportation?
3. What arguments can you think of for nations to develop industry before farming? Do you think they are stronger than the arguments for developing farming first? Give reasons.

D. Things To Do

Suppose you are an official whose job is to persuade villagers to modernize their farming methods. Make a list of five reasons you would present to the villagers. Remember that the villagers will be concerned with practical questions such as the cost involved.

UNIT REVIEW

A. Check Your Reading

1. On a separate sheet of paper write the letter of each description. After each letter, write the number of the term that best matches that description.

(a) Almost two thirds of Indonesia's population lives on this island.	(1) Philippines
(b) It split into two separate countries after a civil war in 1971.	(2) Pakistan
(c) It grows more tea and sugarcane than any other country.	(3) Bay of Bengal
(d) It is the most volcanic country in the world.	(4) Ganges River
(e) It is vital to Indian farmers and highly sacred to Hindus.	(5) India
	(6) Indonesia
	(7) Java
	(8) Borneo

2. Fill in the blanks in the following paragraph by writing the missing term on your answer sheet.

Most of South and Southeast Asia is affected by seasonal rains brought by __(a)__ winds. However, northwest India, Pakistan, and __(b)__ have no wet season, while there is rain all year in Sri Lanka, Indonesia, and __(c)__ . In the wetter areas of the region, the staple food crop is __(d)__ . Most people on the mainland live alongside or close to __(e)__ , because the soil there is so fertile.

B. Think It Over

1. The following statements may be true about any one or more of the following: India (I), Indonesia (IS), Philippines (P). Write the initial letters of the nation(s) to which you think each statement applies.

(a) It is an archipelago.

(b) It is a major oil producer.

(c) Signs of past European influence still remain.

(d) There are still many mineral resources left to be explored.

(e) It was the first nation to give priority to developing farming.

2. Choose any one of the five statements in exercise 1 above. Explain how that statement applies to the nation or nations you chose.

Further Reading

India: Now and Through Time, by Catherine A. Galbraith and Rama Nehta. Houghton Mifflin, 1980. Surveys both history and present-day changes.

Through Indian Eyes, edited by Donald J. and Jean E. Johnson. (two vols). Center of International Training and Education, 1981. Anthology of writings about India by Indians.

Twentieth-Century Indonesia, by Wilfred T. Neill. Columbia University Press, 1973. On Indonesian environment, development, and present-day life.

9

East Asia and the Pacific

Chapter 36

A Distant Region Grows Nearer

It was a strange and fascinating story that Marco Polo brought back from Asia. He described a land known today as China, of which few Europeans at that time had even heard. Polo was a teenager when he set out from Italy in 1271 A.D. He was in his forties when he finally returned home.

Vast distances and natural barriers separated Europe from China. Polo had to cross mountain passes, deserts, and broad rivers. But it was worth it. The China he found was a vast empire with a civilization that was already ancient. It was also the most advanced civilization in the world, with marvels unknown in Europe — gunpowder, paper money, the magnetic compass. China even had noodles, which — thanks to Polo — inspired Italian spaghetti.

Polo's reports fascinated Europe. But they did not set off a rush of tourists to see China. There was too much hostile land (and too many hostile peoples) in the way. East Asia and Europe remained separate worlds.

Two-and-a-half centuries after Polo set out, three small ships rounded South America and sailed into the Pacific Ocean on a voyage around the world. The ships were commanded by Ferdinand Magellan (muh-GEHL-uhn), a Portuguese in the service of Spain. For four months, the ships sailed westward across what seemed an empty and endless ocean. Food and water ran low. Sailors killed and ate the ship's rats to stay alive. At long last, in the year 1521, they reached land — an island now known as Guam (gwahm).

Magellan did not know it, but Guam is one of hundreds of islands that dot the Pacific. Together these islands are called Oceania (oh-shee-AN-ee-yuh). They include one island — Australia — that is so large it is called a continent. Most of the islands are tiny, some even smaller than a football field; but many are populated.

While East Asia is cut off from much of the world by land barriers, Oceania is cut off by water. Yet the two areas have long had much in common. For example, the peoples who lived in Oceania when Magellan passed through were descended from early peoples of Asia. However, they had been isolated from the mainland for so long that they had become distinct cultural groups.

Times have changed. Today you can visit East Asia and Oceania without the hardships faced by Polo and Magellan. Travel times by air are measured in hours, not months or years. Thus a modern traveler on Polo's route might see U.S. oil experts examining rigs pumping oil from beneath China's deserts. A visitor to Guam might hear the scream of jets taking off from U.S. bases.

In the past, it took grueling journeys by land or sea to reach East Asia and the Pacific from other parts of the world. Today visitors to this Tahiti hotel arrive by jet.

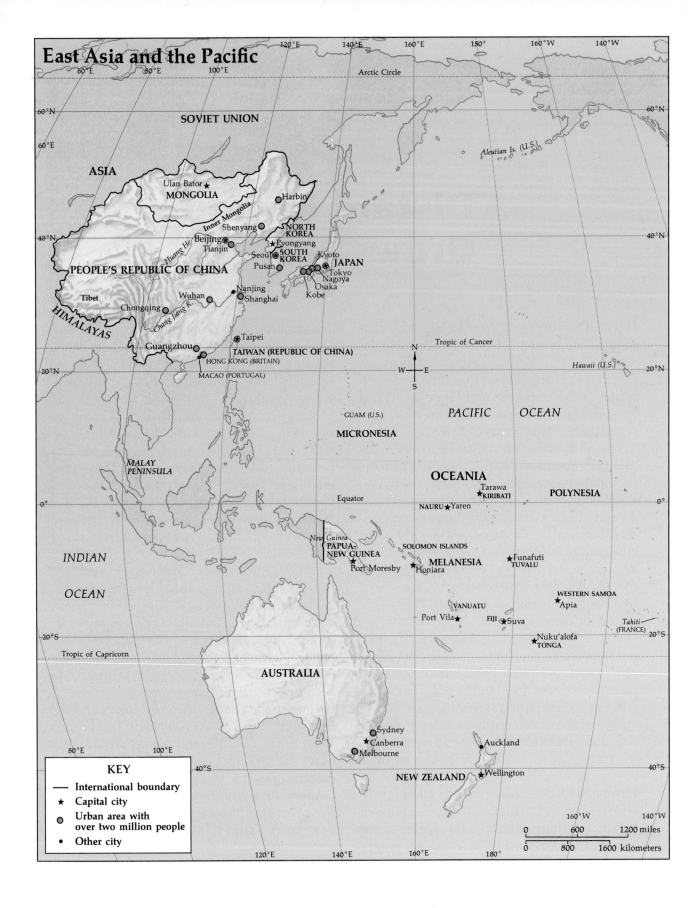

East Asia and the Pacific

SOVIET UNION

ASIA

Aleutian Is. (U.S.)

Ulan Bator ★
MONGOLIA
Harbin

Inner Mongolia
Shenyang
NORTH
Beijing KOREA
Tianjin Pyongyang
Seoul SOUTH
PEOPLE'S REPUBLIC OF CHINA KOREA Kyoto
Pusan Tokyo JAPAN
Nagoya
Tibet Osaka
Nanjing Kobe
Wuhan Shanghai
Chongqing
Chang Jiang R.
Tropic of Cancer
Taipei
Hawaii (U.S.)
Guangzhou TAIWAN (REPUBLIC OF CHINA)
HONG KONG (BRITAIN)
MACAO (PORTUGAL)

Huang He
HIMALAYAS

N
W   E
S

PACIFIC   OCEAN

GUAM (U.S.)

MICRONESIA

MALAY
PENINSULA

OCEANIA

Tarawa
★ KIRIBATI   POLYNESIA

Equator
NAURU ★ Yaren

INDIAN

New Guinea
PAPUA
NEW GUINEA
Port Moresby

SOLOMON ISLANDS

MELANESIA
Honiara

Funafuti
TUVALU

OCEAN

WESTERN SAMOA
Apia

VANUATU
Port Vila ★   FIJI
Suva

Tahiti
(FRANCE)

Tropic of Capricorn

Nuku'alofa
TONGA

AUSTRALIA

Sydney
★ Canberra
Melbourne

Auckland

Wellington
NEW ZEALAND

KEY
— International boundary
★ Capital city
⊙ Urban area with
  over two million people
• Other city

0   600   1200 miles
0   800   1600 kilometers

480

Overview of the Region

East Asia and the Pacific cover more of Earth's surface than any other region discussed in this book. The region includes the world's most populous nation — China. It includes one of the world's industrial leaders — Japan. And it includes tiny island nations like Vanuatu (vah-noo-AH-too).

The Pacific Ocean is the region's most prominent natural feature. Broader and deeper than any other ocean, the Pacific takes up one third of Earth's entire surface. All the world's landmasses together do not equal the Pacific in area.

However, the bulk of the people in the region live on the Asian continent, not in Oceania. Look at the Checklist of Nations on page 587. You will see that most of the region's people are in China, where the population is more than one billion persons. Other populous nations of the region are Japan (more than 120 million), South Korea (more than 40 million), and Taiwan (TIE-wahn) and North Korea (about 20 million each). Although Australia fills a continent of its own, it has only about 16 million people.

While Japan and Taiwan are island nations and Japan is physically on the Pacific rim, they are considered part of East Asia. Most of their peoples and their cultures are closely related to those of the mainland, especially of China. Because China developed a major civilization at an early date, the influence of Chinese culture runs all through East Asia.

The peoples of the Pacific are much more of a mixture. Europeans are a majority in some nations, such as Australia and New Zealand. Peoples of the Pacific include Polynesians, Melanesians, and Micronesians

(see page 486). Many Asians have also settled in the Pacific islands.

Nearly every kind of government can be found in the region. Some nations, like Australia and Japan, are democracies. Some, like China and Mongolia (mon-GOHL-yuh), are under Communist rule. South Korea is under non-Communist military rule. Still other places, like the French island of Tahiti (tah-HEE-tee) and the U.S. island of Guam, are ruled by outside powers.

Physical Geography

On a tropical island in the Pacific, warm ocean breezes blow all year long. People need few clothes, and babies run around naked. But in Mongolia, north of China, fur parkas and warm leggings may be needed for many months of bitter cold each winter. The region has a broad range of climates.

The key population centers of East Asia and the Pacific lie in middle latitudes, as does the United States. Parts of Japan are as far north as Maine; parts of China, as far north as Hudson Bay. But other parts of China are farther south than any portion of the U.S. mainland. For example, the southern city of Guangzhou (gwahng-zhoh) lies almost on a line with Havana, Cuba.

Australia and New Zealand are mostly in middle latitudes, south of the Equator. Parts of Australia lie in the tropics (as close to the Equator as Nicaragua). Other parts lie in middle latitudes (as far from the Equator as Boston). New Zealand's latitudes would stretch from Maine to North Carolina in the Northern Hemisphere.

As you know, the nearer a place is to the Equator, the warmer and more even its climate is likely to be. But warmth also depends on nearness to ocean waters. An island nation like Japan has milder winters and cooler summers than inland areas of China at the same latitudes. A third factor affecting warmth is elevation. On Tibet's high plateaus, which are no farther from the

---

The majority of people in the region live close to the Pacific Ocean. Nations in East Asia (such as China and Japan) have large numbers of people. The other nations, even huge Australia, are thinly populated.

Equator than Los Angeles, winter temperatures resemble those of Minneapolis.

Many of the inland areas of the region are too dry for growing crops. Like the U.S., East Asia faces an ocean to the east. But unlike the U.S., East Asia does not have an ocean to the west. In these latitudes, the prevailing winds are westerlies (see the diagram on page 42). The prevailing winds from the west have lost most of their moisture when they reach China. There are vast deserts in western China — an extension of the deserts that lie to the west in Soviet Central Asia.

Deserts also cover much of the inland area of Australia. As you read in Unit 1 (see pages 42–43), air rising at the Equator loses its moisture and falls back to Earth around 30°N and 30°S. Australia is on the latitudes around 30°S. The Australian deserts are the same distance south of the Equator as the Sahara in Africa is north of the Equator.

Moisture from Monsoons

The coastal areas of the region receive much of their moisture from monsoons. These are the same kind of shifting winds that bring wet and dry seasons to South and Southeast Asia (see Unit 8). However, the monsoons of this region bring rain not from the Indian Ocean but from the Pacific.

As the summer sun heats the air, causing it to rise, cool, moist winds blow inland from the Pacific. The result is heavy summer rains. In winter the ocean is warm, and dry winds blow outward from the continent toward the sea. The result is dry, cold winters on the continent.

South of the Equator in Australia, the seasons are the reverse of those north of the Equator, but the same causes bring summer monsoons to coastal areas. Japan, which is small and surrounded by water, gets a more even distribution of rain year round.

Not to be confused with monsoons are the violent late-summer storms known as **typhoons** (in the northwest Pacific) or **cyclones** (in the southwest Pacific). Like Atlantic hurricanes, these storms may cause great destruction and loss of life. Typhoons often threaten areas near Taiwan and Japan around July. Cyclones threaten eastern Australia and nearby islands around January.

Landforms — Large and Small

When the people of Chongqing (choong-CHEENG) in central China want to cross the Chang Jiang (chang jyahna) River, they must trudge down hundreds of stone steps to reach the riverbank. The city of two million sprawls over steep hills, through which the river has cut a channel. Chongqing is not unusual. Much of East Asia—including Japan, North Korea, and Taiwan—is hilly or mountainous.

The Himalayas, which divide East Asia from South Asia, contain the world's tallest mountains. These are simply the greatest of many great mountain chains around the edges of East Asia. On China's southwestern edge, much of Tibet lies on high plateaus that average 15,000 or 16,000 feet (5,000 meters) above sea level. For comparison, the highest peak in the U.S. Rockies is 14,431 feet (4,399 meters).

East Asia has no interior lowland as big as the U.S. Great Plains. Even along the East Asian coasts, lowlands are limited. The major coastal plain stretches north in China, from Shanghai to Beijing (bay-JING).

By contrast, most of Australia is fairly flat — either plain or plateau. The main mountain chain is about as low as the Appalachians in the U.S.

Two very different types of islands are found in the Pacific. One type you have met before — the submerged (sunken) mountain, with just the peaks poking out of the water. This type is found around the edges of the Pacific. These mountainous islands include Japan, New Guinea, the Solomon Islands, New Zealand, and also the Aleutians

The World's Volcanoes

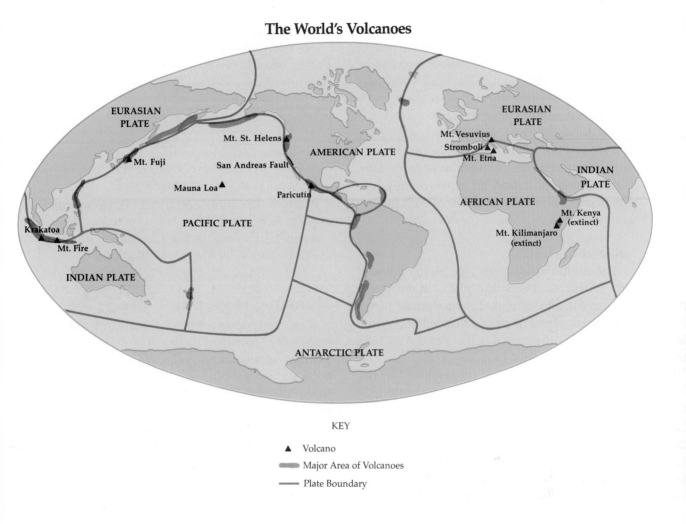

KEY

▲ Volcano

▬ Major Area of Volcanoes

— Plate Boundary

(uh-LOO-shuhnz), which are part of Alaska.

Because many of these islands contain active volcanoes, they are said to form part of a Pacific "ring of fire." Earthquakes are a constant threat in much of the region. No other part of Earth's surface is so unstable. Scientists believe the reason is that this is a zone where segments of Earth's crust are shifting against one another inch by inch. The shifting edges form weak spots where the molten rock below can burst through.

The second kind of island is called a **coral island**. While coral islands also rest on undersea land, their structure is quite different from volcanic islands.

Earth's surface is divided into areas called plates, which move slowly against each other (see page 57). Volcanoes and earthquakes are common where plates meet. The boundaries are most active around the Pacific Ocean, forming what is called a "ring of fire."

Much or all of the land above water has been formed from coral (the shell-like homes of tiny sea animals). These animals live only in warm tropical waters. The animals do not move about but attach themselves to rocks beneath shallow waters, which are kept warm by the sun. Over many years, the coral builds layer upon layer to form **reefs** (ridges close to or above the surface of the water).

Coral reefs usually circle an existing island. Wind and waves batter the surface of the reefs, and bits of coral break away to be ground into fine sand. In time there are sandy beaches that can support trees, such as coconut palms.

Coral islands vary in size, but many are quite small. Sometimes coral islands form a circle around a shallow area of ocean water. Such a circle is called an **atoll** (AT-ol), and the body of water inside it is called a **lagoon** (luh-GOON).

Mineral Resources

The region as a whole has plenty of mineral resources, but some places have fewer resources than others. For instance, Japan lacks most of the minerals needed for industry. On the other hand, China and Australia are rich in minerals.

The most widespread mineral fuel in the region is coal, which is found in abundance in China, Australia, North Korea, and a few other places. Oil once seemed scarce throughout the region. Since the 1960's, however, oil deposits have been discovered in China, Australia, and elsewhere. There also is oil off China's coasts, where active exploration has recently begun.

Major iron deposits occur in China and Australia, and the region has a wide variety of other minerals. For example, copper is found in South Korea and Australia. Bauxite is found in Australia. Other minerals range from lead and zinc to silver and tin.

Rivers that flow out of the mountains of East Asia, Australia, and New Zealand can be used for hydroelectric power. These rivers also provide water for irrigation in many of the drier parts.

Like Japan, the islands of the Pacific tend to be poor in minerals. One exception is Papua-New Guinea (PAP-yoo-uh new GIN-ee), which has gold and copper. (Papua-New Guinea and Indonesia's Irian Jaya share the large island of New Guinea, north of Australia.) Papua-New Guinea also has some hydroelectric power. Most of the other Pacific islands are too small to have strong-flowing rivers.

SECTION REVIEW

1. Name two Chinese inventions that were unknown in Europe at the time of Marco Polo.

2. What is the "ring of fire"?

3. How do China and Japan compare for mineral resources?

4. In what way is the climate of inland China similar to that of inland Australia?

Human Geography

It is rush hour. You are standing on a busy street corner in Beijing, China's capital. Trucks, buses, and bicycles are whizzing by. Bicycles are so popular, in fact, that more than 100 of them zip past your corner in a minute.

Before you know it, you are swept up by the crowds that swarm out of office buildings and shops. There are always crowds in East Asia. Except in deserts and on lonely mountain peaks, masses of people seem to be everywhere. This is understandable, since one out of four people in the world live in East Asia — most of them in China.

Cities are especially crowded. The metropolitan area of Tokyo, Japan's capital, is one of the biggest in the world. Shanghai, China, contains more than 10 million people. Beijing has over eight million.

Yet the bulk of East Asia's people live in rural areas. For example, four out of every five Chinese live in the countryside. However, rural East Asians do not live on isolated farms. Many rural villages have far more people than a U.S. urban community of equal area. And these villages lie close together.

By contrast, Oceania tends to be lightly populated. On many islands, people num-

China has had a strong cultural influence on much of the East Asian mainland. For example, both Korea and Japan borrowed Chinese writing. Note the signs on these buildings in Seoul (sohl), South Korea.

ber in the dozens or hundreds rather than millions. Even Australia, with its population of 15 million, has great empty spaces. Most Australians live in cities, leaving the countryside sparsely settled.

The Influence of China

East Asia is the center of one of the world's oldest and most dynamic cultures — that of the Chinese. The Chinese had a written language before 1000 B.C. Chinese philosophers such as Confucius (kuhn-FEW-shuhs), who lived around 500 B.C., were known far and wide for their teachings. For centuries China led the world in technology — as Marco Polo was to discover. It is said that in 1800 A.D., when the U.S. was an infant republic, more books were published in China than in all other countries combined.

Many aspects of Chinese culture spread to other parts of East Asia. For example, the Koreans and the Japanese took up the Chinese system of writing.

This system is sometimes called "picture-writing." It has no alphabet but instead uses symbols (or characters) that stand for whole words. For example, the character 山 stands for the Chinese word *shan*, meaning "mountain." Chinese school children must learn more than 2,000 such characters, and advanced Chinese dictionaries use far more.

You should note that the Koreans and the Japanese borrowed Chinese writing — not the Chinese language. The Korean and Japanese languages are quite different from Chinese. (Korean and Japanese are related to one another, however.)

Chinese influence gave East Asia a cultural unity far beyond any to be found in South and Southeast Asia. Even nations like Mongolia, which did not adopt the Chinese style of writing, borrowed ideas and customs from the Chinese. The only parts of this region not affected by the ancient Chinese were the Pacific islands. The peoples of these islands were quite isolated until the arrival of Europeans.

At one time or another, much of East Asia — including Korea, Mongolia, Taiwan, and the small territories of Macao (muh-KOW) and Hong Kong — were part of the Chinese empire. That empire began about 1500 B.C. and ended in 1912 A.D., when the last emperor was overthrown. Today the political situation in the area is as follows:

- Korea is split into two independent nations. North Korea has a Communist government; South Korea, a non-Communist government.
- Mongolia is a Communist nation wedged between China and the Soviet Union. But Mongols also live across the border in the Soviet Union and in a section of China known as Inner Mongolia.
- Macao, a tiny port on China's south coast, is ruled by Portugal.
- Hong Kong, a somewhat larger port just east of Macao, is a British colony. The bulk of the colony is held on a lease from China that expires in 1997.
- The status of Taiwan, a sizable island off China's southeast coast, is in dispute. In 1949 a Chinese civil war ended with Communists controlling the mainland and anti-Communists fleeing to what was then the Chinese province of Taiwan. While the mainland became the People's Republic of China, Taiwan kept the name of Republic of China — in effect, becoming a separate nation. But neither side has given up hope of reuniting Taiwan and China.

Ethnic Groups

Many of the people of East Asia have broadly similar backgrounds. Most Chinese, Mongolians, Koreans, and Japanese belong to a broad ethnic group known as Asian. Ethnic minorities do exist, however. They include Caucasians (or Europeans) and the Ainu (IE-noo), a light-skinned people who live in northern Japan.

The peoples of the Pacific are much more varied. Many are European. Many others are people with brown or black skins. Oceania can be divided into three geographic areas, each with its own distinct ethnic group:

- Melanesia includes New Guinea and other islands in western Oceania. The Melanesians have dark skins. They are related to the **aborigines,** the original inhabitants of Australia. Melanesians are believed to be the people with the oldest roots in Oceania, dating back as far as 30,000 years.
- Polynesia lies farther east, in the central and south Pacific. The Polynesians have brown skins. They appear to have migrated from southern Asia, perhaps Malaya, about 15,000 years ago. Many of the people in Hawaii are Polynesians.
- Micronesia, north of Melanesia, takes its name from the tiny (although not "microscopic") coral islands sprinkled through the area. The Micronesians have yellowish skins. They appear to have arrived from Asia sometime after the Polynesians.

The peoples of East Asia and the Pacific follow a variety of religions. Traditional religions include Buddhism (see page 432) in China, Shintoism (see page 512) in Japan, and nature worship in Oceania. Some peoples of China and Mongolia are Moslems. However, Communist governments in the region now discourage religion. Christianity is strong in Australia and New Zealand, and has won many converts in East Asia, especially in South Korea.

Ways of Living

If you traveled from nation to nation through the region, you would find many different ways of living. Some differences are based on physical geography; some, on culture.

In drier areas, which are unsuitable for growing crops, many peoples are nomadic. For example, Mongolians have traditionally been herders of sheep, goats, and camels. When their livestock eats up the scanty grass in one place, these herders move on to another. The typical Mongolian home is a *yurt* (yoort), a type of tent with a heavy covering of felt and animal hides. Even in Mongolia's capital of Ulan Bator (oo-LAHN bah-TOR), a city of 350,000, many of the people live in yurts.

In Papua-New Guinea, home is likely to be a thatched hut in a jungle village. People here keep pigs instead of camels or sheep.

Modern ways are more common in cities than in rural areas. Cities such as Auckland, New Zealand, and Seoul (sohl), South Korea, look much like modern cities anywhere. In Ulan Bator, alongside the yurts, one sees prefabricated apartment buildings.

Some parts of the region are just starting to enter the modern world. In Papua-New Guinea, for example, people in each jungle village have traditionally lived on their local resources. Mountains, forests, and swamps cut the people off from other villages. As a result, villagers often speak a language all their own. More than 700 languages are spoken in New Guinea — one sixth of the world's total languages.

In many isolated villages, people live as their ancestors did. They grow their own food — mainly pigs and a few crops. They make clearings by using stone adzes (sharp tools) to chop down trees. They dig holes with sticks and plant crops such as sweet potatoes, taro (TAH-roh), and cucumbers. (**Taro** is a plant with an edible root that is a staple food on many Pacific islands.) When the soil wears out, the villagers move to a new site. They have little need for cash, and use pigs as a form of money.

Traditional ways of life remain in many of the islands of the Pacific. Below, in Fiji (FEE-jee), boys are collecting straw to thatch the roof of a village school.

At the same time, many villages in Papua-New Guinea now have modern links with Port Moresby, the capital city. Instead of building railroads or extensive highways, the nation jumped into the air age. It has a network of routes served by small planes.

Koina (KOY-nuh) lives in one of the villages that has contacts with the modern world. His village is on the coast. His home is a traditional one, made of the leaves and stalks of the sago (SAY-goh) palm. But his tools include power-driven machinery that is owned jointly by many villagers. Koina raises soybeans and sells them for cash.

As a boy, Koina enjoyed hearing his village's folktales and learning traditional dances. Once the villagers put on a skit about a man caught stealing from another's garden. Koina played the part of a taro root. Now, about once a year, Koina can see such skits performed by a professional drama troupe that tours around the country in a minibus.

The audience gathers in a half circle around the bus. Admission is free, but people are asked to donate something — a sweet potato, a crab, a bunch of bananas. Besides folktales, the troupe puts on its own versions of classics such as *Hamlet*.

The drama troupe has its home in Port Moresby, which is an outpost of the modern world. Port Moresby has offices, factories, schools, and a university. There you can pay your hotel bill with a credit card.

SECTION REVIEW

1. How do the rural areas of Oceania differ from those of East Asia in density of population?

2. Describe one example of China's cultural influence on other parts of East Asia.

3. Give one reason why so many different languages are spoken in Papua-New Guinea.

4. Taiwan used to be a province of China. What is Taiwan's status today?

Economic Geography

Li Yuk Chun (LEE yook CHOON), a young woman in Hong Kong, works on an assembly line. Her job is to fasten the bases on 35-millimeter cameras. One day these cameras may be sold in the United States. Cameras are among many products now made in East Asia and the Pacific. Such products have become a key part of world trade.

The countries of the region are at many different stages in building up industry. Japan is an industrial giant, producing more than any nation in Western Europe. Australia has a wide range of modern industries, from steel to textiles. Hong Kong, South Korea, and Taiwan specialize in light industries. Besides cameras, they make goods such as electronics, textiles, and jogging shoes. Communist nations such as China and North Korea first stressed heavy industry. Recently they have placed more emphasis on consumer goods. Few of the smaller islands of the Pacific have much industry at all.

Farming

You read earlier about the steep hills on which the Chinese city of Chongqing is built. Such hills also cover the surrounding countryside. If you visited the area, you would see what appear to be giant steps cut into the hillsides. These are terraces built by farmers to make the land more level for growing rice. Since rice is grown under water, its fields must be flat.

The farmers of East Asia grow crops wherever they can. The good farmlands have been used for thousands of years and are densely settled. Thus farmers today use the hillsides as well as the plains and valleys.

Changes have come more rapidly to East Asia than to South and Southeast Asia. Some changes are in technology. In Japan, for instance, tractors are used more and

East Asia is densely populated. The most crowded nation is Taiwan, where these TV sets (top) are being made. Pacific nations have much more space. In New Zealand (bottom), sheep far outnumber people.

more. Other changes are in social policy. The greatest policy changes have been in Communist nations like China and North Korea, where collective and state farms have been set up, similar to those in the Soviet Union and Eastern Europe.

In most nations of East Asia, at least some areas have a growing season that is long enough to permit **double cropping** (planting two crops on the same plot in one year). Rice is by far the most important crop. Rice lends itself to double cropping. Its seedlings can be started in a small seedbed and set out in the main field when the land is free. Double cropping has helped give East Asia some of the highest yields per acre in the world. Without such high yields, it would be hard to feed the large populations in nations like China.

Farmers can produce more food from a given area of land by growing crops than by raising livestock. Thus in crowded East Asia, nearly all usable farmland is planted with crops. Only in areas that are too dry for crop growing is there much stock raising. There are nomadic herders in Mongolia, as you read earlier, and also in northwestern China.

By contrast, Australia and New Zealand have small and scattered populations. Thus there is enough land to spare for livestock. In these two countries, large herds of sheep and cattle are raised on farms and ranches similar to those in the United States and Canada.

Before Europeans came, the peoples of the Pacific had a self-sufficient economy based on such local products as coconuts and fish. Coconuts served as a source of "milk," "meat," and oil. The empty shells could be used as cooking vessels. Leaves from the coconut palm went to make shelters, clothes, and woven containers.

Today coconuts are still the chief agricultural product of many Pacific islands. *Copra* (the dried "meat" of coconuts) is exported to the U.S. and other nations. Oil pressed from copra is often used in margarine, cosmetics, and soaps.

Fishing is a major industry in most of the region. The Japanese fleet is especially large, and accounts for more than one fourth of the total world catch of fish.

Australia and China are the region's only countries with large resources of minerals for industry. This Australian plant processes bauxite (aluminum ore).

The Importance of Transport

Today it's easy to catch a plane to Tokyo or Hong Kong. There are frequent flights from the U.S. and Europe. But some parts of East Asia and the Pacific are much harder to reach. Certain islands in the South Pacific get mail service by ship only once a year.

Transportation is vital to any region's economy. Great ocean distances, mountains, and deserts hinder travel and trade throughout the region. Over the ages, travelers have overcome these barriers.

In Marco Polo's time, camel caravans carried silks and spices across China's western deserts to the Middle East. Chinese navigators sailed to India and beyond. Arab ships visited the China coast.

Today fleets of modern ships carry the region's trade. Japan leads the world in shipbuilding. Its ships bring in raw materials vital to Japanese industry and carry away finished products.

Overland travel has also changed. Most countries of the region have built networks of roads and railroads. But some of these networks are still inadequate. For example, China has seven times as many miles of highways today as it had 30 years ago, and more roads are being built. Yet China's total highway mileage is still less than that of Japan, a much smaller country.

Regular airline routes crisscross the Pacific. Most flights carry passengers *across*, rather than *to*, Oceania. But large numbers of tourists visit islands such as Tahiti. Other islands are trying to build up tourism too.

What's Ahead

This unit will now take you to three nations big in area or population or both.

Chapter 37 focuses on China, the biggest nation of the region — both in area and in population. Chapter 38 turns to Japan, the region's most modern nation. Chapter 39 looks at Australia, a large country with a small population.

These three nations have developed, or are developing, in different ways. Chapter 40 compares those ways and shows what they may mean for the rest of the region and for the world.

SECTION REVIEW

1. Why are terraces cut into many hillsides in East Asia?

2. How does industry in Hong Kong and Taiwan differ from industry in Japan and Australia?

3. What kind of farming is most common in the drier parts of the region? Give one specific example.

4. What is copra? In what part of the region is copra a leading product?

YOUR LOCAL GEOGRAPHY

1. Geography Skills 45 (page 492) compares pie graphs showing how the labor force is distributed in two countries. Prepare and compare your own pie graphs showing how the labor force is distributed in (a) the United States as a whole and (b) your state. Use an almanac to find the information. The breakdown of different types of employment will be simpler than in the skills lesson, and not all categories may be listed. Any percentage left over should be marked "Other." When you have prepared the two pie graphs, compare the U.S. and your state for each category. In what ways is your state typical of the U.S. in its pattern of employment? In what ways is it different?

2. From an atlas, find the latitude of your community. Then using the map on page 480, find a place on or closest to the same latitude (north) in East Asia and the Pacific. Write down the name of the place and the territory in which it is located. Then find another place on the same latitude *south* of the Equator. Look at the world climate map on page 550. Do the two places have the same kind of climate as your community? If not, can you suggest why?

Comparing Pie Graphs

You learned in Geography Skills 13 (page 90) that a pie graph divides the whole amount of something into parts or percentages. Each pie graph below represents the whole work force of one country. Slices of each pie graph show what shares or percentages of workers in that country are employed at certain kinds of jobs. By comparing the two pie graphs, you can pick out differences in the employment profiles of China and Japan.

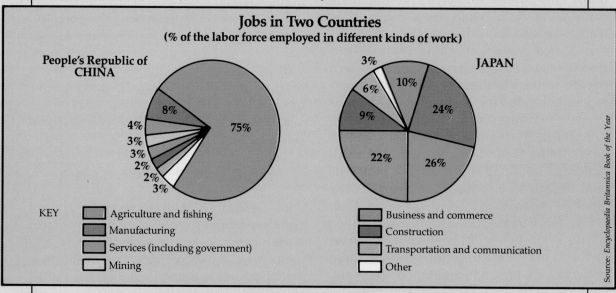

Jobs in Two Countries
(% of the labor force employed in different kinds of work)

People's Republic of CHINA

JAPAN

KEY

Agriculture and fishing
Manufacturing
Services (including government)
Mining

Business and commerce
Construction
Transportation and communication
Other

Source: *Encyclopaedia Britannica Book of the Year*

Look at the key and at each pie graph. Then answer the following questions.

1. According to the key, what color would represent your job category if you were: (a) a farmer? (b) a government postal worker?
2. From these pie graphs, can you find out: (a) how many people are in each work force? (b) how many construction workers are in each work force? (c) the percentages of workers in construction?
3. In China the vast majority of workers earn a living in what area of employment? All of the other categories of employment together make up what percentage of the total Chinese work force?
4. In Japan what three employment categories *each* account for over 20 percent of the work force? What percent of Japan's work force is in agriculture?

Use both graphs together to make some comparisons and answer the following questions.

5. In which country is business and commerce a fairly large area of employment?
6. Which of the two countries has a more industrial society?

CHAPTER REVIEW

A. Words To Remember

Three of the following terms are defined by the numbered phrases below. On a separate sheet of paper, write each term next to the number of its definition. Then write the other two terms and give a definition for each.

aborigine   double cropping   typhoon
atoll   taro

1. A circle of coral islands that enclose a lagoon.
2. A plant with edible roots.
3. A violent storm in the northwest Pacific Ocean.

B. Check Your Reading

1. Give one reason why there are vast deserts in Western China.
2. How does the land surface of Australia differ from that of the East Asian mainland?
3. East Asia and the Pacific have a wide range of climates. Where in the region would you find: (a) a tropical climate? (b) bitterly cold winters?
4. What are the two most populous nations in the region? Are they in East Asia or in Oceania?
5. Why is rice such an important crop in most of East Asia?

C. Think It Over

1. What is picture-writing? How is it different from writing in English?
2. Describe two ways in which the environment of the Pacific islands has influenced their culture or economy.
3. The text describes Port Moresby as "an outpost of the modern world." What does this mean? Do you agree? Explain.

D. Things To Do

Turn to the world time zone map on page 573. Note how the International Date Line runs through the middle of the Pacific Ocean. At this line, by international agreement, each calendar day begins. Any time you cross from west to east, you move into the previous day. Any time you cross from east to west, you move into the following day. The time of day does not change, however. If it is 3 P.M. Monday when you reach the line from the east, it will be 3 P.M. Tuesday after you cross. If you cross back one hour later, it will be 4 P.M. Monday.

Suppose you are flying from Auckland, New Zealand, to Tahiti, with a stopover at Western Samoa (check the map on page 480 for these places). Your plane leaves at 11 A.M. tomorrow. Each leg of the flight takes two hours, and the stopover is one hour. Auckland is west of the International Date Line, while both Western Samoa and Tahiti are east of the line. Western Samoa is one hour later than Auckland, and Tahiti is two hours later. Work out the timetable of your flight, with times and dates at each stopping place.

Chapter 37

China
An Ancient Land Enters the Modern World

If you have ever eaten in a Chinese restaurant, you may have definite ideas about the Chinese diet. For instance, you may think it consists of rice, small bits of pork, chunks of vegetables, and more rice.

Well, there's some truth in this. But your ideas of Chinese food are incomplete — and misleading.

Until recently most of the Chinese who moved to the United States came from the Guangzhou area (also known as Canton) in south China. So the main style of cooking in the United States is called Cantonese.

But China is a giant country and has many different climates. In southern China, including Guangzhou, people eat lots of rice because it grows well in the warm, moist climate. In northern China, where the climate is drier and cooler, wheat grows better than rice. So the people of northern China eat many wheat products — such as noodles.

Some food differences are related to cultural differences. Take meat, for example. Pork is popular in most of China, since many families find it easy to keep pigpens in their backyards. But in northwest China, many people are Moslems, whose religion forbids eating pork. So lamb plays a larger role in the diet of this area.

The point to remember is that China is vast and contains a great variety of peoples. Naturally diet and other cultural factors differ from place to place.

However, with all China's variety, it has much more cultural unity than a nation like India. For much of India's history, it was divided into many separate states. By contrast the most densely populated parts of China have been united under a single government for most of the past 3,000 years. The civilization developed by China smoothed out many differences among its far-flung peoples.

Physical Geography

The city of Guangzhou lies in what is known as China Proper. This part of China forms the nation's core and holds most of its people. The Moslems of the northwest live outside China Proper. More than half of China's present territory is made up of outlying areas such as Tibet, Xinjiang (sheen-JAHNG), Inner Mongolia, and Manchuria (man-CHOO-ree-uh).

The main distinction between the two parts of China is cultural. China Proper is the homeland of the Chinese people themselves. Outlying areas contain a variety of minority groups, as you will see in the next section.

China's Great Wall was built to defend its northern border some 2,200 years ago. This ancient nation, now under Communist rule, is struggling to build modern industry.

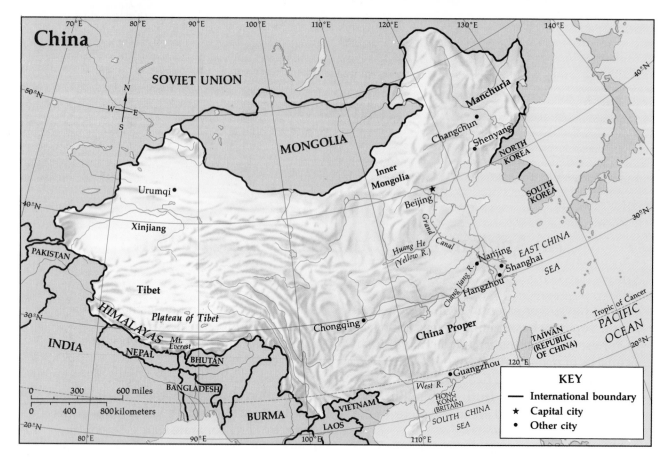

China

The distinction is also based on physical geography. China Proper contains most of China's good farmland — where land and climate create the best growing conditions.

In its size and environment, China is more like the United States than any other country in the world. South China has a climate similar to that of the U.S. South, while North China is similar to the U.S. Northeast. West China is like the U.S. West without the Pacific coast.

China Proper

If you wanted to see flat plains and green valleys in China, the best place to look would be China Proper. China's chief coastal lowland is there (see map above). So are China's three great river valleys—the Yellow, Chang Jiang, and the West. These rivers flow eastward to the sea.

But much of the land in China Proper is not flat. Large areas of the interior are hilly or mountainous. In the southeast, opposite the island of Taiwan, even the coast is mountainous.

In area and latitude, China is similar to the United States. Also like the U.S., China has a densely populated east coast. However, China does not have a west coast with another center of population. Most of the near-billion Chinese live in the crowded east.

The climate in China Proper varies greatly from south to north. The south is both wetter and warmer. Conditions often resemble those along the U.S. Gulf Coast. The growing season is long enough for two rice crops in one year. Annual rainfall runs from 40 to 80 inches (*100 to 200 centimeters*) — mainly from summer monsoons.

North of the Chang Jiang, yearly rainfall drops off. Most of northern China Proper gets from 20 to 40 inches (*50 to 100 centimeters*). Some areas get even less. The growing season there is shorter; the winters are colder. North of the Chang Jiang, it is difficult to grow rice. There, wheat is the main crop. In drier areas, barley and millet are grown.

The rivers of China Proper have long been both a blessing and a curse to the people living along them. The blessing is the silt they deposit in their valleys, providing rich soil for farming. The curse is frequent flooding. In some 3,000 years of recorded history, the Yellow River, or Huang He (hoo-AHNG HUH) in Chinese, has flooded at least 1,500 times. Because this flooding has killed so many people, the Yellow River has the nickname of "China's Sorrow."

Drought is also a problem in the Yellow River Valley. The upper part of the valley is almost too dry for farming. In the years when rain is meager, crops wither and die. The valley's fertile yellow soil is also quite powdery, so it is easily blown away in dust storms. Much of this soil returns to the river. In places the river water resembles a thick yellow soup.

As the silt settles to the riverbed, it raises the water level. Over the centuries, the Chinese built dikes along the river to control its floods. As the riverbed rose, the Chinese built the dikes higher. Today the riverbed lies above nearby farmland. Constant efforts are needed to keep the river from breaking out of the dikes.

Outlying Areas

If you travel outside China Proper, you will find mountains, deserts, and — in some places — fertile valleys. In most areas the climate is dry or cold or both.

To the west are the Tibetan highlands. Here a high windswept plateau borders the Himalayas. These mountains include the highest point on Earth, Mount Everest, which stands on Tibet's southern border. Bitter cold and drought make much of this area barren, except for scattered tufts of grass where yaks (long-haired Asian cattle) can graze. In a few protected valleys, cool-weather crops like barley can grow.

North of the Tibetan highlands is a vast basin surrounded by mountains. This area is called Xinjiang. It is very dry, since it receives less than 15 inches (40 centimeters) of rain a year. Deep wells and runoff

from mountain snows provide water for crops such as cotton and rice. Sheep, goats, and camels graze on mountain slopes.

East of Xinjiang and north of China Proper is a dry highland plateau known as Inner Mongolia. Skimpy steppe grasses support flocks of camels, cattle, sheep, and goats. Hot, dry summers and bitterly cold winters make farming difficult.

Northeast of China Proper is an area known as Manchuria. However, the Chinese themselves refer to this area simply as the Northeast. Part of the Northeast is a rolling plain that resembles the northern Great Plains of the United States. Soils are quite fertile, and crops include corn, soybeans, and millet.

Natural Resources

In China, wrote Marco Polo, "there is found a sort of black stone, which they dig out of the mountains, where it runs in veins. When lighted, it burns like charcoal, and retains the fire much better than wood." Of course, this black stone was coal. Europeans did not use it in Polo's time.

China's coal reserves are huge. The main deposits lie in the north — in China Proper and in Manchuria. These areas are also rich in iron. Thus China has the resources for a major steel industry.

China's oil reserves seem to be much greater than was thought just a few years ago. Oil fields have been found in Manchuria, in the deserts of Xinjiang, and in China Proper. More oil seems to lie under the seas off China's shores.

A great variety of other minerals is found mainly on the fringes of China Proper. Some of these minerals are metals such as tungsten and manganese, which are used in high-grade steels. Others include gold, silver, copper, tin, uranium, and bauxite.

The waters of China are another key resource. China has many rivers that can produce hydroelectric power. The Chang Jiang and West rivers are deep and navigable over great distances. Many of China's rivers and the seas near its coasts are rich in fish.

SECTION REVIEW

1. How does the southern part of China Proper differ in climate from the northern part?

2. Name two areas of China outside China Proper. Where are they?

3. Why are China's rivers considered both a blessing and a curse?

4. Is China rich in fuel resources? Explain.

Human Geography

In the city of Urumqi (oo-ROOM-chee) in Xinjiang is a bright red billboard bearing the latest slogan of China's Communist party. The slogan is not written in Chinese but in Arabic, the language of many of Xinjiang's people. For them Chinese is a second language.

About 60 million of China's one billion inhabitants belong to ethnic minority groups. Many are ethnically similar to people who live across China's borders. For instance, many in Xinjiang are Moslems who resemble people in nearby Soviet Asia. Elsewhere in China are people of Korean, Mongol, and Thai ancestries. Outlying areas such as Tibet and Manchuria have their own distinct peoples (Tibetans and Manchus). Combined, the ethnic minorities of China would form a nation more populous than most countries of the world.

But minorities make up only six percent of China's people. The rest are Chinese. Actually the name the Chinese use for themselves is "people of Han" (hahn). The terms *China* and *Chinese* are Western ones, taken from the name of an early Chinese dynasty, the Qin (chin). A **dynasty** (DIE-nuhs-tee) is a series of rulers who belong to one family.

The Han people live both in China Proper and in outlying parts of China. Millions more, known as "overseas Chinese," live in nearby countries of Asia. They all speak Chinese — but use different dialects. Thus a Han from Guangzhou might have trouble understanding a Han from Chongqing. But all Han people use the same written language, and thus can read the same newspapers, books, and billboards. (This is much the same as the way you can read numbers. If you see the number *20* in a Spanish book, you can read it as "twenty," while a Spanish speaker reads it as "veinte." It means the same to both of you.)

An Ancient Culture

China is one of the oldest areas of human settlement in the world. Scientists have found traces of humans who lived near modern Beijing (Peking) about half a million years ago. Other people later moved into the area and became ancestors of today's Han Chinese. By about 3000 B.C., they were farming along the Yellow River.

All along the river, there was a need to control floods and provide water for irrigation. Only large-scale organization could do these tasks. Partly as a result of this need, a strong Chinese government was formed before 1500 B.C. Other features of civilization developed. Chinese writing appeared, and a unique Chinese culture evolved. Today China has the oldest continuing civilization on Earth.

Chinese achievements over the centuries have been many. Paper, gunpowder, the water-powered mill, the wheelbarrow, and movable printing type were all Chinese inventions. The Chinese also set new trends in literature and art. You have read already how Chinese writing spread to nearby Korea and Japan.

Traditional Chinese culture placed great stress on respect for one's elders and for earlier generations. Many families had shrines in their homes where they honored their dead ancestors.

Learning was highly valued. Government officials had to pass tough examinations in the works of such sages (wise ones) as Confucius. For centuries collections of Confucius' sayings were the basic texts for pupils in Chinese schools.

For more than 3,000 years, the Chinese empire survived. At times it was taken over by outside conquerors, like the Mongols who ruled China when Marco Polo was there. Through all such periods, however, the Chinese kept their form of government and their culture. To the Chinese, all outsiders were "barbarians." China remained *Zhongguo* (joong-GWOH) — "the Central Kingdom."

By the 19th century, China was no longer a leader in technology. The rest of the world was changing rapidly, while China kept to its old ways. "Barbarians" from Europe now had better weapons. Europeans forced China to open its ports to outside goods. Soon outside powers demanded and got more — including special trading rights and special treatment for foreigners within China. China's government grew weaker.

A revolution in 1911–1912 overthrew the last emperor. Later came civil war and invasion by Japan (1937–1945). In 1949 a Communist victory put an end to civil war. China became a Communist state.

In many ways, modern China reflects its ancient heritage. But Communist rule has brought extensive changes. The following section presents two families — one in a village and one in a city — so that you can see what sort of lives they lead.

In a Farming Village

The Liu (lew) family lives in a farming village near Beijing. They have a small brick home with a tile roof. Their home is grouped with about 20 similar homes, on a rocky hill with fields all around. The village was built on the rockiest ground so as not to waste good farmland.

The Liu family has two rooms for four people. But there is an extra room, if you

Three fourths of China's people live on farms, which are controlled by the government. This large farm is near Hangzhou (hahng-JOH) in the coastal plain.

There are no private cars in China, but people do buy bicycles. Here, cyclists wait for a green light during the morning rush hour in Beijing, China's capital.

count the patio on one side, shaded by a grape arbor. In summer the family eats its meals around a low table under the arbor.

Almost all meals include noodles or rice, served in individual bowls, with a big platter of stir-fried vegetables. These vegetables may include Chinese cabbage, snow peas, and carrots. Liu Yingli, who is the mother, usually prepares the meal. (In China the family name is put first, followed by the given name.)

Mrs. Liu chops the vegetables into small chunks. This feature of Chinese cooking developed for a practical reason. In densely settled parts of China, wood or coal was always scarce, so cooking techniques had to save fuel. Small chunks of food cook quickly. Mrs. Liu stirs the vegetables in a small amount of oil in a round-bottomed pan called a *wok*. The round bottom allows heat to spread rapidly to all parts of the wok.

Since the vegetables are in bite-sized pieces, there are no knives at the table. The Lius eat with chopsticks (two sticks held between the fingers of one hand). They use chopsticks as handily as you use a fork.

Several times a week, Mrs. Liu cooks small bits of pork or chicken along with the vegetables, but meat is not a major part of the Chinese diet. The Liu family has its own small plot where it raises vegetables, a few chickens, and a pig. However, the main fields are managed and worked by the villagers as a group.

Electricity has come to the Liu family's village only in recent years. Now the family is working to get the "three rounds and a sound" — that is, a watch, a bicycle, a sewing machine, and a radio. These are the goods that most Chinese today would like to own. So far the Lius have only the "sound."

Life in Shanghai

Zhang Yanlan (ZHAHNG yahn-LAHN) lives in Shanghai, China's largest city. He owns one of the "three rounds" — a bicycle. Each morning he pedals his bike from his apartment building to the factory where he works. During the rush hour in Shanghai, hundreds of thousands of bikes clog the streets—just as in Beijing. There are no private cars in China. The only cars in use are those assigned by the government to holders of top jobs.

Yanlan lives with his parents and a sister in a three-room apartment. Soon the apartment will be more crowded, for Yanlan is about to get married. His bride Yuzhen (yoo-ZHEHN) will move in with his family.

Yanlan saves a good deal each month from his wages of 75 *yuan* (yoo-AHN), about 42 dollars. Both his parents work (as is common in China). Rent is only about five dollars a month, food bills are low, and medical care is free. A night out is cheap too — there are free movies at the factory.

But money is not all that is needed to buy things in China. Some things, such as cotton clothing, are **rationed** (made available only in restricted amounts). If Yanlan wants a Chinese-made polyester shirt, he can buy it readily. But if he wants a cotton shirt, he must save special ration coupons that are passed out each month. When he has enough coupons, he can take his money and his coupons to a store and get the shirt.

You have already read that cotton grows in China. In fact China is the world's third-largest producer of cotton, after the U.S. and the Soviet Union. But China's one billion people use more cotton than China produces. So China must import cotton from the U.S. and other nations. In order to cut down on these imports, which are expensive, China has rationed cotton goods.

SECTION REVIEW

1. Name three early Chinese inventions.

2. Why would you rarely see cars in China?

3. China is the world's third-largest producer of cotton. Why does it import cotton?

4. Describe one thing that Liu Yingli and Zhang Yanlan have in common in their lives.

Economic Geography

You read earlier that the Liu family lives in a village where the homes are built on rocky ground, to save the best soil for farming. Like many other Chinese farmers, the Lius also keep a pig, because a pig needs very little room, can be fed on scraps, and produces manure for the fields as well as meat.

Throughout history the Chinese have had to make the greatest possible use of available resources. Today China's farmers feed 24 percent of the world's people on just seven percent of the world's cropland.

Although China has been building industry, it remains a nation of farmers. Three fourths of all workers are farmers.

Modern machines are coming into use on Chinese farms, but many areas still have few tractors. Even in more advanced areas, much work is done by hand — or foot. For example, in southern China's rice fields, women stand in bamboo frames, pumping pedals with their feet. These machines look a little like American exercise machines, but they are used to pump water out of irrigation ditches into fields.

An electric pump could do the work much faster. Yet in Chinese terms, the pedaling makes sense. First, few electric pumps are available so far. Second, human energy is one resource that China has in abundance, so why not use it to the fullest? The U.S. gets higher crop yields *per worker*. However, China often gets higher yields *per acre*.

In some ways, the Chinese still follow their age-old ways of farming. Yet farm life has changed drastically since the Communists came to power. To see how farming is organized under communism, let's look again at the Liu family's village near Peking.

Farming Today

The 20 families in the Liu village make up a *production team*. Their members work together in the fields around the village. Land,

501

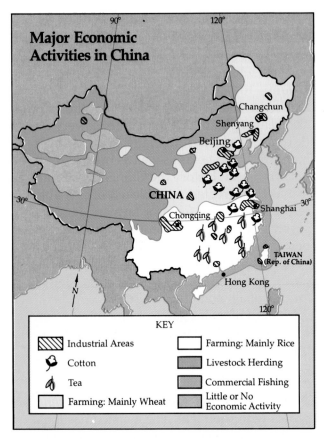

KEY

Industrial Areas	Farming: Mainly Rice
Cotton	Livestock Herding
Tea	Commercial Fishing
Farming: Mainly Wheat	Little or No Economic Activity

Changchun

Shenyang

Beijing

CHINA

Chongqing

Shanghai

TAIWAN (Rep. of China)

Hong Kong

Mountains and deserts make the western half of China unsuitable for most farming. In the eastern half, rice is the staple food crop in the south; wheat, in the north. Most mineral resources and industrial centers are also found in this northeastern section.

tools, and work animals are owned by the team as a whole. Any profits are divided at the end of the year.

The Liu family's village is grouped with other villages in a *production brigade*. Several such brigades make up a *commune*. The commune to which the Liu family belongs has some 20,000 members.

Under the Communist system, each commune must meet quotas (set amounts) decided by government officials. As in the Soviet Union, members of the Communist party play key roles in running the farm.

In the past, family plots were so small that it was difficult to rotate the crops. With **crop rotation,** the same piece of land is planted

with different crops from year to year. Different crops need different nutrients, so that the land has a chance to become fertile again. Today plots have been merged to make bigger fields, and crops are rotated to help raise crop yields.

Communes vary greatly from one place to another. Some have set up industries, such as shirt factories and craft workshops. These boost family incomes, since earnings from industry exceed those from farming.

In some areas, such as Xinjiang, the Chinese have set up *state farms* on the Soviet model (see page 286). Such farms tend to be on new lands around the fringes of China's traditional farm areas. Many cover large expanses and depend heavily on modern machinery. On a state farm, workers get a wage rather than a share of the profits.

Building Up Industry

If you were a Chinese farmer, you might run a giant Red Flag tractor, or a medium-sized Iron Ox tractor, or more likely the two-wheeled Worker-Peasant hand tractor. These are Chinese tractors. They are manufactured in Chinese factories, using steel made from Chinese iron and coal. The Chinese take great pride in this. Up to the 1950's, China produced no motor vehicles of any sort — no cars, no trucks, no tractors.

China's new leaders have made great efforts to build up industry. They have concentrated on heavy industries such as iron and steel. These industries are building blocks for other industries (such as tractor making).

China's heavy industries are concentrated in two types of areas. Some are in large port cities like Shanghai, where barges and ships can deliver raw materials. Others are in areas like Manchuria, where raw materials are mined.

Shanghai is China's leading industrial city. Besides iron and steel works, Shanghai has hundreds of light industries. Products range from cotton and silk cloth to cigarettes, paper, and cement.

Much of China's mining takes place in mineral-rich Manchuria. The leading products of the mines are iron and coal, which are used in factories at cities such as Changchun (chahng-CHOON) and Shenyang (shehn-YAHNG). Changchun is sometimes called China's "Detroit"; Shenyang, its "Pittsburgh."

For many years, China's industries produced few consumer goods — sewing machines, bicycles, and the like. The Chinese government held down consumption, in order to invest about 30 percent of China's output each year in building up industry. (By comparison, U.S. investment is roughly 18 to 20 percent of output.)

However, opposition grew among the Chinese people to the consumer shortages. As a result, the government has eased this policy in recent years. Chinese factories now turn out goods such as television sets. In large cities like Shanghai, as many as half of all homes have a TV set. But in rural areas, TV sets are still rare.

As in the Soviet Union, major industries are owned and operated by the government. A few private enterprises do exist, however. A tailor shop might be privately owned, for example, or a small restaurant.

Building Up Transportation

China has great distances and rugged terrain, which make transportation a real problem. In ancient times, a major road network linked different parts of the empire. The network's main purpose was to carry taxes (in the form of goods such as grain) to the empire's capital. Marco Polo was impressed by the roads of 13th-century China.

However, China's transportation network

For years China's government pushed the development of heavy industry. Now more consumer goods are being made—as in this radio factory in Nanjing.

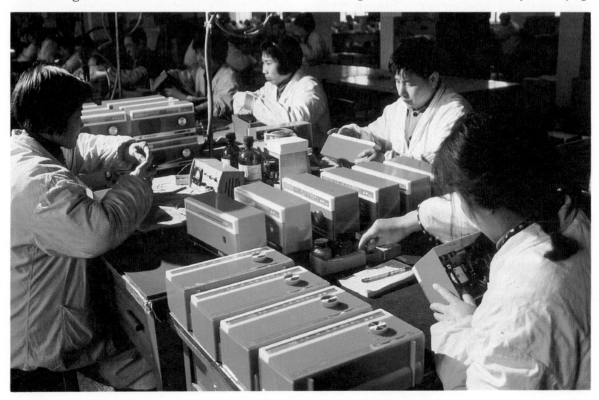

failed to keep up with modern times. Railway building, started in the 19th century by foreign companies, was not planned to meet China's needs. The aim was to carry goods to ports to be shipped overseas.

In recent years, railroads have been built as part of a broader plan to spread industry around the country. Most freight is now carried by railroads, and most travel between cities is by rail. A system of highways is being extended slowly.

Water transport is also important. China contains the world's longest canal — the Grand Canal — which runs for some 1,000 miles (*1,600 kilometers*) from Hangzhou to Beijing. The canal dates back to the 13th century. It fell into disuse for many centuries but has been rebuilt since 1958. Barges use it mainly to carry coal to the south and grain to the north.

Trade: From Silks to Oil

To early Europeans, China was a source of fine silks and rare spices. Chinese trade played a major role in world commerce. In more recent times, China became cut off from the world — especially after 1949. The U.S. and its allies broke off trade with China's Communist government, and China seemed to lose interest in Western trade.

In the 1970's, relations between China and the U.S. became closer again. Now Chinese silks, Oriental rugs, and canned lichee (LEE-chee; a fruit) are found in U.S. stores. And U.S. soft drinks are found in China.

Through trade China wants to make more products available to its consumers. Imports now include Japanese television sets and U.S. wheat, corn, and cotton. A second Chinese goal is to build up industry and agriculture. Thus China has bought massive steel mills, oil-field equipment, and fertilizers. Like the U.S., China has a vast range of resources. If those resources were fully developed, China could meet most of its own basic needs.

Japan and China are close neighbors, but they are far apart in size and development. China, the larger nation, has lagged behind in development. It is just entering the modern world. Although Japan lacks China's vast resources, it has plunged headlong into the modern world. Its industries are among the world's leaders. You will take a closer look at Japan in the next chapter.

SECTION REVIEW

1. Why do many Chinese farmers keep pigs?

2. The world's longest canal dates back to the 13th century. What is it called? How is it used today?

3. How has the production of consumer goods changed in China in recent years?

4. In what kinds of areas are China's heavy industries located? Why?

YOUR LOCAL GEOGRAPHY

1. You have read that China's rivers have been called both a blessing and a curse. What feature of the environment in or near your community has both advantages and disadvantages for the people who live there? (The effect may be smaller in scale than the flooding of China's rivers.) Explain your choice.

2. Because China Proper has such a great population density, the Chinese have found various ways to make use of all available space. Consider how space is used in your community. Are there any ways in which space is used as economically as possible? (In urban areas, for example, you might note high-rise office buildings, underground parking garages, etc.) Are there any ways in which space is used lavishly? (For example, you might note extensive shopping malls.) Draw up two lists, one of economical use of space and the other of lavish use. Then write a brief summary of the use of space in your community. Describe any improvements you can suggest.

Interpreting Pie Symbols on a Map

Does China produce all the energy it uses? Does it use a lot more energy than a smaller but more industrialized nation such as Japan? What kinds of energy does it use?

You can find the answers to such questions in the two maps below, on energy production and energy consumption in East Asia and the Pacific. Both maps use pie symbols that look like pie graphs. Like a pie graph, each symbol represents the whole amount — or 100 percent — of something, and shows how that whole amount is divided into shares or parts. For example, each symbol on the left map represents the whole amount of energy produced by one country in a year. What does each symbol stand for on the right map?

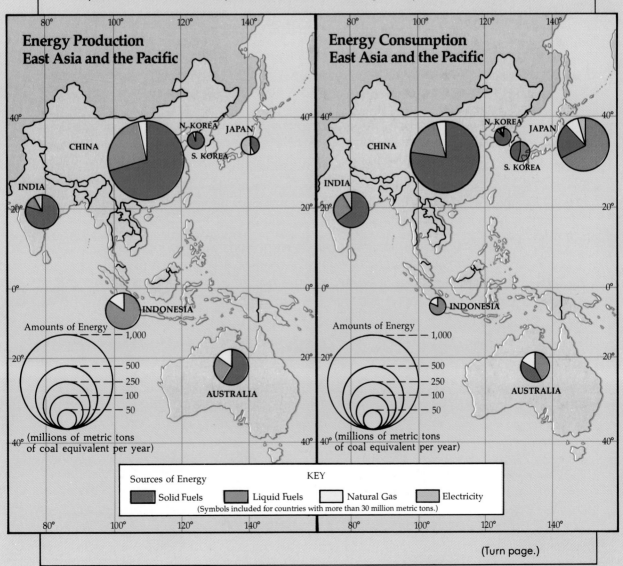

Energy Production
East Asia and the Pacific

CHINA

N. KOREA

JAPAN

S. KOREA

INDIA

INDONESIA

Amounts of Energy
— 1,000
— 500
— 250
— 100
— 50
(millions of metric tons of coal equivalent per year)

AUSTRALIA

Energy Consumption
East Asia and the Pacific

CHINA

N. KOREA

JAPAN

S. KOREA

INDIA

INDONESIA

Amounts of Energy
— 1,000
— 500
— 250
— 100
— 50
(millions of metric tons of coal equivalent per year)

AUSTRALIA

KEY
Sources of Energy

■ Solid Fuels   ■ Liquid Fuels   □ Natural Gas   ■ Electricity
(Symbols included for countries with more than 30 million metric tons.)

(Turn page.)

Unlike most pie graphs, however, pie symbols on maps usually come in different sizes to represent different total amounts. A special key with each map on page 505 tells you the quantities that pies of different sizes represent. Study this key now, and the color key too. Write the best choice for each of the following statements on a separate sheet of paper.

1. Energy amounts are given in millions of:
   (a) barrels.   (b) tons.   (c) metric tons.
2. Energy from electricity is represented by the color:
   (a) red.   (b) green.   (c) yellow.
3. Energy from coal would be represented by the color:
   (a) red.   (b) green.   (c) purple.

Use both keys to study and compare symbols on the energy *production* map. Then tell whether each of the following statements is true or false.

4. China produces between 500 and 1,000 million metric tons of energy per year.
5. More than three quarters of the energy produced by China comes from liquid fuels.
6. Liquid fuels make up a larger percentage of total energy production in China than in any other area shown.
7. Japan produces about the same total amount of energy in a year as North Korea.

Compare symbols on the energy production map with those on the energy consumption map. Answer the following questions.

8. Which countries consume much more energy than they produce?
9. Which country produces more than twice as much energy as it consumes?
10. Which countries seem to consume about the same amount of energy as they produce?

CHAPTER REVIEW

A. Words To Remember

From the following list, select the term that best completes each sentence below. Write your answers on a separate sheet of paper.

crop rotation	pie symbol	rationed
drought	production team	state farm
navigable	quota	steppe

1. _____ is a problem in the upper Yellow River Valley, which is almost too dry for farming.

2. Cotton clothing is _____, because China uses more cotton than it produces.

3. _____ helps to keep farmlands fertile and raise crop yields.

4. The Yangtze and West rivers are deep and _____ over great distances.

5. A _____ is a set amount.

B. Check Your Reading

1. What makes the Yellow River yellow?

2. Why do many families living in rural areas build their houses on rocky ground?

3. Name three consumer goods that Chinese people are especially eager to own.

4. Give one reason why Chinese foods are planned so that they can be cooked quickly.

5. Briefly describe the size, location, and industry of Shanghai.

C. Think It Over

1. "Human energy is one resource that China has in abundance." What does this mean? How does it affect China's economy?

2. What are the major differences between China Proper and the outlying areas?

3. In what ways does China have cultural unity?

D. Things To Do

You have read that Americans may get a misleading idea of the Chinese diet from Chinese restaurants in the U.S. In fact, it's easy to get a distorted picture of another nation from any limited view of it. Test this for yourself. Imagine you are a Chinese student who knows nothing about life in the United States. Watch any television program of your choice. On the basis of this program alone, write a brief description of the physical geography, human geography, and economic geography of the United States. Now become an American again and read your description. In what ways is the description misleading?

Chapter 38

Japan
A Blend of Two Worlds

When Kimie Shibata (KEEM-yeh shee-BAH-tuh) works at a department store in Tokyo, Japan's capital, she wears a Western-style skirt and blouse. At home helping her mother serve dinner, she wears a traditional Japanese *kimono* (kee-MOH-nuh). A kimono is a loose robe with wide sleeves, tied around the waist with a sash.

Kimie sometimes eats lunch at a fast-food restaurant that serves American-style fried chicken. But for supper she always has rice, perhaps with a Japanese delicacy called *sushi* (SOO-shee). Sushi is raw fish.

The Shibatas own a color television set and a refrigerator. But in Japanese tradition, they sleep on floor mats that are stored in closets during the day. On a wall nearby is a shrine dedicated to a god of nature.

On weekends Kimie may go dancing at a disco with her fiancé Ichiro (EE-chuh-roh). The two first met in a traditional Japanese manner, on a visit arranged by their parents.

Kimie's life seems to be divided into two parts — the modern and the traditional. In this she is like many other Japanese young people. Old ways are still strong in Japan, but they exist side-by-side with new ways.

Japan is trying to fit together two different worlds. Since World War II, Japan has become one of the top industrial countries, selling its goods all over the world. Japanese autos zip along the streets of Chicago and London. Japanese video recorders sit beside TV sets in Miami and Paris.

Yet while factories and tall apartment buildings spring up in Japan's crowded cities, other parts of the country look much the same as they did long ago. Tiny rice fields dot the land. Ancient temples draw crowds of worshipers and tourists. In many families, women are still expected to remain "in their place" — serving men and raising children — even though half of all Japanese women hold jobs.

With its mixture of old and new, Japan is a real force in today's world. Of all the nations with non-European cultures, Japan is the only one that has industrialized on a large scale. Its economy ranks third in the world, behind only the United States and the Soviet Union. Yet Japan is a tiny place compared with those nations. And Japan has few of the natural resources needed for modern industry. How has it come so far, so fast?

Physical Geography

In the 19th century, another "tiny place" developed into a great industrial power. That place was Britain (Chapter 16). In many ways, the geography of Japan resembles that of Britain. Here are some of the similarities:

Japan is the only country with a non-Western culture that has become a leading industrial nation. The worlds of the *kimono* and the T-shirt meet in a Tokyo department store.

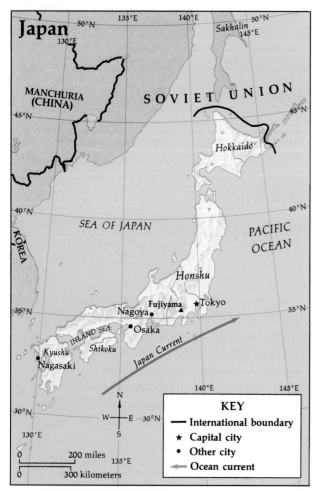

Japan's climate ranges from cool in the north to mild or warm in the south. Mountains cover most of the land in all of the major islands. Thus most of the population is crowded along the coasts. The biggest cities run from northern Kyushu (kee-OO-shoo) along the southern coast of Honshu (HON-shoo) to Tokyo.

■ Both are island nations that lie close to a major continent.

■ Both are archipelagos (groups of islands). Japan has four large islands and hundreds of smaller ones.

■ Both benefit from warm ocean currents that flow past their shores. Just as the North Atlantic Drift warms Britain, the Japan Current flows north to warm Japan.

Japan and Britain have many differences as well. For example:

■ Japan is bigger by half, but has fewer mineral resources.

■ Japan is much more rugged. Mountains take up 85 percent of its land, leaving little room for farms and cities.

■ Japan lies in lower latitudes and has a more varied climate. As you can see from the map here, parts of Japan are as close to the Equator as northern Florida. The main population centers are near the latitudes of North Carolina and Virginia.

Land, Water, and Climate

You are never far from the sea in Japan. No spot is more than 100 miles (*160 kilometers* from the coast. You are also never far from mountains. Japan's mountains give the country a striking landscape that has inspired painters and poets.

The most spectacular mountain is the tallest, Fujiyama (foo-jee-YAH-muh), or Mount Fuji, which rises to 12,388 feet (*3,776 meters*). One look at Fuji's cone-shaped peak tells you it is a volcano, as are many other Japanese mountains. Fuji last erupted in 1707. Today you could climb its snowy slopes and look down into the quiet crater.

Japan's islands are part of the Pacific "ring of fire," and many volcanoes are still active. Earthquakes are frequent. A big quake in 1923 killed 143,000 people in and around Tokyo. To save lives in future earthquakes, Japanese building codes now set strict standards for tall buildings.

Because Japan has so many mountains, large areas of land are unfit for farming or settlement. Most people are crowded into a few lowland areas along the coasts or a short way inland. The chief lowland areas are on Honshu (HON-shoo), Japan's largest island.

Between Honshu and two other main islands, Kyushu (kee-OO-shoo) and Shikoku (shee-KOH-koo), lies the Inland Sea. Some people call this sea the "Japanese Mediterranean." The Inland Sea is dotted with hundreds of small islands, and early Japanese civilization grew up on its shores.

Japan's fourth major island, the one farthest north, is Hokkaido (hoh-KIE-doh). Only a narrow strait separates Hokkaido from the island of Sakhalin (SAH-kuh-leen) to the north. Sakhalin, once part of Japan, now belongs to the Soviet Union.

Japan has a fine climate for farming. Rainfall is plentiful, and temperatures are mild. In southern areas, farmers can grow two crops of rice each year. Even in northern Hokkaido, the growing season can be five months long — about the same as in much of Iowa and Nebraska.

Warm sea currents from the south add to the mildness of the climate in southern Japan. Cold sea currents from the north cause fog and chill in Hokkaido.

In the south, summers are warm and sunny — good growing weather for rice, wheat, mulberries, and vegetables like peas and beans. In some places, conditions are right for such subtropical crops as oranges and tea. Winters in central and southern Japan tend to be mild. January temperatures near Tokyo average around 40°F (4°C) —

about the same as in Little Rock, Arkansas.

In the north, summers are mild and moist. Grains such as wheat, barley, oats, and — in some places — rice can grow there. Northern winters are cold and snowy. Winter storms tend to blow in from the northwest. People who live along the northwestern coast of Honshu often place large rocks on the roofs of their houses to keep the roofs from being blown off.

Natural Resources

Japan has some of the minerals needed for industry — but only in small amounts. Japan had enough resources at first to start building industry. Now the nation must import most of the raw materials it uses.

Coal is found in a number of places, but it lies deep in the ground and thus is costly to mine. There is little iron ore. Oil is present only in small amounts. Streams tumbling down from Japan's mountains make it possible to turn out hydroelectric power. But Japan uses far more electricity than it can supply by such means.

One other resource the mountains provide is timber. Forests grow over about 60 percent of Japan. Even so, Japan must import timber to fill out its needs.

The oceans and seas around Japan supply an important food resource. Fish are plentiful where the warm and cold ocean currents meet. The mixture of water temperatures enables tiny ocean creatures to thrive, and many kinds of fish feed on such creatures. The Japanese eat far more fish than meat.

Only 15 percent of Japan's land is flat enough for farming and settlement. On the island of Hokkaido (hoh-KIE-doh), snowy mountains loom over plowed fields.

SECTION REVIEW

1. What natural dangers are a threat to Japan?

2. Is Japan rich in mineral resources? Explain.

3. In what way is Japan suited for farming? In what way is it unsuited for farming?

4. How do sea currents affect Japan?

511

Human Geography

Japan is one of the most densely settled lands on Earth. It has squeezed 120 million people into an area smaller than California (which has about 24 million people). For each square mile of Japanese land, there are more than 800 people (*for each square kilometer, 300 people*). If the United States were as crowded as Japan, it would have a population of three billion!

But Japan is even more tightly packed than such numbers show. You have read that 85 percent of Japan's land is taken up by mountains. Almost all the people live on the remaining 15 percent. Counting only this usable land, Japan's population density becomes 5,500 per square mile (*2,100 per square kilometer*).

Crowding is a fact of life in Japan. Homes and apartments are tiny. Farms are barely as big as the yards in some U.S. suburbs. In some places, streets may be only the width of one Japanese car.

The Japanese have learned to make the best of their crowded conditions. Inside their homes, they save space by doing without much furniture. People sit on the floor, which is covered with mats called *tatami* (tuh-TAH-mee). As you have read, bedding is placed on the floor at night, allowing the same room to be used for daytime activities and sleeping. Instead of walls between rooms, most houses have paper-covered screens. These screens may be removed to create a larger room for special events.

Japan has a homogeneous (hoh-muh-JEE-nee-uhs) population. (**Homogeneous** means all basically alike — in this case, belonging to one ethnic group.) The main minority groups are a few hundred thousand Koreans and some 200,000 people known as Ainu.

No one knows when the first people settled in Japan, or what they were like. The Ainu seem to have been living there when the ancestors of today's ethnic Japanese showed up around the third century B.C. These new settlers came from the Asian mainland.

By 400 A.D., Japan was united under one government, headed by an emperor. For centuries people worshiped the emperor as a god. They believed he was descended from a Sun goddess. This goddess was the central figure of a religion known as Shinto (SHEEN-toh), a form of nature worship. Shinto became Japan's official religion.

The Sun is a key symbol in Japan. The Japanese call their land *Nippon* (nee-PON), "source of the Sun." In English this is sometimes worded "land of the rising Sun." The Japanese flag has a red Sun on a white background.

In the late 19th and early 20th centuries, the Japanese raised their flag over other lands. Japanese armed forces took control of Korea, Taiwan, and Manchuria. A key purpose was to get resources that Japan lacked.

But Japan overreached itself. In 1941 it attacked the United States at Pearl Harbor, in Hawaii. When World War II ended four years later, Japan had lost its empire. U.S. troops occupied Japan.

The U.S. insisted on changes in Japanese life. The emperor became a figurehead, with little power. Shinto lost its place as a state religion. Today Japan, like Britain, is a democratic monarchy. Elected leaders run the government.

A Family in Tokyo

At the start of this chapter, you met Kimie Shibata of Tokyo. (Although Japanese names are "translated" with the family name last, the Japanese place it first. Kimie calls herself Shibata Kimie.) Her father Hisao (hee-SOW) works in a factory. Her mother Kei (kay) stays home to keep house. Kimie, who is 18, has no brothers or sisters.

The Shibatas live in a six-story apartment building. From the outside, it looks like a

Because Japan is so densely populated, most homes are quite small and contain very little furniture. This family has a TV and stereo, but no chairs or sofa. It is a Japanese tradition to sit on mats on the floor.

modern building anywhere. Inside it is distinctly Japanese. As you enter the Shibatas' apartment, you must remove your shoes, to protect the *tatami* mats that cover the floors from wall to wall.

If it is winter, you will find the Shibatas' rooms chilly. As in most Japanese homes, there is no central heating. A small oil heater, which can be moved from room to room, is the main source of heat. The Shibatas wear padded clothing to keep warm.

At meal time, the Shibatas sit on the floor around a low table. They stick their feet under a blanket that hangs to the floor. On the underside of the table is an electric heating unit. The Shibatas often sit around the table after supper to keep their legs warm as they watch television or talk to visitors.

The Shibatas try not to waste heat because electricity and fuel are costly in Japan. Electricity costs five times and oil two times as much as in the United States. The main reason is that Japan must depend on outside sources for most of its energy. Japan uses

less than half as much energy per person as does the United States.

Kimie works as a store hostess. Her job is to greet shoppers with a polite bow and help them to find what they want. Bowing and other outward shows of respect are important in Japanese society.

Kimie does not try to "stand out" on her job, but rather to fit in with the dozens of other hostesses at the store. Teamwork is all-important in Japan. Kimie does not expect to get a promotion until all the hostesses her age are promoted together.

Kimie finds many things to do in Tokyo. Sometimes she and her fiancé Ichiro go to a play, perhaps a *Kabuki* (kuh-BOO-kee) drama. This is a lively show with music, dancing, and elaborate costumes. It also follows an old Japanese tradition of having male actors play all the parts. Sometimes Kimie goes to a game of *beisuboru,* which is the Japanese word for baseball. Baseball has been one of Japan's most popular sports since Christian missionaries introduced the game almost a century ago. There are two professional baseball leagues.

The Shibatas are among the three out of four Japanese people who live in towns and cities. Tokyo is part of a big metropolitan area that has some 22 million people — close to the total population of California. Several other Japanese cities have more than a million inhabitants.

A Farm Family

Yoichi (yoh-EE-chee) and Matsuko Kawakami (maht-SOO-koh kah-wuḥ-KAH-mee) own a small farm north of Tokyo. They raise rice and cherries, plus a variety of garden vegetables. They have two sons, Tamizo (tah-MEE-zoh) and Shuichi (shoo-EE-chee), who are now in high school. Tamizo plans to be a farmer. Shuichi wants to become an accountant.

The Kawakamis live in a one-story frame house with four small rooms. Their home is in a village with other houses. Only the garden is near the house. The farmland is made up of four plots scattered outside the village.

When Yoichi and Matsuko were first married, their farm was in 10 tiny, separate plots. After 1945 farmlands were **consolidated** (pulled together) to form larger plots. In all, the Kawakami farm has 2½ acres (*one hectare*). That is typical for a Japanese farm, but less than one hundredth of the size of an average U.S. farm.

The Kawakamis set out about 200,000 rice seedlings every spring. They used to do it all by hand, but now they use transplanting machines.

Yoichi, the husband, combines farming with a second job — fishing. He is often away from home. Sometimes he brings back seaweed that he has collected along the shore. Matsuko dries the seaweed and adds it to soups and other dishes.

Rice, fish, and vegetables like seaweed are a major part of the Kawakamis' diet. But as the family's income has risen in recent years, they have added other foods such as beef. Matsuko makes a small amount of beef go a long way, because beef is costly in Japan — far more so than in the U.S. Most beef is imported, since there is so little room in Japan to grow hay or grain for livestock.

Like many Japanese, Tamizo is studying English in school. English is a popular second language in Japan, and many English words have entered the Japanese language. If you went to buy clothes in a Japanese store, you might ask for *shuraku* (slacks), a *shatsu* (shirt), or a *nekutai* (necktie).

Shuichi has one year to go in high school, and already he is studying for a college entrance exam. College education is free in Japan, but only the top 50 percent of scorers on a national exam are admitted. Because Japan has such a large population, its college places are limited.

Shuichi wants to make sure he is in the top half in his exam. He studies at least two hours every night. On weekends he attends a special school that prepares students for exams.

As recently as 1945, half of all Japanese workers were on farms. Now barely 10 percent are on farms. For young people like Shuichi, city jobs seem more rewarding.

SECTION REVIEW

1. What events changed Japan from an empire to a democratic monarchy?

2. Since fuel and electricity are costly in Japan, how do many Japanese families conserve energy?

3. Why do the Japanese eat less meat than fish?

4. Describe two ways in which the Kawakamis' farm is typical of many Japanese farms.

Economic Geography

At Nagoya (nah-GOY-yuh) in southern Honshu, there is a factory that runs day and night. No matter what time you might pass by, you would hear the throbbing of machinery. Normally a plant of such size would need a work force of 200. But this plant has only ten workers — five on first shift, five on second shift, and none on third shift.

The factory is an automated machine-tool plant. It uses computer-controlled machines called robots to make more computer-controlled machines for use in other factories. As the robots make robots, human workers do little more than check the machines for wear.

Japan has led the way into the age of industrial robots. In the early 1980's, Japan had seven out of every ten working robots in the world. This is one sign among many of the leaps taken by Japanese industry in recent years.

Japan's industrial revolution began late in the 19th century. Until the middle of that century, Japan had almost no contact with the outside world. It lived in a traditional world of its own.

Then in 1853, U.S. Navy ships under the command of Matthew Perry visited Japan. Shortly afterward a new emperor rose to power, and it was decided that Japan would become a modern industrial nation like the U.S.

At first Japan concentrated on light industry, such as textiles — cotton, silk, and wool. By the 1930's, Japan was turning to heavy industry — producing goods such as iron and steel, ships, and machinery.

World War II broke off Japan's growth. Destruction caused by bombing required Japan to rebuild many plants. However, the rebuilt plants included the latest equipment. This meant they could produce goods more cheaply than older plants — for instance, plants in the United States. Moreover, factory workers in Japan were paid less than those doing similar jobs in the United States. Japanese industry boomed as its products found customers all over the world.

Soon Japan branched out into new industries. Japanese autos surged onto the world market in the 1960's. In 1980 Japan became the world's number one auto maker.

But already autos were being overshadowed by other products — like electronics. Japan is a top maker of such consumer goods as cameras, TV sets, and stereos. It also excels in industrial products like computer chips and robots.

How has Japan been able to do all this with so few natural resources? One answer is that it has made the most of the resources it *does* have — above all, a skilled labor force of people with a tradition of teamwork. Japan has also traded extensively with other nations for the goods it lacks.

Japan has only half the population of the U.S. and is much smaller in area. Yet in some ways its economy rivals that of the U.S. In this graph, Japanese figures are shown as percentages of U.S. amounts.

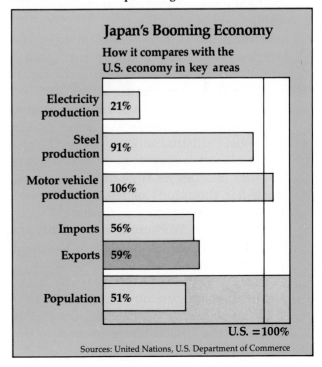

Japan's Booming Economy

How it compares with the U.S. economy in key areas

Electricity production	21%
Steel production	91%
Motor vehicle production	106%
Imports	56%
Exports	59%
Population	51%

U.S. = 100%

Sources: United Nations, U.S. Department of Commerce

How Industry Works

Hisao Shibata, Kimie's father, works for an auto company. He is part of a team of a dozen workers — 11 men and a woman — who bolt fenders onto cars. Today the workers take a break at 10 A.M. to attend a meeting about quality control.

Hisao's company, like other Japanese firms, urges its workers to think up ways to cut costs, speed work, and make better products. In the 1950's, Japanese firms sent observers to U.S. plants to get tips on quality control. The Japanese learned so well that now U.S. firms study Japanese methods. A reputation for careful design and quality work has won a growing share of the world market for many Japanese products.

Hisao's pay is about 350 dollars a week — less than he would earn in a U.S. factory. But he gets benefits that most U.S. workers don't get. Twice each year, Hisao gets a bonus of a month's extra pay. He pays a low rent (about 45 dollars a month) because the apartment the Shibatas live in is owned by the firm that employs him. At vacation time, the family can go to a company-owned resort for a special low rate.

Perhaps the biggest benefit of all is the company's "lifetime employment" policy. The firm almost never lays off a worker. If business drops off, the firm cuts back on work hours, but it keeps paying full wages. One result is that Japan has fewer strikes than most industrial nations.

Not all workers get such advantages, however. Besides its large industrial firms, Japan has thousands of small companies. Most large manufacturers "farm out" work to small plants. Workers at such plants get fewer benefits, and are more likely to be laid off when times are tough.

Wages in Japan have risen rapidly in recent decades. Japan now has the highest standard of living in East Asia. By 1990 it may rival the United States in income per person.

In building up industry, Japan has had to look outside its borders for raw materials. This has affected the location of major industries. Most are in coastal cities where ships can deliver materials such as iron and oil. A belt of big industrial cities stretches from Tokyo west to Nagasaki (nah-guh-SAH-kee).

The Importance of Trade

Japan buys raw materials from many nations. From the U.S. and Canada, it buys things such as coal and timber. From Australia it buys coal and minerals like iron and bauxite. Most of Japan's oil comes from the Middle East, though Japan has been trying to buy more from China and other places.

To pay for its many imports, Japan tries to export as much as possible. It had great success in the 1970's and early 1980's — too much success, in the view of some nations.

The products Japan had for sale were mostly manufactured goods — such as autos, ships, and television sets. But the main markets for such goods were in industrial nations like the U.S., West Germany, and Britain. These nations wanted to sell their own manufactured goods in Japan. But Japan wasn't buying. It wanted raw materials — not finished goods.

The result was that Japan's balance of trade with many nations became lopsided. (The balance of trade is the difference between money spent on imports and money received from exports.) Japan was draining money out of Western Europe and the United States.

Many governments protested to Japan. In response Japan agreed to put limits on some of its exports (such as autos). Japanese companies also stepped up investments in other nations. For instance, a Japanese auto firm built a plant in Tennessee to make pickup trucks. While the profits go to Japan, the wages go to American workers.

Japan buys more U.S. farm products than any other nation. These products include

wheat, corn, grapefruit, cotton, poultry, pork, and beef. Japan also buys many farm goods from Canada, China, and Australia. Japan imports about one fifth of its food.

Growing and Catching Food

Farmers like the Kawakamis do a first-rate job of raising food. They get 1½ times as much rice from a piece of land as do growers in the U.S. But they still fail to grow all the food Japan needs. There is too little space.

Japan's farms are run more like big gardens than like U.S. farms. Half of all farmland is irrigated, mainly because rice needs lots of water, and half of all farmland is used for growing rice.

Actually Japan grows more rice than it needs. It must sell some to other nations. Japan's government had encouraged rice growing by buying rice for three or four times the world price. Lately the government has been urging farmers to switch to other crops like wheat and soybeans.

Fishing is another way Japan fills its food needs. The Japanese eat more fish per per-

Japan has a modern transportation system, with railroads offering some of the world's fastest runs. This train averages 100 miles (*160 kilometers*) an hour on its runs between Tokyo and Osaka (oh-SAH-kuh).

son than the people of any other nation. Most fish are caught near the shores of Japan, but modern Japanese fishing fleets go to distant waters too. Huge "mother ships" go along to can or freeze the catch.

A Time of Change

Japanese life has been changing in many ways. As industry has grown and incomes have risen, new ways of living have taken hold.

You saw that the Kawakamis have changed some of their eating habits. Rice and vegetables play a smaller part in the Japanese diet than they did 20 years ago. People now eat far more meat and other protein-rich foods. One result is that many of today's young adults stand a head taller than their parents (who tend to be shorter than Americans).

Changing standards have made it hard to keep up the tradition of extended families, with many generations living together. Today's young couples feel cramped in small apartments. More and more, the generations are living apart.

Still, the Japanese are trying to keep old traditions, while switching to modern ways. They are seeking the best of both worlds. Just as some people are bilingual, says one Western observer, the Japanese are "bicultural." Kimie Shibata may wear skirts and *shatsu*, but she's keeping her *kimono* too.

Japan is a world leader in the design and use of industrial robots and other methods of computer control. Instead of checking machines, this plastics worker simply glances over a computer-controlled panel.

SECTION REVIEW

1. Even though Japan lacks most natural resources, it has been able to develop an advanced industrial economy. Describe one resource that it could rely on for that development.

2. Describe two benefits that are common for Japanese workers but not for U.S. workers.

3. Why is most of Japan's industry located in coastal cities?

4. In what way did Japan's trade cause problems for some industrialized nations? What happened as a result?

YOUR LOCAL GEOGRAPHY

1. You have read that Japanese homes are small. The average Japanese dwelling in 1980 had an area of 600 square feet (*55 square meters*). In Tokyo the average dwelling had only 400 square feet (*35 square meters*). Compare these figures with the area of your classroom. Measure the width and length of the classroom in feet or meters. Next multiply width by length. The result will be the number of square feet or square meters in the room. Compare the size of your classroom with that of an average Japanese home.

2. You have read how the Japanese diet has changed in recent years. Rice and vegetables are eaten less now than they were 20 years ago. Meat and protein-rich foods are eaten more. Has the American diet also changed in the past few decades? To find out, look in a current almanac under the heading "Food" or "Diet" for information on consumption of different types of foods in the United States. Most almanacs will supply this information for several years or decades. From the list of foods, find three whose consumption has changed noticeably since the earliest date. What does this tell you about changes in Americans' eating habits? What do you think are the main reasons for these changes?

A Visit to Tokyo

Here are some scenes from a Japanese film called Tokyo Story, *which dates from 1953. At that time, Japan was adapting to changes brought by its defeat in World War II, and it was fast rebuilding its economy. The film is about an elderly couple, Shukichi and Tomi, who make their first visit to Tokyo to see their grown-up children. As you read the following scenes, notice how modern ways exist side by side with traditional ways.*

Shukichi and Tomi are staying with their eldest daughter Shige, who runs a beauty parlor. They have just gone out.

Scene 58. Shige picks up the telephone.

SHIGE: Hello, is this the Yoneyama Company? May I speak to Mrs. Hirayama? Thank you . . . Noriko? It's me. Would you do something for me? Mother and Father haven't been anywhere yet, and I wonder if you could take them out some place tomorrow. I really ought to, but I'm just too busy here at the shop.

Scene 59. The Yoneyama Company. Noriko is on the telephone. She is the widow of Shukichi's and Tomi's dead son.

NORIKO: Will you wait just a moment? (*She puts down the receiver and goes to her superior's desk.*) Excuse me. I know it's short notice, but could I have the day off tomorrow?

MAN: That would be all right. How about the Asahi Aluminum?

NORIKO: I'll finish it today. (*She bows and returns to the telephone.*) I'll be at your place at nine tomorrow.

Scene 60. A sightseeing bus. Shukichi, Tomi, and Noriko are riding in it. The explanation of the woman bus guide is heard.

GUIDE: The Imperial Palace was built originally some 500 years ago. In its quiet setting with green pine trees and the moat — what a contrast to the bustle of Tokyo today.

Later Shukichi goes to visit an old friend who moved to Tokyo from their hometown.

Scene 96. Inside the home of Hattori and his wife Yone. Shukichi is talking about old times.

SHUKICHI: It's already been 17 or 18 years.

HATTORI: Really? And here you've been sending me a New Year's card every year.

SHUKICHI: And so have you.

YONE: I suppose that Onomichi has changed.

SHUKICHI: Well, fortunately the city wasn't bombed during the war. The neighborhood where you lived is still just like it used to be.

YONE: Is that so? Well, it was a nice place. We used to like the view from the temple.

HATTORI: And after the cherry-blossom season, the price of sea bream would always drop. All these years, we've missed the taste of those delicious fish. (*From upstairs a young man comes down.*)

MAN: Tell my friends I'll be over there playing pinball, will you? (*Yone nods.*)

HATTORI: We rent the upstairs room to that man. He's really a playboy. Says he's a law student, but spends all his time at pinball. I feel sorry for his father back home.

> — From *Tokyo Story*, by Yasujiro Ozu with Kogo Noda, translated by Donald Richie and Eric Klestadt (in *Contemporary Japanese Literature*, edited by Howard Hibbett). New York: Alfred A. Knopf, 1977.

Ask Yourself . . .

1. Onomichi is the town where the visiting parents live. How does it compare with Tokyo?

2. In these scenes, what clues are there to Japan's economic development?

3. Traditionally, family ties in Japan are very close. What changes does the movie show in such ties?

Understanding a Population Profile

What percent of Japan's population is under age 15? Are there more boys or girls? A population profile like the one below shows how a country's population is distributed by age and sex.

A **population profile** is a kind of bar graph. Each bar represents a particular age group. The bars are stacked on top of one another, with the youngest age group on the bottom. The central vertical line divides the bars by sex. You can compare the sizes of different age groups by comparing lengths of the bars.

For a country with a rapidly growing population, the profile would look like a pyramid, with the widest bars at the bottom. What is the shape of Japan's population by age and sex? Examine the profile for Japan and answer these questions on a separate sheet of paper.

1. From ages 5 to 84, how many years of age does each bar represent?
2. Which side of the center line represents: (a) males? (b) females?
3. On the number scale, what number does the vertical line represent?

Numbers on either side of the "0" line are percentages. Here is how to use them. Find the bar for the 30–34 age group. Notice that the bar on the male side of the "0" line extends to just beyond "4" on the number scale, showing that males age 30–34 make up just over four percent of the total Japanese population. About what percent of the population are females in this age group? Add the two amounts to find the percentage of population represented by the whole bar. Use the population profile to answer the following questions.

4. What is the largest age group? This group makes up what percentage of the population?
5. Teenagers, both males and females, age 15–19, make up what percentage of the Japanese population?
6. How does the shape of the profile change: (a) upward from 29? (b) downward from 25?
7. What will happen to the "under 5" age group on another profile drawn five years after this one?

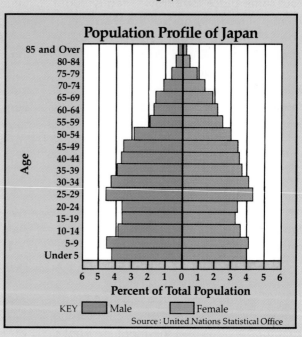

Population Profile of Japan

Age

Percent of Total Population

KEY ☐ Male ☐ Female

Source: United Nations Statistical Office

CHAPTER REVIEW

A. Words To Remember

In your own words, define each of the following terms.

archipelago   consolidated   industrial robots
balance of trade   hydroelectric power

B. Check Your Reading

1. Describe two ways in which Honshu differs from the other major islands of Japan.

2. What does Japan buy from the U.S. in large quantities?

3. How has the surface of Japan's land affected population density? How does Japan's population density compare with that of the U.S.?

4. Name two natural resources that Japan does have in fairly large supply.

5. Describe one way the Japanese make the most of limited space in their densely populated land.

C. Think It Over

1. Why doesn't Japan want to import more manufactured goods?

2. How does teamwork play an important part in Japanese life?

3. In what way is Japan today both traditional and modern? Give three examples each of traditional and modern features.

D. Things To Do

Look through the advertisements in two different national magazines. Make a list of the Japanese-made goods that are advertised. Then write a brief paragraph on the kinds of Japanese goods that are imported by the United States.

Chapter 39

Australia

A Country That Is Also a Continent

It's midsummer on an Australian sheep farm, and there's a hot sun overhead. Some 3,500 sheep are bleating as they graze on wheat stubble. Earlier, giant combines clattered through to harvest the grain.

This farm stretches as far as the eye can see. A few months ago, a team of workers sheared all the sheep. They got about 12 pounds (*five kilograms*) of wool from each sheep. This wool has now been shipped to distant parts of the world. It may end up in a tweed suit worn in Scotland or a wool sweater worn in Maine.

Wool, like most of Australia's farm products and natural resources, finds its way to many distant markets. Like Canada, Australia is a supplier of raw materials to other nations.

Also like Canada, Australia is a big land with few people. It sprawls over an entire continent — an area as big as the United States without Alaska. But much of Australia is dry and lonely. Sheep outnumber people eight to one.

The entire human population, 15 million, is less than the population of the Tokyo metropolitan area. Most Australians live in a few main cities along the east and south coasts. In the *outback* (the Australian interior), a rancher's next-door neighbor may live 20 miles (*30 kilometers*) away.

Yet Australia is not just a land of farms. It is a modern industrial country — again like Canada. More Australians work in manufacturing than in either farming or mining. And whatever work they do, large numbers of Australians are well off. Australia's standard of living is among the highest in the world — slightly above that of Japan.

Physical Geography

Every Monday a train leaves the city of Sydney, on the east coast, for Perth, on the west coast. If you took this train, you would not arrive in Perth until Thursday.

During much of the train trip, you would see little except scrubby bushes, sparse grass, and telephone poles. The land looks endlessly flat and empty. Along one stretch, railroad tracks run straight as an arrow for 297 miles (*475 kilometers*). In this stretch, the tracks do not cross a single bridge or river.

Australia is the smallest and driest of the world's seven continents, and the only one besides Antarctica that lies wholly in the Southern Hemisphere. Australia is sometimes called the "land of the living fossils." **Fossils** are rocks that show the remains or traces of ancient life.

Many scientists believe that Australia was once connected to other continents but broke away hundreds of millions of years

Most of Australia's people live in the southeast of their island continent. At the port of Sydney, the nation's biggest city, a freighter passes the modern opera house.

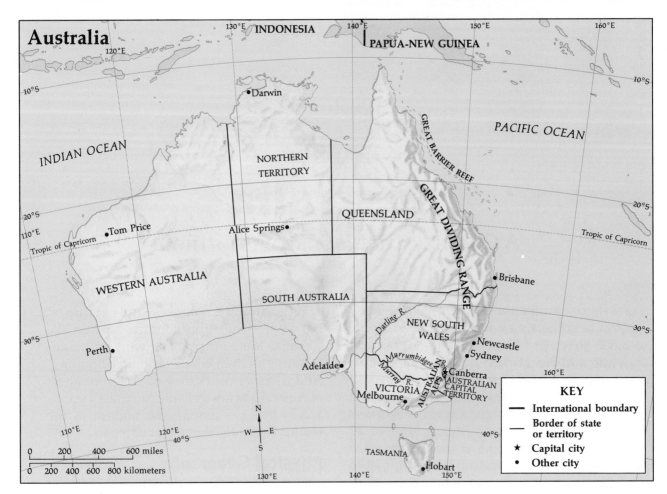

Australia

130°E INDONESIA

140°E PAPUA–NEW GUINEA

PACIFIC OCEAN

•Darwin

INDIAN OCEAN

NORTHERN TERRITORY

GREAT BARRIER REEF

QUEENSLAND

GREAT DIVIDING RANGE

•Tom Price

Tropic of Capricorn

Alice Springs•

WESTERN AUSTRALIA

SOUTH AUSTRALIA

Tropic of Capricorn

•Brisbane

Darling R.

NEW SOUTH WALES

Perth •

Murrumbidgee R.

•Newcastle
•Sydney

Adelaide•

Murray R.

AUSTRALIAN ALPS

Canberra
AUSTRALIAN CAPITAL TERRITORY

VICTORIA

160°E

Melbourne•

N
W—E
S

TASMANIA

| 0 | 200 | 400 | 600 miles |

| 0 | 200 | 400 | 600 | 800 kilometers |

•Hobart

KEY
— International boundary
— Border of state or territory
★ Capital city
• Other city

ago. Isolated from the rest of the world, many of Australia's forms of life are different from those elsewhere. Today in Australia, you can see living ferns with trunks like those of trees. On other continents, such plants are known only as fossils.

Because of Australia's isolation, it has many unique animals. Some, such as the kangaroo, wombat, and koala, are marsupials (mar-SOO-pee-uhls). (**Marsupials** are mammals that carry their young in a pouch.) Another unique animal is the platypus (PLAT-i-puhs), one of the few mammals that hatches its young from eggs. It has fur like an otter and a beak like a duck.

Land and Climate

Most of the land is flat. A low plateau, broken in a few places by highlands, covers the western two thirds of the country. This plateau drops off to a lowland along its eastern edge. Farther east the lowland rises to a ridge of mountains that parallels the Pacific

Australia has a population the same size as that of Texas in an area as big as the U.S. without Alaska. The interior is too dry for crop farming. The main centers of population are along the coast between Adelaide and Brisbane and at Perth in the southwest.

coast. This ridge is about as high as the U.S. Appalachians, and is called the Great Dividing Range. The range extends to Tasmania (taz-MAY-nee-uh), a large island off the south coast.

Between the Great Dividing Range and the Pacific coast lies a narrow coastal plain. Here are many sandy beaches, where conditions are fine for surfing and boating.

Off the northeast shore is the longest coral reef in the world — the Great Barrier Reef. It stretches for 1,250 miles (*2,000 kilometers*).

In most of Australia, you would not have to worry about rain spoiling a picnic. Rainfall tends to be spotty or seasonal. Any year with more than 20 inches (*50 centimeters*) of rain is a "good" year.

Deserts cover one third of the country, mainly on the western plateau. Bordering the desert edges are steppe lands. Here rainfall can support skimpy grasses and shrubs. Much of the central lowland contains such steppe vegetation. Sheep can graze here, but they need large areas of land to find enough grass.

The rainiest parts of Australia are along the south and east coasts. Some places in the northeast and in Tasmania receive 100 inches (*250 centimeters*) or more a year. A short way inland, as in the Great Dividing Range, the total drops to about 25 inches (*65 centimeters*). Only quite narrow strips of land along the east, southeast, and southwest coasts are suitable for growing crops.

Because Australia is south of the Equator, its seasons are the reverse of those in the U.S. Also, the warmest areas are in the north, not the south. Northern Australia lies in the tropics, so temperatures remain fairly high all year. Where rainfall is heavy, as near the northeast coast, crops like sugarcane, pineapples, and bananas can grow. Where rainfall is lighter, tropical grasses can grow and provide grazing for cattle. This area is too warm for sheep.

Southern Australia lies in the cooler middle latitudes. In the southeast, there is rainfall throughout the year. Growing conditions are like those in the U.S. Midwest. Crops include fruits, vegetables, and fodder (plants grown to feed livestock) for sheep and cattle. Elsewhere near the south coast, the rain falls mainly in winter, as in California. Crops include wheat, barley, citrus fruits, and grapes.

West of the Great Dividing Range, a broad stretch of pastureland lies between the rainy areas and the dry interior. The distance is about the same as the distance from Boston to Chicago. This area is mainly sheep country. On many farms, sheep and wheat are raised together, as you saw on page 523.

Mineral Resources

If you dream of being a prospector, Australia may be the place for you. Some Australians spend their weekends out in the *bush* (the countryside), looking for gold and other precious metals. They pay a small fee for a permit to prospect on public lands.

Australia has many valuable minerals, and new finds are still being made. Gold attracts most prospectors, but diamonds, silver, copper, bauxite, tin, lead, zinc, tungsten, and other minerals are present as well.

Two kinds of minerals — fossil fuels and iron — are the keys to modern industry, and Australia has both. Major deposits of high-quality coal lie beneath the eastern highlands. One of the world's richest deposits of iron ore exists near the west coast.

Australia has oil and natural gas as well. Until recently deposits of these fuels seemed small; but in the late 1970's, a major natural gas field was found off the northwest coast. Oil is found under the sea between the mainland and Tasmania, and in a few other spots. The search for more oil goes on.

In the north, Australia has some of the world's biggest deposits of uranium. This metal is a key fuel for atomic reactors.

Unlike Canada, Australia can produce only a moderate amount of hydroelectric power. Most Australian rivers are dry at least part of the year. Many "lakes" that show on maps are little more than mud flats. They hold water only in the rainy season.

SECTION REVIEW

1. Why shouldn't you be surprised to find yourself in the middle of a heat wave if you traveled to Australia in January?

2. Where would you find: (a) the driest parts of Australia? (b) the wettest parts?

3. What do scientists mean when they call Australia the "land of the living fossils"?

4. Is Australia rich in mineral resources? Explain.

Human Geography

If you visit Australia, don't cross any street without thinking twice. Traffic in Australia drives on the left — not the right. Unwary Americans risk being struck by a car coming the "wrong" way.

Australian driving habits are one sign of the nation's British heritage. Another sign is the use of the English language. You'd have no trouble making yourself understood in Australia. You might have trouble understanding some Australian expressions, though. For example, a *station* is a ranch in the bush. A *billabong* is a stream that is dry part of the year. A *matilda* is a backpack — and "waltzing matilda" is what a *swagman* (hobo) does as he roams the land.

British influence in Australia dates from the 1770's. At that time, Australia was populated only by some 300,000 aborigines. The aborigines did not plant crops. They hunted, fished, and gathered foods that were growing wild.

Dutch explorers had visited Australia's arid west coast more than a century before. They were not impressed. In 1770 James Cook, a British navigator, explored the more inviting east shore. He claimed it for Britain.

Britain set up a number of separate colonies, as it had done earlier in America. Some of these began as **penal colonies**. They were places to send people who had been convicted of crimes both large and small. Free settlers also came. The discovery of gold and other minerals helped attract immigrants, and European settlement slowly spread inland.

Like the U.S., Australia has a federal form of government. This was set up when six colonies united in 1901. As in the U.S., a separate federal territory was created to house the national capital, Canberra (KAN-buh-ruh). Canberra is Australia's sixth largest city, with a population of 220,000, and is the only major city that is not on a coast.

In much of Australia's dry interior, there is enough grass to support sheep farms and cattle ranches. Above, an amphibious buggy helps with a roundup.

Can you find Canberra on the map on page 524? One large section of Australia is still not organized as a state. Can you find the Northern Territory on the map?

Today Australia's biggest cities are the capitals of the major states. The most important cities are Sydney, with a population of 3.2 million; Melbourne (MEL-burn), 2.8 million; and Brisbane, one million. All of these cities are in the east and southeast.

The people of present-day Australia are almost all whites, and a majority are of British or Irish descent. A minority trace their ancestry to Germany or to southern European countries such as Italy, Greece, and Yugoslavia. This minority has grown since World War II. Australia has discouraged immigration by Asians, although after the Vietnam War, it took in many refugees from

Indochina. The government has encouraged immigration by whites.

Aborigines now make up less than one percent of the population. Some live in cities. Others live on reservations, much like many Indians in the United States. Australian law calls for equal treatment for aborigines, but many live in poverty.

The chief religion in Australia is Christianity. Large numbers of people are Protestants (mainly members of the Church of England) or Roman Catholics. Smaller groups follow the Jewish or Moslem faiths.

Life on a Sheep Station

To get a look at life in the bush, let's visit the family of Rod and Mary Masterton. They own a sheep station in the eastern highlands, about 150 miles (*250 kilometers*) northwest of Sydney. The station covers about 2,000 acres (*800 hectares*).

The Mastertons live in a two-story house near one edge of their property. They have electricity and a telephone. Around the house are pens and buildings needed for working the sheep and storing equipment. A windmill towers over the pens. It pumps underground water into a pond where the sheep can drink.

The gently rolling landscape looks much like parts of Texas, with scattered trees and tufts of grass. Wire fences divide the land into *paddocks* (pastures), each one enclosing many acres. The Mastertons' two children, Geoff (jehf) and Gretel, help move the sheep to different paddocks from time to time so the grass has a chance to grow back.

The Mastertons are trying to grow better grasses so the land can support more sheep. They sometimes hire an airplane to spread fertilizer and grass seed. Unlike many other sheep graziers, they do not grow wheat. (A grazier is a farmer who raises livestock on pasture for market.)

Gretel and Geoff have horses that they ride about the property. They keep an eye out for holes that kangaroos break in the fences. Kangaroos like the same grasses that sheep like. Some graziers shoot kangaroos as pests.

The children ride a school bus to school. Each weekday morning, they walk to the end of the lane and wait at the edge of a dirt road. Australia has many paved roads to connect major towns and cities, but most rural roads are unpaved. Some sheep and cattle stations in the outback are too far from schools to have bus service. There children are taught by two-way radio over "schools of the air." They send their homework by mail.

The food the Mastertons eat is much like American food. Of course they eat a good deal of lamb and mutton. Lamb is meat from a young sheep; mutton is meat from an old sheep. The children especially like the meat pies their mother makes out of mutton, onions, carrots, and flour.

Life in Melbourne

If you visited Melbourne on Australia's south coast, you might meet the family of Nigel (NIE-juhl) and Barbara Huff. They live in a suburb of neatly kept homes on the city's outskirts. Twice each day, the Huffs fight city traffic to drive to their jobs in the business district. Nigel is a bank executive. Barbara works for an insurance firm.

Car ownership is as common in Australia as in the United States. The Huffs own two cars — one American, the other Japanese. Both cars were assembled at factories in Australia, and have their steering wheels on the right.

Melbourne is closer to the South Pole than any other big city on the Australian mainland. Yet Melbourne is no farther from the Equator than Richmond, Virginia. Snow almost never falls in Melbourne, but that does not keep the Huffs and their children from enjoying winter sports. A few hours' drive from home, they can ski in a part of the

Great Dividing Range known as the Australian Alps.

The Huffs' children are Sandra, age 16, and Roger, age 12. They go to a church-related school in Melbourne, one of many private schools that exist alongside the public school system. Both are active in after-school sports. Sandra is on the school's tennis team. Roger plays cricket and the Australian version of football, with 18 players on a team.

Australia's sunny climate is just right for outdoor sports. Water sports are especially popular, since most Australians live near the coast. Sandra and Roger spend much of the summer at beaches near Melbourne.

The Huffs like to eat out, and they find a variety of foods to choose from. Melbourne has many ethnic neighborhoods with restaurants run by immigrants or the descendants of immigrants. Sandra and Roger like Italian cooking. Barbara likes Greek dishes, and Nigel favors Vietnamese food. At home they often cook steaks on a backyard grill.

SECTION REVIEW

1. What people lived in Australia before the British arrived? What percentage of the population does this people make up today?

2. How has the makeup of Australia's population changed since World War II?

3. In what way is Australia's capital city, Canberra, similar to Washington, D.C.?

4. Give two examples of British influence that you would notice in Australia.

Economic Geography

If you took a surfboard to the beaches at Newcastle, on Australia's east coast, you could ride the big waves as they rolled in off the Pacific. You could also see Australia's economy at work. A short way out to sea, dozens of freighters would be sitting at anchor, ready to be loaded with coal at the nearby docks.

Newcastle was named after an English city that became world-famous as a coal-shipping port. The Australian city not only exports coal from nearby mines but also uses it in industry. Local blast furnaces and steel mills turn out products for Australian factories. Giant power plants burn coal to run machines that make electricity for homes and offices and for industries like aluminum smelting. (To **smelt** a metal is to separate it from its ore by heat.)

Australia's coal has fueled both local industry and foreign trade. In 1979 for the first time, coal beat out wool and wheat to become Australia's leading export.

Australia is one of the few industrial nations to produce more energy than it uses. At a time of soaring energy prices, this is good news for Australia. It can sell coal and gas to other nations and use the profits to develop its own industries.

There's only one hitch: The profits don't belong just to Australians. Like Canada, Australia lacks the **capital** (money) it needs to develop its natural resources. So it "imports" capital. Companies in the United States, Japan, and other nations have made large investments in Australian mines and industries. Some of the profits go to them.

Like the Old West in the United States, Australia has its boom towns that sprang up around mineral strikes (discoveries). It also has its ghost towns that died when the mines played out. These towns are scattered about the outback, often in lonely spots.

Today's boom towns have schools, libraries, and golf courses. But mine workers may live in cheap boxlike houses, unshaded by trees. Usually the road to the nearest city is unpaved and bumpy. Iron miners in Tom Price, a town in Western Australia, take three spare tires for the half-day drive to the nearest beach.

Australia has built roads and railroads to reach its mines, and is building more. But the distances are great, and costs are high.

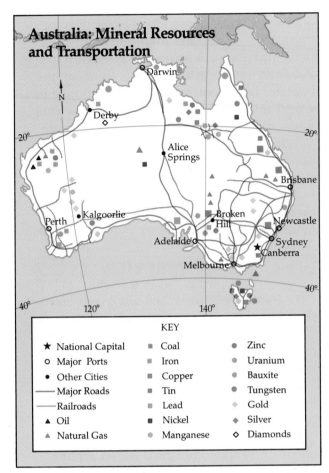

Australia: Mineral Resources and Transportation

KEY

★ National Capital	■ Coal	● Zinc
○ Major Ports	■ Iron	● Uranium
● Other Cities	■ Copper	■ Bauxite
— Major Roads	■ Tin	● Tungsten
— Railroads	■ Lead	◆ Gold
▲ Oil	■ Nickel	◆ Silver
▲ Natural Gas	● Manganese	◇ Diamonds

Australia is rich in mineral resources, but most of them are located far from the main population centers in the southeast and southwest. As a result, many short highway and railroad routes have been built to connect mining areas with small ports around the coast.

There is still no north-south railroad across the continent.

In the settled areas near the east and south coasts, the road and rail network is much thicker. Trucks and trains haul farm goods to market. They carry manufactured goods to customers in the cities, towns, and country.

Most consumer goods are made at factories in and near the largest cities. Besides autos these factories make products like boots, clothing, razor blades, radios, television sets, and plastic baby rattles. Heavy industries include the manufacture of ships and railroad locomotives.

Unlike most industrial nations, Australia exports few of the goods it manufactures. These goods are sold mainly in Australia.

Agriculture

An American farmer visiting an Australian farm would probably feel at home. There are plenty of modern machines — pickup trucks, tractors, milking machines, even airplanes. By using such machines, Australian farmers can get a lot done in a short time. This means that one family can work a large farm or station without much hired help. Only seven percent of Australia's work force is in agriculture.

The animals and plants raised commercially in Australia are almost all **exotic** (ehg-ZOT-ik) — that is, they have been brought from other parts of the world. Sheep, cattle, and wheat, the leading farm products, were all introduced by Europeans. Kangaroos are one of the few **indigenous** (in-DIJ-i-nuhs; native) animals to be put to use. Their meat goes into pet food, the hides into leather products such as belts, and the fur into fur coats and toys. However, some people are very concerned that kangaroos are in danger of dying out, and would like to see them protected.

Many breeds of sheep are raised — some for wool, some for meat, and some for both. Australia is the world's leading producer and exporter of wool.

When graziers like the Mastertons have their sheep sheared, they pack the fleeces into bales weighing about 300 pounds (*135 kilograms*). They send the bales to port cities to be sold at auction. Agents representing buyers all over the world attend such wool auctions and bid on the bales. If the wool is especially fine or if the supply is short, they will bid high. If the wool is coarse or if there is an oversupply of wool, they will bid low. A farmer can't be sure how much the wool is worth until the sale is made.

Some farmers raise cattle instead of

sheep. In the outback, these are likely to be beef cattle to be sold to meat packers. Near the coast, where cities provide a market for milk and cheese, dairy cattle are more common. Australia exports both beef and cheese.

Wheat is another key export. In some years, Australia earns more from wheat than from wool.

In areas of the northeast where sugarcane grows, dozens of small mills process the cane into sugar. This also is exported.

Because of the dryness of many areas, Australian farmers use irrigation to grow crops such as grapes, citrus fruits, and melons. About one tenth of all cropland is irrigated.

Australia and the World

If you haven't been to Australia, you needn't feel alone. Few Americans have. Australia is twice as far from the U.S. as Europe is.

Australians would like to attract tourists to their beaches and resorts, but distance is a major drawback, especially for Americans and Europeans. Many tourists do come to Australia from closer lands, such as New Zealand and Japan. But tourism is not yet a leading industry.

Fortunately for Australia, distance has little effect on foreign trade. Modern ships can carry bulky goods for great distances at fairly low cost. Frozen meat can travel in refrigerated ships.

At one time, Britain was Australia's main trading partner. Britain furnished manufactured goods and bought farm products in return. Lately Australia has stepped up its exports of minerals, and has found new customers in Japan and the U.S. These two nations have become Australia's leading trading partners. They sell Australia manufactured goods such as trucks, appliances, and large machines. Australia also has done a brisk trade with New Zealand.

In some ways, Australia is like a developing nation. It imports many finished products and exports mainly raw materials. Between one tenth and one third of the bauxite, iron, coal, nickel, lead, and zinc in world trade are Australian. With new fuel finds, mineral exports will grow even more.

Yet Australia is also like a modern, developed nation. It has enough wealth to bring in the modern machinery it needs. It manufactures many of its own products. And its people enjoy one of the highest standards of living in the world.

SECTION REVIEW

1. How do Australia's exports of wool compare with those of other nations?

2. How does Australia's energy production compare with that of most other industrial nations?

3. Although many farms in Australia cover huge expanses of land, they require very little hired help. Why is that?

4. What kinds of goods does Australia export to and import from the U.S. and Japan?

YOUR LOCAL GEOGRAPHY

1. For Australians it is normal to travel north for a warmer climate and to ring in the New Year in midsummer. Imagine that an Australian student is visiting your community on his or her first trip north of the Equator. Describe three things that he or she would find unfamiliar about living "in reverse."

2. You have read that Australia's distance from other parts of the world makes it hard to attract many tourists. Using the maps on pages 574–580, measure the approximate air distance from your nearest airport to Sydney, Australia, via Honolulu. Then measure the distance from your nearest airport to Tokyo, Japan, and to London, England. About how many times farther would you have to travel to Sydney than to Tokyo or London?

Interpreting a Landsat Image

Space technology has given people new ways of looking at Earth's resources. Since the mid-1970's, for example, two human-made satellites called *Landsat* 1 and 2 have been in orbit around Earth. A *satellite* is an object that circles a planet in outer space, rather like a tiny moon. Aboard each Landsat satellite is a sensor. A *sensor* is an instrument that can measure the intensity of light over large areas of Earth's surface. (Turn page.)

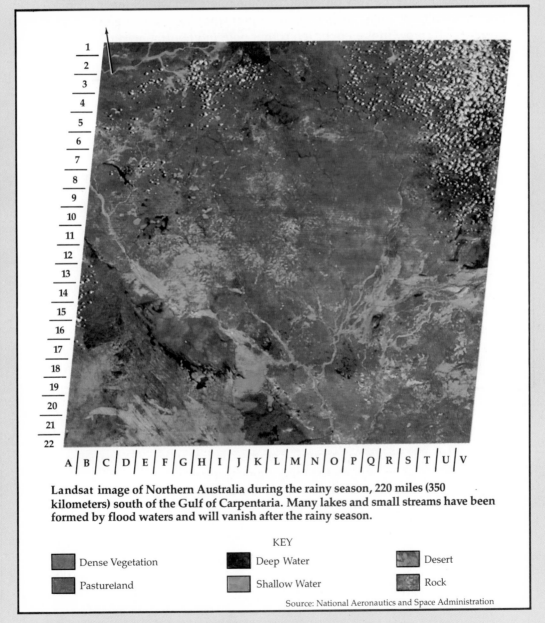

Landsat image of Northern Australia during the rainy season, 220 miles (350 kilometers) south of the Gulf of Carpentaria. Many lakes and small streams have been formed by flood waters and will vanish after the rainy season.

KEY

Dense Vegetation   Deep Water   Desert

Pastureland   Shallow Water   Rock

Source: National Aeronautics and Space Administration

From the Landsat sensors, data or information is relayed to ground stations on Earth and recorded as numbers on magnetic tape. From these numbers, a computer can create a colorful maplike image such as the one of Northern Australia on page 531.

If you compare the Landsat image with the map of the same area below, you will notice several differences. The Landsat image does not show borders or other political features that can be shown on a map. It does show natural features more accurately and in more detail than a map can.

The Landsat image has features similar to some maps. For example, colors of the Landsat image are very different from the colors you would see in a photograph. A key is provided to help you interpret what the colors mean. You will also find an arrow in the upper left corner pointing north. Coordinates along the edges can help you pinpoint locations.

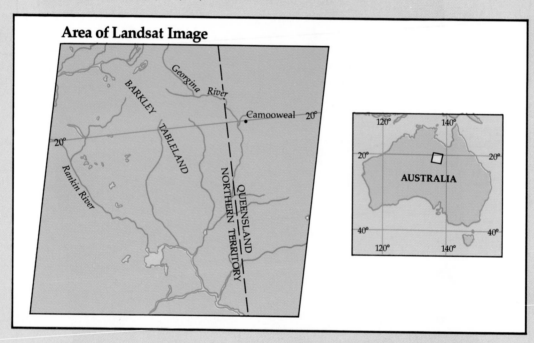

Area of Landsat Image

Compare the Landsat image and the map. Study the caption and the color key on page 531. Then do the following exercises on a separate sheet of paper.

1. What town is located at about P-7 on the Landsat image? Is this town located in the polar regions, the midlatitudes, or the tropics?
2. Which river is located at about F-14 on the Landsat image?
3. Give the letter-number coordinates of at least one place on the Landsat image where you would find: (a) deep water; (b) shallow water; (c) dense vegetation.
4. Give the letter-number coordinates of at least one place on the Landsat image where you might expect to find: (a) cattle grazing; (b) a dry lake or stream bed after the rainy season.

CHAPTER REVIEW

A. Words To Remember

From the following list, select the term that best completes each sentence below. Write your answers on a separate sheet of paper.

aborigines   grazier   marsupial
exotic   indigenous   outback
fossils   investment

1. A _____ is a mammal that carries its young in a pouch.
2. The inhabitants of Australia when the British arrived were the _____ .
3. A(n) _____ animal is one that is native to the area.
4. A _____ is a farmer who raises livestock on pasture for market.
5. The Australian interior is known as the _____ .

B. Check Your Reading

1. In what parts of Australia do the majority of the people live?
2. In what part of Australia are sugarcane, pineapples, and bananas grown? Why can they grow there?
3. Why does Australia produce little hydroelectric power?
4. What are two of Australia's leading exports? Are they manufactured goods or raw materials?
5. Why does Australia have difficulty attracting many tourists?

C. Think It Over

1. Do you think the life of the Mastertons on a sheep station has a lot in common with the life of the Huffs in Melbourne? Explain.
2. How does Australia's environment influence the way its population is distributed?
3. In what ways is Australia like a developing nation? In what ways is it like a modern nation? Which do you think it resembles more? Why?

D. Things To Do

Imagine you have been chosen to fly to Australia to collect a variety of animals for a new zoo. Using an encyclopedia or other library source, make a list of the animals you would like to see in your zoo. Tell where in Australia you expect to find them. Then write after each animal's name the kind of environment and food you will need to provide.

Chapter 40

Different Paths to the Modern World

The Region in Perspective

More than a century ago, a young Japanese student set out to do an experiment in science. He wanted to take a piece of iron and coat it with tin.

The student, whose name was Fukuzawa Yukichi (foo-koo-ZAH-wah yoo-KEE-chee), had read about tin-plating in a school book. A thin coat (or plating) of tin prevents iron from rusting. In modern Western countries, tin-plated iron was in common use — in things such as metal trays and tin boxes. However, the Industrial Revolution had not yet reached Japan. There were no factories making tin objects.

Fukuzawa read that a chemical called zinc chloride could be used to make the tin coat the iron. But there was no place in Japan to buy zinc chloride. So Fukuzawa and some friends had to make their own. They bought the materials and borrowed some equipment. After much work, they finally succeeded in plating iron with tin.

Fukuzawa went on to become a famous educator who helped to modernize Japanese teaching methods. Japan went on to become a modern industrial nation. Today all kinds of Japanese foods, from soup to soda pop, are packaged in tin-plated cans.

In order to become a modern nation, Japan needed the same kinds of resources as Fukuzawa needed — on a much larger scale, of course. Fukuzawa needed *physical resources* — iron, tin, and other materials. He also needed *human resources* — his own skill and determination to succeed. In addition, he needed *capital* (money) in order to pay for or borrow physical resources. Nations also need money to develop their human resources — by paying workers' wages and providing schools to teach skills.

In this unit, you have studied three nations that have reached or are approaching the modern world. They are on three different paths. They differ in their human and physical resources, and in the ways they seek to develop those resources. They also differ in the ways they have chosen to pay for their development.

China, Japan, and Australia all have the same goal. They want to build up industry and create a modern economy. Most other nations in East Asia and the Pacific also share the goal and are taking one or other of the three paths.

Human and Physical Resources

How can you measure the human resources of a nation? You might start by counting the people. China has a huge population, Japan has more people than most other nations, and Australia has relatively few people. Some countries in the

---

Most nations in the region have had to struggle with various difficulties to develop their resources. This railroad carries freight across the barren interior of Australia.

region have far fewer even than Australia.

A nation with a small population finds it hard or impossible to create a modern economy on its own. There are probably too few people to supply all the labor or capital needed. Thus Australia by itself has been unable to raise enough capital to develop and expand its economy.

However, human resources depend on more than numbers. In fact, a large population can be a handicap. People cannot help their nation to develop unless they produce more than their basic needs. In addition, there must be enough people with the technical skills needed in a modern society. So a nation with a small but well-educated population may enter the modern world more readily than a populous nation with a low level of education.

China, Japan, and Australia all have built extensive school systems. You read on page 514 that Shuichi Kawakami sees education as a way to get ahead.

Education is also a way for a nation as a whole to get ahead. It widens people's horizons, from their locality to the rest of the world. It trains people for new types of jobs. And it prepares people to take a more active part in politics and goverment.

The level of education, then, is one measure of a nation's human resources. Another measure is the unity (or lack of unity) of a nation's people. China, Japan, and Australia all have a strong sense of unity. Each has one ethnic group that makes up a large majority, speaks a common language, and shares a common culture. There are minority groups that want to preserve their own culture, but they do not expect it to become the culture of the majority.

To develop on its own, a nation needs strong human resources. This graph compares the populations of China, Japan, and Australia in three ways—by *size*, *skills* (literacy and education), and ethnic *unity*. (See the text on this page for further explanations.) Bar areas are drawn in proportion to population sizes.

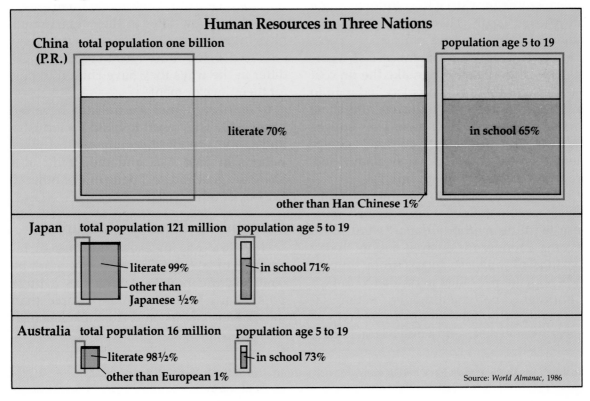

Human Resources in Three Nations

China (P.R.) total population one billion — population age 5 to 19

literate 70%
in school 65%
other than Han Chinese 1%

Japan total population 121 million — population age 5 to 19

literate 99%
other than Japanese ½%
in school 71%

Australia total population 16 million — population age 5 to 19

literate 98½%
other than European 1%
in school 73%

Source: *World Almanac*, 1986

A nation that lacks such unity finds it harder to enter the modern world. The different groups may put more effort into struggling for power than building a common future. Sometimes people in such a nation must learn a new language before they can get an education. In Papua-New Guinea, as you have read, hundreds of local languages are spoken. Koina speaks one of those languages at home, but he would have to learn English if he wanted to go to the main college in the nation's capital.

A nation trying to modernize needs various kinds of physical resources. It needs raw materials such as coal and iron. It also needs machinery to help turn those raw materials into finished products.

However, most nations do not have nearly enough of all the physical resources they need. For example, Japan has few raw materials within its borders. On the other hand, China is well supplied with raw materials, but it cannot produce enough machinery. Therefore, both Japan and China must import physical resources from other nations. To do this, they need some form of money.

Australia has plenty of physical resources but it needs money too. As you read earlier, Australia has too few people to raise the capital needed for development.

In the next section, you will see how the nations in the region have found different ways of raising the money they need.

SECTION REVIEW

1. What three kinds of resources does a nation need in order to develop?

2. Can a large population be a handicap to a nation's development? Explain.

3. Why do many nations in the region see education as an important step toward development?

4. How does the unity of a nation affect its development?

Paying for Development

If you want to buy a new stereo, you can do it in two ways. One way is to save your money and pay for the stereo out of your savings. The other way is to borrow money and pay it back later.

Nations seeking to develop modern industry face the same choice. Japan chose to use its own savings. Australia, with much fewer people than Japan, borrowed from outside sources. China has swung between the two alternatives.

Japan's choice is sometimes called the "bootstrap" way, because it requires "pulling oneself up by one's own bootstraps " — an old saying that means struggling to succeed without help from others.

The Bootstrap Way

Japan began building modern industry in the 1860's. To pay for this development, it placed high taxes on the Japanese people — especially farmers. In this way, people were forced to "save" money they could otherwise have spent on their own needs. The Japanese government used these "savings" to start new industries or to help private firms do so.

High taxes are not popular with voters, so this method is hard for many nations to choose. In Japan's case, it was made easier by the fact that voting was limited to the wealthier classes. Not until 1925 did all adult males get the right to vote. Women had to wait until 1947.

China used a variety of the bootstrap way, especially in the early years of Communist rule that began in 1949. The government took control of existing industry. Instead of setting high taxes, China's leaders ordered industry to limit the output of consumer goods. People *had* to "save" — there was little they could buy with their money.

Moreover, the Chinese government controlled both wages and prices throughout

the economy. Therefore the government set low prices for the farm products that it "bought" from the communes. In this way, it cut down the amount of money that farmers were able to spend. (Farmers make up the vast majority of China's population.) By such methods, the Chinese government was able to channel money into building up industry.

Outside Investment

To some extent, all nations use taxes or other methods to ensure "savings." However, most nations that want to build up industry also look for outside sources of help. Australia is a good example.

You have read of Australia's vast mineral resources. Australia wants to develop these resources rapidly, which requires large sums of money. Australian investors put up

Some nations in the region have raised money to develop by attracting outside investors. South Korea is one country that has taken this course. Above, workers in Pusan (POO-sahn) build a nuclear plant.

some money, and some money comes from taxes — but much more is needed. So Australia welcomes investments from outsiders. When Australia began tapping its offshore reserves of natural gas in the 1980's, investors included major firms in Europe and the United States.

Many other modernizing nations also welcome outside investments. U.S. firms have invested money into copper mines in Papua-New Guinea, auto assembly plants in Taiwan, and a new banking system in Micronesia. Even Communist nations like China sometimes use this method. When China decided to explore for oil in its offshore waters recently, it invited U.S. and European oil firms to take part. China offered a share of any oil found to the firm making the discovery.

Only nations with stable governments can attract much foreign investment. (A stable government is one that is unlikely to be overthrown. It may be almost any form of government from a democracy to a dictatorship.) Investors don't like to put money into a project that might be destroyed in a civil war or seized by new leaders. In recent years, most governments in East Asia and the Pacific have been stable.

Japan and Australia have gone a long way on their chosen paths from the traditional to the modern world. China has made a strong start. Other countries in the region, notably the two Koreas and Taiwan, are also rapidly building industry.

In Chapter 38 you met the Shibata family, who live in modern Tokyo. The Shibatas enjoy a modern life today partly because of the "savings" of their rural ancestors a few generations ago. They also owe a debt to people like Fukuzawa Yukichi who worked to move their nation along the path to the modern world.

In many parts of the region, the modern life of the Shibatas may not become widespread for a long time. But there are many people like Fukuzawa and the "savers" who are paving the way to that future.

SECTION REVIEW

1. What are the two major ways in which a nation can raise money to pay for modern industry?

2. How did Japan obtain money from its people in the 1860's? How did China obtain money after 1949?

3. How did China recently raise money for offshore oil exploration?

4. What kind of nation would have trouble attracting outside investors? Why?

YOUR LOCAL GEOGRAPHY

1. In Geography Skills 49 (page 540), you study a cartogram in which places are drawn to the relative size of their population. You can make a simple cartogram showing the relative populations of both your state and any neighboring state. First find the population figures in an almanac. Round out the figures to the nearest hundred thousand. Divide the number of hundred thousands in the smaller state into the number of hundred thousands in the larger state. If the remainder is less than half of the divider, ignore it. If the remainder is half or more, use the next higher number. For example, suppose your state has 4,400,000 people and another state has 1,700,000 people. You divide 44 by 17, obtaining 2 with a remainder of 10. Since 10 is more than half of 17, you use 3 as the figure.

Now draw the state with the smaller population as a square with sides of one inch. The other state will also be a square. The length of its sides will be the same number of inches long as the figure you obtained by dividing. In the example above, your state would have sides three inches long. Mark an arrow on your sheet of paper to show north. Then draw the second state so that it touches the first on the side nearest its actual compass position. You now have a cartogram.

Compare the relative areas of the two states in your cartogram with those on the map on page 118. What can you tell about the density of population in each state? What other figures in the almanac could be made into a state cartogram?

2. Think of a school or community project that is under way or being planned. It may be anything from a litter drive to a costly item of construction. List the major physical and human resources needed to carry out the project. Then describe the way or ways in which most of the cost will be funded. Is the method of obtaining money most like that used by Japan, Australia, or China?

Reading a Cartogram

How "big" is a country? On most maps, the sizes of places are based on their actual land area. But what if a map were drawn to compare countries by the sizes of their populations?

The map below does just that. It is a special type of map called a cartogram. On a **cartogram**, the locations of places are shown as accurately as possible, but sizes are based on energy use, population, or some quantities other than areas. Use the cartogram below to answer the following questions on a separate sheet.

1. What do the four colors on the map and key stand for?

2. What happens on this map to Greenland, the world's largest island in area? Which continent is missing from the map?

3. In land area, the two largest countries in the world are the Soviet Union and Canada. What are the two largest countries on this map?

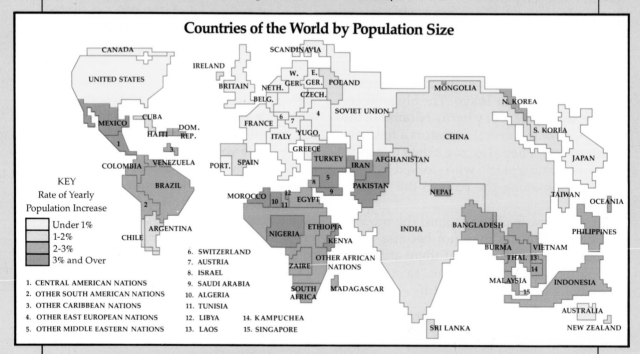

Countries of the World by Population Size

KEY
Rate of Yearly Population Increase

- [ ] Under 1%
- [ ] 1-2%
- [ ] 2-3%
- [ ] 3% and Over

1. CENTRAL AMERICAN NATIONS
2. OTHER SOUTH AMERICAN NATIONS
3. OTHER CARIBBEAN NATIONS
4. OTHER EAST EUROPEAN NATIONS
5. OTHER MIDDLE EASTERN NATIONS

6. SWITZERLAND
7. AUSTRIA
8. ISRAEL
9. SAUDI ARABIA
10. ALGERIA
11. TUNISIA
12. LIBYA
13. LAOS
14. KAMPUCHEA
15. SINGAPORE

The world clearly looks different when countries are sized on a map by population instead of area. What can such a map tell you? Use the cartogram to complete these statements.

4. The smallest of these nations in population is (Japan; Indonesia; Australia).

5. The largest of these nations in population is (Spain; South Africa; Bangladesh).

6. Of these nations, the one with the highest rate of yearly population increase is (the United States; India; Mexico).

CHAPTER REVIEW

A. Words To Remember

In your own words, define each of the following terms.

bootstrap method   human resources   physical resources
foreign investment   minority group

B. Check Your Reading

1. What kind of resource does Fukuzawa Yukichi represent?

2. Is it easy for a democracy to use the bootstrap method of economic development? Why, or why not?

3. In what way does Papua-New Guinea lack unity?

4. What is a stable government? Why is a stable government important for attracting foreign investment?

5. Describe two ways that the U.S. has been involved in the development of nations in the region.

C. Think It Over

1. Can nations overcome a shortage of physical and human resources? Explain.

2. Why does a nation with many different language groups face difficulties in developing?

3. The chapter says that too many people may be a handicap to a developing nation. Are too few people a handicap? Explain.

D. Things To Do

Select any two countries from the region you have just studied. Using the Checklist of Nations on page 587, draw up a table comparing the area, population, average personal income, and life expectancy for each nation. Do those data tell you whether one country is more developed than the other? If so, how? If not, what further statistics would you need?

A. Check Your Reading

1. On your answer sheet, write the letter of each description. After each letter, write the number of the place that best matches that description.

(a) The most modern industrial nation in the region.

(b) A Pacific island that is a U.S. possession.

(c) An island that is also a continent.

(d) A dry plateau that borders the Himalayas.

(e) The largest and leading industrial city in China.

(1) Beijing
(2) Shanghai
(3) Tibetan highlands
(4) Australia
(5) Guam
(6) Japan
(7) New Zealand

2. Fill in the blanks in the following paragraph by writing the missing term on your answer sheet.

Two countries in the region are rich in mineral resources. For fuels they have plenty of coal and some oil. These two countries are __(a)__, north of the Equator, and __(b)__, south of the Equator. __(c)__, an island nation with few minerals, has made use of its human resources. Other resources a nation needs in order to develop are __(d)__ and __(e)__ (money).

B. Think It Over

1. The following statements may be true of any one or more of the following: Australia (A), China (C), Japan (J). Write the initial letter of the nation(s) to which you think each statement applies.

(a) It is one of the most populous nations in the world.

(b) It has an advanced modern industrial economy.

(c) A large majority of its people belong to the same ethnic and cultural group.

(d) It is a democracy.

(e) It relies heavily on outside investment for its development.

2. Choose any one of the five statements in exercise 2 above, and write its letter on your answer sheet. Then explain how this statement applies to the nation or nations you chose.

Further Reading

Holding Up the Sky: Young People in China, by Margaret Rau. Lodestar, 1983. Chinese youth in modern China.

The Pacific Navigators, by Oliver E. Allen. Silver Burdett, 1980. Describes early exploration of the Pacific Ocean and its islands.

Red Earth, Blue Sky: The Australian Outback, by Margaret Rau. Crowell Junior Books, 1981. Traveler amasses first hand information on Outback life.

10

A Global View

Chapter 41

More People—Enough Food?

How long would it take you to visit all the states of the U.S.? It could take several weeks — even if you traveled in a private plane. Now think about visiting every major city in the U.S., or every one of the more than 3,000 counties. And after finishing that trip, suppose you decided to explore other countries in the same detail. You would be traveling for the rest of your life. Earth is a big place.

Yet from another viewpoint, Earth is small. To the alien visitors you met in Unit 1, Earth was just a speck in the vastness of space. The part of Earth that supports human life is even smaller — the outer layer of the crust. Except for energy from the Sun, all the resources we need must come from that thin layer and its wrapping of air.

In Units 2 through 9, you studied the separate regions of Earth. You looked at some nations that are rich in resources and others that have few resources. You saw how some nations make the most of their resources, while others have barely started to develop. You met some people who are struggling to survive, some people who enjoy a high standard of living, and others at many stages in between.

This unit deals with all of those nations and all of those different people together. The unit takes a global view of some of the basic issues involved in our attempt to make a better life out of Earth's resources.

The rest of this chapter focuses on three linked issues: the rising number of people on Earth, the need to provide food for all these people, and the natural hazards that often threaten the food supply. These issues are important for all of us, but are especially urgent for people in the traditional world.

The following chapter deals with another group of linked issues: conserving our environment and its resources, and finding enough energy sources to meet people's needs. These issues become more urgent as more and more nations develop and enter the modern world.

More and More People

The world holds more people today than it has ever held before. Earth will hold more people next year, and still more the year after.

Some see this increase as a burden — more mouths to feed. In this view, a "population explosion" has occurred and must be checked. Others see each new person as an asset — two more hands to work the land, or another brain to add to human knowledge. In this latter view, a growing population offers opportunities for putting Earth's resources to fuller use.

Can food production be improved to keep pace with the world's growing population? These Peruvian Indians have to grow their food in the thin soil of the chilly Andes Mountains.

People on Earth

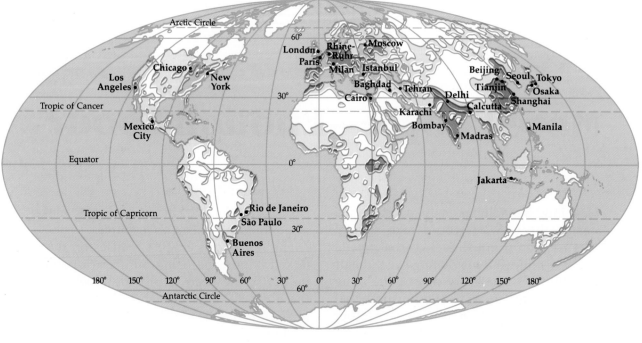

KEY

Number of People

per square mile		per square kilometer
Under 2		Under 1
2 to 50		1 to 20
50 to 100		20 to 40
100 to 500		40 to 200
Over 500		Over 200

• Urban Areas with over 5 Million People

While the *meaning* of population growth is disputed, the *fact* of this growth is not. The world's population has grown from an estimated five *million* in 8000 B.C. to almost five *billion* today.

A few landmarks over the centuries stand out. These are turning points when population took a jump upward. One turning point was the introduction of farming around 8000 B.C. This advance in technology allowed the land to feed more people.

You can see another turning point if you look at the graph on page 79. In the late 18th century, Earth's total population was still less than one billion. However, the Industrial Revolution was about to bring a rapid change.

You can see at a glance that large areas of Europe, South Asia, and East Asia are densely populated. There are smaller crowded areas in the Americas and Africa. To see why other places have few people, compare this map with the climate map on page 550.

The Industrial Revolution was based on machines powered by fuels such as coal and oil. These machines could develop Earth's resources faster than human muscles. Thus Earth could be made to support more people. Moreover, the extra resources could meet more needs than in the past. More and more people began to enjoy better diets, clothing, homes, and means of transportation. Not only were there more people, but they were living longer.

Advances in medicine and public health helped to lengthen people's lives even more. For example, millions of people used to die every year from smallpox. Today vaccination appears to have wiped out the disease. Vaccines have also curbed many other diseases, such as typhoid and polio.

Even in the most advanced nations, it was common in the past for many babies to die in their first year of life. Today better diets and improved health care have lowered infant mortality rates. However, these rates are still higher in the traditional world than in the modern world.

How Much Increase and Where?

As you can see from the Checklist of Nations on page 581-587, population growth varies widely in different parts of the world. The population of some European countries is no longer increasing. In other industrial countries like the United States, the population increases at a slow rate of about one percent each year. In many developing countries of Latin America, as you read earlier (see page 208), the population grows at a faster rate of two percent or more. In several African nations the population growth rate is more than four percent.

For most countries, the growth rate has been slowing down. Experts say that world population growth will ease off further as more and more nations build industry and as more people leave farms for cities. In the traditional world, human muscles do much of the work on farms. Therefore a farm family needs many people to work the land. In the city, however, space is limited and food is bought rather than grown. Thus a farm family may see a new baby as a future worker on the farm, while a city family may see a new baby as "another mouth to feed." As a result, city families tend to be smaller than farm families.

Although the growth rate is going down, the actual numbers of people on Earth are still rising rapidly. Even the low 0.7 percent increase in the population of the U.S. means more than 1.5 million new people per year. All these people have needs that must be met from Earth's resources. In the next section, you will read about one of the most important of these needs — food.

SECTION REVIEW

1. How did the Industrial Revolution affect world population growth?

2. What two factors have helped to lower infant mortality rates around the world?

3. What kind of country is likely to have a population growth rate of one percent or less per year?

4. Why are families living in rural areas more likely to want more children than families living in cities?

Producing More Food

Back in 1920, the world's population was about 1.6 billion. At that time, farmers in Iowa were doing very well if they got 35 bushels of corn to the acre (*2.2 metric tons to the hectare*). The average yield in the U.S. was about 25 bushels (*1.6 metric tons*).

Today the world's population is over five billion. The average yield of corn on U.S. farms is now more than 90 bushels per acre (*six metric tons per hectare*). Some farmers claim yields as high as 200 bushels (*13 metric tons*). Thus the same amount of farmland produces four times as much food as it did in 1920. Meanwhile the world's population has grown only about three times as big.

Of course, this does not mean that the world's food problems have been solved — far from it. The yields of different crops on different farmlands around the world have not all grown as fast as corn yields in the U.S. Midwest. Even when crop yields do rise, there are many accidents and dangers that may set them back again. (You will read more about these problems later in this chapter.) All the same, it has been possible to increase food production to keep step with or even outpace population growth.

Better Crops

There are several different ways of increasing food production. In the U.S. corn belt, much of the increase began with the development of hybrid corn in the 1930's. A **hybrid** plant is a cross between two or more varieties of plants. For example, a fast-growing plant that yields little grain might be crossed with a slow-growing plant that yields a lot of grain. The resulting hybrid might grow fast *and* yield a lot of grain.

Hybrid corn was an early example of plant breeding to boost crop yields dramatically. In Chapter 35, you read about other examples — the new varieties of wheat, rice, and corn that formed part of the green revolution of the 1960's and 1970's. These three crops are staple foods in large areas of the traditional world. Corn is a basic food in much of Latin America; wheat is basic in North Africa, the Middle East, north India, and north China; and rice is basic in much of Asia. Thus the new crop varieties have helped to increase food production in many areas with high population growth.

Breeding new crop varieties takes time and a lot of trial and error. Scientists are hoping one day to be able to improve plants in the lab by a technique known as "gene splicing." Genes are microscopic parts of living cells, and they pass on the qualities of a plant or an animal from one generation to the next. Scientists hope they can identify the plant genes that carry qualities they want. Such qualities include not only fast growth and high yield but also resistance to drought and to poor soils. Then the scientists will try to splice (connect) those genes together within one plant.

A simpler way to develop new food plants is to domesticate (tame) "wild" plants — which is what happened during the Agricul-

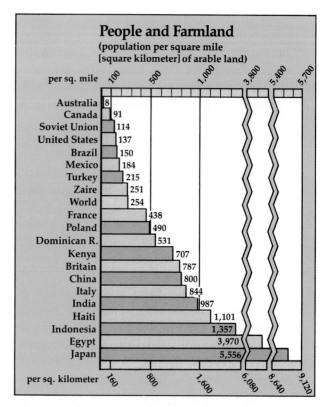

This graph is based not on total area but on the area that can produce food. It gives a more accurate picture of population density. Thus Egypt's total density is only 113 per square mile—but most of its people live crowded together in the Nile Valley.

tural Revolution around 8000 B.C. At that time, wheat, corn, and rice grew wild. Early farmers learned to pick the best seeds for planting and to till the soil.

Today there are still many wild plants that might be developed as food crops. For instance, a type of wild palm that grows in the Amazon River Valley bears a purplish fruit that is high in protein. Local people use the fruit to make a rich drink. The palm might make a useful crop for the tropics, where it could be cultivated quite easily.

Much wild food also grows in the oceans. One example is a tiny shrimplike creature called *krill*, which swarms by the billion. A pound of krill provides as much protein as a sirloin steak. In fact, many huge whales live almost entirely on krill. Not long ago, a scientific expedition found a single swarm of krill that was estimated to weigh 11 million tons (*10 million metric tons*) — one sixth the weight of the world fish catch for one year.

However, new food sources such as krill cannot be put to use easily or quickly. Harvesting krill is difficult and expensive. There is the problem of distributing krill to the people who need it most. There is also the problem of persuading people to change their eating habits — which can be very difficult. For example, you probably enjoy drinking milk but would be disgusted if someone offered you a protein-rich insect. Yet there are many people in the world who feel exactly the opposite about these two foods.

More Farmlands

One method of increasing the supply of existing foods is to open new lands to farming. At present some 1.75 billion acres (*700 million hectares*) are farmed around the world, including almost all the best land. But experts say an equal area of **marginal** (not-so-good) land might be developed.

This development is being done in many places, especially in drier parts of the world. In Israel, Egypt, and other places, irrigation projects are helping open desert lands for crops. At one project in Iran, workers have sprayed a fine mist of oil over shifting sand dunes. The oil forms a crust that keeps the sand from blowing away. Then grasses (for livestock) or trees can grow on the dunes.

In Mali, one of the poorest nations in the Sahel (see page 336), school children have been organized to plant trees to "hold back the Sahara." Some day this program of **reforestation** (planting trees) could create a belt up to 60 miles (*100 kilometers*) wide. This new land for trees, crops, and grasses could be a big help in feeding Mali's hungry population.

These are just some of the methods that have been used to increase Earth's food sup-

ply. There are others, but the two methods that have proved most effective so far are increasing crop yields with new varieties of plants and opening new lands to farming.

Both methods have involved cooperation between the modern world and the traditional world. Modern techniques have been adapted for use on traditional farm villages, and traditional villagers have adapted their old farming ways. At the same time, age-old techniques such as irrigation and tree-planting have been improved by modern methods so that they can be practiced on a much larger scale.

SECTION REVIEW

1. Describe one age-old technique that is being used to increase food production.

2. Give an example of marginal land.

3. Give two reasons why a new food such as krill does not offer a quick solution to food shortages.

4. Describe one modern way in which farmers in different parts of the world have successfully increased food production.

Natural Hazards

Five hundred miles (*800 kilometers*) above Earth's surface, two U.S. Landsat satellites slip silently through space. Special "eyes" scan the ground, studying the crops being grown in major farming areas all over the world. Using data from these eyes, scientists can make color pictures that show what crops are growing and how many acres have been planted. More important, the pictures show whether the crops are suffering from droughts or floods. They can even show whether plant diseases have damaged the crops.

Thus the satellites provide vital information to the U.S. Department of Agriculture. They serve as an early warning system to watch for crop failures, especially those caused by natural hazards.

Earth's Climates

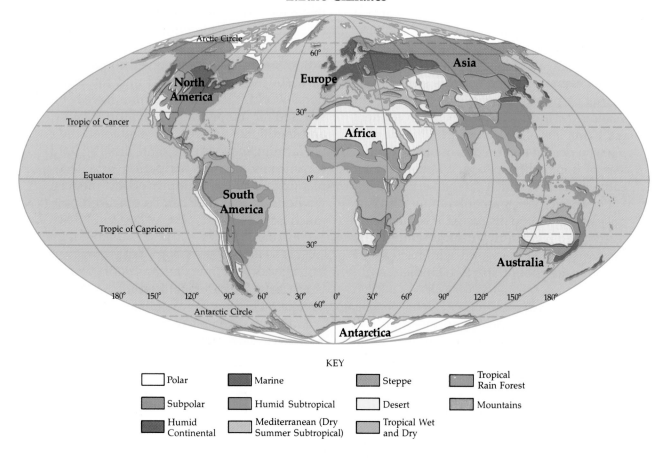

KEY

☐ Polar   ■ Marine   ■ Steppe   ■ Tropical Rain Forest

■ Subpolar   ■ Humid Subtropical   ☐ Desert   ■ Mountains

■ Humid Continental   ■ Mediterranean (Dry Summer Subtropical)   ■ Tropical Wet and Dry

A natural hazard is any kind of event or change in the natural environment that threatens human life or property. Earthquakes and volcanic eruptions are hazards that come from Earth's crust. Storm winds and lightning are among the hazards that come from movements of the air. Floods and droughts are hazards of the water cycle. Diseases and pests are hazards of living creatures.

Some of these hazards are more obviously destructive than others. Floods and storm winds (hurricanes, tornadoes, typhoons) can spread death and destruction over vast areas. A severe earthquake may kill hundreds of thousands of people.

Ways have been found to reduce the death toll from these hazards. Satellite photos can

Compare this with the maps on pages 546 and 561. You can see the way climate affects where and how people live. Polar and Subpolar areas are too cold for farming; Deserts, too dry. The most populous areas have Humid, Marine, or Mediterranean climates.

give warning of a storm wind, so that people in its path can move to take shelter. In areas where earthquakes often occur (see the map on page 483), buildings can be constructed more solidly, so that they are less likely to collapse. (Most earthquake deaths are caused by collapsing buildings.) Scientists are still trying to find reliable ways of predicting *when* earthquakes will occur.

Some natural hazards strike much more slowly or quietly than earthquakes and storms, yet can be just as destructive. This is especially true of hazards that threaten human food supplies. The rest of this section focuses on three such hazards.

Drought

In Chapter 32, you read about Indian farmers waiting eagerly for the monsoon rains to arrive in June each year. You also read that in some years the rains do not come. In the past, no rain meant no crops, and hundreds of thousands of people died of starvation.

Almost everywhere on Earth, the weather varies from year to year. Where you live, you know that in some years there is more rainfall than in others. In places where the rainfall is moderate or light, a change in the weather may mean too little rain (drought). Such places include much of the U.S. West and the steppe lands of Africa and Asia. If the drought is severe, crops may wither and die. With a mild drought, crops may survive — but with a lower yield.

India survived its recent drought with a surplus of food built up in previous years.

However, to survive two dry years in a row, India must find a way of growing some crops even in a drought. One solution is to treat farmlands where drought may occur as if they were always dry — like farmlands reclaimed from desert. A proposed canal (see page 450) would make water from India's rivers available, if needed, to large areas of farmland.

Many farmlands around the world rely on river water for irrigation. In time of drought, however, a river may carry too little water to supply all the nearby farms, or the river may run completely dry. One widespread solution to this problem is to build a dam on the river. The dam stores water in the form of an artificial lake. In time of drought, this water can be released to keep the river flowing.

When the topsoil is dry, there may still be plenty of groundwater below the surface. In some North African nations, deep wells

---

Irrigation of some kind has been used for thousands of years to grow food in dry lands. Below, water pumped from a simple well supplies a farm in Iran.

have been drilled to tap groundwater below the Sahara. Such deep wells are expensive, however. Also, in arid places such as the Sahara, there is too little rain to replace the groundwater. In other words, the water is being used up, like oil.

Fortunately many areas where drought can occur have groundwater closer to the surface. There is usually enough rainfall to keep the level high. In these areas, shallow and inexpensive wells can be used. Thousands of such wells now dot the farmlands of Pakistan, India, and Bangladesh.

Pests

Insects far outnumber humans on Earth. So do birds. So do mammals. Many of these creatures eat the same kinds of food plants as humans do. Humans often find it difficult to stop them.

In Chapter 34, you read about a system of bells and strings that is used in Indonesian fields to scare off birds. Similar means are used in small farms around the world. When the numbers of bird or mammal pests become too large, farmers may kill them with poison or gunshot.

Insects are usually the most difficult pests to control. Insects are smaller and far more numerous than birds or mammals, and cannot be scared off or shot. Poison, in the form of a chemical insecticide, is the most common defense. It has to be used carefully, however. Too much poison may affect humans and livestock. It may also kill all except a few tough insects, whose descendants will be resistant to (unaffected by) the poison.

One of the worst insect pests is the desert locust, which lives in much of North Africa and the Middle East. The locust, similar to a grasshopper, sometimes gathers in swarms of tens of millions. Such a swarm arrives as a huge cloud that blackens the sky, descends on a farming area, and strips the crops bare. Once the swarm arrives, there is no defense.

Modern technology has helped to reduce

Insects threaten much of the world's food supply. One of the worst pests is the desert locust. Above, in Ethiopia, a swarm of locusts descends on a village.

the destruction caused by locusts. For example, in northeast Africa, airplanes report on the movements of any locust swarms. Planes are then used to spray insecticide on farmlands on the locusts' route.

Scientists have developed techniques of insect control that do not involve the widespread spraying of poisons. One technique uses a chemical produced by the insects' own bodies to attract the insects to a spot where they can easily be killed. Another technique is to breed insects that are sterile (unable to produce offspring). By releasing these insects to mate with the normal insects, it is hoped that the insect population will be cut down.

So far such techniques have been used mainly in the modern world. They have had some success. For example, there is a kind of fly that lays its eggs in the skin of cattle. The cattle become diseased and often die. In Texas the use of sterile flies has cut down the number of these pests and practically wiped out the disease.

Plant Diseases

You read earlier that many diseases affecting humans have been brought under control. Many such diseases are still widespread, especially in the poorer parts of the traditional world. One reason is that public health measures do not reach many people who live far from cities. Another reason is that many people are undernourished — their diet is too poor to keep them healthy.

Diseases attack plants too. When a disease attacks a plant used by humans, it becomes a natural hazard. When a disease attacks a staple food plant in part of the traditional world, it can be a disaster.

Humans have learned many ways of fighting plant diseases. Sometimes chemicals can be used, much as drugs are used to fight disease in humans. For example, grape growers may spray their vines to control diseases such as mildew.

Another way to fight plant diseases is to develop plants that have built-in disease resistance. In the early 1970's, a hybrid corn used widely in the U.S. showed a weakness to a fungus that attacked its leaves. As a result, corn output dropped 15 percent. Scientists then developed a new variety of hybrid corn that was resistant to the fungus. In a short time, corn output rose again.

In developing improved varieties of crops for the green revolution, scientists try to make sure that the new plants are hardy and disease-resistant. But they may not always succeed in avoiding all possible hazards. For example, the hybrid corn that suffered from fungus in the early 1970's was a new and supposedly improved variety. In this case, human actions played a part in creating a hazard — a destructive plant disease.

In the following chapter, you will read about other hazards that have arisen as humans have developed Earth's resources and tried to improve their lives. You will also learn how the challenge posed by those hazards is being met today.

YOUR LOCAL GEOGRAPHY

1. As in many other nations, the pace of population growth in the United States has slowed in the past few decades. Discover whether this trend holds true for your community. Look through an up-to-date almanac for population statistics for your county or city or the large city nearest to you. You should also be able to find population figures for one year in each decade from 1940 through 1980. Check these figures to see if the number of people in your area has been increasing, decreasing, or fluctuating from decade to decade in recent years. Estimate or calculate the percentage of population increase or decrease in each of the four or five decades. How does the population trend in your city compare with that of the nation as a whole?

2. Speak with your parents or another adult member of your community who has lived there for some time to learn what natural hazards have occurred in your community over the years. Were the hazards linked to changes in Earth's crust, the air, the water cycle, or to plant and animal life? Ask a local librarian for help in locating newspaper or magazine articles about one of these hazards. After reading several of these articles, answer the following questions: Was this hazard entirely natural or was it caused in part by human activities? Could a similar hazard be prevented today? If so, how? If not, are there ways to reduce the damage?

A Village in Trouble

The novel Afrika Ba'a, written by Rémy Mvomo of Cameroon, describes the struggle for survival in a West African village. The village, named Afrika Ba'a, is imaginary, but its problems are real. The hero is a young villager named Kambara (kahm-buh-RAH). As you read what follows, compare his problems with those of the village as a whole, and note how he tries to solve them.

"You talk about marriage," said Kambara to his mother, "but where will I get the bride wealth? I know Ada's parents like me, but where they live it's the custom to give both cash and goods. Today people expect luxury goods like kerosene refrigerators and even cars. . . . Our plot of land will never bring in enough money for me to marry."

"Couldn't you work at the sawmill?"

Kambara shrugged. He had little desire to go and work in the sawmill. All things considered, he'd rather go to the city, where he could surely find a better-paid job. After all, he'd gone through high school.

Kambara goes to check his plot of land. On the way, he sees smoke rising from a field. He stops to talk with a woman who is slashing weeds and burning them to prepare the ground for planting.

"Your field is so far from the village, you'll have trouble bringing in the crops."

"What can I do?" she said. "There isn't any other land. There are more and more people all the time, and everyone wants to live close to the roads."

When Kambara reached his plot of land, he could see that it was in terrible shape. He examined each palm tree in turn, hoping against hope to find one healthy branch. But insects and rot had destroyed everything.

"Now I'll have to cut everything down and start over," he said to himself. "And then wait."

It had been a bad season for everyone in the village. The men, who sold cacao as a cash crop, had earned very little because the price had dropped. The women, who raised the food crops, had suffered various misfortunes. The rainy season had arrived very late, and when it did come, there was a deluge. The Kavao River had overflowed its banks and flooded the fields right up to the edge of the village.

To survive, the people of Afrika Ba'a had to eat plantains that were barely ripe. Some even ate the roots.

When the heavy rains finally stopped, it took a long time for village life to return to normal. People were thinner, weaker, and depressed. They had got into the habit of going to bed early in the evening, since the best remedy for hunger is sleep.

Kambara goes to the city but cannot find work. He decides to return to Afrika Ba'a and help the villagers to make a better life. He persuades the government to lend them a truck, tools, and seed. The people work together to clear the land near the village and farm it in cooperation. As the book ends, the villagers have new hope for the future.

— From the French of *Afrika Ba'a*,
by Rémy Médou Mvomo.
Editions Clé, Yaoundé, Cameroon, 1969.

Ask Yourself . . .

1. What was the main source of money for the people of Afrika Ba'a? Why was it hard for them to earn much?

2. Why was the flood such a disaster for the villagers?

3. What main problems does Kambara tackle at the end of the book?

Solving Problems

As you start to feel hungry late one afternoon, you find there is no food in the kitchen. It seems you and your family have a problem tonight. Meanwhile, 7,500 miles *(12,000 kilometers)* away in a farm village of Zaire, Mbombo and her family have finished their one daily meal of cassava and may be feeling hungry too (see Chapter 30, page 409). How would you and your family solve your problem? How could Mbombo and her family solve theirs?

Solving any problem — about food, population, or something else — should be a thoughtful step-by-step process, if the results are to turn out well. The basic steps of problem-solving are illustrated in the flowchart below. At each step, the problem-solvers ask themselves a key question that requires an answer.

To start the process, for example, the problem-solvers ask, "What's the problem?" and then try to define the problem clearly. In the first situation above, you and your family might describe your problem by saying, "There's no food in the house for our evening meal." The next step is to ask, "What caused the problem?" Was there no money? Did someone forget to shop? Understanding the cause of a problem may help to find a solution.

Next you would think of several possible solutions. Maybe you can eat at a restaurant tonight. Or you might be able to get to the supermarket before it closes. You may need to find out more facts before picking out the possible alternatives — for example, you might need to find out exactly what time the supermarket closes.

Finally, before deciding which solution is best, you would compare the advantages and disadvantages of each possibility. In other words, you would try to predict its consequences.

Study the steps of the problem-solving process shown in the flow chart below. Then read the story about Mbombo's problem on the next page. As you read, think how the problem-solving steps might be used to solve this problem.

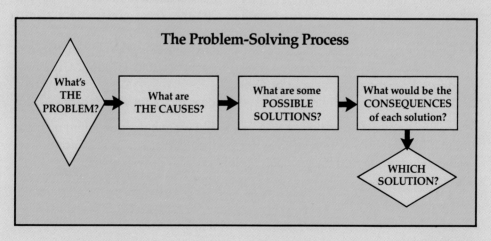

The Problem-Solving Process

What's THE PROBLEM? → What are THE CAUSES? → What are some POSSIBLE SOLUTIONS? → What would be the CONSEQUENCES of each solution? → WHICH SOLUTION?

(Turn page.)

Mbombo's Problem

Like three out of four people in Zaire, Mbombo and her family lead a rural life. They eat just once a day and have a limited diet. Part of the meal is usually a starchy root called cassava. Most meals contain no meat.

Mbombo's family uses simple farm tools and raises only subsistence crops — that is, just enough food for their own needs. Cassava is the main crop in Mbombo's village because of the poor soils. If chemical fertilizer could be used to improve the soil, the farmers could grow corn and other crops to feed themselves and still have some left to sell. But fertilizer is expensive, and most farmers in Mbombo's village cannot afford it. Even if they could grow extra food, they would have a hard time getting their crops to distant markets, because roads are poor.

Mbombo and her family sometimes think of moving to the city, but they know city life can be hard for unskilled workers. It can be difficult for them to find jobs or adequate housing. Still, by hard bargaining at the food market, city people may be able to eat better by adding fish to their cassava or rice, for more protein.

Each question below matches a step in the problem-solving process. Refer back to the story to answer each question. Write your answers on a separate sheet of paper.

1. What is the problem of Mbombo and her family? Describe it clearly.
2. What are some causes of this problem? Explain at least two.
3. What are some possible solutions? Try to think of some alternatives which: (a) Mbombo and her family might choose for themselves; (b) the whole village might choose together; (c) the Zaire government might choose to help them. Explain the alternatives.
4. Can you predict some *consequences* (results) of each alternative? Explain the advantages and disadvantages of each possible solution. Give evidence from the story for your answers.
5. Which solution or solutions would you choose? Explain your reasons.

Are there any additional facts that you might want before deciding on a solution to Mbombo's problem? If so, what are they? What sources of information would you check to find these facts? Explain on your answer sheet.

A. Words To Remember

In your own words, define each of the following terms.

domesticate	groundwater	reforestation
drought	hybrid (plant)	

B. Check Your Reading

1. Why have population growth rates been declining recently in most countries?

2. What was the main advance made during the Agricultural Revolution? What similar change might be made today?

3. What is gene splicing? How has it affected world food production so far?

4. Describe one modern method that has been used effectively against a natural hazard.

5. Why are insects often a greater threat to food crops than other animals are?

C. Think It Over

1. Why do you think that most efforts to develop new varieties of food crops were focused on corn, rice, and wheat?

2. You read that there are two ways to view the world's growing population. Which view do you agree with? Explain your answer.

3. "In my opinion, Earth's worst natural hazard is. . . ." Choose a hazard to complete the statement, and explain your choice.

D. Things To Do

World population has increased rapidly in the 20th century. To gain insight into one reason for this increase, find examples of the way infant mortality has changed. Make a list of six famous American men and women who died before 1900. Using an encyclopedia or biographical dictionary, choose three of the six who were married and had children. For each person, note the total number of children born and the number who died in infancy. Now make a list of six famous Americans who were born after 1900. Using an encyclopedia or current biography, choose three who were married and had children. (If you choose living Americans, they should be over 40.) Again note the total number of children born and the number who died in infancy. Compare the two sets of infant mortality figures. What changes, if any, do they suggest?

Chapter 42

Using Earth's Resources — For Better or Worse?

In some places, you can see scars on Earth that look like craters on the moon. In other places, smoke in the air sometimes blots out the Sun. Chemicals mix with moisture in the air to form "acid rain," which kills fish in the lakes where it falls.

These three effects can be traced to human activities. All three can result from one cause — the digging and burning of coal. Coal is a natural resource — a fossil fuel that is widely used as a source of energy to run industry or heat homes. But the three effects mentioned above are hazards to human and other forms of life. They are human-caused hazards, not natural ones.

All through this book, you have read of the many ways in which humans use the resources they find around them. Earth's resources are central to life itself. They provide us with food, shelter, and tools — the means of living. Ever since humans first walked on Earth, they have used nature's resources.

In recent centuries, new uses have been found for old resources, and new resources have been discovered. The pace of life has quickened. So has the pace at which people are using resources. This increasing use of resources has led to higher standards of living. But it has also had hazardous side effects.

So long as human use of resources was limited, most side effects were also limited. On a still day, smoke from cooking fires might hover over a small village. But when the wind picked up, the smoke blew away. Humans added so little smoke to the air that it could disperse easily.

Today factories, automobiles, and power plants pump vast amounts of smoke and fumes into the air. Under certain weather conditions, the polluted air may not be able to move. Then this air may become dangerous to breathe. People may suffer headaches, coughing spells, shortness of breath. Some may even die. One of the worst crises to result from polluted air took place in 1930 in a river valley in Belgium. A poisonous cloud of smog settled over the valley. More than 1,000 people became ill, and 60 died.

In recent years, vigorous efforts have been made to deal with human-caused hazards like air pollution. For example, each of the hazards mentioned at the start of this chapter has been tackled. The scars from some coal mines have been filled in, and grass and flowers have hidden much of the damage. Special devices have been put on some smokestacks to take the soot out of smoke and keep air cleaner. In some cases, these devices also remove acids from smoke, so that rain no longer carries these acids to lakes and streams.

In developing Earth's resources, humans often damage the environment they live in.
At left, smoke from an industrial complex pollutes the air and dims the Sun.

If people cause hazards, they can also correct them. In this chapter, you will read about many human-made hazards and the efforts that are being made to reduce these hazards. You will also read about efforts to find new sources of energy to lessen our dependence on fossil fuels such as coal.

Human-Caused Hazards

Most human-caused hazards involve a trade-off between risks and benefits. To get the benefit, one must risk a hazard. For example, you know that dirty air is a hazard — but does this mean you would *never* dirty the air? What if you were on a camping trip and wanted to cook a meal, or just keep warm? Would you light a campfire, knowing that the fire will pollute the air with smoke? Which would be more important to you — the meal and the warmth, or avoiding pollution? You would have to make a choice.

The choice is not always easy to make, since the benefits and risks may be hard to compare. One reason is that the risk is sometimes hidden. A hazard may become apparent only after many years. Usually benefits are easier to see than risks, and their value is easier to calculate.

Another reason is that the benefits and risks may not be spread equally. One group of people may get the benefits, while a different group may face the hazards.

Still another choice is often involved in judging risks. Think again of coal-burning and of the devices added to smokestacks to cut pollution. These devices are a means of reducing the hazard caused by burning coal. But they cost money. Do they lessen the hazard enough to make the cost worthwhile? People often disagree over this question.

In recent years, people have become more aware than ever before of the hazards that human activities may cause. They have stepped up efforts to control these hazards. The following section will look at some of the types of hazards that are under attack.

Protecting Air and Water

In 1978 a U.S. oil tanker struck a reef during a storm off the coast of France. Into the English Channel poured 1.6 million barrels of oil. The oil spread out across the surface of the water in an oil slick 60 miles (*100 kilometers*) long. Fish, oysters, and birds died by the thousands. Oily slime coated French beaches, and France's tourist industry suffered large losses. Millions of dollars had to be spent to clean up the mess.

Vast amounts of oil are moved across the world's oceans every year. Most oil is used in developed countries. But much of it starts out in developing nations of the Middle East, Africa, Asia, and Latin America. To move this oil, massive supertankers have been built. When disaster strikes such a ship, the resulting oil spill can be immense.

When oil is spilled in the oceans, it does more than make the water dirty. It also spreads damage and destruction. The damage caused by oil spills affects humans directly — for instance, by killing fish that could have been used for food. The damage also harms wildlife such as birds. A bird that lands on oily water may become coated with oil so that it can neither fly nor swim.

What is being done to control oil spills? For one thing, rules that govern shipping have been tightened. Crews must now receive training in handling emergencies. Tankers must be equipped with radar to spot

The World: Economic Activities

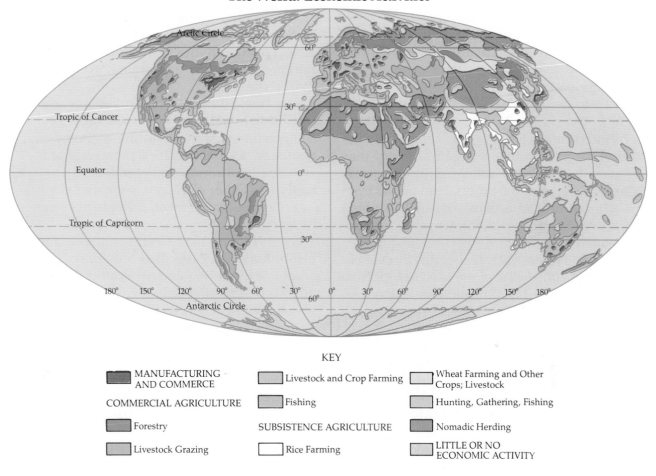

KEY

MANUFACTURING AND COMMERCE

COMMERCIAL AGRICULTURE

Forestry

Livestock Grazing

Livestock and Crop Farming

Fishing

SUBSISTENCE AGRICULTURE

Rice Farming

Wheat Farming and Other Crops; Livestock

Hunting, Gathering, Fishing

Nomadic Herding

LITTLE OR NO ECONOMIC ACTIVITY

reefs, icebergs, and other ships in time to avoid collisions. In addition, new tools have been developed for cleaning up spills. Gene splicing (see page 548) may even play a part. Scientists have been working to develop a type of bacteria that could "eat" oil and thus get rid of the problem.

Oil spills are an example of water pollution. Smoke and smog are examples of air pollution. Both can affect wide areas, for water and air do not stand still. A **pollutant** (something that pollutes) dumped into water or air at one place may cause a hazard far away. For example, acid waters in Sweden have been blamed on coal smoke from factories in West Germany and elsewhere. Thus efforts to deal with pollution may involve cooperation among different nations.

Most people live in the best farming areas (see the map on page 546). Thus manufacturing tends to compete with farming for the best land. Areas too dry for crops are used for grazing or herding. Hunting, Gathering, Fishing also includes occasional crop farming.

Industrial nations have taken the lead in trying to halt pollution. The early 1970's were a time of strong antipollution drives in the United States. New laws were passed, often after sharp debate. Some of these laws were later made even tougher. Other laws were eased, after complaints that the costs sometimes outweighed the benefits.

Developing nations have been slower to take action against polluted air and water. One reason is that these nations have fewer factories and oil supplies to cause heavy pol-

561

lution on the spot. In the late 1970's, however, a river in China caught fire because it was coated with oil that had been dumped into the river by five factories upstream. China passed its first antipollution law in 1979 to deal with hazards such as this.

Protecting the Soil

Ever since humans began farming, they have had to battle soil erosion. Of course, erosion is a natural process and occurs even without human activity. But by disturbing the soil to plant crops, farmers speed up erosion drastically.

In one recent year, U.S. farms lost 6.4 billion tons (*5.8 billion metric tons*) of topsoil to erosion. The lost topsoil was enough to cover all of the state of Missouri one inch (*2½ centimeters*) deep. If such losses continued, U.S. crop yields would begin to drop.

Fortunately there are farming techniques that can keep erosion to a minimum. One technique is to rotate crops — that is, to change crops from field to field rather than to plant the same crop in the same place every year. However, some farmers don't do this because they specialize in one crop.

Another way to fight erosion is to do no plowing at all. This is called the **no-till method** of farming. It was made possible by recent advances in chemistry and by the development of more powerful tractors.

In the no-till method, seed is planted in a narrow channel ripped in the stubble of the last crop. Chemicals are spread to keep weeds down. However, in some places, these chemicals might be washed into streams and cause water pollution. Here risks and benefits have to be balanced.

Protecting Animals and Plants

Seven hundred years ago, Kampuchea was the center of an empire. At the time, the empire contained more than 200,000 elephants. Many of the elephants were tame and trained to haul heavy loads.

Today only 17,000 elephants remain in all of Southeast Asia. Two human activities have wiped out the rest. One activity is hunting. The other is farming, which requires the clearing of forests. By cutting down trees, humans destroyed the elephants' **habitat** (homes and food supplies).

Elephants play a role in the economy of Southeast Asia. Some 2,500 tame elephants are used by Burma's timber industry. They haul logs. Other elephants are used by villagers as a source of meat. If elephants were to die out, Asians would pay a big price.

Over the ages, many forms of animal and plant life have indeed died out. They have become **extinct**. An ancient example is the dinosaur. A more recent one is the passenger pigeon, which thrived in North America until it was wiped out by hunters less than a century ago.

Scientists warn that as many as one million species of animals and plants are threatened with extinction between now and the end of this century. Threatened animals include the bald eagle (see photo opposite). Threatened plants include many primitive forms of wheat, corn, and rice. Once a species dies out, humans cannot bring it back. The genes that give that species its precise form may be lost forever.

Many people believe that we humans have a duty to protect the other life forms with which we share Earth. But there is also a practical reason for protecting these life forms. There is no telling when a plant or animal that seems of no use to humans will turn out to be a valuable resource.

For example, you read in Chapter 41 that scientists have developed new varieties of food plants. To create each new variety, scientists may cross many different types or species. A species that seems useless by itself may have a quality that is vitally needed for crossbreeding.

To save animal species, attempts are being made to stop the destruction of habitats

The habitat of many animals is threatened by human activities. The bald eagle (above), the California condor, and the alligator are three endangered species in the U.S. In the world, the elephant, the tiger, and various whales are among animals facing extinction.

around the world. This is not easy, for Earth's growing population needs more and more land for living on and for growing food. There is continual debate about risks, benefits, and costs.

To save plant species, scientists have found a different method. They are collecting seeds from plants in all parts of the world. With these "seed banks," they can still draw on the genes of plant species that have died out on the land.

Protecting Workers on the Job

PVC is one of the "miracle plastics" that have been invented in recent decades. The letters stand for polyvinyl chloride. PVC is used in a wide range of modern products,

from plastic pipe to food wrappings. Unfortunately vinyl chloride — the chemical from which PVC is made — can kill people. Workers exposed to vinyl chloride dust or gas in factories face a higher-than-normal chance of getting liver cancer.

In a chemical plant in Malaysia, workers must enter the vats in which PVC is made and clean them out after each batch. The company provides cloth masks, but no special clothes. The workers increase their risk of early death every time they go to work.

Some 70,000 different chemicals are in use in industries and homes around the world. Each year about 1,000 more chemicals are added. Most of these chemicals are believed to be safe. But some, like vinyl chloride, may be deadly. They are human-caused hazards, just like oil spills.

Chemicals like vinyl chloride don't *have* to kill. Measures can be taken to protect workers and others who might come in contact with them. In many lands, including the United States, such measures are being taken.

In the mid-1970's, reports of cancer deaths among PVC workers caused concern in the U.S. Strict new safety measures were ordered by a U.S. government agency. The agency said factories must stop vinyl chloride from leaking into the air. It said workers must wear special masks that supply pure air for them to breathe.

There was a price to pay for the new safety measures. It cost money to plug all leaks and provide masks. The cost of manufacturing PVC rose — and so did the prices of many plastic products.

Malaysia has its own safety rules. However, as in most developing countries, the rules are not very strict. Business and government leaders try to hold down costs so newly begun industries can compete on the world market.

Job-related hazards are a worldwide problem. Some have existed for centuries. One such hazard is a crippling disease nick-named "black lung," which may affect miners who breathe coal dust day after day. Other hazards are linked directly to modern products and techniques. For example, radiation may be a threat to hospital workers who work near X-ray machines. As more and more nations develop modern industry, worker health becomes a bigger issue.

In most cases, measures can be taken to provide protection. But cost is an ever-present concern. As with products made of PVC, safety costs add to the price of many goods, from autos to designer jeans.

SECTION REVIEW

1. What kind of damage is caused by an oil spill? Describe one step being taken to reduce such damage.

2. What necessary human activity causes soil erosion?

3. Describe two methods being taken to save plant and animal species threatened by extinction.

4. What difficulty is often involved in making working conditions safe for workers?

Energy for the Future

In Chapter 30, you read about Mbombo, a farm girl in Zaire. Since she lives in the tropics, her home needs little heating. Her food is cooked over a small wood fire. When she has to go somewhere — to school, for instance — she walks. As a result, Mbombo makes few demands on the world's energy resources.

In Chapter 17, you read about Benjamin Lesage, a farm boy in France. Benjamin uses more energy resources than Mbombo. He rides a motorbike, which takes gasoline and oil. He lives in a large house that is heated in winter.

If you are like most Americans, you use even more resources than Benjamin. You "consume" a wide variety of goods, most of which require large amounts of energy to manufacture and transport. You use fossil fuels to cook, heat, and travel.

You might draw two conclusions from these examples. First, all people use energy resources. Second, people in modern nations like France and the U.S. use more energy resources than those in developing nations like Zaire.

The growth of the modern world since the Industrial Revolution has depended heavily on fossil fuels. Not counting food, almost 90 percent of the energy used in the United States today comes from fossil fuels. But fossil fuels are no longer cheap and readily available. In particular, petroleum has shot up in price since the early 1970's. As a result, a search is on for new forms of energy. One goal is to find new ways to use fossil fuels such as coal. A second is to develop alternative sources of energy.

More from Fossil Fuels

Coal is the most abundant fossil fuel, and is found on every continent. But coal has a major drawback: It is a solid. Liquid fuels are much better for powering vehicles like cars, trucks, and airplanes. (Liquid fuels can be piped to the engine a little at a time, as needed.) Gases are easier to transport and to use for purposes like home heating.

In the past, coal was turned into liquid and gaseous fuels. But it cost more to do this than to use oil or natural gas. Efforts are under way to reduce the cost so that coal products can compete with natural gas and oil.

Gaseous and liquid coal are human-made, or synthetic, fuels — **synfuels**, for short. They are among a variety of synfuels that researchers hope to make available at reasonable cost in the near future to supplement natural fuels.

Another possible synfuel is **shale oil**, which can be extracted (removed) from certain rocks that occur in different parts of the world. The richest sources in the U.S. are in

The World: Energy Sources

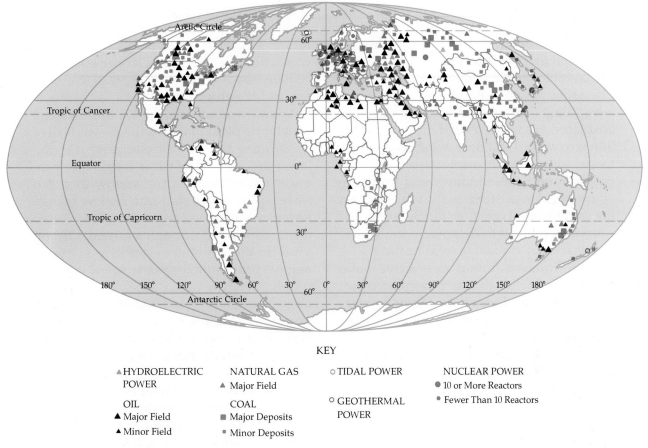

KEY

△ HYDROELECTRIC POWER

OIL
▲ Major Field
▲ Minor Field

NATURAL GAS
▲ Major Field

COAL
■ Major Deposits
▪ Minor Deposits

○ TIDAL POWER

○ GEOTHERMAL POWER

NUCLEAR POWER
● 10 or More Reactors
• Fewer Than 10 Reactors

the Rocky Mountains. But the rocks must first be mined, crushed, and heated — an expensive process and one that might harm the environment.

Shale oil was produced in Scotland on a large scale as long ago as World War I. In the U.S., a dozen small plants have produced shale oil in experiments since then. Two large plants are to be in operation in Colorado by the mid-1980's. If shale oil can be made at a low enough cost, it may gain great importance in the future. U.S. reserves of shale oil are said to be 22 times as plentiful as U.S. reserves of ordinary oil.

Other Sources of Energy

There are many sources of energy that could supplement — or replace — fossil

Most energy still comes from fossil fuels, which cannot be renewed. Other energy sources are renewable. Hydroelectric and tidal power make use of water; geothermal power takes heat from underground; and some nuclear power plants produce their own fuel.

fuels. Some have been known and used for centuries. Others are still in early stages of development. Here are a few examples of what's being done.

■ *Solar Energy.* All energy on Earth comes directly or indirectly from the Sun. How much use can humans make of the Sun's direct energy, known as **solar energy**?

There is evidence that people have been making good use of it for a long time. In Greece archaeologists have unearthed a town where people lived some 24 centuries

565

ago. Every house contained a living room that faced south onto a courtyard.

It seems certain that the reason houses were built this way was to soak up the greatest possible warmth from the winter Sun. The same principle is being used today in newly built "solar homes." Large windows are placed in southern walls. In summer the Sun is overhead and does not shine through the windows. But in winter, the Sun's rays pass through the windows and strike an object such as a stone wall, which stores the Sun's heat. Even after the Sun sets, the wall gives off heat to help warm the house. This is known as **passive solar heating.**

Some modern homes use **active solar heating**. Collection panels, often placed on a roof, take in the Sun's heat. Water or air is passed through the panels to absorb the heat, and then passed into the house. In some cases, heated air is passed through large bins of rocks beneath the house. At night or on cloudy days, the rocks give off heat to warm the house.

Since active and passive solar heating do not need advanced technology, they can be used easily in traditional parts of the world. Another form of solar energy, the use of solar cells, requires space-age technology. **Solar cells** are metal strips that can use the Sun's rays to make electricity. Such cells have turned out power for an Indian reservation in Arizona and astronauts in space. The cost of solar cells has dropped sharply in recent years. If it drops further, solar cells may some day produce much of the electricity used in homes and industries.

■ *Energy from Air and Water.* In southern California, winds whistle through a mountain pass and spin what looks like a giant airplane propeller. The three-blade propeller is as tall as a 16-story building. This propeller is a wind turbine, put into operation in 1980. It was set up by a power company as part of a planned "wind park" to produce electricity for California homes.

In places with steady winds, such turbines may help produce the electricity of the future. However, most places do not have winds that are steady and strong enough to be harnessed. A more widespread source of energy is Earth's water.

As you read in Chapter 17, France has a generating plant that uses the rise and fall of ocean tides to turn out electricity. Off the Japanese coast is a yellow barge that produces electricity with the energy of ocean waves. The waves slap into hollow orange towers. Each time a wave comes, it forces air up the tower and through a nozzle, turning a turbine that generates electricity. The technique is promising, but still experimental.

Compared to ocean tides and waves, rivers have been generating electricity for a long time. What is new is the number of hydroelectric plants, which has increased rapidly in recent years. One advantage of hydroelectric power is that the dam used to harness the flow of water can also be used to control the river for irrigation. However, dams may also harm the environment (see Geography Skills 51 on page 569). In any case, hydroelectric power cannot be used in large areas of the world that lack sizable or fast-flowing rivers.

■ *Nuclear Energy.* In France's Rhône Valley is a massive plant to produce electricity from heat given off in nuclear reactions. Nuclear reactions involve the breakup of the nucleus (core) of atoms. The breakup releases energy in the form of heat.

In 1986 the nuclear reactor at Chernobyl in the Soviet Union had an uncontrollable reaction which resulted in the deaths of many persons. Radioactivity resulting from the accident was carried in the atmosphere and detected in Western Europe within hours. The long-term effects will not be known for many years. This and the earlier incident at Three-Mile Island in the U.S. has stirred a lot of debate. Critics say that nuclear reactors are hazardous to run and leave waste

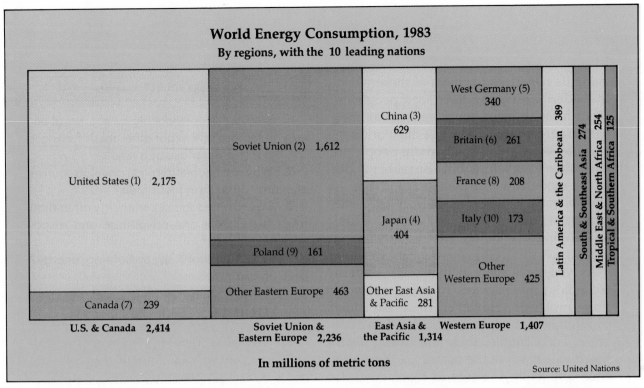

World Energy Consumption, 1983
By regions, with the 10 leading nations

United States (1) 2,175

Soviet Union (2) 1,612

China (3) 629

West Germany (5) 340

Latin America & the Caribbean 389

South & Southeast Asia 274

Middle East & North Africa 254

Tropical & Southern Africa 125

Britain (6) 261

France (8) 208

Japan (4) 404

Italy (10) 173

Poland (9) 161

Other Western Europe 425

Canada (7) 239

Other Eastern Europe 463

Other East Asia & Pacific 281

U.S. & Canada 2,414

Soviet Union & Eastern Europe 2,236

East Asia & the Pacific 1,314

Western Europe 1,407

In millions of metric tons

Source: United Nations

products that are almost impossible to dispose of safely. Supporters say that the hazards have been exaggerated. They say such hazards as do exist can be brought under control. At present, 10 to 15 percent of the electricity in modern nations comes from nuclear power plants.

Existing nuclear plants are based on nuclear **fission** (the splitting of the atoms of a heavy metal such as uranium). Scientists are experimenting with another type of nuclear energy. This is called nuclear **fusion**, and it produces energy by combining atoms of the gas hydrogen to form another gas — helium. The hazards of nuclear fusion are much smaller than those of nuclear fission. However, scientists do not expect to have a fusion plant in operation in this century.

■ *Conserving Energy.* Energy conservation, or cutting down on energy waste, would have the same effect as developing a major new source of energy. By some estimates, up to one fourth of the energy used in the United States serves no useful purpose. For example, much of the heat used in our homes leaks out wastefully. This heat could be kept in — and energy bills cut — if our homes were better insulated.

All energy that was used in the world in this year added up to the amount that would have come from more than eight billion metric tons of coal. The bar shows relative amounts used by regions (vertical bands) and by major energy-producing nations.

Have you ever left a light burning for hours when you were out of your room? Turning off this light would have saved energy. What other ways can you think of to cut energy waste?

Our World Tomorrow

Like the aliens you met in Unit 1, you have not been able to visit every place on Earth or all of its people. However, you have had a broad view of each of Earth's regions and a closer look at representative areas within each region. You have discovered how the various parts of Earth differ from one another — and also what they have in common. You have been able to compare other parts of Earth with your own nation and also with your own community.

Throughout this text, you have seen people interacting with the environment. In traditional areas, you have seen people interacting with their local environment, using its resources for most of their needs. In modern areas, you have seen people interacting with many environments as they use resources from all around the world.

Today there are not only more people on Earth than ever before, but also more people in the modern world then every before — and still more moving into it every year. The modern world holds promise of a better life, but finding that better life involves many problems.

As you have read in this unit, modern technology offers ways to solve many of these problems. However, there are some problems that technology alone cannot solve. For example, there are the conflicts that divide many of the peoples of the world from one another. Some of these conflicts are violent, like the war that broke out between neighboring Iran and Iraq in the early 1980's. Others involve suspicion and mistrust, like the political division between Western and Communist nations. All of these conflicts can make it more difficult to solve the other problems of the world today.

Thomas Jefferson said that "every nation prospers by the prosperity of others." Today the nations of Earth are linked even more closely than in Jefferson's time. They trade in goods and resources. They share the problems of dealing with natural and human-made hazards, and meeting their energy needs. Individual nations must find a way to balance their own aims and ideals with the need for international cooperation.

Thus making a better life tomorrow depends on far more than finding resources and making use of technology. People themselves must first decide what kind of life they want. They must then have not only the skills to create that kind of life but also the desire and dedication to work for it.

This text has shown you how the world of today is changing. Like the other billions of people on Earth, you are part of that change, and are shaping our world of tomorrow.

SECTION REVIEW

1. Why are people looking for sources of energy to replace oil? What other fossil fuels are coming into greater use as a result?

2. Why are people trying to develop sources of energy other than fossil fuels?

3. Describe two sources of energy other than fossil fuels. Give one advantage and disadvantage of each.

4. What is meant by conserving energy? Give an example.

YOUR LOCAL GEOGRAPHY

1. What kind of energy resources are available in or near your community? Are there deposits of coal, oil, or natural gas? Is there a hydroelectric dam or a nuclear power plant, or is electricity generated by fossil fuels? Is there any use of solar energy or wind energy? When you have completed your list, consider for each energy source whether (a) it is running out, or (b) it causes any environmental hazards. If there are few or no energy resources, where does your community's energy come from? If you cannot answer these questions, look through the local government listings in the telephone book and select the departments most likely to have answers. (Look for headings such as "Commerce," "Environment," "Power"; or for a general information service that can refer you to the right departments.) Prepare a letter that your class could send to the appropriate departments asking for information.

2. Has your community had to make any recent decision about the use of resources? For example, there may have been plans to develop an area that was used for recreation or was the habitat of wildlife. When you have chosen an issue, describe it briefly on a sheet of paper. Then draw two columns, one labeled "Risks"; the other, "Benefits." List the risks and benefits, describing each one as clearly as you can. Do the benefits outweigh the risks? Write your conclusion, with an explanation.

Evaluating Decisions

As oil prices rose in the early 1970's, the U.S. began looking for less-expensive sources of energy. Many people saw nuclear energy as the ideal solution, so more nuclear power plants were started in various parts of the U.S. Then in 1979, a radioactive leak was found in a plant in Pennsylvania. Building of other new plants was halted, as people tried to find out whether the possible hazards of this "solution" were really outweighed by the benefits.

It often happens that the solution chosen for one problem may cause new problems or have other unexpected results. For this reason, an important step in problem-solving is to *evaluate* (judge) a decision or solution after it is put into action. This way, if a solution has worked well, it can be continued. But if it has not worked, or has created new problems — as in the case noted above — the same steps used to solve the first problem can be used again to search for a better solution.

The flowchart below is similar to the chart in Geography Skills 50. Instead of asking questions, however, each box of this chart describes a major step of problem-solving. This chart also shows steps before and after a decision is made. Study the steps in order. Which four steps happen before deciding on a solution? Which two steps happen afterward?

In Egypt farmers like Halim Amin depend completely on the Nile River for their livelihood (see page 346). As you read the problem of the Nile on the next page, think how the steps of problem-solving might have been applied to it. Then on a separate sheet of paper, follow the steps listed below the story.

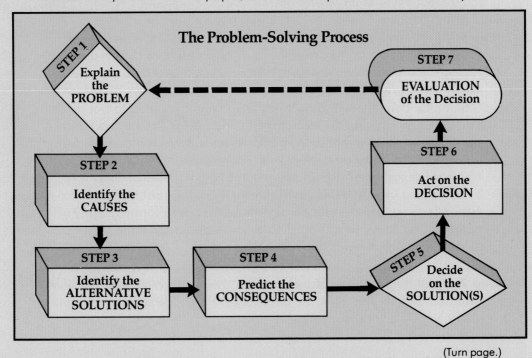

The Problem-Solving Process

STEP 1 — Explain the PROBLEM

STEP 2 — Identify the CAUSES

STEP 3 — Identify the ALTERNATIVE SOLUTIONS

STEP 4 — Predict the CONSEQUENCES

STEP 5 — Decide on the SOLUTION(S)

STEP 6 — Act on the DECISION

STEP 7 — EVALUATION of the Decision

(Turn page.)

The Problem of the Nile

In the late 20th century, Egypt's fast-growing population has required an ever-increasing food supply. Yet the valley of the Nile — where Halim Amin's village lies — contains virtually all of Egypt's farmland. Other areas of Egypt are too dry for growing food crops.

For centuries the Nile River spread out across the valley during the flood season between June and September. When the flood waters dried away, the fields would have a new layer of alluvium — the perfect fertilizer. Egyptian farmers would dig basins in the earth to hold water from the Nile after the flooding was over. From these basins, it was possible to irrigate just one good crop during the dry season.

The Aswan High Dam was proposed to prevent the annual floods and help increase Egypt's food production. Excess waters of the Nile — and the rich fertilizer of alluvium — would be backed up behind the dam in a huge, human-made body of water — Lake Nasser. But downstream from the dam, Egyptian farmers like Halim Amin would be able to plant crops in the valley all year round — using irrigation water from the lake and river. And hydroelectric power from the dam would help Egypt develop modern industry as well.

Step 1: What was the problem of the Nile?

Step 2: Why was it difficult for Egyptians to increase their food supply?

Step 3: The story suggests one solution. What was another alternative?

Step 4: What good consequences might result from building the dam? What bad consequences could result from it?

Step 5: Which alternative do you think would have been best? Why?

Step 6: As you read in Unit 6, the Aswan High dam was completed in 1971. Have benefits of the dam outweighed any new problems? Think about this question as you read the passage below. Then do Step 7.

Since the dam opened, new lands have been converted to cropland through irrigation, and the effort to reclaim more land continues. Crops can be grown all year round. More hydroelectric power for use by industry is generated by the dam every year. However, since the annual floods no longer deposit their rich alluvium in the Nile Valley, farmland is not as fertile as it used to be. Today Halim Amin's sons must buy chemical fertilizers to replace the alluvium. Without the river's former supply of nutrients, moreover, Egypt's sardine catch — also a valuable source of food — was all but wiped out.

Step 7: What benefits were achieved from the dam? What new problems were created? Do you think the benefits outweigh the problems? Explain why on your answer sheet.

A. Words To Remember

From the following list, choose the term that best completes each of the sentences below. On a separate sheet of paper, write your answer next to the number of each sentence.

conservation   habitat   solar energy
extinct   pollutant   synfuels
fossil fuels   soil erosion   topsoil

1. Artificial fuels that will help to supplement natural fuels in the future are called _____ .

2. Coal is the most abundant of all _____ and is found in most countries.

3. Efforts are being made to protect certain animal species and save them from becoming _____ .

4. Energy _____ could have the same effect as developing a new source of energy.

5. Farmers have developed new techniques to keep _____ at a minimum in order to prevent a drop in crop yields.

B. Check Your Reading

1. Most modern countries import billions of barrels of oil every year. Describe one risk that is involved in this action.

2. What is the no-till method of farming? What is its advantage? What possible drawback does it have?

3. Why have scientists created plant seed banks?

4. What is the difference between active and passive solar heating? What are solar cells? Why are they not yet in widespread use?

5. Why has nuclear energy stirred debate in recent years?

C. Think It Over

1. Why are people concerned with saving animal species from extinction?

2. Give one reason why it is often difficult to judge the benefits and risks of a new technological improvement. Give one example.

3. What are the most important requirements of any energy source that is to be used widely?

D. Things To Do

Make a log of your energy use for one day. List all the forms of energy you use from morning until bedtime. Don't forget to include energy you rely on indirectly, such as refrigeration, hot water on tap, or home heating. Include natural forms of energy if you use them — for example, if you hang washing out to dry in the sun instead of using an electric dryer. Also include your muscles, if you walk or ride a bike instead of using a car or bus, or wash dishes by hand instead of using a dishwasher. After completing the list, check those items where you could most easily conserve Earth's energy.

A. Check Your Reading

1. On your answer sheet, write the letter of each description. After each letter, write the number of the term that best matches that description.

(a) Wearing away of soil by wind or water.
(b) The complete dying out of a species of animal or plant.
(c) A strip of metal that uses the Sun's rays to make electricity.
(d) Liquid coal is an example of this.
(e) It may provide energy one day.

(1) extinction
(2) solar cell
(3) nuclear fusion
(4) synfuel
(5) erosion
(6) no-till method
(7) passive solar heat
(8) insecticide

2. Fill in the blanks in the following paragraphs by writing the missing term on your answer sheet.

The rapid increase in world population is due in part to improvements in public health, which have lowered infant __(a)__ rates. Because of the rising population, great efforts have been made to increase food production. One major way of doing this has been to develop new __(b)__ of food plants. Another major way is to open more land for farming. This means developing __(c)__ land — land that is too dry or otherwise not so good for growing crops.

Modern nations rely heavily on __(d)__ fuels as their main sources of energy. One way of increasing available energy is to get more out of these fuels. Another way is to develop other sources of energy. A third way is to __(e)__ energy.

B. Think It Over

1. For each item listed below, decide whether it provides a resource (R), creates a hazard (H), or does both (B). On your answer sheet, write the appropriate capital letter beside the letter of each item.

(a) Drought.
(b) Coal mining.
(c) Plowing.
(d) Solar cell.
(e) Nuclear reactor.

2. Choose any two of the items in exercise 1 above, and write their letters on your answer sheet. Then for each item, explain the answer you gave.

Further Reading

Energy Isn't Easy, by Norman F. Smith. Coward-McCann, 1984. The author explores new energy sources the world may have to rely on in the future.

Energy: The New Look, by Margaret O. Hyde. McGraw-Hill, 1981. A survey of such energy sources as solar energy, ocean energy, and synfuels.

Our Hungry Earth: The World Food Crisis, by Laurence Pringle. Macmillan, 1976. A report on the world hunger crisis and its possible cause or causes.

Stand on Zanzibar, by John Brunner. Ballantine, 1976. A science-fiction novel about overpopulated Earth in the future.

The World

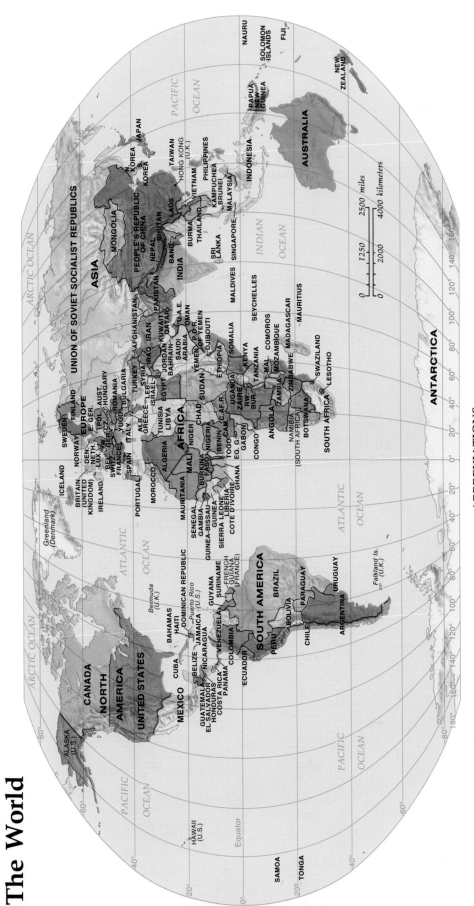

ABBREVIATIONS

ALB. Albania
AUST. Austria
BANG. Bangladesh
BEL. Belgium
BUR. Burundi
CAM. Cameroon
C.AF.R. Central African Republic

CZ. Czechoslovakia
DEN. Denmark
E. GER. East Germany
EQ. G. Equatorial Guinea
LUX. Luxembourg
MAL. Malawi
NETH. Netherlands

P.D.R. OF YEMEN People's Democratic Republic of Yemen
POL. Poland
RW. Rwanda
SWITZ. Switzerland
U.A.E. United Arab Emirates
W. GER. West Germany
YUGO. Yugoslavia

573

North America

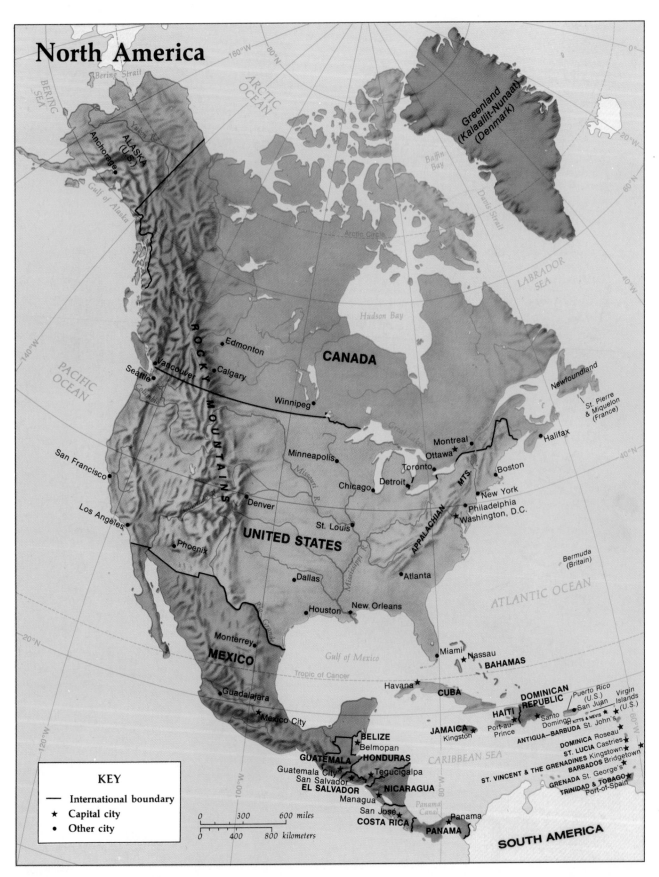

BERING SEA

Bering Strait

ARCTIC OCEAN

160°W

80°N

Greenland
(Kalaallit-Nunaat)
(Denmark)

0°

20°W

Baffin Bay

60°N

Davis Strait

40°W

ALASKA
(U.S.)

Anchorage

Yukon R.

Gulf of Alaska

140°W

Arctic Circle

Hudson Bay

Great Bear Lake

LABRADOR SEA

PACIFIC OCEAN

Edmonton

Calgary

Seattle

Vancouver

CANADA

Winnipeg

ROCKY MOUNTAINS

Montreal

Ottawa ★

Toronto

Great Lakes

Newfoundland

St. Pierre & Miquelon (France)

Halifax

40°N

San Francisco

Minneapolis

Missouri R.

Chicago

Detroit

Boston

APPALACHIAN MTS.

New York

Los Angeles

Denver

St. Louis

UNITED STATES

Philadelphia

Washington, D.C.

Phoenix

Mississippi R.

Bermuda (Britain)

ATLANTIC OCEAN

Dallas

Atlanta

Houston

New Orleans

Rio Grande

120°W

Monterrey

MEXICO

20°N

Gulf of Mexico

Miami

Nassau ★

BAHAMAS

Tropic of Cancer

Guadalajara

Havana ★

CUBA

DOMINICAN REPUBLIC

Puerto Rico (U.S.)

San Juan

Virgin Islands (U.S.)

★ Mexico City

HAITI

Port-au-Prince

Santo Domingo

KITTS & NEVIS

ANTIGUA–BARBUDA St. John's

JAMAICA ★

Kingston

DOMINICA Roseau ★

ST. LUCIA Castries ★

ST. VINCENT & THE GRENADINES Kingstown ★

BARBADOS Bridgetown ★

BELIZE

Belmopan

GUATEMALA

HONDURAS

Guatemala City

Tegucigalpa

San Salvador

EL SALVADOR

Managua

NICARAGUA

San José

COSTA RICA

Panama

PANAMA

Panama Canal

CARIBBEAN SEA

GRENADA St. George's ★

TRINIDAD & TOBAGO

Port-of-Spain

100°W

80°W

60°W

SOUTH AMERICA

KEY
— International boundary
★ Capital city
• Other city

| 0 | 300 | 600 miles |

| 0 | 400 | 800 kilometers |

South America

NORTH
AMERICA

CARIBBEAN SEA

• Maracaibo   ★ Caracas

VENEZUELA

• Medellín

• Bogotá

COLOMBIA

GUIANA
HIGHLANDS

GUYANA   Georgetown ★
★ • Paramaribo
SURINAME
• Cayenne
FRENCH GUIANA
(FRANCE)

Equator

Galápagos Islands
(Ecuador)

★ Quito

ECUADOR

Guayaquil •

Iquitos •

Manaus •

• Belém

Amazon

PACIFIC
OCEAN

PERU

A N D E S

★ Lima

Arequipa •

• La Paz ★

BOLIVIA

• Sucre ★

★ Brasília

BRAZIL

Recife •

BRAZILIAN
HIGHLANDS

• Salvador

20°S

M O U N T A I N S

• Belo Horizonte

Tropic of Capricorn

PARAGUAY

Antofagasta •

CHILE

Asunción ★

• Tucumán

São Paulo •
• Rio de Janeiro

ATLANTIC
OCEAN

• Córdoba

Santiago ★

Buenos Aires ★

URUGUAY
★ Montevideo

ARGENTINA

Río de la Plata

40°S

Falkland Islands

Punta Arenas •

Strait of Magellan

• Tierra del Fuego

Cape
Horn

| 0 | 300 | 600 miles |
| 0 | 400 | 800 kilometers |

80°W

60°W

40°W

KEY

⎯⎯ **International boundary**

★ **Capital city**

• **Other city**

Europe

ATLANTIC
OCEAN

ICELAND
★ Reykjavik

Arctic Circle

Faroe Islands
(Denmark)

Shetland Islands
(Britain)

NORTH
SEA

FINLAND

NORWAY

SWEDEN

Bergen ●

Oslo ★

★ Helsinki

● Leningrad

● Goteborg

BALTIC
SEA

★ Stockholm

● Moscow

Glasgow ●

Belfast ★

Dublin ★

IRELAND

UNITED
KINGDOM
(BRITAIN)

Manchester ●

London ●

Channel Islands
(Britain)

DENMARK
Copenhagen ★

Hamburg ●

NETHERLANDS
Amsterdam ●
The Hague ★

West Berlin ★   East Berlin ★

EAST
GERMANY

Warsaw ★

POLAND

● Kiev

U.S.S.R.
(SOVIET UNION)

Brussels ★

BELGIUM

LUXEMBOURG

Luxembourg ★

Paris ★

Bonn ★

WEST
GERMANY

Prague ★

● Krakow

CZECHOSLOVAKIA

CARPATHIAN MTS.

Munich ●

Vienna ★

AUSTRIA

★ Budapest

HUNGARY

ROMANIA

FRANCE

SWITZERLAND

LIECHTENSTEIN

Bern ★

A L P S

Milan ●

Belgrade ★

Bucharest ★

BLACK
SEA

PYRENEES

ANDORRA

Marseille ●

MONACO

SAN MARINO

YUGOSLAVIA

Sofia ●

BULGARIA

PORTUGAL

SPAIN

Madrid ★

Corsica
(France)

VATICAN
CITY

ITALY

Rome ★

Tirana ★

ALBANIA

Istanbul ●

TURKEY

ASIA

Lisbon ★

Tagus R.

Balearic Islands
(Spain)

Naples ●

Sardinia
(Italy)

GREECE

Athens ★

AEGEAN
SEA

Strait of
Gibraltar

GIBRALTAR
(BRITAIN)

AFRICA

Sicily
(Italy)

ADRIATIC
SEA

MALTA
Valletta

Crete
(Greece)

MEDITERRANEAN   SEA

KEY

— International boundary

★ Capital city

● Other city

| 0 | 200 | 400 miles |

| 0 | 300 | 600 kilometers |

576

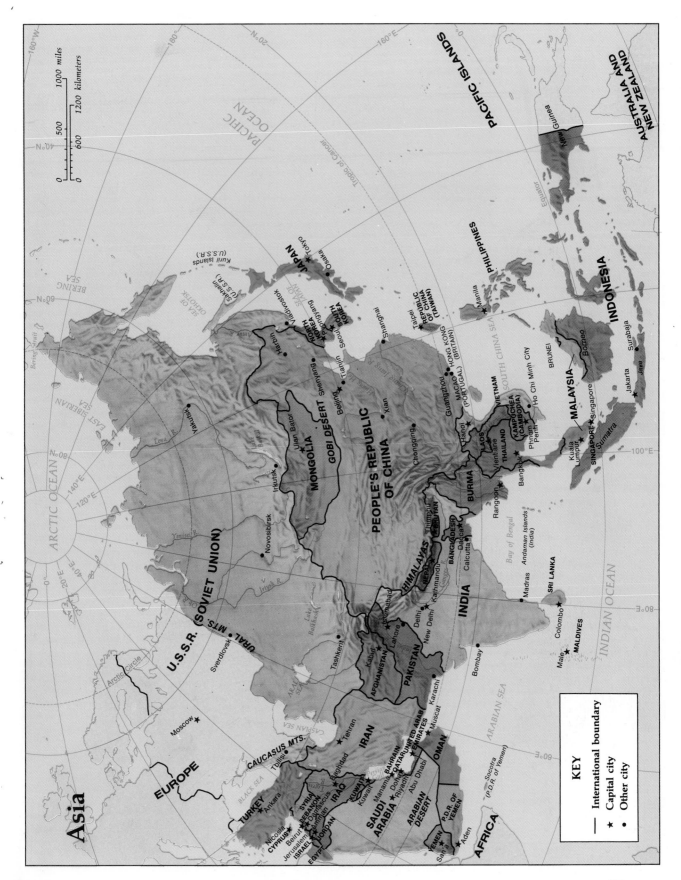

Asia

EUROPE

AFRICA

PACIFIC ISLANDS

AUSTRALIA AND NEW ZEALAND

PACIFIC OCEAN

ARCTIC OCEAN

INDIAN OCEAN

BERING SEA

SEA OF OKHOTSK

EAST SIBERIAN SEA

ARABIAN SEA

Bay of Bengal

SOUTH CHINA SEA

SEA OF JAPAN

BLACK SEA

CASPIAN SEA

ARAL SEA

Lake Baikal

Lake Balkhash

Tropic of Cancer

Equator

Arctic Circle

1000 miles
1200 kilometers
1000
500
600
0
0

U.S.S.R. (SOVIET UNION)

URAL MTS.

CAUCASUS MTS.

Moscow ★

Sverdlovsk •

Novosibirsk •

Tashkent •

Irkutsk •

Yakutsk •

Ulan Bator •

Vladivostok •

Harbin •

Sakhalin (U.S.S.R.)

Kuril Islands (U.S.S.R.)

Ob R.

Irtysh R.

Yenisei R.

Lena R.

Amur R.

MONGOLIA

GOBI DESERT

PEOPLE'S REPUBLIC OF CHINA

Shenyang •

Beijing ★

Tianjin •

Shanghai •

Xian •

Chongqing •

Guangzhou •

Yellow R.

Yangtze R.

JAPAN

Tokyo ★

Osaka •

NORTH KOREA

Pyongyang ★

SOUTH KOREA

Seoul ★

HONG KONG (BRITAIN)

MACAO (PORTUGAL)

Taipei ★

REPUBLIC OF CHINA (TAIWAN)

PHILIPPINES

Manila ★

VIETNAM

Hanoi ★

Ho Chi Minh City •

LAOS

Vientiane ★

THAILAND

Bangkok ★

KAMPUCHEA (CAMBODIA)

Phnom Penh ★

BURMA

Rangoon ★

BRUNEI

Borneo

MALAYSIA

Kuala Lumpur ★

SINGAPORE ★

Sumatra

Jakarta ★

Surabaja •

Java

INDONESIA

New Guinea

HIMALAYAS

NEPAL

Kathmandu ★

BHUTAN

Thimphu ★

BANGLADESH

Dacca ★

INDIA

New Delhi ★

Delhi •

Calcutta •

Madras •

Bombay •

Ganges R.

Andaman Islands (India)

SRI LANKA

Colombo ★

MALDIVES

Male ★

AFGHANISTAN

Kabul ★

PAKISTAN

Islamabad ★

Lahore •

Karachi •

Indus R.

IRAN

Tehran ★

IRAQ

Baghdad ★

KUWAIT

Kuwait ★

BAHRAIN

Manama ★

QATAR

Doha ★

UNITED ARAB EMIRATES

Abu Dhabi ★

OMAN

Muscat •

Riyadh ★

SAUDI ARABIA

ARABIAN DESERT

P.D.R. OF YEMEN

Aden •

YEMEN

Sana ★

Socotra (P.D.R. of Yemen)

TURKEY

Ankara ★

CYPRUS

Nicosia ★

SYRIA

Damascus ★

LEBANON

Beirut ★

ISRAEL

Jerusalem ★

JORDAN

Amman ★

EGYPT

Tbilisi •

Euphrates R.

Tigris R.

KEY

— International boundary
★ Capital city
• Other city

160°W
180
160°E
140°E
120°E
100°E
80°E
60°E
40°E
20°E
20°N
40°N
60°N
80°N

577

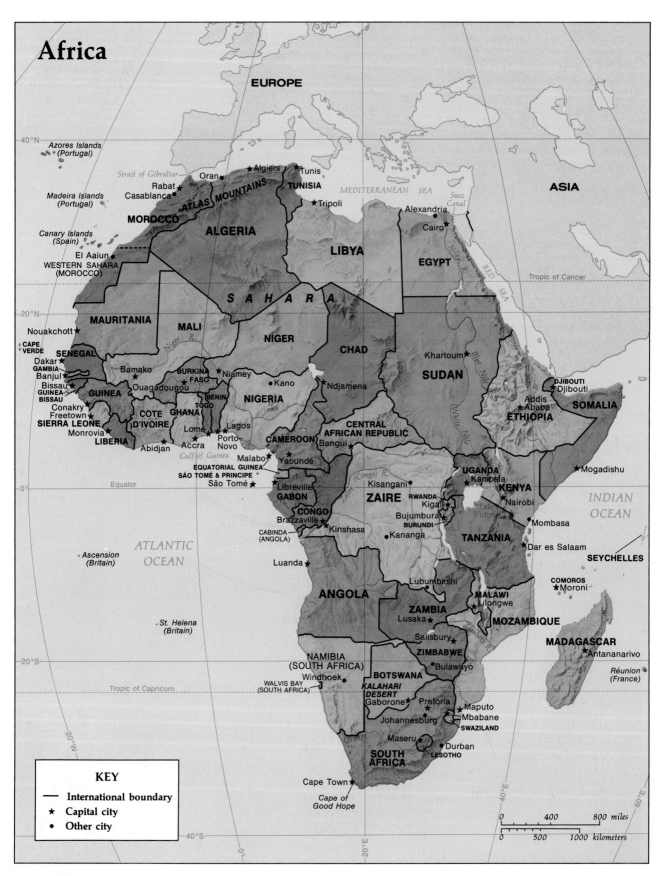

Africa

EUROPE

ASIA

Azores Islands
(Portugal)

Madeira Islands
(Portugal)

Canary Islands
(Spain)

Strait of Gibraltar

Oran
Rabat
Casablanca

Algiers
Tunis

MEDITERRANEAN SEA

Suez Canal

TUNISIA

Tripoli

Alexandria
Cairo

ATLAS MOUNTAINS

MOROCCO

ALGERIA

LIBYA

EGYPT

El Aaiun
WESTERN SAHARA
(MOROCCO)

Tropic of Cancer

RED SEA

S A H A R A

MAURITANIA

MALI

NIGER

CHAD

Nouakchott

CAPE
VERDE

SENEGAL

Dakar

GAMBIA

Banjul

Bissau

GUINEA-
BISSAU

Bamako

Niger R.

BURKINA
FASO

Niamey

Kano

Khartoum

SUDAN

Blue Nile

DJIBOUTI
Djibouti

Addis
Ababa

SOMALIA

GUINEA

Conakry
Freetown

SIERRA LEONE

Monrovia

LIBERIA

Ouagadougou

COTE
D'IVOIRE

BENIN
TOGO

GHANA

Lomé

Abidjan

Accra

Porto-
Novo

Lagos

Ndjamena

NIGERIA

ETHIOPIA

CAMEROON

CENTRAL
AFRICAN REPUBLIC

Bangui

White Nile

Gulf of Guinea

Malabo

EQUATORIAL GUINEA

SÁO TOMÉ & PRINCIPE

São Tomé

Yaoundé

Libreville

GABON

CONGO

Brazzaville

Kinshasa

CABINDA
(ANGOLA)

(Congo) R.

Zaïre R.

Kisangani

ZAIRE

RWANDA
Kigali

Bujumbura

BURUNDI

Kananga

Mogadishu

UGANDA
Kampala

KENYA

Nairobi

Lake
Victoria

Mombasa

Equator

INDIAN
OCEAN

ATLANTIC
OCEAN

Ascension
(Britain)

Luanda

TANZANIA

Dar es Salaam

SEYCHELLES

Lubumbashi

COMOROS
Moroni

St. Helena
(Britain)

ANGOLA

ZAMBIA

Lusaka

MALAWI
Lilongwe

MOZAMBIQUE

MADAGASCAR

Salisbury

Antananarivo

NAMIBIA
(SOUTH AFRICA)

Windhoek

ZIMBABWE

Bulawayo

Zambezi R.

Réunion
(France)

Tropic of Capricorn

WALVIS BAY
(SOUTH AFRICA)

BOTSWANA

KALAHARI
DESERT

Gaborone

Pretoria

Johannesburg

Maputo

Mbabane

SWAZILAND

Maseru

Durban

LESOTHO

SOUTH
AFRICA

Orange R.

Cape Town

Cape of
Good Hope

KEY

— International boundary

★ Capital city

• Other city

0   400   800 miles

0   500   1000 kilometers

578

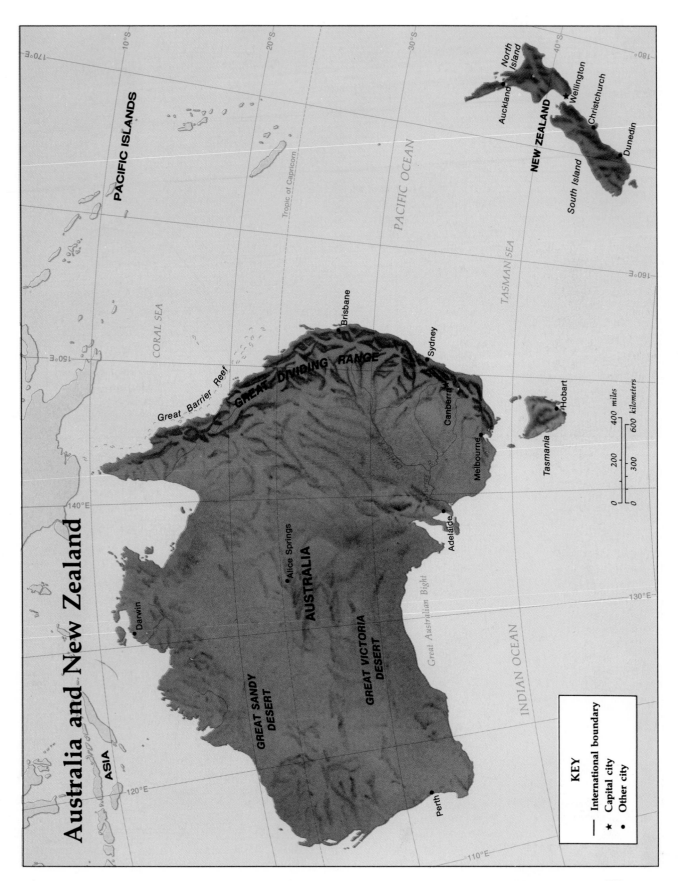

Australia and New Zealand

PACIFIC ISLANDS

ASIA

PACIFIC OCEAN

Tropic of Capricorn

CORAL SEA

TASMAN SEA

NEW ZEALAND

North Island

Auckland

Wellington

Christchurch

Dunedin

South Island

Brisbane

Sydney

GREAT DIVIDING RANGE

Great Barrier Reef

Canberra

Hobart

Melbourne

Tasmania

Adelaide

Alice Springs

AUSTRALIA

Great Australian Bight

Darwin

GREAT SANDY DESERT

GREAT VICTORIA DESERT

INDIAN OCEAN

Perth

10°S
20°S
30°S
40°S

170°E
180°
160°E
150°E
140°E
130°E
120°E
110°E

400 miles
600 kilometers
200
300
0
0

KEY
— International boundary
★ Capital city
• Other city

579

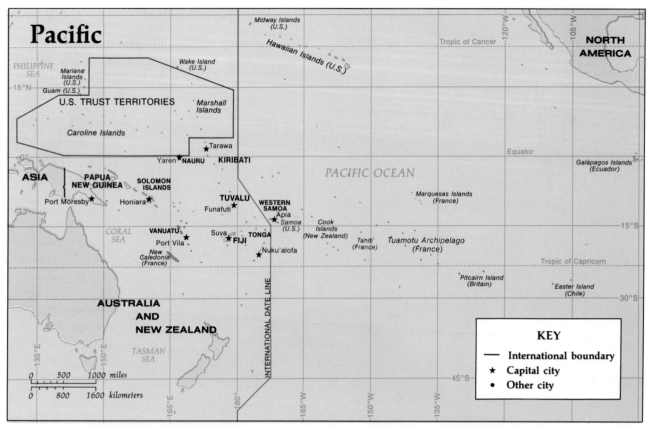

Pacific

PHILIPPINE
SEA

Midway Islands
(U.S.)

Hawaiian Islands (U.S.)

Tropic of Cancer

NORTH
AMERICA

Wake Island
(U.S.)

Mariana
Islands
(U.S.)

Guam (U.S.)

15°N

U.S. TRUST TERRITORIES

Marshall
Islands

Caroline Islands

120°W

105°W

Equator

Galápagos Islands
(Ecuador)

★ Tarawa

Yaren ★ NAURU   KIRIBATI

PACIFIC OCEAN

ASIA

PAPUA
NEW GUINEA

SOLOMON
ISLANDS

Marquesas Islands
(France)

Port Moresby ★

Honiara ★

TUVALU

WESTERN
SAMOA

Funafuti ★

Apia ★
Samoa
(U.S.)

Cook
Islands
(New Zealand)

15°S

CORAL
SEA

VANUATU

Suva ★ FIJI

TONGA

Tahiti
(France)

Tuamotu Archipelago
(France)

Port Vila ★

New
Caledonia
(France)

Nuku'alofa ★

Tropic of Capricorn

Pitcairn Island
(Britain)

Easter Island
(Chile)

30°S

AUSTRALIA
AND
NEW ZEALAND

TASMAN
SEA

135°E

150°E

165°E

180°

INTERNATIONAL DATE LINE

165°W

150°W

135°W

45°S

KEY
── International boundary
★ Capital city
● Other city

0   500   1000 miles

0   800   1600 kilometers

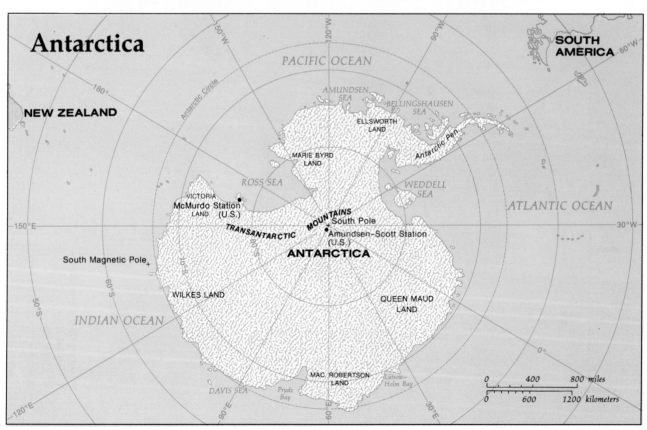

Antarctica

150°W

120°W

90°W

SOUTH
AMERICA

60°W

PACIFIC OCEAN

NEW ZEALAND

180°

Antarctic Circle

AMUNDSEN
SEA

BELLINGSHAUSEN
SEA

ELLSWORTH
LAND

Antarctic Pen.

MARIE BYRD
LAND

ROSS SEA

WEDDELL
SEA

ATLANTIC OCEAN

150°E

VICTORIA
LAND

McMurdo Station
(U.S.)

TRANSANTARCTIC

MOUNTAINS

South Pole

Amundsen–Scott Station
(U.S.)

30°W

South Magnetic Pole +

60°S

ANTARCTICA

QUEEN MAUD
LAND

WILKES LAND

0°

INDIAN OCEAN

50°S

60°E

120°E

90°E

MAC. ROBERTSON
LAND

Lützow-
Holm Bay

30°E

DAVIS SEA

Prydz
Bay

60°E

0   400   800 miles

0   600   1200 kilometers

580

Sources:
Britannica World Data, 1987
Heath Student Atlas, 1987
Information Please Almanac, 1987
Population Reference Bureau, Washington, D.C., 1987

These population figures will change. Please refer to the *Encyclopaedia Britannica Book of the Year* World Data or any of the above.

Checklist of Nations

Nation	Area (sq. miles) (sq. km.)	Population	Population Growth Rate (% yearly)	Life Expectancy (years)	Capital City	Per Capita Personal Income (U.S. $)
Unit 2: The United States and Canada						
Canada	3,851,309 9,974,890	25,640,000	1.0	74	Ottawa	7,572
United States	3,628,150 9,396,909	241,489,000	1.0	74	Washington, D.C.	8,612
Unit 3: Latin America and the Caribbean						
Antigua-Barbuda	171 443	81,000	1.2	NA	St. John's	NA
Argentina	1,072,157 2,776,889	31,030,000	1.6	69	Buenos Aires	1,900
Bahamas	5,380 13,935	235,000	1.9	69	Nassau	3,510
Barbados	166 431	253,000	0.2	70	Bridgetown	1,840
Belize	8,866 22,963	171,000	2.7	NA	Belmopan	NA
Bolivia	424,163 1,098,581	6,611,000	2.8	51	La Paz, Sucre	730
Brazil	3,286,473 8,511,965	138,403,000	2.2	64	Brasília	1,523
Chile	292,256 756,945	12,278,000	1.7	67	Santiago	970
Colombia	439,734 1,138,914	28,231,000	1.7	62	Bogotá	830
Costa Rica	19,575 50,700	2,534,000	2.3	70	San José	1,370
Cuba	44,218 114,524	10,194,000	1.0	72	Havana	840
Dominica	290 752	86,000	2.5	58	Roseau	460
Dominican Republic	18,816 48,734	6,390,000	2.4	60	Santo Domingo	880
Ecuador	109,482 283,561	9,651,000	2.9	60	Quito	780
El Salvador	8,260 21,393	5,461,000	2.4	63	San Salvador	610
Grenada	133 344	97,000	1.7	63	St. George's	500
Guatemala	42,042 108,889	8,191,000	2.9	58	Guatemala City	846
Guyana	83,000 215,000	786,000	0.8	68	Georgetown	510

NA=not available

Nation	Area (sq. miles) (sq. km.)	Population	Population Growth Rate (% yearly)	Life Expectancy (years)	Capital City	Per Capita Personal Income (U.S. $)
Haiti	10,714 27,750	5,427,000	1.7	51	Port-au-Prince	230
Honduras	43,277 112,088	3,938,000	2.9	57	Tegucigalpa	490
Jamaica	4,411 11,424	2,351,000	1.7	70	Kingston	1,610
Mexico	761,600 1,972,500	80,472,000	2.5	65	Mexico City	1,244
Nicaragua	53,938 139,699	3,384,000	3.4	55	Managua	980
Panama	29,208 75,650	2,227,000	2.2	70	Panama	1,716
Paraguay	157,047 406,752	3,531,000	3.6	64	Asunción	750
Peru	496,222 1,285,216	20,207,000	2.6	55	Lima	748
St. Christopher and Nevis	102.9 266.6	46,000	0.7	65	Basseterre	NA
St. Lucia	238 616	140,000	2.1	67	Castries	698
St. Vincent and the Grenadines	150 389	111,000	1.3	67	Kingstown	250
Suriname	63,251 163,820	395,000	2.3	66	Paramaribo	1,240
Trinidad-Tobago	1,979 5,126	1,202,000	1.7	69	Port-of-Spain	3,040
Uruguay	68,536 177,508	3,035,000	0.7	71	Montevideo	1,612
Venezuela	352,143 912,050	17,791,000	2.8	66	Caracas	2,590

Unit 4: Western Europe

Nation	Area (sq. miles) (sq. km.)	Population	Population Growth Rate (% yearly)	Life Expectancy (years)	Capital City	Per Capita Personal Income (U.S. $)
Andorra	175 453	46,000	5.3	NA	Andorra	NA
Austria	32,374 83,848	7,552,000	−0.0	72	Vienna	6,739
Belgium	11,781 30,512	9,856,000	−0.0	73	Brussels	8,040
Britain (United Kingdom)	94,209 244,001	56,679,000	0.1	72	London	4,360
Cyprus	3,572 9,251	674,000	1.2	71	Nicosia	1,580
Denmark	16,619 43,043	5,112,000	−0.0	74	Copenhagen	9,663
Finland	130,119 337,009	4,927,000	0.5	72	Helsinki	9,900
France	211,207 547,026	55,427,000	0.5	73	Paris	8,759
Greece	50,944 131,944	9,987,000	0.5	73	Athens	3,209
Iceland	39,768 102,999	246,000	1.3	76	Reykjavik	5,450

NA = not available

Nation	Area (sq. miles) (sq. km.)	Population	Population Growth Rate (% yearly)	Life Expectancy (years)	Capital City	Per Capita Personal Income (U.S. $)
Ireland	27,136 70,282	3,547,000	0.5	73	Dublin	3,000
Italy	116,303 301,224	57,298,000	0.3	73	Rome	3,470
Liechtenstein	62 160	27,000	1.1	NA	Vaduz	16,864
Luxembourg	999 2,586	367,000	0.1	71	Luxembourg	6,900
Malta	122 316	336,000	0.8	71	Valletta	1,971
Monaco	0.73 1.89	29,000	1.4	NA	Monaco	NA
Netherlands	14,125 36,583	14,561,000	0.4	75	Amsterdam, The Hague	7,597
Norway	125,181 324,218	4,166,000	0.3	75	Oslo	8,316
Portugal	35,510 91,970	10,250,000	0.8	70	Lisbon	1,509
San Marino	24 62	23,000	1.2	NA	San Marino	NA
Spain	194,883 504,746	38,818,000	0.6	72	Madrid	3,190
Sweden	173,654 449,763	8,358,000	0.1	75	Stockholm	12,020
Switzerland	15,941 41,288	6,566,000	0.4	75	Bern	11,606
Vatican City	0.17 0.44	NA	NA	NA	Vatican City	NA
West Germany	95,937 248,476	60,861,000	10.3	72	Bonn	9,278

Unit 5: The Soviet Union and Eastern Europe

Nation	Area (sq. miles) (sq. km.)	Population	Population Growth Rate (% yearly)	Life Expectancy (years)	Capital City	Per Capita Personal Income (U.S. $)
Albania	11,100 28,750	3,023,000	2.1	68	Tirana	520
Bulgaria	42,729 110,668	8,974,000	0.2	72	Sofia	2,100
Czechoslovakia	49,370 127,869	15,552,000	0.3	70	Prague	3,985
East Germany	41,659 107,897	16,636,000	−0.1	72	East Berlin	4,120
Hungary	35,918 93,028	10,624,000	−0.2	70	Budapest	2,750
Poland	120,664 312,519	37,456,000	0.9	71	Warsaw	2,740
Romania	91,699 237,500	22,809,000	0.4	70	Bucharest	2,360
U.S.S.R. (Soviet Union)	8,647,172 22,396,175	280,038,000	0.9	70	Moscow	3,990
Yugoslavia	98,766 255,804	23,289,000	0.7	69	Belgrade	2,210

NA=not available

Nation	Area (sq. miles) (sq. km.)	Population	Population Growth Rate (% yearly)	Life Expectancy (years)	Capital City	Per Capita Personal Income (U.S. $)

Unit 6: The Middle East and North Africa

Nation	Area (sq. miles) (sq. km.)	Population	Population Growth Rate (% yearly)	Life Expectancy (years)	Capital City	Per Capita Personal Income (U.S. $)
Algeria	919,591 2,382,673	22,564,000	3.2	56	Algiers	1,100
Bahrain	231 598	435,000	4.1	63	Manama	2,500
Chad	495,753 1,284,000	5,139,000	2.3	44	Ndjamena	70
Djibouti	8,800 22,800	456,000	4.5	NA	Djibouti	360
Egypt	386,100 1,000,000	48,007,000	2.8	60	Cairo	280
Ethiopia	471,779 1,221,908	48,850,000	2.9	40	Addis Ababa	100
Iran	636,293 1,648,000	46,097,000	3.3	57	Tehran	2,170
Iraq	167,924 434,923	15,946,000	3.3	55	Baghdad	1,561
Israel	7,993 20,699	4,381,000	2.1	74	Jerusalem	3,200
Jordan	37,500 97,125	2,749,000	3.9	56	Amman	870
Kuwait	6,880 17,819	1,791,000	4.5	70	Kuwait	11,431
Lebanon	4,015 10,400	2,707,000	0.4	65	Beirut	730
Libya	679,358 1,759,537	3,953,000	4.4	55	Tripoli	6,260
Mali	464,000 1,201,760	8,457,000	3.0	42	Bamako	96
Mauritania	419,231 1,085,808	1,689,000	2.0	40	Nouakchott	376
Morocco	172,834 447,640	22,455,000	2.5	55	Rabat	520
Niger	489,189 1,267,000	6,423,000	2.7	42	Niamey	100
Oman	82,000 212,000	1,288,000	4.4	47	Muscat	4,880
Qatar	4,400 11,400	311,000	3.7	55	Doha	25,320
Saudi Arabia	870,000 2,253,300	11,670,000	3.9	48	Riyadh	6,089
Somalia	246,201 637,660	5,992,000	2.8	43	Mogadishu	110
Sudan	967,500 2,505,825	24,603,000	3.9	46	Khartoum	165
Syria	71,500 185,200	10,612,000	3.3	62	Damascus	780
Tunisia	63,170 163,610	7,327,000	2.2	57	Tunis	930
Turkey	301,380 780,574	52,419,000	2.9	61	Ankara	1,070

NA=not available

Nation	Area (sq. miles) (sq. km.)	Population	Population Growth Rate (% yearly)	Life Expectancy (years)	Capital City	Per Capita Personal Income (U.S. $)
United Arab Emirates	32,300 83,700	1,700,000	3.2	60	Abu Dhabi	18,500
Yemen	75,290 195,000	2,365,000	3.1	45	San'a	180
Yemen, P.D.R. (People's Democratic Republic)	111,074 287,681	7,046,000	2.8	45	Aden	290

Unit 7: Tropical and Southern Africa

Nation	Area (sq. miles) (sq. km.)	Population	Population Growth Rate (% yearly)	Life Expectancy (years)	Capital City	Per Capita Personal Income (U.S. $)
Angola	481,351 1,246,690	8,823,000	2.9	41	Luanda	440
Benin	43,483 112,620	4,126,000	2.9	46	Porto-Novo	200
Botswana	231,804 600,372	1,126,000	3.7	56	Gaborone	480
Burkino Faso	105,869 274,200	8,126,000	3.5	42	Ouagadougou	113
Burundi	10,747 27,835	4,830,000	2.8	45	Bujumbura	70
Cameroon	183,581 475,474	9,873,000	2.5	44	Yaoundé	328
Cape Verde	1,557 4,033	342,000	2.4	58	Praia	170
Central African Republic	240,534 622,983	2,706,000	2.5	46	Bangui	177
Comoros	863 2,235	409,000	3.4	46	Moroni	153
Congo	132,046 341,999	2,097,000	3.9	46	Brazzaville	490
Cote d'Ivoire	127,520 330,276	10,694,000	4.2	46	Abidjan	821
Equatorial Guinea	10,852 28,107	322,000	1.6	46	Malabo	240
Gabon	103,346 267,666	1,187,000	1.7	43	Libreville	250
Gambia	4,005 10,373	765,000	3.2	41	Banjul	210
Ghana	91,843 237,873	13,144,000	2.6	49	Accra	790
Guinea	94,925 245,855	6,225,000	2.4	58	Conakry	140
Guinea-Bissau	13,948 36,125	891,000	2.1	41	Bissau	230
Kenya	224,960 582,646	21,148,000	4.0	53	Nairobi	270
Lesotho	11,716 30,344	1,586,000	2.6	50	Maseru	270
Liberia	43,000 111,370	2,303,000	3.3	48	Monrovia	600
Madagascar	228,000 590,520	10,294,000	2.8	46	Antananarivo	250
Malawi	45,483 117,800	7,279,000	3.2	46	Lilongwe	154

NA=not available

Nation	Area (sq. miles) (sq. km.)	Population	Population Growth Rate (% yearly)	Life Expectancy (years)	Capital City	Per Capita Personal Income (U.S. $)
Mauritius	720 1,865	1,034,000	1.3	67	Port Louis	640
Mozambique	303,373 785,736	14,143,000	2.6	46	Maputo	220
Nigeria	356,669 923,772	98,112,000	2.5	48	Lagos	500
Rwanda	10,166 26,330	6,336,000	3.5	46	Kigali	106
São Tomé and Principe	372 963	110,000	2.7	NA	São Tomé	270
Senegal	75,750 196,190	6,699,000	2.7	44	Dakar	353
Seychelles	107 277	66,000	0.6	65	Victoria	710
Sierra Leone	27,699 71,740	3,733,000	1.9	46	Freetown	176
South Africa	471,442 1,221,034	33,704,000	2.5	60	Cape Town, Pretoria	1,450
Swaziland	6,704 17,363	682,000	3.3	46	Mbabane	664
Tanzania	363,708 942,003	22,463,000	3.2	50	Dar es Salaam	200
Togo	21,860 56,617	3,072,000	2.8	46	Lomé	300
Uganda	91,134 236,037	15,638,000	3.4	52	Kampala	240
Zaire	905,563 2,345,408	31,079,000	2.1	46	Kinshasa	132
Zambia	290,584 752,612	6,896,000	3.4	48	Lusaka	480
Zimbabwe	150,333 389,362	8,553,000	3.0	52	Salisbury	520

Unit 8: South and Southeast Asia

Nation	Area (sq. miles) (sq. km.)	Population	Population Growth Rate (% yearly)	Life Expectancy (years)	Capital City	Per Capita Personal Income (U.S. $)
Afghanistan	249,999 647,497	16,892,000	1.3	42	Kabul	130
Bangladesh	55,126 142,776	103,084,000	2.6	46	Dacca	85
Bhutan	18,147 47,001	1,146,000	2.0	43	Thimphu	70
Brunei	2226 5765	233,000	3.7	71	Bandar Seri Begawan	NA
Burma	261,789 678,033	38,493,000	2.1	53	Rangoon	120
India	1,261,597 3,267,536	777,230,000	2.1	52	New Delhi	140
Indonesia	735,865 1,905,890	168,662,000	2.2	50	Jakarta	310
Kampuchea (Cambodia)	69,898 181,035	7,469,000	2.9	45	Phnom Penh	NA
Laos	91,429 236,801	3,703,000	2.5	42	Vientiane	70

NA=not available

Nation	Area (sq. miles) (sq. km.)	Population	Population Growth Rate (% yearly)	Life Expectancy (years)	Capital City	Per Capita Personal Income (U.S. $)
Malaysia	128,430 332,633	16,090,000	2.6	66	Kuala Lumpur	990
Maldives	112 290	189,000	3.4	NA	Male	150
Nepal	54,362 140,797	16,863,000	2.3	44	Kathmandu	108
Pakistan	310,403 803,943	102,878,000	3.0	51	Islamabad	200
Philippines	115,707 299,681	56,004,000	2.5	61	Manila	460
Singapore	225 583	2,588,000	1.2	71	Singapore	2,800
Sri Lanka	25,332 65,610	16,087,000	1.5	68	Colombo	220
Thailand	200,148 518,383	52,654,000	1.9	61	Bangkok	444
Vietnam	127,251 329,580	61,218,000	2.2	41	Hanoi	140

Unit 9: East Asia and the Pacific

Nation	Area (sq. miles) (sq. km.)	Population	Population Growth Rate (% yearly)	Life Expectancy (years)	Capital City	Per Capita Personal Income (U.S. $)
Australia	2,967,909 7,686,884	15,912,000	1.3	73	Canberra	7,515
China, People's Republic of	3,692,000 9,562,000	1,053,703,000	1.2	68	Beijing	390
China, Republic of (Taiwan)	13,886 35,965	19,439,000	1.6	71	Taipei	1,170
Fiji	7,055 18,272	710,000	1.9	70	Suva	1,100
Japan	142,726 369,660	121,470,000	0.7	75	Tokyo	6,010
Kiribati	263 681	65,000	2.0	58	Tarawa	NA
Korea, North	46,540 120,538	20,543,000	2.3	62	Pyongyang	590
Korea, South	38,004 98,430	41,569,000	1.4	66	Seoul	880
Mongolia	592,664 1,534,999	1,938,000	2.5	63	Ulan Bator	750
Nauru	8.2 21.2	8,000	−0.3	NA	Yaren	17,140
New Zealand	103,736 268,676	3,288,000	1.0	73	Wellington	4,303
Papua-New Guinea	178,260 461,690	3,400,000	2.1	50	Port Moresby	453
Solomon Islands	17,600 45,600	277,000	3.5	57	Honiara	300
Tonga	269 697	98,000	0.9	NA	Nuku'alofa	430
Tuvalu	10 26	8,000	1.5	58	Funafuti	NA
Vanuatu	5,700 14,750	137,000	3.3	NA	Port Vila	NA
Western Samoa	1,097 2,841	160,000	0.6	65	Apia	290

NA=not available

Glossary

aborigine. An original inhabitant of an area.

adobe. A building material made of clay mixed with straw. It is shaped when wet and dried in the sun.

agribusiness. The complex of giant farms that help to grow, process, and market U.S. farm goods.

agriculture. Farming.

alluvium. A rich soil left behind by running water; for example, the fine mud left by a flooding river.

altitude. The height above sea level.

Antarctic Circle. The parallel of latitude at 66½°S. In midsummer the period of daylight lasts 24 hours on all parts of Earth south of this parallel.

apartheid. The official policy of separation of the races in the Republic of South Africa.

arable land. Land suitable for farming.

archipelago. A large group of islands.

Arctic Circle. The parallel of latitude at 66½°N. In midsummer the period of daylight lasts 24 hours on all parts of Earth north of this parallel.

arrow symbols. The symbols used on maps to indicate motion and direction; for example, the motion and direction of winds or ocean currents.

atlas. A series of maps, bound together, which show different parts of the world.

atmosphere. The layer of gases that surround some planets. Earth's atmosphere is vital to life.

atoll. A ring-shaped coral island.

axis (of Earth). An imaginary rod through the center of Earth from the North Pole to the South Pole. Earth rotates around this axis.

balance of trade. The relation between a nation's imports and exports. If imports exceed exports, the balance is negative or unfavorable. If exports exceed imports, the balance is positive or favorable.

band graph. A kind of line graph that shows how two or more quantities change over time.

bar graph. A graph that uses bars of various lengths to compare different quantities.

basin (of a river). The area drained by a river and its tributaries.

bauxite. Aluminum ore.

bilingual. Speaking two languages.

bloc. A tightly knit group of nations.

blue-collar job. A job that involves manual labor.

bride wealth. In some societies, the money or goods given by a man to the family of his future wife.

cacao. A tropical tree whose seeds are used to make chocolate (see illustration on page 437).

capital. The money used to start a business, an industry, or other economic activity.

caravan. A group of travelers banded together for protection on difficult routes; for example, to cross a desert.

cartogram. A type of map in which the sizes of nations are based on some quantity other than area.

cash crop. A crop grown for sale or export.

cassava. An edible root, also known as **manioc** (see illustration on page 437).

caste. In India one of the 3,000 groups into which Hindus are divided.

cataracts. Low waterfalls and rapids.

census. The counting of a population.

centralized government. A system in which most decisions (for example, about economic production and prices) are made by the national government. The making of these decisions is known as **centralized planning.**

chernozem. A type of rich black soil (the Russian word for "black earth").

citrus fruits. A family of tropical and subtropical fruits that include oranges and lemons.

civilization. An organized society with advances in technology and art.

clan. A group of families who have a common ancestor, and who live and work together.

climate. The kind of weather a place has over a period of time.

climatic graph. A combined bar graph and line graph that shows the precipitation and average temperature of a place for each month of the year.

collective farm. In Communist countries, a government-owned farm where the workers share in the day-to-day management.

Common Market, or **European Economic Community (EEC).** An economic grouping of many West European nations.

communications. Ways of exchanging information.

condense. To change from a vapor to a liquid.

coniferous tree. A type of tree that bears seeds in the form of cones. Most such trees are also evergreens.

consolidated. Put together; for example, forming one large farm out of several small farms.

constitutional monarchy. A type of government in which the head of state is a king or queen but the real power lies with elected officials.

consumer goods. Products such as clothes, radios, pens, and cars that are used directly by people.

contiguous. Joined together.

continent. One of Earth's largest landmasses.

continental climate. A type of climate with hot summers and cold winters.

continental shelf. The underwater edge of a continent. A continental shelf slopes under the oceans before dropping away to the ocean deeps.

contour line. A line joining all points on a map that have the same elevation on Earth.

coordinates. See **letter-number coordinates.**

copra. The dried "meat" of the coconut.

coral island. An island formed from many layers of coral, the shell-like homes of tiny sea animals.

cosmopolitan. International; containing a variety of different ethnic groups or cultures.

cottage industry. A small industry in which the workers produce goods in their own homes.

crop. A plant grown for food or other use.

crop rotation. Planting the same piece of land with different crops from year to year.

crude oil. Oil that is not yet treated or refined.

crust (of Earth). The outermost layer of Earth.

cultivation (of the land). Plowing and planting crops.

cultural geography. See **human geography.**

cultural region. An area of land where most people have enough in common to be considered as a group that is different from other peoples.

culture. Customs and ways of doing things.

current (in oceans). One of the paths in which ocean waters usually move, or the moving water itself.

cycle. A series of events that is repeated over and over.

cyclone. A name for various air movements involving spiral motion. A **tropical cyclone** is a violent late-summer storm in the southwest Pacific Ocean.

deciduous tree. A tree that drops its leaves in the fall.

degree (of latitude and longitude). A unit of measure for latitude and longitude. There are 90 degrees of latitude between the Equator and each pole, and 180 degrees of longitude east and west.

delta. A broad, usually triangular, area of land where a river divides into channels near its mouth.

democracy. A system of government in which the people rule.

desert. An extremely dry area where ordinary plants cannot grow.

developing area. A nation whose government is trying to bring modern farming, industry, transportation, and communications into traditional areas.

diagram. A drawing.

dialect. A local language that differs from the official language of a nation.

dike. An earth barrier, used to hold back waters.

diversify (an economy). To build up a range of different industries.

divide. The separation between two river systems.

divided-bar graph. A graph that compares not only total amounts but also parts of those totals.

domesticated animal. An animal that has been tamed to do work (also, a pet).

double cropping. Planting two crops on the same plot of land in one year.

downstream. The direction of a river's flow; toward the river's mouth.

drainage. The running off of rain water from the land.

drought. An extra-long period without precipitation.

dynasty. A series of rulers who belong to one family line.

earthquake. A movement along weak sections in Earth's surface caused by pressures in the crust.

economic geography. The study of the way people use Earth's resources.

edible. Fit to eat.

ejidos. Community-owned lands in Mexico.

elevation. The height of land above sea level.

embassy. The office of one nation's representatives in another nation.

energy. The power to do work. **Energy resources** are resources (such as fuels) that people use to make things work or move.

environment. The surroundings in which people and animals live.

Equator. The 0° line of latitude on Earth, halfway between the North and South poles.

equinox. One of the two days every year (in spring and fall) when there are 12 hours of daylight and 12 hours of darkness on all parts of Earth.

erosion. The wearing away of the land by natural forces (such as wind and water) or by human activities (such as plowing).

ethnic group. A large group of people who have more in common with each other than they do with other peoples.

evaluating. Judging the usefulness of (information).

evaporate. To change from a liquid to a gas.

exotic. Brought in from another part of the world.

exports. The goods sold by one nation to other nations.

extended family. A household that includes not only parents and children but also grandparents, uncles, and aunts.

extinct. No longer in existence; having died out.

extracted. Taken out.

fallow. Land that is out of use for growing crops. A field may be left fallow for a year so that its soil can build up nutrients.

fault (in Earth's crust). A weak section in Earth's surface. Earthquakes often take place along faults.

ferroalloy. A mineral that is added to iron to produce high-quality metals such as stainless steel.

fertilizer. Manure or chemicals that contain nutrients for crops.

fission (nuclear). The splitting of the atoms of a heavy metal such as uranium to produce energy.

fjord. A steep-sided inlet in a coast.

flowchart. A chart showing what happens in a process. Different parts of the process are described in separate boxes.

fodder. Food for livestock.

fossil. A rock that shows traces of ancient life.

fossil fuel. A fuel that comes from dead plants or tiny animals that have been changed by lying underground for millions of years. Coal, petroleum, and natural gas are the major fossil fuels.

fuel. A resource used to produce energy. In addition to fossil fuels (see above), there are fuels such as wood and animal oils.

functional diagram. A diagram that shows how something works or moves.

fusion (nuclear). Producing energy by combining atoms of hydrogen to form another gas called helium.

gazetteer. A geographical dictionary; a list of places in alphabetical order, with facts about each place.

generalization. A general statement.

general-purpose map. A map showing both physical and political features. Also known as a **geopolitical map**.

geographical dictionary. See **gazetteer**.

geopolitical map. See **general-purpose map**.

geyser. A spout of water heated by molten rock underground.

glacier. A slowly moving mass of ice.

graph. A drawing that shows quantities of things by means of lines, shapes, colors, and other symbols.

grassland. An area where the major type of natural vegetation consists of grasses (rather than trees, desert plants, or tundra plants).

grid (global). The pattern made by parallels of latitude and meridians of longitude crossing each other around the globe.

Gross National Product (GNP). The total value of all the goods and services produced in a country during one year. The GNP measures a country's overall economic activity and is one way of comparing the economies of different countries.

groundwater. The water below the topsoil.

growing potential. The possibility of growing crops (in an environment); the kind and quantity of crops that could be grown.

growing season. The period when it is warm enough for a plant to begin and complete its growth.

habitat. An animal's home and food sources.

hacienda. In Latin America, a large estate owned by a wealthy family.

head of navigation. The farthest point up a river that oceangoing ships can reach.

headwaters. The small streams that come together in a highland area to make up the source of a river.

heartland. The central, important area (of a nation or region).

heavy industry. The manufacture of steel, machinery, cars and trucks, weapons, construction materials, and other massive items.

hemisphere. One of the halves into which Earth can be divided. The Equator divides Earth into the **North-** **ern Hemisphere** and the **Southern Hemisphere**. The Atlantic and Pacific oceans divide Earth into the **Eastern Hemisphere** and the **Western Hemisphere**.

high latitudes. The areas more than about 60° north or south of the Equator.

hill. A raised area of land, smaller than a mountain.

homogeneous (population). All basically alike; belonging to one main ethnic or cultural group.

human geography. The study of people and their ways of life in different places on Earth. Also known as **cultural geography**.

humid (air). Containing a lot of moisture. Moisture in the air is known as **humidity**.

hurricane. A violent late-summer storm in the Atlantic Ocean.

hybrid (plant). A cross between two or more varieties of plants.

hydroelectric power. The electricity created by the flow of water in rivers or from lakes.

ice age. A lengthy cold period on Earth.

illiterate. Unable to read and write.

immigrant. A person who migrates, or moves, *into* an area.

imports. The goods bought by one nation from other nations.

incentive. A motive such as a reward promised in advance.

inclined. Tilted.

independent nation. A nation that governs itself and does not officially depend on the government of any other nation.

index. The alphabetical listing at the back of a book of all the things included in the book.

indicator line. A line that connects the points drawn on a graph.

indigenous. Native; belonging to or growing naturally in a place.

industrialization. Building up industries.

Industrial Revolution. A huge advance in technology that occurred about 250 years ago. The basic change was the use of new sources of energy.

infant mortality. The percentage of children who die at or shortly after birth.

interdependence. Relying on one another (as with nations that need to trade with one another).

International Date Line. A line based on the 180° meridian that separates one calendar day from the next. The date west of the line is always one day later than the date to the east.

interrupted projection. A map projection that breaks up the oceans in order to present the continents more accurately.

invest. To use money or capital to bring in more money; for example, by starting an industry or drilling for oil. Someone who invests money is an **investor**; the money used is an **investment**.

irrigation. Bringing water for crops from wells,

rivers, or other sources, in areas with too little rain.

isolated. Cut off; having little or no contact with other peoples or nations.

isotherm. On a map, a line connecting places that have the same average temperature.

jute. A tropical plant containing a fiber used for making burlap.

key (of a map). A list of the symbols used on a map, with an explanation of their meaning.

lagoon. The body of water inside an atoll (a ring-shaped coral island).

landlocked. Having no outlet to the sea.

latitude. The position of a place north or south of the Equator.

leeward. Sheltered from the wind; on the side away from the direction from which the wind is blowing.

letter-number coordinates. Letters or numbers marked along the edges of a map to help locate places on the map.

life expectancy. The average number of years that persons in a nation (or other group) live.

light industry. The manufacture of any goods that do not require massive equipment; for example, clothing, processed foods and drinks, and paper goods.

line graph. A graph that shows how a quantity varies over time.

lingua franca. A common language used by people who speak different languages.

literate. Able to read and write.

loess. Wind-spread soil that is excellent for crops.

longitude. The position of a place east or west on the globe.

loss/gain graph. A graph that shows quantities less than, as well as more than, zero.

lowland. A low-lying area.

low latitudes. Areas less than about 30° north or south of the Equator.

malnutrition. Lack of a proper diet.

manioc. See **cassava**.

marble. A fine stone used in building.

marginal land. Land that is only just good enough for farming; usually it needs irrigation or other special treatment.

marine (climate). Mild and wet.

maritime. Close to and influenced by a sea or an ocean.

marsupial. A mammal that carries its young in a pouch.

Mercator projection. A map projection in which meridians of longitude are shown as parallel lines.

meridian. A line of longitude.

mestizo. A Latin American of mixed Indian and Spanish descent.

metal. A mineral that is smooth, tough, and easily shaped.

meteorological. To do with **meteorology**, the study of the weather.

metropolitan area. A city and its suburbs.

midlatitudes. The areas between about 30° and 60° north and south of the Equator.

migrate. To move from a homeland.

mineral. A rock or rocklike material that can be dug up and put to use.

minority people. People who belong to a different ethnic group from most of a nation's population.

minute (of latitude and longitude). One sixtieth of a degree.

monsoon. A wind that changes direction at regular times of the year. Monsoons bring wet and dry seasons to parts of South Asia and Africa.

moon. A planetlike mass that revolves around a planet.

moraine. A hill made up of rocks carried by glaciers.

mountain. A raised area of land, usually with steep, rocky sides. A **mountain range** is a chain or group of mountains.

mouth (of a river). The part of a river that joins a larger body of water.

mulatto. A person of mixed African and European descent.

nation. A group of people who share the same political system or type of government.

nationalized. Taken over and run by the government.

natural harbor. The part of a coast where the land itself protects ships from the open sea.

natural vegetation. The land's original plant life.

navigable (waterway). Wide and deep enough to be used by ships.

newsprint. The paper made from timber and used for newspapers.

nomad. A person who moves from one place to another as a way of life. Most people who follow a **nomadic** way of life are herders in dry areas, where too little grass grows to support their livestock in one place year round.

no-till method (of farming). A method of farming without plowing. Its purpose is to avoid soil erosion.

nutrient. Food; for example, the food that plants need from the soil.

oasis. In a desert, a source of water that allows some crops to grow.

ocean. A large body of water separating continents.

ore. Rock from which a mineral can be extracted.

oxygen. A gas, found in Earth's atmosphere, that is necessary for life on Earth.

parallel. Running in the same direction and always at the same distance. A **parallel of latitude** is a line on a map or globe joining places on the same latitude.

peat. Decayed plants used as fuel and fertilizer.

penal colony. An overseas place where people convicted of crimes were sent.

per capita personal income. The average income of all individuals in a nation.

percentage. A part of share expressed in hundredths. Thus 10 percent equals 10/100 or 1/10.

permafrost. Permanently frozen ground.

petrochemical. A chemical made from petroleum.

petroleum. The oil used most widely as a fuel.

physical feature. A land or water form on Earth's surface.

physical geography. The study of Earth's land, water, and atmosphere.

pictograph. A graph that uses picture symbols to represent amounts.

pie graph. A circular graph that shows how a whole amount is divided into segments.

plains. Broad lowlands, flat or gently rolling.

planet. A huge mass, cooler than a star, that revolves around a star.

plantation. A large farm that employs many workers and usually specializes in one crop.

plateau. A broad area of high, mostly flat land.

plate (of Earth's crust). One of the hard sections of Earth's crust on which the continents lie.

polar projection. A map centered on either the North or South Pole.

pole. One of the points at the end of Earth's axis. The **North Pole** is the point at the northern end; the **South Pole**, the point at the southern end.

political feature. A feature, shown on a map, of the way humans have divided Earth's surface; for example, a city, state, or national boundary.

pollutant. Any waste matter that causes **pollution**, which is dirt in the air or water, or on the land.

polygamy. The marriage of a person to more than one wife or husband. **Polygyny** specifies the marriage of a man to more than one wife.

population profile. A kind of bar graph that shows the size of different age groups in a population.

prairie. A grassy plain. The term is used mainly for the plains of Canada.

precipitation. Any of the forms in which water falls on Earth's surface (rain, snow, etc.).

prevailing wind. The direction of the wind that blows most often in a place.

Prime Meridian. The 0° line of longitude, which runs through Greenwich, England.

priority. The order of importance of things to be done.

projection. One of the many ways in which Earth's curved surface can be represented as a flat map.

quota. A set amount; for example, the amount of goods that must be produced in a certain time.

rain forest. A tree-covered area that remains hot and wet throughout the year.

rain shadow. The side of a mountain that gets very little rainfall. This occurs when prevailing winds cause clouds to drop their moisture on the other side.

rationed. Made available only in restricted amounts.

raw materials. Substances from which manufactured goods are made.

reef. A ridge in a sea or ocean bed, close to or above the surface of the water.

refined. Treated to remove impurities; for example, impurities from crude oil.

reforestation. Planting of trees in areas where trees have been cut down.

relief. Changes in elevation. An area of **high relief** has big changes in elevation. An area of **low relief** has only small changes in elevation, and is flat.

republic. A government of elected representatives, headed by a president.

resource. Any element of the environment that humans can put to use.

revolution (of Earth). The broadly circular movement of Earth around the Sun.

river system. A river and its tributaries.

rotation. Turning or spinning, like Earth on its axis.

savanna. Tropical grassland with scattered trees.

scale. The proportion between the size of a place on a map and its actual size on Earth. A map's **scale line** shows what distances on the map represent miles and kilometers on Earth.

seaboard. The land along a coast.

self-sufficient. Meeting all of its own needs; as a nation that produces all of the resources it uses.

separatists. People who want their part of a nation to form a separate nation.

shale oil. The type of petroleum that is imbedded in rocks.

sisal. A tropical plant containing a fiber used in making rope.

slag. The waste left when a metal is extracted from its ore.

slash and burn. A farming system in which farmers cut down the trees in a forest area and burn the underbrush. The ashes provide some fertilizer for the cleared land. However, the soil is quickly used up and more forest is then destroyed.

smelt. To separate a metal from ore by heat.

soil. A thin layer found on much of Earth's land surface. Soil consists of crushed rock, often mixed with dead plant and animal matter.

solar energy. Energy from the Sun. At present this energy is used in three main ways. **Passive solar heating** stores the Sun's heat in walls and other objects. **Active solar heating** uses special panels to store the Sun's heat. **Solar cells** are devices that convert solar energy into electricity.

solar system. A group of planets and other bodies revolving around a star.

solstice. One of the two days of the year when the period of daylight reaches a limit. The **summer solstice** marks the longest daylight period; the **winter solstice** marks the shortest daylight period.

source (of a river). The beginning of a river.

specialized (farming). Concentrating on just one or two kinds of crops.

special-purpose map. A map giving information on a topic other than physical or political features; for example, climate, vegetation, or population.

spring wheat. Wheat that is planted in the spring and harvested in the summer.

stability. Calm; the condition of a nation that has no political upheavals.

staple. Basic (food).

state farm. In Communist countries, a government-owned farm run directly by the government.

steppe. Flat, dry land that gets just enough rainfall to grow some grass.

stock. A share in the ownership of a business. People who buy stock pay money now in exchange for a share in any profits later. A place where stock is traded is known as a **stock exchange.**

structural diagram. A drawing that shows how the inside of something (such as Earth) is structured or put together.

stucco. A plaster wall covering.

subcontinent. A major division of a continent.

subsidiary. A branch of a company.

subsidy. A payment made by a government toward the cost of a certain item in order to keep that item's price low for the buyers.

subsistence farming. Raising just enough crops to feed the farmer's own family.

subtropical. Nearly tropical in latitude or climate.

Sunbelt. The group of U.S. states that extend from coast to coast across the Southeast and Southwest, from California to Florida.

surplus. An amount left over; for example, crops left over to sell after a farming family has taken enough for its own needs.

synfuels. Synthetic fuels, especially those made from coal.

synthesize. Combine; fit together.

table. A listing of facts in rows and columns so that they can be compared easily.

taiga. Cool, forested land.

tariff. A tax on imports.

taro. A plant with an edible root, grown widely in the Pacific islands.

technology. Any skill or tool that people have created to help them live better.

temperate. Moderate; without extremes of heat or cold. The **temperate zone** is a name given to the midlatitudes.

tenant farmer. A farmer who grows crops on land owned by someone else.

thatched (roof). Covered with straw, reeds, or other plant material.

Third World. The group name for all developing nations that are not allied with either the U.S. or the Communist powers.

tidal wave. A huge, destructive ocean wave caused by an earthquake.

timber line. The highest elevation at which trees can grow.

time zone. One of 24 divisions of Earth based on longitude. The official time in each zone differs by one hour from the neighboring zones.

topographic map. A large-scale map of an area, showing physical and cultural features in detail.

topsoil. The uppermost and richest layer of soil.

trade winds. The prevailing winds of the tropics. In the days of sailing ships, traders relied on these winds for much of their travel.

traditional. Following old ways.

transportation system. A network of roads, highways, waterways, railroads, and airports.

tree chart. A type of chart that shows many people or things branching off from a common source.

tremor. A shaking; one of the shock waves caused by an earthquake.

trend. An overall increase or decrease in a quantity over a period of time. Depending on the length of the period, it may be called a **short-term trend** or a **long-term trend.**

tributary. A stream that flows into a larger stream.

tropical wet and dry. A low-latitude climate that is wet in the summer and dry in the winter.

tropics. The area of Earth closest to the Equator. The area lies between the **Tropic of Cancer,** at latitude 23½°N, and the **Tropic of Capricorn,** at latitude 23½°S.

trust territory. An area placed under the temporary control of a nation by the United Nations.

tundra. A type of vegetation that consists of mosses and small flowering plants. It is found in the cold high latitudes.

typhoon. A violent late-summer storm in the northwest Pacific.

unit cost. The cost of making a single product, such as one car or one pen. To find the unit cost, manufacturers divide the cost of making all of the products (cars, pens, etc.) by the number produced.

valley. An area of low land, usually longer than it is wide, between hills or mountains.

vegetation. Plant life.

volcano. An opening in Earth's crust from which molten rocks erupt. The rocks usually form a mountain around the opening.

wadi. In the Middle East and North Africa, a temporary stream in a desert.

water cycle. The process by which water goes from Earth's surface to its atmosphere and back again.

westerly (wind). A wind that blows from the west.

white-collar job. An office job.

wind. Air moving across Earth's surface.

windward. The side toward which the wind blows.

winter wheat. Wheat that is planted in the fall and ripens the next summer.

Index

In this index, each foreign place-name is followed by its pronunciation in parentheses ( ). The key below shows how the letters used in the pronunciation guide are to be pronounced. For example, the letters *ow* should always be pronounced as in *cow*, never as in *low*. Letters such as *b* and *m*, which are not listed below, are pronounced as usual.

a (as in cat)	ehr (as in errand)	oh (as in old)	th (as in thin)
ah (as in odd)	ew (as in few)	oo (as in too)	uh (as in first and
ar (as in art)	g (as in go)	oor (as in lure)	last "a" of banana)
aw (as in lawn)	h (as in hit)	or (as in for)	ur (as in urn)
ay (as in aim)	i (as in it)	ow (as in out)	y (as in yet)
ch (as in chair)	ie (as in ice)	oy (as in boy)	zh (as in vision)
ee (as in eat)	j (as in joke)	s (as in sit)	
eer (as in hear)	ng (as in sing)	sh (as in ship)	
eh (as in end)	o (as in hot)	t (as in tin)	

Sources:
Where not provided in the text, material was obtained from one of the following sources:

Encyclopaedia Britannica Book of the Year
Population Reference Bureau
Europa Yearbook
Heath Student Atlas
Information Please Almanac
Statesman's Yearbook
World Almanac
Joint Economic Committee of the U.S. Congress
Population Reference Bureau, Washington, D.C.
Statistical Office of the United Nations
U.S. Department of Commerce
U.S. Department of Energy

Acknowledgements

Cover and Title Page
C.P. Jones

Illustrations: 18, 41, 56, 59, 437: Howard Friedman.

Introduction
10: *l* George Holton (Photo Researchers); *r* Bruno J. Zehnder. 11: *l* W. Steinmetz (Image Bank); *r* Haas (Magnum).

Unit 1
16, 19, 25: NASA. 32: Waugh (Peter Arnold). 34: Emil Schulthess (Black Star). 44: *tl* Dieter Blum (Peter Arnold); *bl* Bruno J. Zehnder; *tr* Stephen J. Krasemann (Peter Arnold); *br* Max Tortel (DPI). 48: P. Thomann (Image Bank). 50: H.R. Uthoff (Image Bank). 52: Guido Mangold (Image Bank). 61: *t* J. Alex Langley (DPI); *b* J. DiMaggio/ J. Kalish (Peter Arnold). 66: Yoram Lehmann (Peter Arnold). 68: *l* Chuck Fishman (DPI); *r* John Zoiner (Peter Arnold). 72: *l* Stephanie FitzGerald (Peter Arnold); *r* Jacques Jangoux (Peter Arnold). 76: David Burnett (Woodfin Camp/ Contact). 82: Marc & Evelyne Bernheim (Woodfin Camp). 85: *t* Peter Larsen (Photo Researchers); *b* Roland Michaud (Woodfin Camp). 86: *t* J. Alex Langley (DPI); *b* Jerry Frank (DPI). 89: Jacques Jangoux (Peter Arnold). 92: *t* Jerry Frank (DPI); *b* Marvin E. Newman (Woodfin Camp).

Unit 2
100: J. Alex Langley (DPI). 102: Steve Wilson (DPI). 104: Ginger Chih (Peter Arnold). 105: Bryan Hitchcock (Photo Researchers). 107: Bill Bridge (DPI). 113: John Todaro (DPI). 116: Joe Munroe (Photo Researchers). 119: J. Alex Langley (DPI). 121: Norman Myers (Bruce Coleman). 122: Tom McHugh (Photo Researchers). 125: Craig Aurness (Woodfin Camp). 126: Adam Woolfitt (Woodfin Camp). 130: Craig Aurness (Woodfin Camp). 135: Edna Douthat (Photo Researchers). 137: Earl Roberge (Photo Researchers). 144: Jim Brandenburg (Woodfin Camp).

Unit 3
154: Nicholas DeVore III (Bruce Coleman). 157: Malcolm S. Kirk (Peter Arnold). 163: Richard W. Wilkie (Black Star). 165: Ed Drews (Photo Researchers). 170: Bob Schalkwijk (Black Star). 174: Jacques Jangoux (Peter Arnold). 176: Carl Frank (Photo Researchers). 182, 185: Jacques Jangoux (Peter Arnold). 189: J. Alex Langley (DPI). 194: Andrew Holbrooke (Black Star). 197: Manu Sassoonian (Editorial Photocolor Archives). 199: Gerhard Gscheidle (Image Bank). 206: Richard W. Wilkie (Black Star).

Unit 4
216: Helmut Gritscher (Peter Arnold). 221: Marvin E. Newman (Woodfin Camp). 225: J. Alex Langley (DPI). 226: Adam Woolfitt (Woodfin Camp). 230: John Moss (Camera Press London). 233: Adam Woolfitt (Woodfin Camp). 236: Manley Photo (Shostal). 242: Bernard Beaujard/Parimage (Camera Press London). 245: Dennis K. Purse (Photo Researchers). 247: Guy Gillette (Photo Researchers). 248: M. Serraillier (Rapho/Photo Researchers). 251: *t* Collection, State Museum Kröller-Müller, Otterlo, The Netherlands; *b* Private collection. 254: Thomas D.W. Friedmann (Photo Researchers). 259: Tony Howarth (Woodfin Camp). 260: A. Vergani (Image Bank). 262: John Bryson (Rapho/Photo Researchers). 266: Herb Levart (Photo Researchers). 268: J.G. Ross (Photo Researchers). 271: Thomas Hopker (Woodfin Camp).

Unit 5
278: Paolo Koch (Photo Researchers). 283: Karales Look (Peter Arnold). 285: D. Waugh (Peter Arnold). 286: World Film Enterprise (Black Star). 292: Dieter Blum (Peter Arnold). 296, 298, 300: John Launois (Black Star). 304: Chris Niedenthal (Black Star). 307: Peggy Kahana (Peter Arnold). 309: Chuck Fishman (DPI). 312: Rudi Frey (Black Star). 316: Henry G. Jordan (DPI).

Unit 6
326: Derek Bayes (DPI). 330: Craig Aurness (Woodfin Camp). 332: M. Biber (Photo Researchers). 333: Robert Azzi (Woodfin Camp). 334: Gordon Gahan (Photo Researchers). 336: D. Waugh (Peter Arnold). 342: J.G. Ross (Photo Researchers). 345: Stern (Black Star). 347: Farrell Grehan (Photo Researchers). 348: Diane Rawson (Photo Researchers). 349: Kelly B. Langley (DPI). 354: Roland Michaud (Woodfin Camp). 356, 359: Lenore Weber (Omni-Photo Communications). 361: Farrell Grehan (Photo Researchers). 366: Anthony Howarth (Woodfin Camp). 369: Diane Rawson (Photo Researchers). 370: Olivier Rebbot (Woodfin Camp).

Unit 7
376: Paolo Koch (Rapho/Photo Researchers). 381: Michael Lee (Design Photographers International). 383: Slim Aarons (Photo Researchers). 384: Marc & Evelyne Bernheim (Woodfin Camp). 386: Pamela Meyer (Photo Researchers). 389: *l* Rod Borland (Bruce Coleman); *r* Harvey Lloyd (Peter Arnold). 392: Yoram Kahana (Peter Arnold). 395: George Holton (Photo Researchers). 396: Katrina Thomas (Photo Researchers). 398: Marc & Evelyne Bernheim (Woodfin Camp). 400: Michael Hardy (Woodfin Camp). 404: Thomas D.W. Friedmann (Photo Researchers). 407: John & Bini Moss (Photo Researchers). 409: Lionello Fabbri (Photo Researchers). 410: George Holton (Photo Researchers). 416: M. Lee (Design Photographers International). 421: Marc & Evelyne Bernheim (Woodfin Camp).

Unit 8
426: Hans Hoefer (Woodfin Camp). 433: David Alan Harvey (Woodfin Camp). 434: Marvin E. Newman (Woodfin Camp). 435: J. Alex Langley (Design Photographers International). 436: Kal Muller (Woodfin Camp). 442: J. Alex Langley (Design Photographers International). 445: Lenore Weber (Omni-Photo Communications). 447: Vic Cox (Peter Arnold). 448: Jacques Jangoux (Peter Arnold). 449: Penny Tweedie/Daily Telegraph Magazine (Woodfin Camp). 451: Robert W. Young (Design Photographers International). 456: Artfotopolke (Image Bank). 461: Kal Muller (Woodfin Camp). 462, 464, 468: J. Alex Langley (Design Photographers International. 471: Joseph F. Viesti (Design Photographers International).

Unit 9
478: Malcolm S. Kirk (Peter Arnold). 485: R. Bunge (Black Star). 487: Jack Fields (Photo Researchers). 489: *t* J.P. Laffont (Sygma); *b* G.R. Roberts (Omni-Photo Communications). 490: David Moore/Fortune (Black Star). 494, 499, 500, 503: J. Alex Langley (Design Photographers International). 508: Jason Lauré (Woodfin Camp). 511: Mike Yamashita (Woodfin Camp). 513: J.P. Laffont (Sygma). 517, 518, 522: J. Alex Langley (Design Photographers International). 526: Tom Nebbia (Design Photographers International). 531: NASA. 534: Rick Smolan (Contact Press Images). 538: Jim Pickerell (Black Star).

Unit 10
544: Nik Wheeler (Black Star). 551: J.C. Stevenson (Peter Arnold). 552: Gianni Tortoli (Photo Researchers). 558: Peter Arnold. 563: Stephen J. Krasemann (Peter Arnold).